Aluminum Alloys

Special Issue Editor
Nong Gao

MDPI • Basel • Beijing • Wuhan • Barcelona • Belgrade

MDPI

Special Issue Editor
Nong Gao
University of Southampton
UK

Editorial Office
MDPI AG
St. Alban-Anlage 66
Basel, Switzerland

This edition is a reprint of the Special Issue published online in the open access journal *Metals* (ISSN 2075-4701) in 2016 (available at:
http://www.mdpi.com/journal/metals/special_issues/aluminum_alloys).

For citation purposes, cite each article independently as indicated on the article page online and as indicated below:

Author 1; Author 2. Article title. *Journal Name* **Year**, *Article number*, page range.

First Edition 2018

ISBN 978-3-03842-474-1 (Pbk)
ISBN 978-3-03842-475-8 (PDF)

Cover photo courtesy of Laixiao Lu, Jie Sun, Xiong Han and Qingchun Xiong

Table of Contents

About the Special Issue Editor

Nong Gao is a Lecturer at the Materials Research Group and Associate Director of the Centre for Bulk Nanostructured Materials at the University of Southampton, United Kingdom. He obtained his PhD from the University of Sheffield in 1994. After several years research at the University of Strathclyde and the University of Sheffield as a Post-Doctoral Research Associate, he joined the University of Southampton in 2000. He is a Fellow of the Higher Education Academy in the UK. Nong Gao has many years research experience in materials characterization, mechanical property evaluation, creep-fatigue, rolling contact mechanism, tribological behavior, ultrafine grained and nanostructured materials and additive manufacturing. At the moment, he is on Editorial Board of Journal of Metals, and Associate Editorial Board of Materials Letters. To date, he has published over 170 papers in journals and conferences, with over 1500 citations and h-index: 25.

Preface to "Aluminum Alloys"

Aluminium is the world's most abundant metal and is the third most common element, comprising 8% of the Earth's crust. The versatility of aluminium makes it the most widely used metal after steel. By utilising various combinations of their advantageous properties such as strength, lightness, corrosion resistance, recyclability, and formability, aluminium alloys are being employed in an ever-increasing number of applications. In the recent decade, a rapid new development has been made in production of aluminium alloys, and new techniques of casting, forming, welding, and surface modification, have been evolved to improve the structural integrity of aluminium alloys.

This Special Issue covers wide scope of recent progress and new developments regarding all aspects of aluminium alloys, including processing, forming, welding, microstructure and mechanical property, creep, fatigue, corrosion and surface behavior, thermodynamics, modeling, and application of different aluminum alloys.

I am really grateful for the contributions from all the authors around world, whose support and effort make this Special Issue particularly successful, so we have a total of 29 papers included in this book—the largest number of the papers among all the Special Issues from the Journal of Metals.

Nong Gao

Special Issue Editor

Article

Metallographic Index-Based Quantification of the Homogenization State in Extrudable Aluminum Alloys

Panagiota I. Sarafoglou, John S. Aristeidakis, Maria-Ioanna T. Tzini
and Gregory N. Haidemenopoulos *

Department of Mechanical Engineering, University of Thessaly, Volos 38334, Greece;
pa.sarafoglou@gmail.com (P.I.S.); iaristeidakis@gmail.com (J.S.A.); margiannatz@gmail.com (M.-I.T.T.)
* Correspondence: hgreg@mie.uth.gr; Tel.: +30-242-107-4061

Academic Editor: Nong Gao
Received: 10 April 2016; Accepted: 16 May 2016; Published: 21 May 2016

Abstract: Extrudability of aluminum alloys of the 6xxx series is highly dependent on the microstructure of the homogenized billets. It is therefore very important to characterize quantitatively the state of homogenization of the as-cast billets. The quantification of the homogenization state was based on the measurement of specific microstructural indices, which describe the size and shape of the intermetallics and indicate the state of homogenization. The indices evaluated were the following: aspect ratio (AR), which is the ratio of the maximum to the minimum diameter of the particles, feret (F), which is the maximum caliper length, and circularity (C), which is a measure of how closely a particle resembles a circle in a 2D metallographic section. The method included extensive metallographic work and the measurement of a large number of particles, including a statistical analysis, in order to investigate the effect of homogenization time. Among the indices examined, the circularity index exhibited the most consistent variation with homogenization time. The lowest value of the circularity index coincided with the metallographic observation for necklace formation. Shorter homogenization times resulted in intermediate homogenization stages involving rounding of edges or particle pinching. The results indicated that the index-based quantification of the homogenization state could provide a credible method for the selection of homogenization process parameters towards enhanced extrudability.

Keywords: homogenization; aluminum alloys; extrudability; metallographic indices

---

1. Introduction

The process chain of extrudable Al-alloys of the 6xxx series involves direct-chill casting followed by a homogenization cycle, prior to hot extrusion. The as-cast billets contain several inhomogeneities, such as elemental microsegregation, grain boundary segregation, and formation of low-melting eutectics as well as the formation of iron intermetallics. The presence of intermetallic phases, in particular, which possess sharp edges, can impair the deformability of 6xxx extrudable alloys especially when located in the grain boundary regions [1–4]. Among the intermetallics the most important are the Fe-bearing intermetallics, α-$Al_{12}(FeMn)_3Si$ and β-Al_5FeSi, from now on called α-AlFeSi and β-AlFeSi respectively. The α-AlFeSi has a cubic crystal structure and globular morphology while the β-AlFeSi possesses a monoclinic structure and a plate-like morphology, limiting the extrudability of the as-cast billet by inducing local cracking and surface defects in the extruded material [5–7]. The above effects are partially removed by the homogenization treatment, which includes the removal of elemental microsegregation, removal of non-equilibrium low-melting eutectics, the transformation of β-AlFeSi to α-AlFeSi and the spheroidization of the remaining undissolved intermetallics [1]. The effect of various parameters of the homogenization treatment, such as the homogenization temperature, time, as well as the cooling rate, have been studied experimentally [8–11]. The dissolution of Mg_2Si during homogenization is a rather fast process while the transformation of β-AlFeSi to α-AlFeSi is a much slower process [12–14]. In industrial practice, the minimum homogenization time is controlled by the completion of the β-AlFeSi to α-AlFeSi transformation. After the transformation β → α-AlFeSi is complete, the α-AlFeSi phase undergoes coarsening and spheroidization, adopting, finally, a "necklace" morphology, which enhances the extrudability of the billet. This explains the fact that the actual homogenization times in industrial practice are longer than the times required for Mg_2Si dissolution and the completion of the β → α-AlFeSi transformation.

The morphological changes of the α-AlFeSi phase have been described mostly qualitatively in the published literature. Studies have been made on the microstructural evolution during the homogenization of AA7020 aluminum alloy concerning the dissolution of detrimental grain-boundary particles, which degrade the hot workability of the alloy [15–17]. In other studies, it was found that the spheroidization of intermetallic phases is a key mechanism in the microstructural evolution during homogenization [1,18,19]. A method to quantify the microstructure with 3D metallography has been applied for a 6005 Al-alloy [20]. The method, which involved serial sectioning and 3D reconstruction techniques, revealed that the connectivity of the intermetallics decreases with homogenization time. Despite the above works, studies on the "quantification" of the homogenization state are still very limited.

The aim of the present paper is the quantification of the homogenization state by means of quantitative metallography, in order to describe the morphological evolution of the intermetallic phases. An index-based methodology has been developed. The aspect ratio, feret, and circularity are metallographic indices, among others, that can be used to characterize the homogenization state. These indices can be determined by quantitative metallography, involving image analysis. A fully homogenized billet, with the potential for high extrudability should have all β-AlFeSi transformed to α-AlFeSi with necklace morphology and appropriate values of aspect ratio and circularity.

2. Materials and Methods

The chemical composition of the 6060 alloy investigated is Al-0.38Mg-0.40Si-0.2Fe-0.03Mn (mass %). Three homogenization heat treatments consisted of holding at 560 °C, for 2, 4, and 6 h followed by air cooling (see also Table 1). These conditions were selected in order to study the morphological changes of the α-AlFeSi phase after the complete transformation of β → α-AlFeSi.

Table 1. Chemical composition and homogenization conditions for the 6060 alloy.

Chemical Composition (wt. %)		Temperature (°C)	Time (h)
Al	Bal.		
Mg	0.38		2
Si	0.4	560	4
Fe	0.2		
Mn	0.03		6

After the homogenization heat treatment, the specimens were prepared for standard metallographic examination involving optical microscopy (Leitz Aristomet, Leica Microsystems, Wetzlar, Germany), SEM-JEOL 6400 (JEOL Ltd, Tokyo, Japan), and image analysis (Image J software, Version 1.50g, 2016, National Institutes of Health, Bethesda, MD, USA). The specimens were subjected to grinding, polishing, and etching with a Poulton's reagent consisting of 1 mL HF, 12 mL HCl, 6 mL HNO_3, and 1 mL H_2O, modified by the addition of 25 mL HNO_3 and 12 g Cr_2O_3 (in 40 mL H_2O). The as-cast as well as the homogenized microstructures were characterized for intermetallic phases and the particles were categorized in three morphological types as rounded particles, pinched particles, and particles exhibiting a necklace formation. It should be noted that pinched particles are those that are in the initial stage of separation to smaller rounded particles towards the formation of a necklace group. The number of images processed and the number of particles measured for each condition appears in Table 2.

Table 2. The number of images processed and the number of particles measured for each condition.

Alloy	Number of Images	Number of Particles
As-cast	58	106
2 h	57	161
4 h	56	150
6 h	58	133

As mentioned above, the quantification of the homogenization state was based on the measurement of indices that describe the size and shape of the intermetallics and indicate the state of homogenization. The indices employed were the aspect ratio (AR), feret (F), and circularity (C) and are defined in Table 3.

Table 3. The indices employed for the quantification of the homogenization state.

Aspect Ratio	Feret	Circularity
A ratio of the major to the minor diameter of a particle, where d_{max} and d_{min} correspond to the longest and the shortest lines passing through the centroid $$AR = \frac{d_{max}}{d_{min}}$$	The longest caliper length $$F$$	Circularity is a measure of how closely a particle resembles a circle. It varies from zero to one with a perfect circle having a value of one $$C = \frac{p^2}{4\pi A}$$

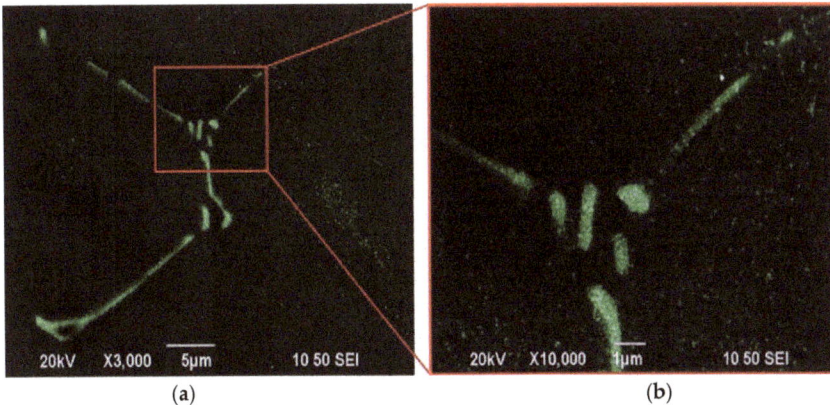

Figure 1. *Cont.*

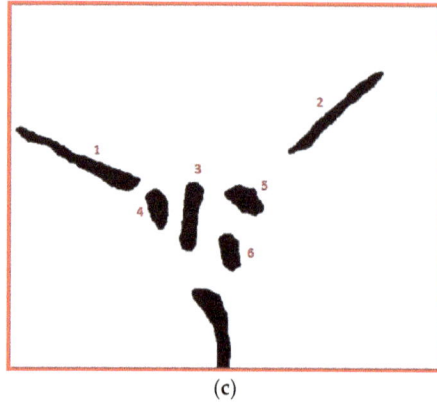

(c)

Figure 1. SEM image used for the measurement of indices: (**a**) low magnification image; (**b**) high magnification isolation of the group of particles; (**c**) image J display used for the measurement of indices.

After the standard metallographic observation, measurement of particle dimensions was carried out in the SEM using the appropriate magnification and a suitable numerical aperture as suggested in [21]. The method is indicated for a group of particles (Figure 1a). The group is isolated (Figure 1b) and transferred to the image analysis program (Figure 1c) where the particles are numbered and their dimensions measured. The respective measurements for each particle in the group are depicted in Table 4. In most cases, the measuring frames contained whole particles. In the cases where the frame passes through a particle, then this particle was not taken into account.

Table 4. Respective measurements for each particle referring to Figure 1c. Indices *AR*, *C*, and *F* correspond to the aspect ratio, circularity, and feret of the measured particles respectively. Accordingly, d_{max} and d_{min} are the major and minor diameters, *p* is the perimeter and *A* is the area of particles (refer to Table 3).

No.	$d_{max}/\mu m$	$d_{min}/\mu m$	*AR*	$p/\mu m$	$A/\mu m^2$	*C*	$F/\mu m$
1	3.655	0.376	9.720	8.405	1.083	5.190	3.656
2	3.289	0.379	8.678	7.427	0.844	5.200	3.308
3	1.792	0.389	4.606	4.395	0.735	2.103	1.793
4	1.069	0.534	2.001	2.820	0.441	1.433	1.068
5	1.123	0.632	1.776	3.080	0.524	1.439	1.160
6	0.976	0.489	1.995	2.712	0.423	1.384	1.000

The area of measurement (scanned area) was kept constant for all homogenization treatments. Statistical analysis is required to assess the data and allow for credible conclusions to be made. In order to examine if the data samples were comparable, the Kruskal-Wallis test [22] was used. It is a non-parametric method for testing whether samples originate from the same distribution and it is used to compare two or more independent samples of equal or different sample size. With a confidence level of 99%, it was proved that the samples derive from different distributions. As a result, the samples are not comparable without further processing. In order to compare between the dissimilar samples, the "Bootstrapped Mean" [23] method was used. Bootstrapping is a non-parametric statistical technique that allows accurate estimations about the characteristics of a population to be made when the examined sample size is limited. As it is non-parametric, the method can be used to compare between samples derived from different distributions, such as Normal and LogNormal distributions. It works by recursively calculating the preferred parameter, like the mean or the median, for a part of the sample and then combining the results to make robust estimates of standard errors and confidence intervals of the population parameter. In this case, a 95% confidence interval was used, while the standard error was kept to a minimum by using a large number of iterations. This process leads to comparable statistical parameters for each measurement.

3. Results and Discussion

The microstructural evolution of the 6060 alloy during homogenization is depicted in Figure 2. The as-cast microstructure is depicted in Figure 2a. Mg_2Si, α-AlFeSi, and β-AlFeSi intermetallics are located at the grain boundaries, while the α-AlFeSi phase exhibits the characteristic "Chinese-script" morphology. The morphological evolution with homogenization time is indicated in Figure 2b,c for 2 h, Figure 2d,e for 4 h and Figure 2f,g for 6 h homogenization time. Connectivity between intermetallics is decreased with homogenization time, in agreement with the observations in [20]. Clear spheroidization of particles and necklace formation is evident only in the micrographs of Figure 2f,g, *i.e.*, after 6 h homogenization. It is clear that optical metallography can supply only qualitative data on the progress of homogenization.

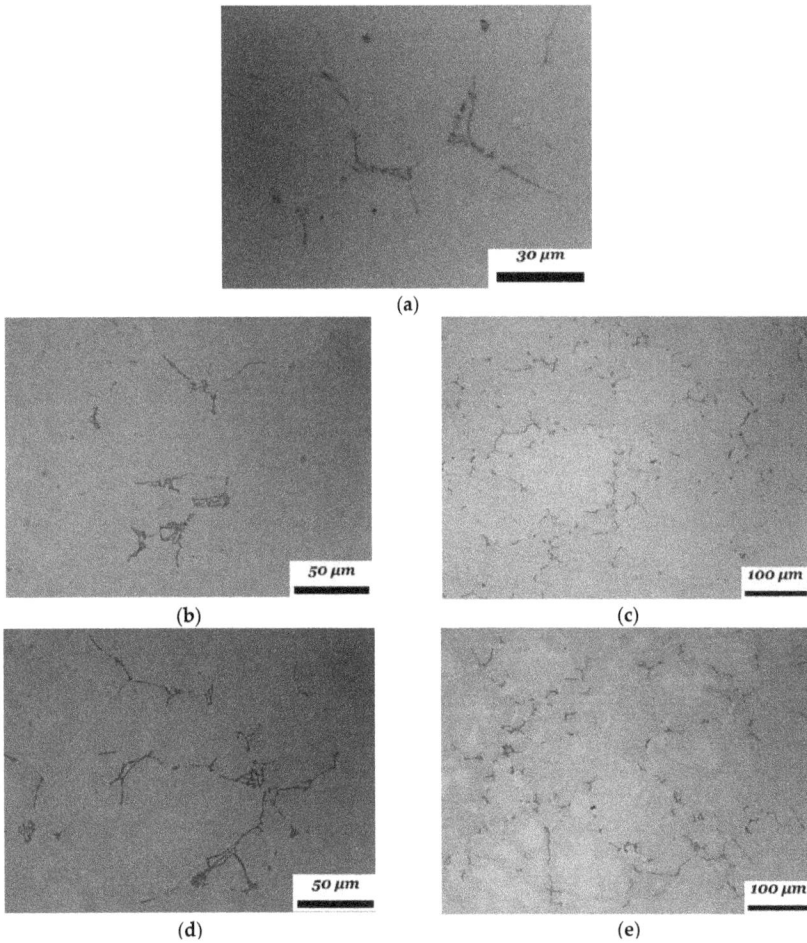

Figure 2. *Cont.*

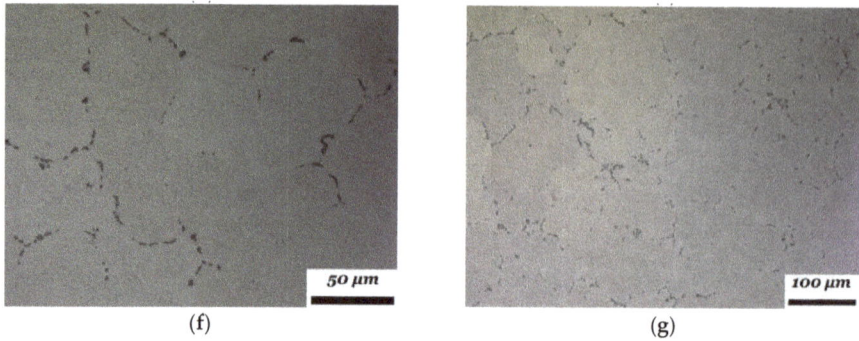

Figure 2. Metallographic images: as-cast (**a**); homogenized at 560 °C (**b**) and (**c**) for 2 h; (**d**) and (**e**) for 4 h; (**f**) and (**g**) for 6 h.

The SEM analysis revealed that the morphological changes of the α-AlFeSi phase during homogenization could be classified in three stages:

First stage, rounding of edges, 2 h homogenization (Figure 3). The β-AlFeSi particles exhibit sharp edges, this being the main reason for their detrimental effect on extrudability. After 2 h, all particles with sharp edges have been transformed and there are no particles with sharp edges in the microstructure. Therefore, we assume that there are no β-AlFeSi particles after 2 h homogenization. As discussed in the previous section after the completion of the β to α-AlFeSi transformation, the intermetallic α-AlFeSi phase undergoes spheroidization. In the first stage of this process the plate-like particles exhibit a slight decrease in their width. Although they do not exhibit complete spheroidization the particles become more rounded at the edges as depicted in Figure 3.

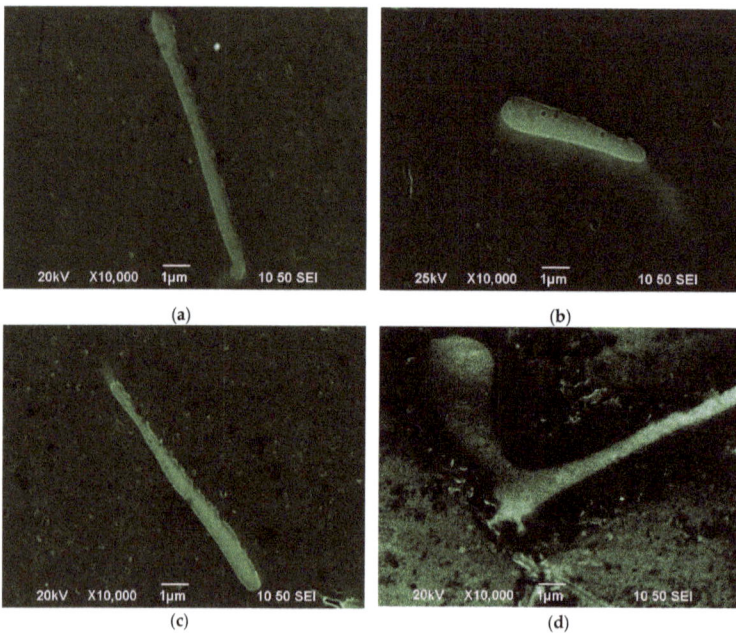

Figure 3. Images indicating the rounding of the edges of the particles after 2 h holding time. (**a**) Long elongated particle; (**b**) short particle; (**c**) elongated particle and (**d**) particle with segment.

Second stage, particle pinching, 4 h homogenization (Figure 4). At the second stage, the rounding of edges is intensified while there is a clear tendency of the particles to be separated into smaller rounded particles by a process called particle pinching. The process has been also observed during homogenization of a 7020 alloy [15] and is indicated by arrows in Figure 4.

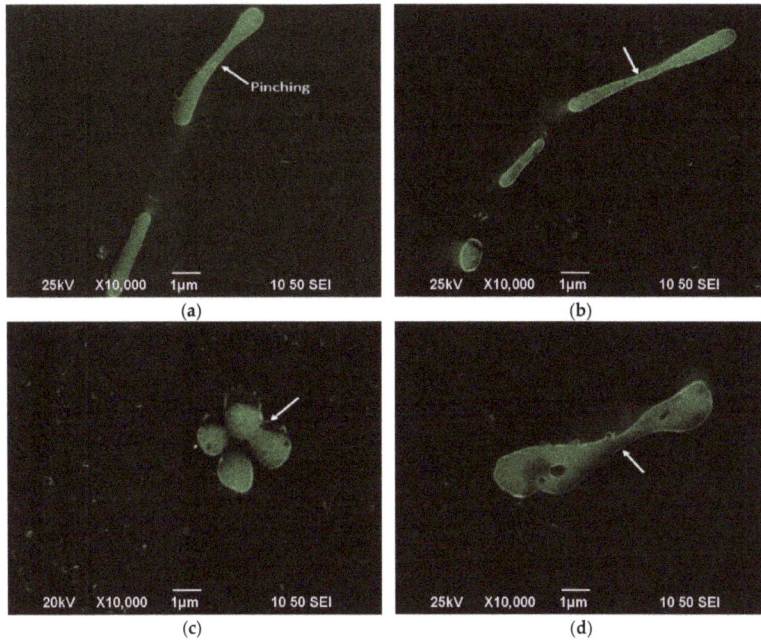

Figure 4. Images revealing the pinching process after 4 h holding time. (**a**) Local reduction of thickness; (**b**) pinching at advanced stage with seperation and local thickness reduction; (**c**) pinching in spherical particles; (**d**) local necking leading to pinching of a particle.

Third stage, necklace formation, 6 h homogenization (Figure 5). The reduction of surface energy of the α-AlFeSi phase is the driving force for spheroidization. With this process, the total interface area between the matrix and the α-AlFeSi phase is reduced. The particles finally adopt a spherical shape and are arranged in a necklace formation during the third stage, as depicted in Figure 5.

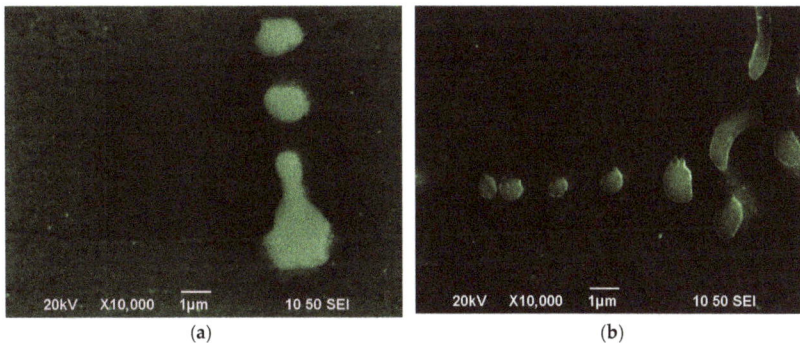

Figure 5. *Cont.*

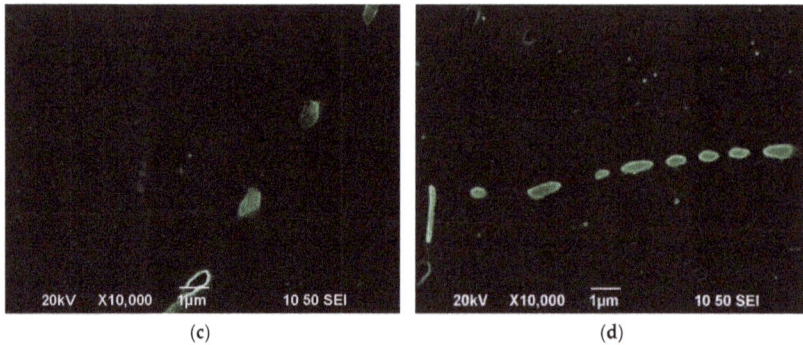

Figure 5. Images revealing the spheroidization and necklace formation after 6 h holding time. (**a**) Pinching leading to particle separation; (**b**) separated particles; (**c**) isolated particles after pinching; (**d**) neclace formation (aligned particles).

The morphological changes of the α-AlFeSi phase described above, include rounding of edges, pinching, and spheroidization. A reduction in surface energy drives the rounding of the edges, since the total interfacial area of the particle is reduced. Particle pinching, *i.e.*, the breakdown of a large plate to smaller particles is driven by the reduction of strain energy, caused by the plate morphology. The spheroidization of the small particles and necklace formation are also driven by the reduction in surface energy. All the above processes are accomplished by the diffusion of alloying elements through the matrix.

The mean values of microstructural indices, aspect ratio, feret and circularity have been determined for the as-cast and homogenized alloys. The 2.5% and 97.5% quantiles were used to define a confidence interval of 95%. The mean index values for the entire population (not just the measured sample), are located inside the confidence interval and have an expected value given by the Bootstrapped Mean. From these data, which are shown in Figure 6a–c, the following remarks can be made.

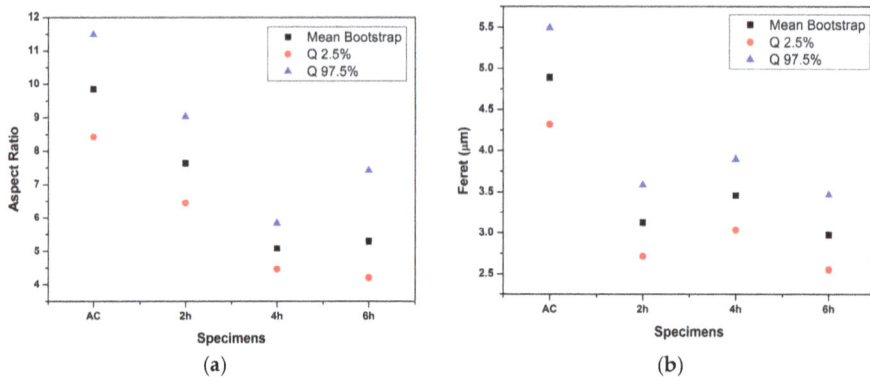

Figure 6. *Cont.*

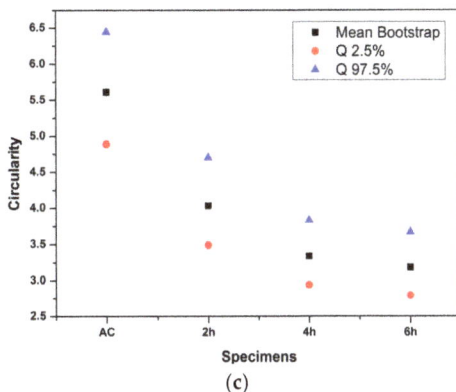

(c)

Figure 6. The values of indices for the as-cast and after homogenization time 2 h, 4 h, and 6 h: (**a**) aspect ratio; (**b**) feret; (**c**) circularity.

The as-cast condition exhibits the highest values of all three indices. Homogenization leads to the reduction in these indices. Regarding the aspect ratio (Figure 6a) the greatest reduction appears up to 4 h homogenization. Extending the homogenization time to 6 h does not change the aspect ratio considerably. Regarding feret, (Figure 6b), the index is reduced appreciably at 2 h homogenization with a further slight reduction at 6 h homogenization. The intermediate slight increase of the feret index between 2 and 4 h homogenization is attributed to the protrusions formed at the particle surface, a process accompanying the pinching process, as suggested in [15,16]. The circularity index, (Figure 6c), exhibits a continuous reduction with homogenization time, with the largest reduction appearing after 2 h homogenization. This is attributed to the initiation of the spheroidization process at the first stage (rounded particles) discussed above. Circularity achieves its lowest value at after 6 h homogenization. This is in agreement with the observation of necklace formation after 6 h homogenization (third stage). The necklace formation is characterized by the lowest value of the circularity index among the conditions examined. The fact that there is no further reduction of the aspect ratio between 4 and 6 h homogenization, discussed above, is attributed to the decreased connectivity of the α-AlFeSi phase, which follows the necklace formation. A continuous decrease of connectivity with homogenization time has been also observed for a 6005 Al-alloy [20]. Spheroidization and in particular, necklace formation, has been considered a key process for increased extrudability [18,19]. It appears that the index exhibiting the more consistent variation with homogenization time is the circularity index, which, as stated above, exhibits a continuous reduction with homogenization time.

4. Conclusions

An index-based method to quantify the homogenization state has been developed. Indices such as the aspect ratio, feret, and circularity have been determined in order to characterize the stage of spheroidization of the α-AlFeSi phase, following the β to α-AlFeSi transformation. The effect of the homogenization time was studied in a 6060 alloy. The major conclusions are the following:

- The α-AlFeSi particles, after the completion of the β to α-AlFeSi transformation undergo morphological changes leading to spheroidization. This process can be divided in three stages: (1) rounding of edges, (2) particle pinching, and (3) necklace formation.
- The evolution of the morphological changes can be described quantitatively by the use of indices, such as aspect ratio, feret and circularity, which are sensitive to homogenization process parameters, such as the homogenization time.

- The circularity index exhibited the most consistent variation with homogenization time. The lowest value of the circularity index (more circular particles) coincided with the metallographic observation for necklace formation. Shorter homogenization times resulted in intermediate stages involving rounding of edges or particle pinching.
- The method requires the measurement of a large number of particles and the implementation of a statistical analysis in order to be credible.

Acknowledgments: Part of this work has been supported by a grant from Aluminium of Greece (AoG).

Author Contributions: P.I. Sarafoglou and G.N. Haidemenopoulos conceived and designed the experiments; P.I. Sarafoglou, M.-I.T Tzini, and J.S. Aristeidakis performed the experiments and analyzed the data; All authors contributed to the preparation of the manuscript.

Conflicts of Interest: The authors declare no conflicts of interest.

References

1. Sheppard, T. *Extrusion of Aluminum Alloys*; Kluwer Academic Publishers: Dordrecht, The Netherlands, 1999.
2. Xie, F.Y.; Kraft, T.; Zuo, Y.; Moon, C.H.; Chang, Y.A. Microstructure and microsegregation in Al-rich Al-Cu-Mg alloys. *Acta Mater.* **1999**, *47*, 489–500. [CrossRef]
3. Robinson, J.S. Influence of retrogression and reaging on fracture toughness of 7010 aluminum alloy. *Mater. Sci. Tech. Ser.* **2003**, *19*, 1697–1701. [CrossRef]
4. Rokhlin, L.L.; Dobatkina, T.V.; Bochvar, N.R.; Lysova, E.V. Investigation of phase equilibria in alloys of the Al-Zn-Mg-Cu-Zr-Sc system. *J. Alloy. Compd.* **2004**, *367*, 10–16. [CrossRef]
5. Liu, Y.L.; Kang, S.B. The solidification process of Al-Mg-Si alloy. *J. Mater. Sci.* **1997**, *32*, 1443–1447. [CrossRef]
6. Saha, P.K. *Aluminum Extrusion Technology*; ASM International: Metals Park, OH, USA, 2000.
7. Mukhopadhyay, P. Alloy designation, processing and use of AA6xxx series aluminum alloys. *ISRN Metall.* **2012**. [CrossRef]
8. Birol, Y. The effect of homogenization practice on the microstructure of AA6063 billets. *J. Mater. Process. Technol.* **2004**, *148*, 250–258. [CrossRef]
9. Cai, M.; Rodson, J.D.; Lorimer, G.W.; Parson, N.C. Simulation of the casting and Homogenization of two 6xxx Series Alloys. *Mater. Sci. Forum* **2002**, *396*, 209–214. [CrossRef]
10. Usta, M.; Glicksman, M.E.; Wright, R.N. The effect of Heat Treatment on Mg_2Si Coarsening in Aluminum 6105 Alloy. *Metall. Mater. Trans. A* **2004**, *35*, 435–438. [CrossRef]
11. Van de Langkruis, J. The effect of thermal treatments on the extrusion behaviour of AlMgSi alloys. Ph.D. Thesis, Technical University of Delft, Delft, The Netherlands, June 2000.
12. Kuijpers, N.C.W.; Vermolen, F.J.; Vuik, K.; van der Zwaag, S. A model of the β-AlFeSi to α-Al(FeMn)Si transformation in Al-Mg-Si alloys. *Mater. Trans.* **2003**, *44*, 1448–1456. [CrossRef]
13. Kuijpers, N.C.W.; Vermolen, F.J.; Vuik, C.; Koenis, P.T.G.; Nilsen, K.E.; van der Zwaag, S. The dependence of the β-AlFeSi to α-Al(FeMn)Si transformation kinetics in Al-Mg-Si alloys on the alloying elements. *Mater. Sci. Eng. A* **2005**, *394*, 9–19. [CrossRef]
14. Haidemenopoulos, G.N.; Kamoutsi, H.; Zervaki, A.D. Simulation of the transformation of iron intermetallics during homogenization of 6xx series extrudable aluminum alloys. *J. Mater. Process. Technol.* **2012**, *212*, 2255–2260. [CrossRef]
15. Eivani, A.R.; Ahmed, H.; Zhou, J.; Duszczyk, J. Evolution of Grain Boundary Phases during the Homogenization of AA7020 Aluminum Alloy. *Metall. Mater. Trans. A* **2009**, *40*, 717–728. [CrossRef]
16. Eivani, A.R.; Ahmed, H.; Zhou, J.; Duszczyk, J. Correlation between Electrical Resistivity, Particle Dissolution, Precipitation of Dispersoids, and Recrystallization Behavior of AA7020 Aluminum Alloy. *Metall. Mater. Trans. A* **2009**, *40*, 2435–2446. [CrossRef]
17. Eivani, A.R.; Ahmed, H.; Zhou, J.; Duszczyk, J. Modelling dissolution of low melting point phases during the homogenisation of AA7020 aluminium alloy. *Mater. Sci. Technol.* **2010**, *26*, 215–222. [CrossRef]
18. Robson, J.D.; Prangnell, P.B. Dispersoid precipitation and process modelling in zirconium containing commercial aluminium alloys. *Acta Mater.* **2001**, *49*, 599–613. [CrossRef]
19. Fan, X.; Jiang, D.; Meng, Q.; Zhong, L. The microstructural evolution of an Al-Zn-Mg-Cu alloy during homogenization. *Mater. Lett.* **2006**, *60*, 1475–1479. [CrossRef]

20. Kuijpers, N.C.W.; Tirel, J.; Hanlon, D.N.; van der Zwaag, S. Quantification of the evolution of the 3D intermetallic structure in a 6005A aluminium alloy during a homogenisation treatment. *Mater. Charact.* **2002**, *48*, 379–392. [CrossRef]

21. *ASTM Specification F1877 Standard Practice for Characterization of Particles*; ASTM International: Philadelphia, PA, USA, 1998. [CrossRef]

22. Wayne, D. *Applied Nonparametric Statistics*, 2nd ed.; Cengage Learning: Andover, UK, 2000; pp. 226–234.

23. Efron, B.; Tibshirani, R.J. *An Introduction to the Bootstrap*, 1st ed.; Chapman & Hall/CRC: London, UK, 1993; pp. 1–16.

metals

MDPI

Article

Effect of Heat Treatment on the In-Plane Anisotropy of As-Rolled 7050 Aluminum Alloy

Huie Hu [1] and Xinyun Wang [2],*

[1] Department of Chemistry and Materials, Naval University of Engineering, Wuhan 430033, China;
     huhuie@163.com
[2] State Key Laboratory of Materials Processing and Die & Mould Technology,
     Huazhong University of Science and Technology, Wuhan 430074, China
*   Correspondence: bigaxun@263.net; Tel./Fax: +86-27-8754-3491

Academic Editor: Nong Gao
Received: 22 January 2016; Accepted: 29 March 2016; Published: 2 April 2016

Abstract: Tensile tests were conducted on both as-quenched and over-aged 7050 aluminum alloy to investigate the effect of heat treatment on the in-plane anisotropy of as-rolled 7050 aluminum alloy. The results showed that the tensile direction has limited effect on mechanical properties of the as-quenched 7050 aluminum alloy. The in-plane anisotropy factors (IPA factor) of tensile strength, yield strength, and elongation in as-rolled 7050 aluminum alloy fluctuate in the vicinity of 5%. The anisotropy of the as-quenched 7050 aluminum alloy is mainly affected by the texture according to single crystal analysis based on the Schmid factor method. Besides, the IPA factor of the elongation in the over-aged 7050 aluminum alloy reaches 11.6%, illustrating that the anisotropy of the over-aged 7050 aluminum alloy is more prominent than that of the as-quenched. The occurrence of the anisotropy in the over-aged 7050 aluminum alloy is mainly attributed to the microstructures. which are characterized by visible precipitate free zones (PFZs) and coarse precipitates in (sub)grain boundaries.

Keywords: aluminum alloy; heat treatment; anisotropy; microstructure; texture

---

1. Introduction

Heat treatable high-strength aluminum alloys with high strength-density ratio and excellent mechanical properties have already become the primary structural materials of aircraft and vehicles [1–4]. Plastic forming is often used to achieve the final shape of high-strength aluminum alloy products, during which the anisotropy of workability often takes place. The anisotropy is defined as the difference between property values measured along different axes, and very likely to result in unpredicted material flow behavior. Hence, it is meaningful to reveal the anisotropy of high-strength aluminum alloys during plastic working, so as to precisely control the material flow pattern during forming. Besides, high-strength aluminum alloys in the peak strength state are known to be highly susceptible to stress corrosion cracking (SCC). However, the susceptibility of T6 temper to corrosion can be alleviated through the utilization of over-aged T73 temper, which provides improved corrosion resistance, but with a 10%–15% reduction in strength [5]. Therefore, a study of effect of heat treatment, especially over-aging, on the anisotropy of high-strength aluminum alloys during plastic forming is necessary. Moreover, it can also help to deepen the understanding of anisotropic deformation behavior of high-strength aluminum alloys.

It is well known that the anisotropy of aluminum alloys is mainly caused by the crystallographic texture which develops during rolling and heat treatment operation, and the effects of crystallographic texture can be classified into direct effects and indirect one [6–9]. Direct effects are attributed to the orientation of crystals and slip systems with respect to applied stresses and grain morphologies.

Engler *et al.* built the correlation of texture and anisotropic properties of the Al-Mg alloy 5005 based on experiment and simulation [10,11]. Crooks *et al.* concluded that the anisotropy of 2195 aluminum alloy was a direct effect of texture, with no significant contribution from precipitates [12]. Indirect effects are suggested to be caused by work hardening and precipitation during plastic processing, which include the orientation of precipitates with respect to slip systems, the distribution of dislocation densities in differently orientated slip systems and the corresponding distribution of precipitates. Yang *et al.* reported that the anisotropy of the extruded 7075 aluminum alloy bar was resulted from the elongated grain microstructure and {112}<111> and {110}<111> crystal textures after extrusion [13]. Bois-Brochu *et al.* suggested that the strength anisotropy of Al-Li 2099 extrusions might be controlled by the volume fraction of precipitates that could itself be related to the intensity of the <111> fiber texture [14]. Additionally, modeling and simulation work finished by Tome *et al.* as well as their viscoplastic self-consistent code implied that the microstructure was influential, but the effect became secondary when it was compared to that of the texture [15].

As mentioned in available reports, the anisotropy of aluminum alloys may also be influenced by microstructures, such as the average grain shape [16,17], the topology of second phase particles [18,19], the substructure topology [20–22], *etc.*, which are all closely related with heat treatment processes. As reported in our previous work [23], the microstructure, which considers precipitates and PFZs, while ignores the crystallographic texture, is the primary cause of anisotropy of the 7050 aluminum alloy during high temperature deformation. Thus, it is suggested that the microstructure is also an important cause of aluminum alloys' anisotropy, and heat treatment has a significant influence on the anisotropy of aluminum alloys.

However, few studies have concerned the relationship of heat treatment, texture, microstructure and anisotropy of aluminum alloys, except some works reported by Engler *et al.*, which considered the correlation of texture, microstructure and anisotropy in 5xxx aluminum alloys during rolling and annealing [10,11,24]. Meanwhile, to the best of our knowledge, reports on the relationship of heat treatment, texture, microstructure and anisotropy of high-strength aluminum alloys are still not available.

Hence, in this paper, tensile tests were carried out to study the effects of heat treatment on the in-plane anisotropy of as-rolled 7050 aluminum alloy sheet, with attention mainly paid to the different heat treatment conditions.

2. Experimental Section

Commercial as-hot rolled 7050 aluminum alloy plates with 80 mm in thickness were used as sample material in this study. The chemical composition of the alloy is Al-(5.7–6.7)Zn-(1.9–2.6)Mg-(2.0–2.6)Cu-0.1Zr-0.15Fe-0.12Si-0.10Mn (in wt. %). Different directions and planes of the as-rolled 7050 aluminum alloy plate are shown in Figure 1. The centerline layer with 2 mm in thickness was cut out from the as-rolled plate parallel to the rolling plane. Tensile specimens with a 5 mm gauge width and a 20 mm gauge length were prepared along the rolling direction (RD), at 45° from the RD and along the long-transverse direction (LD) respectively, as shown in Figure 2. Before tensile test, the tensile specimens were solid solution treated at 477 °C for 1 h, and then quenched into water. Half of the as-quenched tensile specimens were then over-aged at 100 °C for 4 h followed by another 1 h at 160 °C [25,26]. Afterward, the tensile specimens were subjected to tensile tests within 24 h after heat treatment to investigate the effect of heat treatment on the in-plane anisotropy. Tensile tests were carried out on an Instron-5500R universal testing machine (ITW Test & Measurement, Glenview, IL, USA) at a strain rate of 0.5 mm/min. For each heat treatment condition, three samples were tested, with the averaged experimental data considered as the final result. The differences of the three measurements are less than 5%. A Hitachi S-570 scanning electron microscope (SEM, Tokyo, Japan) was used to analyze the fracture surfaces of tested specimens. Electron back scatter diffraction (EBSD, Oxford Instruments plc, Oxford, UK) measurement samples were mounted and electro-polished using 10 vol. % $HClO_4$ acids in alcohol followed by examined and analyzed using HKL Channel 5 software in a JEOL 733 electron probe (Advanced Microbeam, Vienna, OH, USA) with an accelerating voltage of 20 kV [27]. The samples for optical microscope (OM) were mounted, polished and etched by Keller solution (1.5% HCl + 1% HF + 2.5% HNO_3 + 95% distilled water, in vol. %) for observation by a ZEISS HAL100 microscope (Carl Zeiss Microscopy, Thornwood, NY, USA). The transmission electron microscope (TEM) samples were thinned to about 50 μm followed by electropolish in a double-jet polishing unit operating at 15 V and −20 °C with a 30% nitric acid and 70% methanol solution, the disks were observed in a Tecnai 20 microscope (FEI, Hillsboro, OR, USA), operating at 200 kV.

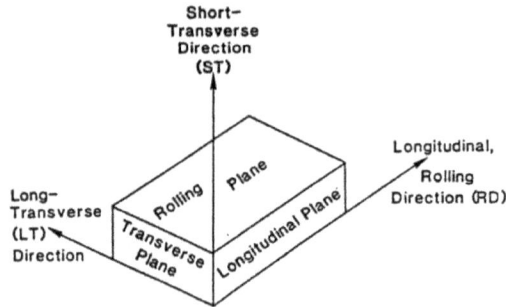

Figure 1. Schematic of different directions and planes in the as-rolled 7050 aluminum alloy plate.

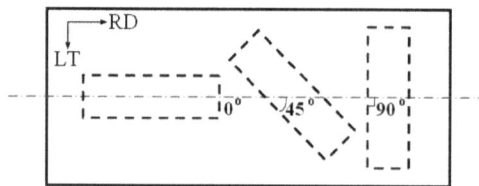

Figure 2. Schematic of tensile specimens with different orientations.

3. Results

3.1. Textures and Grain Microstructures

It is assumed that all the tensile specimens possess of the same texture components before heat treatment because they all were cut out from the centerline layer of an as-rolled 7050 aluminum alloy with 80 mm in thickness. Moreover, over-aging at temperatures lower than 200 °C does not obviously change texture components, thus, both the as-quenched and over-aged tensile specimens also have the same texture components. The variation of orientation densities in α and β fibers of the as-quenched 7050 aluminum alloy indicates that a well-developed fiber consisting of the primary Brass orientation {011}<211> (35°, 45°, 90°), the S orientation {123}<634> (57°, 37°, 63°), and the Copper orientation {112}<111> (90°, 35°, 45°) is evident (see Figure 3). The Brass orientation {011}<211> is found to be the strongest orientation along the β fiber and the maximal intensity of the Brass orientation {011}<211> reaches 35.

Figure 4 is the optical micrographs showing microstructures in the transverse plane and the longitudinal plane of the as-quenched 7050 aluminum alloy. It is demonstrated that the as-quenched 7050 aluminum alloy consists of elliptical grains in the transverse plane, as shown in Figure 4a. The size of grains in the long-transverse direction is about five times of that in the short-transverse direction. The average intercept length measured by random lines drawn parallel to the short-transverse direction is higher than 50 μm. The optical micrograph show that microstructures in the longitudinal plane consist of highly elongated and band-like grains aligned with the rolling direction (see Figure 4b). Figure 5 presents the optical microstructures of the over-aged 7050 aluminum alloy in the transverse plane, which mainly consist of different sized elliptical grains (see Figure 5a). The microstructures of the over-aged 7050 aluminum alloy in the longitudinal plane contain some large elongated grains distributing in small size grains which most likely are sub-grains (see Figure 5b). However, the average grain size of the over-aged 7050 aluminum alloy is around 10 μm and smaller than that of the as-quenched 7050 aluminum alloy.

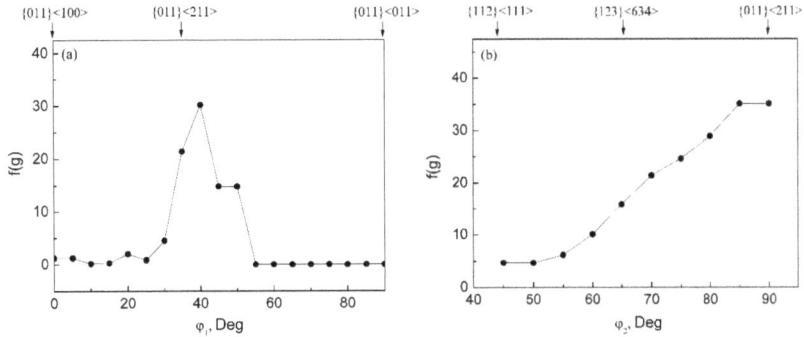

Figure 3. Variation of orientation densities in α and β fibers of the 7050 aluminum alloy (**a**) α fiber; (**b**) β fiber.

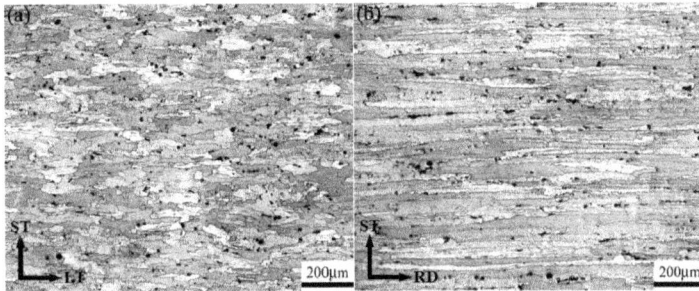

Figure 4. Optical micrographs showing microstructures of the as-quenched 7050 aluminum alloy in (**a**) the transverse plane; (**b**) the longitudinal plane.

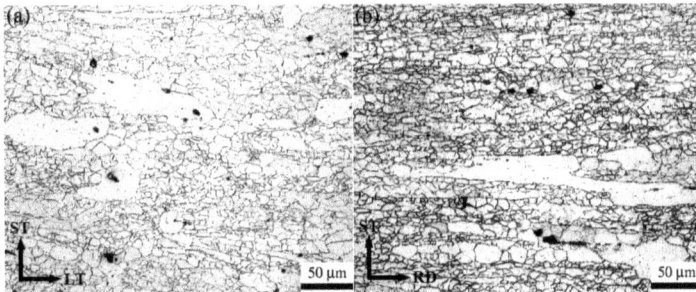

Figure 5. Optical micrographs showing microstructures of the over-aged 7050 aluminum alloy in (**a**) the transverse plane; (**b**) the longitudinal plane.

3.2. Mechanical Properties

Figure 6 shows the true stress-strain curves of both as-quenched and over-aged 7050 aluminum alloys stretched along different directions. Mechanical properties obtained according to Figure 6 are listed in Table 1 including tensile strength (R_m), yield strength ($R_{P0.2}$) and elongation (A). Table 1 shows that over aging can increase the strength while reduce the elongation of the 7050 aluminum alloy in any direction.

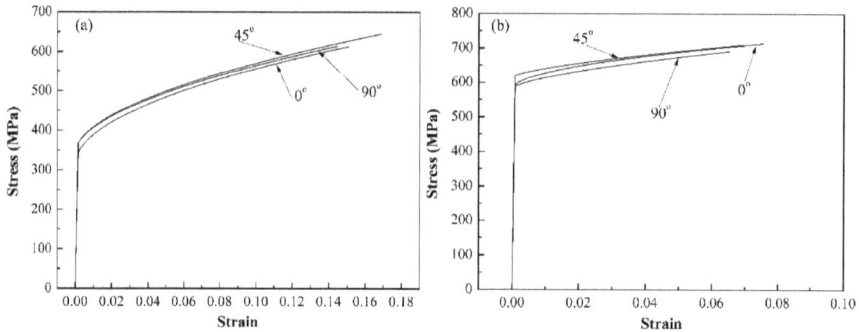

Figure 6. True stress-strain curves of the 7050 aluminum alloy at different tensile directions (**a**) as-quenched; (**b**) over-aged.

Table 1. Mechanical properties of the 7050 aluminum alloy under different heat treatment conditions.

Heat Treatment	As-Quenched			Over-Aged		
Tensile Directions	0°	45°	90°	0°	45°	90°
R_m/MPa	610	641	609	717	705	687
$R_{p0.2}$/MPa	315	345	346	581	616	581
A/%	15.44	15.6	14.08	11.68	11.08	9.56

3.3. In-Plane Anisotropy

The in-plane anisotropy of mechanical properties of the 7050 aluminum alloy is characterized by the IPA factor presented in References [28,29], which is defined as:

$$IPA = \frac{(N-1)X_{\max} - X_{mid1} - X_{mid2} - \ldots X_{mid(N-2)} - X_{\min}}{(N-1)X_{\max}} \times 100\% \tag{1}$$

where, N is the number of specimens' angle along the rolling direction, $X_{\max}$, $X_{\min}$ and X_{mid} are the maximum, the minimum and the rest of mechanical properties respectively. In this study, N is set as 3, since the tensile specimens were prepared along three directions, including the rolling direction (RD), at 45° from the RD and along the long-transverse direction (LD), respectively. So IPA factors of mechanical properties of the 7050 aluminum alloy can be calculated by Equation (2).

$$IPA = \frac{2X_{\max} - X_{mid} - X_{\min}}{2X_{\max}} \times 100\% \tag{2}$$

IPA factors of mechanical properties of as-quenched and over-aged 7050 aluminum alloy are calculated according to Table 1 to illustrate the effect of heat treatment on in-plane anisotropy. Figure 7 shows the IPA factors of tensile strength, yield strength, and elongation of as-quenched and over-aged 7050 aluminum alloy samples. It is shown that the IPA factors of tensile strength, yield strength and elongation of the as-quenched 7050 aluminum alloy fluctuate in the vicinity of 5%. However, both the IPA factors of tensile strength and yield strength are lower than 6%, and while, the IPA factor of elongation reaches 11.6% for the over-aged 7050 aluminum alloy, which is higher than that reported in other 7xxx aluminum alloy [28,29]. The IPA factor results illustrate that tensile direction has greater effect on elongation than tensile strength and yield strength of the over-aged 7050 aluminum alloy. Besides, the over-aged 7050 aluminum alloy shows stronger anisotropy than the as-quenched 7050 aluminum alloy. So the effect of heat treatment on the in-plane anisotropy of the 7050 aluminum alloy was researched by analyzing the relationship between tensile directions and elongations of the 7050 aluminum alloy with different heat treatment conditions.

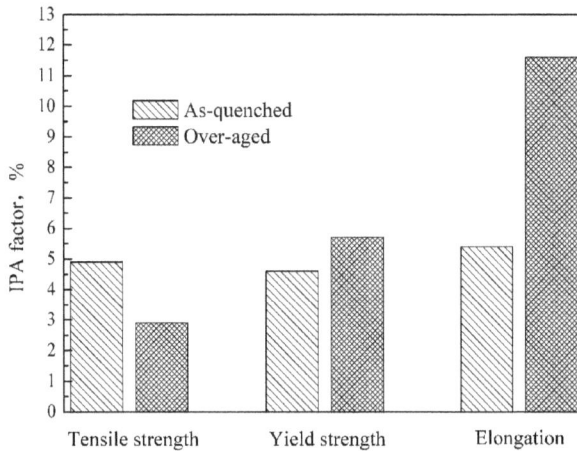

Figure 7. In-plane anisotropy (IPA) factors of mechanical properties of the 7050 aluminum alloy with different heat treatment conditions.

4. Discussion

4.1. In-Plane Anisotropy of the As-Quenched 7050 Aluminum Alloy

Reference [23] reported that the anisotropy of 7050 aluminum alloy was mainly affected by texture components when the alloy elements of the 7050 aluminum alloy are in solution. It is suggested that the texture components are the primary cause of anisotropy of the as-quenched 7050 aluminum alloy.

Changes of orientation densities in α and β fibers of the as-quenched 7050 aluminum alloy imply that the texture components contain the Brass orientation {011}<211> (35°, 45°, 90°), the S orientation {123}<634> (57°, 37°, 63°), and the Copper orientation {112}<111> (90°, 35°, 45°). The intensity of the Brass orientation {011}<211> is 35, much higher than those of the other texture components, indicating that the Brass orientation {011}<211> is the main texture component affecting the in-plane anisotropy of the as-quenched 7050 aluminum alloy. The single crystal analysis, which ignores the rotation of the crystal and the interaction between slip systems, will be conducted on the as-quenched 7050 aluminum alloy based on the Schmid factor ($m = \cos(\varphi)\cos(\lambda)$) method as follows.

It is assumed that the as-quenched 7050 aluminum alloy only comprises the Brass orientation {011}<211> and is considered as a single crystal. The spatial relationship between four possible {111} planes of the as-quenched 7050 aluminum alloy and the Brass orientation {011}<211> is shown in Figure 8. It is demonstrated that $(1\bar{1}1)$ plane and $(\bar{1}11)$ plane are normal to the rolling plane of the as-quenched 7050 aluminum alloy sheet. The angles between the (111) plane, $(11\bar{1})$ plane and the rolling plane are all 35.3°. The deformation behavior of the single crystal with the Brass orientation {011}<211> along the rolling direction (RD) and the long-transverse direction (LD) was analyzed to represent that of the as-quenched 7050 aluminum alloy.

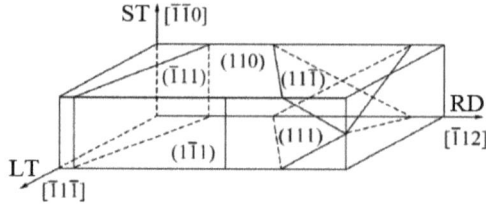

Figure 8. Space relationships of four possible {111} slip planes with the Brass orientation {110}<112>.

Schmid factors of the slip system {111}<110> for a single crystal with the Brass orientation {011}<211> along the rolling direction (RD), at 45° from the RD and along the long-transverse direction (LD) are presented in Table 2. It is shown that the Schmid factors of slip systems (111)[$\bar{1}$01] and (11$\bar{1}$)[011] along the rolling direction (RD) are the biggest and reach 0.408. So the two slip systems with the Schmid factors of 0.408 are the easiest to be activated in the as-quenched 7050 aluminum alloy, which is assumed to be single crystal with the Brass orientation {011}<211>. For the slip system (111)[$\bar{1}$01], the angles between it and the short-transverse direction [$\bar{1}\bar{1}$0], the rolling direction [$\bar{1}$12], and the long-transverse direction [$\bar{1}1\bar{1}$] are 60°, 30° and 90°, respectively. Meanwhile, for the slip system (11$\bar{1}$)[011], the angles between it and the short-transverse direction [$\bar{1}\bar{1}$0], the rolling direction [$\bar{1}$12], and the long-transverse direction [$\bar{1}1\bar{1}$] are 120°, 30° and 90°, respectively. The two slip systems (111)[$\bar{1}$01] and (11$\bar{1}$)[011] are the easiest to be activated during tensile deformation along the rolling direction. Hence, during the deformation of the as-quenched 7050 aluminum alloy with the Brass orientation {011}<211> single crystal along the rolling direction, the sample thickness decreases, the elongation along the tensile direction (RD) increases, while the sample width almost keeps constant. Besides, 86.6% stress acts in the tensile direction to make the as-quenched 7050 aluminum alloy elongate in that direction.

Table 2. Schmid factors of the slip system {111}<110> for various tensile orientations.

Slip Plane	Slip Direction	0°	45°	90°
	[110]	0	0	0
(1$\bar{1}$1)	[011]	0	0.4330	0
	[$\bar{1}$01]	0	0.4330	0
	[110]	0	0	0
($\bar{1}$11)	[0$\bar{1}$1]	0.1361	0.3368	0.2722
	[101]	0.1361	0.3368	0.2722
	[1$\bar{1}$0]	0.2722	0.0962	0.2722
(111)	[0$\bar{1}$1]	0.1361	0.0364	0.2722
	[$\bar{1}$01]	0.4082	0.0598	0
	[1$\bar{1}$0]	0.2722	0.0962	0.2722
(11$\bar{1}$)	[011]	0.4082	0.0598	0
	[101]	0.1361	0.0364	0.2722

It is shown in Table 2 that Schimd factors of slip systems ($\bar{1}$11)[0$\bar{1}$1], ($\bar{1}$11)[101], (111)[1$\bar{1}$0], (111)[0$\bar{1}$1], (11$\bar{1}$)[1$\bar{1}$0] and (11$\bar{1}$)[101] all are the maximal value of 0.2722 when the tensile direction is [$\bar{1}1\bar{1}$] and along the long-transverse direction (LD). Thus, it is indicated that the six slip systems mentioned above in the as-quenched 7050 aluminum alloy with the Brass orientation {011}<211> single crystal were operated at the same time. Table 3 is the shear stress distribution of the six slip systems in the three directions (the rolling direction, the short-transverse direction and the long-transverse direction) in the Brass orientation {110}<112> when the tensile direction is along the long-transverse direction. It is shown that during the tensile deformation along the long-transverse direction in the as-quenched 7050 aluminum alloy with the Brass orientation {011}<211> single crystal, the thickness and width decreases while the elongation of the tensile direction increases. Furthermore, 81.7% shear stress acts in the tensile direction to make the as-quenched 7050 aluminum alloy elongate in the tensile direction. So the strengths of the as-quenched 7050 aluminum alloy along different directions are similar, as shown in Table 1. The difference of elongation along different directions is small and the IPA factor of elongation is only 5.4%.

Table 3. Shear stress distribution of slip systems in the three directions of the Brass orientation {110}<112> when the tensile direction is along the long-transverse direction.

Slip System/Direction	RD[$\bar{1}12$]	ST[$\bar{1}10$]	LT[$\bar{1}\bar{1}1$]
($\bar{1}11$)[$0\bar{1}1$]	$1/\sqrt{12}$	$1/2$	$-2/\sqrt{6}$
($\bar{1}11$)[101]	$1/\sqrt{12}$	$-1/2$	$-2/\sqrt{6}$
(111)[$1\bar{1}0$]	$-\sqrt{1/3}$	0	$2/\sqrt{6}$
(111)[$0\bar{1}1$]	$1/\sqrt{12}$	$1/2$	$-2/\sqrt{6}$
($11\bar{1}$)[$1\bar{1}0$]	$-\sqrt{1/3}$	0	$2/\sqrt{6}$
($11\bar{1}$)[101]	$1/\sqrt{12}$	$-1/2$	$-2/\sqrt{6}$

4.2. In-Plane Anisotropy of the Over-Aged 7050 Aluminum Alloy

The above analysis shows that the texture components in the over-aged 7050 aluminum alloy, which are similar to those in the as-quenched 7050 aluminum alloy, can only result in slight anisotropy. So the effect of tensile direction on the elongation of the over-aged 7050 aluminum alloy is attributed to the microstructure instead of texture, similarly to what was reported in Reference [30].

TEM images show that precipitates can be founded in the over-aged 7050 aluminum alloy, with small size precipitates uniformly distributing inside grains (see Figure 9a). Coarse precipitates in grain boundaries or subgrain boundaries are visible in the over-aged 7050 aluminum alloy (see Figure 9b). As indicated by arrows in Figure 9b, PFZs are very visible and the widths of the PFZs are less 100 nm. The small size precipitates inside grains have limited influence on the elongation of the as-rolled 7050 aluminum alloy. However, coarse precipitates in grain boundaries or subgrain boundaries and obvious PFZs have a significant effect on the plastic deformation behavior and elongation of the as-rolled 7050 aluminum alloy [23].

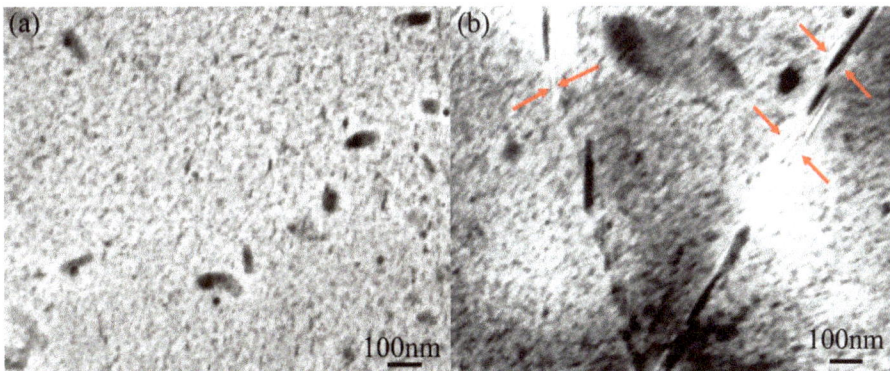

Figure 9. TEM micrographs showing microstructures of the over-aged 7050 aluminum alloy (a) precipitates inside grains; (b) precipitates in grain boundaries and subgrain boundaries.

Figure 10 shows the fracture surfaces of the over-aged 7050 aluminum alloy stretched along different directions. It is implied that the fracture surfaces stretched along different directions are different from ordinary ductile transgranular fracture surface of aluminum alloys characterized by dimples with different sizes. The fracture surfaces are intergranular, in which the initial grain structures and grain boundaries can be clearly distinguished. The grain size of the fracture surface of the over-aged 7050 aluminum alloy stretched along the rolling direction is about 10 μm, which is consistent with the optical microstructure results (see Figures 5a and 10a). Big size elongated grains were observed in the fracture surface of the over-aged 7050 aluminum alloy stretched along the long-transverse direction, which consist of small size grains (see Figure 10b). Figure 9 shows that the size of precipitates in grain boundaries and subgrain boundaries is bigger than that inside grains. It is easy to be eroded for the subgrain boundaries as the grain boundaries. So the small size grains in Figures 5b and 10b are subgrains in nature. Similar research results have also been reported in our previous studies [23,31,32].

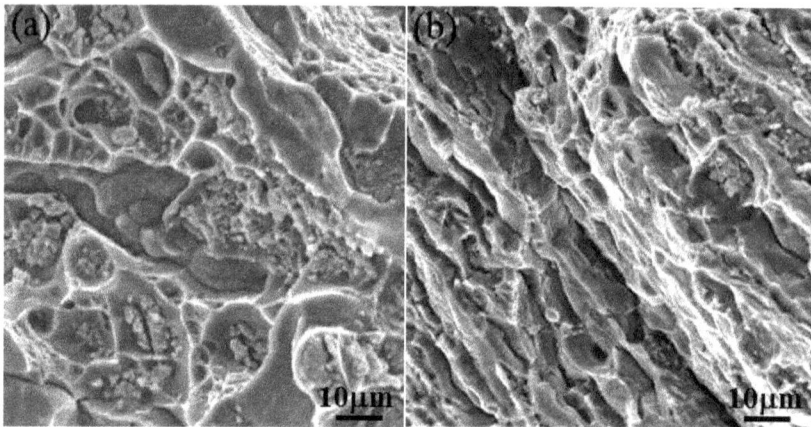

Figure 10. SEM micrographs showing fracture surfaces of the over-aged 7050 aluminum alloy along different directions (**a**) the rolling direction (RD); (**b**) the long-transverse direction (LT).

During plastic deformation of the over-aged 7050 aluminum alloy, dislocations bow around, but do not cut through, the precipitates with big size and high hardness. At the same time, the precipitates in subgrain boundaries pin and inhibit the migration of subgrain boundaries. However, the friction of dislocations movement in PFZs is lower than that inside the grains because there are only a few precipitates in PFZs of the over-aged 7050 aluminum alloy. Meanwhile, alternate slipping is easy to occur in PFZs because of few precipitates in (sub)grain boundaries and property of texture (slip systems are nearly parallel to (sub)grain boundaries). Thus, there are more plastic strains in PFZs than those inside the grains. It is to say that the in-plane anisotropy of the over-aged 7050 aluminum alloy has a primary relationship with PFZs' shapes, *viz.* grains' and subgrains' shapes. The greater difference of grains' and subgrains' shapes in different planes, as shown in Figures 5 and 10 is the primary cause of higher IPA factor in the elongation of the over-aged 7050 aluminum alloy. Figures 5a and 10a show that the grain size in the transverse plane of the over-aged 7050 aluminum alloy is smaller than that in the longitudinal plane. So the PFZs can provide more strains when the over-aged 7050 aluminum alloy is stretched along the rolling direction, than being stretched along the long-transverse direction. Besides, intergranular fractures in Figure 10 indicate cracks in the over-aged 7050 aluminum alloy grow mainly along (sub)grain boundaries. So the smaller the grain size of fracture surfaces, the longer the crack propagation path before failure will be, impling that the elongation of the over-aged 7050 aluminum alloy along the rolling direction is higher than that along the long-transverse direction. The above analyzes also reveal that the elongation and microstructure results are consistent with fracture surfaces.

5. Conclusions

(1) For the as-quenched 7050 aluminum alloy, the tensile direction has little effect on anisotropies of mechanical properties, and the IPA factors of tensile strength, yield strength and elongation fluctuate in the vicinity of 5%.

(2) For the over-aged 7050 aluminum alloy, the difference of IPA factors of mechanical properties is apparent. The tensile direction has a significant effect on the elongation, and the IPA factor of elongation reaches 11.6%.

(3) The intensity of the Brass orientation {011}<211> in the as-quenched 7050 aluminum alloy is much higher than those of the other texture components. The influence of texture on the in-plane anisotropy of the as-quenched 7050 aluminum alloy is revealed by building the relationship between the elongation and the Brass orientation {011}<211> using the single crystal analysis based on the Schmid factor method.

(4) The microstructures of the over-aged 7050 aluminum alloy are characterized by obvious PFZs and coarse precipitates in (sub)grain boundaries. Deformation is easier to take place in PFZs than that inside grains. The shapes of PFZs, *viz.* grains' and subgrains' shapes, are the primary cause of the in-plane anisotropy in the over-aged 7050 aluminum alloy.

Acknowledgments: This work was supported by National Nature Science Foundation of China (NSFC-51575522), and State Key Laboratory of Materials Processing and Die & Mould Technology (P2016-02).

Author Contributions: The preparation of test samples was supported by Hu Huie. The statistical analysis was undertaken by Wang Xin-yun. The paper was written by Hu Huie and Wang Xin-yun.

Conflicts of Interest: The authors declare no conflict of interest.

References

1. Williams, J.C.; Starke, E.A. Progress in structural materials for aerospace systems. *Acta Mater.* **2003**, *51*, 5775–5799. [CrossRef]
2. Deschamps, A.; Brechet, Y. Influence of quench and heating rates on the ageing response of an Al-Zn-Mg-(Zr) alloy. *Mater. Sci. Eng. A* **1998**, *251*, 200–207. [CrossRef]
3. Tajally, M.; Emadoddin, E. Mechanical and anisotropic behaviors of 7075 aluminum alloy sheets. *Mater. Des.* **2011**, *32*, 1594–1599. [CrossRef]
4. Hu, H.E.; Zhen, L.; Chen, J.Z.; Yang, L.; Zhang, B.Y. Microstructure evolution in hot deformation of 7050 aluminium alloy with coarse elongated grains. *Mater. Sci. Technol.* **2008**, *24*, 281–286. [CrossRef]
5. Ranganathal, R.; Anilkumar, V.; Nandi, V.S.; Bhat, R.R.; Muralidhara, B.K. Multi-stage heat treatment of aluminum alloy AA7049. *Trans. Nonferrous. Met. Soc. China* **2013**, *23*, 1570–1575.
6. Crumbach, M.; Neumann, L.; Goerdeler, M.; Aretz, H.; Gottstein, G. Through-process modelling of texture and anisotropy in AA5182. *Modell. Simul. Mater. Sci. Eng.* **2006**, *14*, 835–856. [CrossRef]
7. Engler, O.; An, Y.G. Correlation of texture and plastic anisotropy in the Al-Mg alloy AA 5005. *Solid State Phenom.* **2005**, *105*, 277–284. [CrossRef]
8. Beaudon, A.J.; Dawson, P.R.; Mathur, K.K. A hybrid finite element formulation for polycrystal plasticity with consideration of macrostructural and microstructural linking. *Int. J. Plast.* **1995**, *11*, 501–521. [CrossRef]
9. Van Houtte, P. Treatment of elastic and plastic anisotropy of polycrystalline materials with texture. *Mater. Sci. Forum* **1998**, *273–275*, 67–75. [CrossRef]
10. Engler, O. Texture and anisotropy in the Al-Mg alloy AA5005—Part I: Texture evolution during rolling and recrystallization. *Mater. Sci. Eng. A* **2014**, *618*, 654–662. [CrossRef]
11. Engler, O.; Aegerter, J. Texture and anisotropy in the Al-Mg alloy AA5005—Part II: Correlation of texture and anisotropic properties. *Mater. Sci. Eng. A* **2014**, *618*, 663–671. [CrossRef]
12. Crooks, R.; Wang, Z.; Levit, V.I.; Shenoy, R.N. Microtexture, micro structure and plastic anisotropy of AA2195. *Mater. Sci. Eng. A* **1998**, *257*, 145–152. [CrossRef]
13. Yang, Y.B.; Xie, Z.P.; Zhang, Z.M.; Li, X.B.; Wang, Q.; Zhang, Y.H. Processing maps for hot deformation of the extruded 7075 aluminum alloy bar: Anisotropy of hot workability. *Mater. Sci. Eng. A* **2014**, *615*, 183–190. [CrossRef]

14. Bois-Brochu, A.; Blais, C.; Goma, F.A.T.; Larouche, D.; Boselli, J.; Brochu, M. Characterization of Al-Li 2099 extrusions and the influence of fiber texture on the anisotropy of static mechanical properties. *Mater. Sci. Eng. A* **2014**, *597*, 62–69. [CrossRef]
15. Lebensohn, R.A.; Tomé, C.N.; Maudlin, P.J. A self-consistent formulation for the prediction of the anisotropic behavior of viscoplastic polycrystals with voids. *J. Mech. Phys. Solid* **2004**, *52*, 249–278. [CrossRef]
16. Choi, S.H.; Brem, J.C.; Barlat, F.; Oh, K.H. Macroscopic anisotropy in AA5019A sheets. *Acta Mater.* **2000**, *48*, 1853–1863. [CrossRef]
17. Delannay, L.; Melchior, M.A.; Signorelli, J.W.; Remacle, J.F.; Kuwabara, T. Influence of grain shape on the planar anisotropy of rolled steel sheets—Evaluation of three models. *Compos. Mater. Sci.* **2009**, *45*, 739–743. [CrossRef]
18. Bate, P.; Roberts, W.T.; Wilson, D.V. The plastic anisotropy of two-phase aluminium alloys—I. Anisotropy in unidirectional deformation. *Acta Metall.* **1981**, *29*, 1797–1814. [CrossRef]
19. Choi, S.H.; Barlat, F.; Liu, J. Effect of precipitates on plastic anisotropy for polycrystalline aluminum alloys. *Metall. Mater. Trans. A* **2001**, *32*, 2239–2247. [CrossRef]
20. Peeters, B.; Seefeldt, M.; Kalidindi, S.R.; van Houtte, P.; Aernoudt, E. The incorporation of dislocation sheets into a model for plastic deformation of b.c.c. polycrystals and its influence on *r* values. *Mater. Sci. Eng. A* **2001**, *319–321*, 188–191. [CrossRef]
21. Mahesh, S.; Tome, C.N.; McCabe, R.J.; Kaschner, G.C.; Beyerlein, I.J.; Misra, A. Application of a substructure-based hardening model to copper under loading path changes. *Metall. Mater. Trans. A* **2004**, *35*, 3763–3774. [CrossRef]
22. Li, Z.J.; Winther, G.; Hansen, N. Anisotropy in rolled metals induced by dislocation structure. *Acta Mater.* **2006**, *54*, 401–410. [CrossRef]
23. Wang, X.Y.; Hu, H.E.; Xia, J.C. Effect of deformation condition on plastic anisotropy of as-rolled 7050 aluminum alloy plate. *Mater. Sci. Eng. A* **2009**, *515*, 1–9.
24. Engler, O. Texture and anisotropy in cold rolled and recovery annealed AA 5182 sheets. *Mater. Sci. Technol.* **2015**, *31*, 1058–1065.
25. Rometsch, P.A.; Zhang, Y.; Knight, S. Heat treatment of 7xxx series aluminium alloys—Some recent developments. *Trans. Nonferrous. Met. Soc. China* **2014**, *24*, 2003–2017. [CrossRef]
26. Zheng, Y.; Yin, Z.M.; Zhu, Y.Z. Microstructure investigation of a new type super high strength aluminum alloy at different heat-treated conditions. *Rare Metal.* **2004**, *23*, 377–384.
27. Cao, W.Q.; Godfrey, A.; Liu, Q. Annealing behavior of aluminium deformed by equal channel angular pressing. *Mater. Lett.* **2003**, *57*, 3767–3774. [CrossRef]
28. Singh, R.K.; Singh, A.K.; Prasad, N.E. Texture and mechanical property anisotropy in an Al-Mg-Si-Cu alloy. *Mater. Sci. Eng. A* **2000**, *277*, 114–122. [CrossRef]
29. Jata, K.V.; Hopking, A.K.; Rioja, R.T. The ainsotropy and texture of Al-Li alloy. *Mater. Sci. Forum* **1996**, *217–222*, 647–652. [CrossRef]
30. Melton, K.N.; Cutler, C.P.; Edington, J.W. Anisotropy during superplastic deformation of the Sn-Pb eutectic alloy. *Scr. Metall.* **1975**, *9*, 515–520. [CrossRef]
31. Hu, H.E.; Zhen, L.; Yang, L.; Shao, W.Z.; Zhang, B.Y. Deformation behavior and microstructure evolution of 7050 aluminum alloy during high temperature deformation. *Mater. Sci. Eng. A* **2008**, *488*, 64–71. [CrossRef]
32. Zhen, L.; Hu, H.E.; Wang, X.Y.; Zhang, B.Y.; Shao, W.Z. Distribution characterization of boundary misorientation angle of 7050 aluminum alloy after high-temperature compression. *J. Mater. Process. Technol.* **2009**, *209*, 754–761. [CrossRef]

metals

MDPI

Article

High-Temperature Compressive Resistance and Mechanical Properties Improvement of Strain-Induced Melt Activation-Processed Al-Mg-Si Aluminum Alloy

Chia-Wei Lin, Fei-Yi Hung * and Truan-Sheng Lui

Department of Materials Science and Engineering, National Cheng Kung University, Tainan 701, Taiwan; qqkm0526@gmail.com (C.-W.L.); luits@mail.ncku.edu.tw (T.-S.L.)
* Correspondence: fyhung@mail.ncku.edu.tw; Tel.: +886-6-275-7575 (ext. 31395); Fax: +886-6-234-6290

Academic Editor: Nong Gao
Received: 13 June 2016; Accepted: 21 July 2016; Published: 5 August 2016

Abstract: Even though the high-temperature formability of Al alloys can be enhanced by the strain-induced melt activation (SIMA) process, the mechanical properties of the formed alloys are necessary for estimation. In this research, a modified two-step SIMA (TS-SIMA) process that omits the cold working step of the traditional SIMA process is adopted for the 6066 Al-Mg-Si alloy to obtain globular grains with a short-duration salt bath. The high-temperature compressive resistance and mechanical properties of TS-SIMA alloys were investigated. The TS-SIMA alloys were subjected to artificial aging heat treatment to improve their mechanical properties. The results show that the TS-SIMA process can reduce compression loading by about 35%. High-temperature compressive resistance can be reduced by the TS-SIMA process. After high-temperature compression, the mechanical properties of the TS-SIMA alloys were significantly improved. Furthermore, artificial aging treatment can be used to enhance formed alloys via the TS-SIMA process. After artificial aging treatment, the mechanical properties of TS-SIMA alloys are comparable to those of general artificially-aged materials.

Keywords: aluminum alloy; Strain-Induced Melt Activation (SIMA); mechanical properties

---

1. Introduction

6xxx series Al alloys, a series of precipitation-hardened Al alloys, are widely used. 6066 Al alloy, used in this study, has a strength that is higher than that of the great majority of other alloys in this series due to its Cu and Mn addition and excess Si [1,2]. This alloy is widely applied in the automobile industry, bicycle industry, and architecture components, due to its high strength and low density. Even though Cu and Mn increase strength, they decrease formability. In order to promote formability, the strain-induced melt activation (SIMA) process is used for forming at high temperatures.

The SIMA process is a semi-solid process, in which the materials are manufactured at temperatures of solid-liquid coexistence. The finished products have a near-net shape advantage [3–5]. The SIMA process has great potential due to its low cost and high stability [6–10]. Figure 1a shows the procedure of the two-step SIMA (TS-SIMA) process proposed in this study. The steps are: (1) casting, which produces a dendritic structure; (2) hot extrusion, which disintegrates the initial structure and introduces sufficient strain energy into the alloy; and (3) salt bath, which makes the material recrystallize and partially melt at temperatures of solid-liquid coexistence. TS-SIMA is defined as a two-step process because the casting materials are via only two steps to obtain globular grains. The two major differences between the traditional SIMA process and the TS-SIMA process are: (1) the proposed TS-SIMA process uses severe hot extrusion instead of cold work to introduce a large amount of strain energy; and (2) the proposed SIMA process uses a salt bath instead of an air furnace to improve heating uniformity and reduce heating time. The globular grain evolution for the proposed TS-SIMA process is shown in Figure 1b [11].

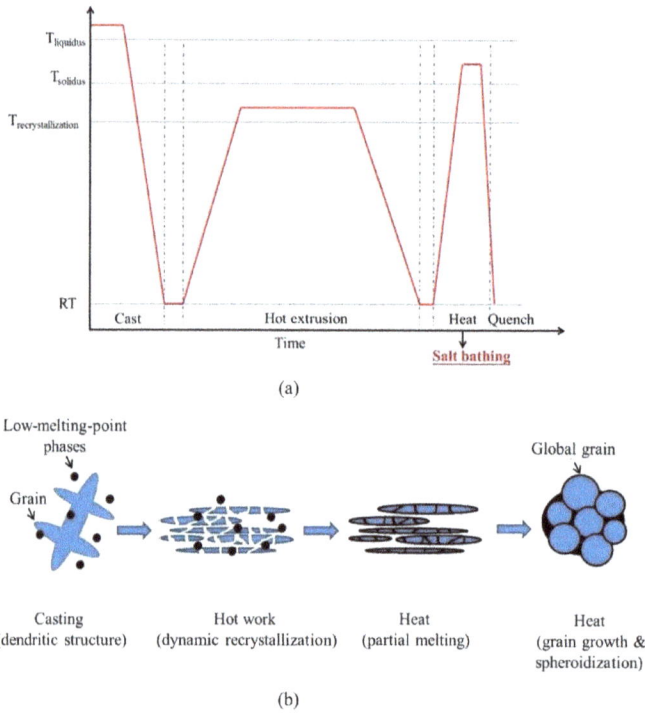

(a)

(b)

Figure 1. (**a**) Procedure of TS-SIMA process and (**b**) formation steps of globular grains in TS-SIMA process.

In our previous study [11], the high-temperature deformation resistance and forming behavior of TS-SIMA alloys were investigated. The improvement in the high-temperature formability of alloys subjected to the TS-SIMA process was confirmed. However, the mechanical properties of the formed alloys after the TS-SIMA process were not examined. In this research, the high-temperature compressibility and the improvement of the mechanical properties of TS-SIMA alloys are investigated. High-temperature compressibility is evaluated using high-temperature compression. The compression deformation mechanism of TS-SIMA alloys is also investigated. The mechanical properties of TS-SIMA alloys are investigated and improved via artificial aging (T6) heat treatment.

2. Materials and Methods

The material used in this study was extruded 6066 Al alloy. Its composition, determined using a glow discharge spectrometer, is shown in Table 1. Six-inch (15.24 cm) diameter casting materials were extruded with dimensions of 52 mm (width) × 3 mm (thickness) and 75 mm (width) × 9 mm (thickness). The extrusion ratio was 27:1 and the true strain was 3.3. The as-extruded alloy is denoted as "F".

Table 1. Composition of 6066 Al alloy.

Element	Mg	Si	Cu	Mn	Fe	Cr	Al
Mass %	1.02	1.29	0.98	1.02	0.19	0.18	Bal.

The salt bath for spheroidized grain formation was conducted at 620 °C for 10 min and then cooled down by quenching in water. The grains were spheroidized uniformly and the fraction of liquid phases was high with these salt bath settings. The material deformed severely, or was partially melted severely, when the temperature was higher than 620 °C. The TS-SIMA alloy subjected to this salt bath is denoted as "S10".

Aluminum alloys are often fully annealed for subsequent manufacturing. Therefore, the test alloy in this study was fully annealed for comparison with TS-SIMA-processed specimens. In the full annealing treatment, F was heated to 420 °C for 2 h, cooled to 220 °C at a cooling rate of 25 °C/min, and then cooled in a furnace to room temperature. The fully annealed 6066 Al alloy is denoted as "O".

The microstructural characteristics and grain size were analyzed using optical microscopy (OM). The specimens were polished using SIC papers from 80# to 5000# (the number before # means how many hard particles in per square inch), Al_2O_3 aqueous suspension (1.0 and 0.3 μm), and SiO_2 polishing suspension and etched using Keller's reagent. The liquid fraction of the lower-melting-point second phases was measured using ImageJ (National Institetes of Health, Java 1.8.0_60, New York, NY, USA) software. Two shape parameters, x and z, were defined for the degree of spheroidization [5]. In Figure 2, a, b, c, and A represent the major axis, minor axis, perimeter, and area of a grain, respectively. According to the definitions $x = (b/a)$ and $z = (4\pi A)/c^2$, x is the ratio of the minor axis to the major axis and z becomes closer to 1 as the shape becomes more circular. As x and z become closer to 1, the grains become more equi-axial and the degree of spheroidization increases.

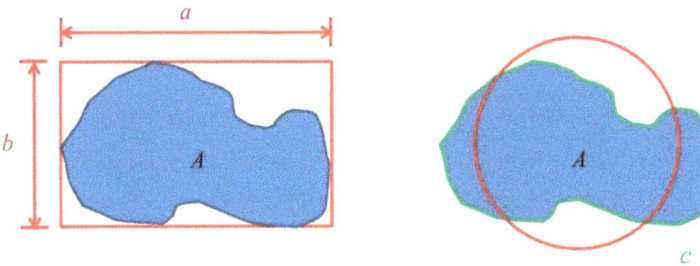

Figure 2. Parameters of spheroidization degree definition.

The hardness of the matrix and globular grain boundaries were evaluated using nano-indentation to understand the hardness distribution in the TS-SIMA alloys. A triangular pyramidal diamond probe was used for nano-indentation. The measurement conditions were a drift velocity of 0.25 nm/s and a depth of 800 nm. The space between measurement points was 5 μm.

In the high-temperature compression test, as-extruded alloys, fully-annealed alloys, and TS-SIMA alloys were tested to compare their high-temperature formability. The compression ratio is defined as $R\% = (t_0-t_f)/t_0$, where t_0 is the thickness of the initial sheet (9 mm) and t_f is the thickness after compression. The specimens for compression had dimensions of 40 mm (length) × 20 mm (width) × 9 mm (thickness). The compression temperature was set as 600 °C and the compression rate was set as 20 mm/min. The compressive loadings of the different materials were estimated and compared as the compression ratio reached 50%. When the compression ratio is higher, the deformation resistance is lower, which indicates better high-temperature formability [12]. The specimens compressed to a compression ratio of 50% were used in further experiments for improving the mechanical properties. Specimens subjected to compression are marked with the prefix "C-".

Finally, in order to confirm that the mechanical properties of the compressed TS-SIMA alloys can be enhanced, T6 heat treatment was adopted. T6 heat treatment includes solution heat treatment and artificial aging. Solution temperatures of 530 °C and 550 °C were used in this research. Specimens subjected to T6 heat treatment are marked the suffix "T6$_{530}$" or "T6$_{550}$". The hardness of the specimens was measured using a Rockwell hardness tester and the tensile properties were tested using a universal tester. The dimensions of the tensile test specimen are shown in Figure 3. The tensile specimen was prepared by a milling machine for thinning and wire cutting for shaping. The tensile initial strain velocity was 1.67×10^{-3} (crosshead velocity of 1 mm/min). Each hardness and tensile datum was the average from at least three testing samples.

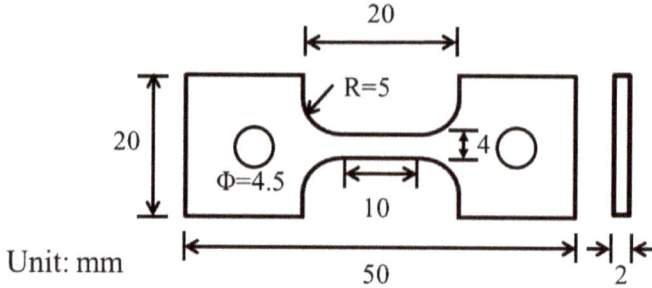

Figure 3. Dimensions of tensile test specimen.

3. Results and Discussion

3.1. Microstructure Characteristics

Figure 4 shows the microstructures of as-extruded alloys and TS-SIMA alloys. The typical extrusion microstructure can be seen in the metallography of the as-extruded alloys, as shown in Figure 4a. Dynamic recrystallization only occurred in parts of F; the recrystallized grain size was about 5–8 μm. Grains were spheroidized uniformly after a salt bath for 10 min. The average globular grain size was about 78 μm. The shape parameters x and z were 0.62 and 0.65, respectively.

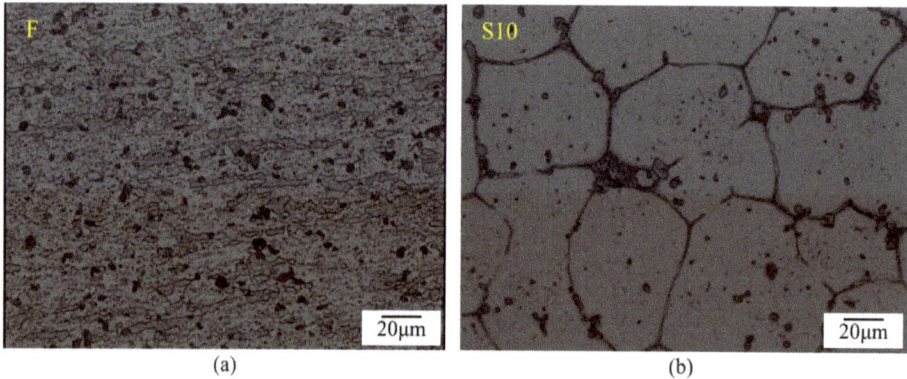

Figure 4. Microstructures of (**a**) as-extruded alloys (F) and (**b**) TS-SIMA alloys (S10).

The distribution of elements in S10 was analyzed using electron probe microanalysis (EPMA) (JEOL, Peabody, MA, USA). The results are shown in Figure 5. After a salt bath, Mg, Si, and Cu were located at the grain boundaries and formed a network structure, but Mn, Fe, and Cr just aggregated and formed a particle-shaped phase due to the melting point of the Mn-rich phase being higher than 620 °C [13]. The phases at globular boundaries are composed of the eutectic phase of Al and Al_2Cu, the eutectic phase of Al and Mg_2Si, and the eutectic phase of Al and Si. The melting points of these eutectic phases are below 620 °C and, thus, they melted and penetrated into the globular grain boundaries.

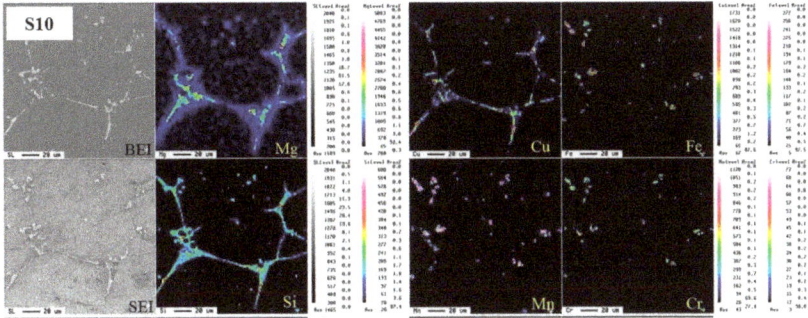

Figure 5. Elemental distribution of TS-SIMA alloy (S10) obtained using EPMA.

The nano-indentation data for S10 are shown in Figure 6. The same results were obtained for five samples. The spheroidized grain boundaries, abundant in Cu, Mg, and Si, are much harder than the internal grains. This proves that the grain boundaries of the TS-SIMA alloy are the hard and brittle parts of the material. When a TS-SIMA alloy is defomed, the deformation should be where stress concentration occurs.

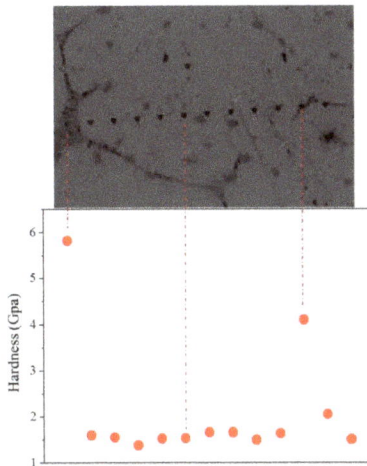

Figure 6. Hardness distribution in TS-SIMA alloys evaluated using nano-indentation.

3.2. High-Temperature Compressive Resistance of TS-SIMA Alloy

For the compression test at 600 °C, Figure 7 shows the compression loading at a 50% compression ratio for various materials. It can be seen that S10 has the lowest compression loading. Full annealing reduced compression loading by only about 9% but the TS-SIMA process reduced it by about 35% compared with that of the as-extruded alloys. This proves that the TS-SIMA process is beneficial for enhancing high-temperature compressibility. The compressive resistance of the TS-SIMA alloy was the smallest.

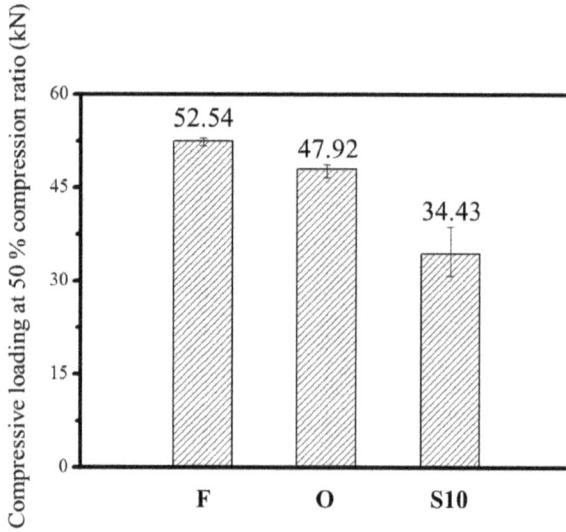

Figure 7. Deformation resistance of several materials.

Figure 8 shows the microstructures of compressed TS-SIMA alloys. It can be seen that after high-temperature compression, globular grains became flat and oval-shaped, as shown in Figure 8a. Under large magnification (Figure 8b), it can be seen that the original broad grain boundaries of the TS-SIMA alloys vanished after compression. Only Mn-rich particle phases existed at the grain boundaries and in the internal grains. This resulted from the low-melting-point phases at grain boundaries melting at 600 °C and flowing during high-temperature compression.

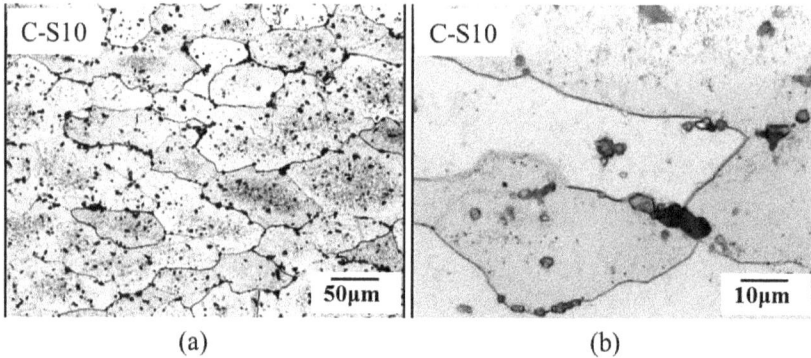

(a)   (b)

Figure 8. Microstructure of TS-SIMA alloy at (**a**) small and (**b**) large magnification.

Figure 9 shows the elemental distribution of compressed TS-SIMA alloys. It shows that Cu, Mg, and Si were no longer located at the globular grain boundaries after high-temperature compression. They diffused and solid-soluted into the matrix during hot-temperature compression. In contrast, Mn, Fe, and Cr still aggregated and formed a particle-shaped phase.

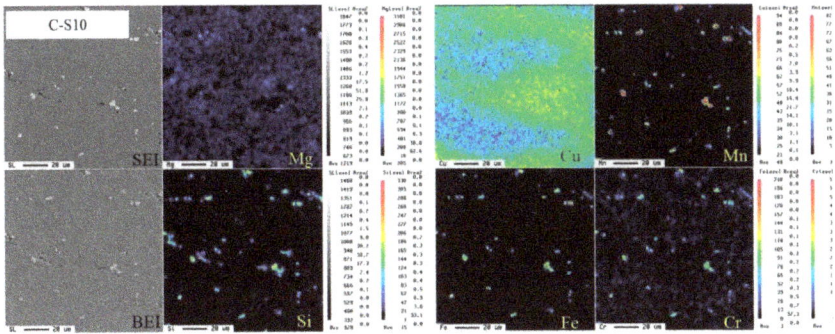

Figure 9. Elemental distribution of compressed TS-SIMA alloy (C-S) obtained using EPMA.

3.3. Mechanical Properties Improvement of TS-SIMA Forming Alloys

In order to ensure that the formed products are suitable for applications, the mechanical properties of compressed TS-SIMA alloys were investigated. T6 heat treatment, the most commonly used method for strengthening 6xxx series Al alloys, is used in this study. Figure 10 shows the hardness data and Figure 11a shows the tensile properties data of compressed and heat-treated materials. Hardness data show that the hardness values of as-extruded alloys and TS-SIMA alloys are similar. High-temperature compression significantly enhanced hardness. After T6 heat treatment, the hardness of all specimens increased obviously. Hardness increased with increasing solution heat treatment temperature due to the solution limit being enhanced by increased solution temperature. The hardness of compressed TS-SIMA alloys is slightly lower than that of as-extruded alloys after T6 heat treatment. Strength data trends are similar to those of hardness data. The strength of specimens increased after T6 heat treatment. The strength of compressed TS-SIMA alloys was slightly lower (by about 10–20 MPa) than that of as-extruded alloys. The ultimate tensile strength (UTS) of TS-SIMA alloys reached about 430–440 MPa. This shows that the strength of TS-SIMA forming materials after T6 heat treatment is high enough for common applications.

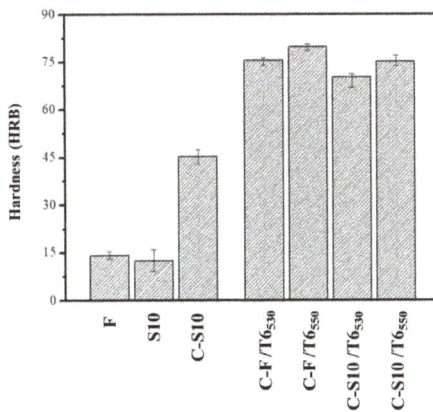

Figure 10. Hardness data of specimens.

Elongation data are shown in Figure 11b. The tensile elongation of TS-SIMA alloys is much lower than that of as-extruded alloys. Elongation can be improved to about 23% uniform elongation (UE) and 27% total elongation (TE) after compression at 600 °C. The enhancement of elongation is majorly due to the hard and brittle phases located at globular boundaries composed of Al, Mg, Si, and Cu diffused into matrix After T6 heat treatment, elongation increased with decreasing solution heat treatment temperature. Uniform elongation reached about 12% and total elongation reached 16% when the solution heat treatment temperature was 530 °C. Even though the elongation of TS-SIMA alloys was quite low, high-temperature compression improved it. The mechanical properties of TS-SIMA alloys can thus be improved by high-temperature compression and T6 heat treatment. Strength can reach more than 400 MPa and elongation can reach more than 10% after appropriate heat treatment.

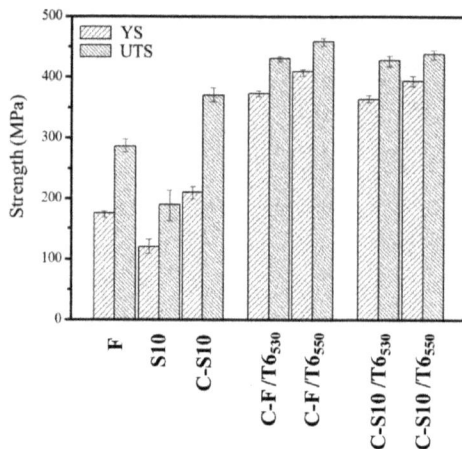

(a)

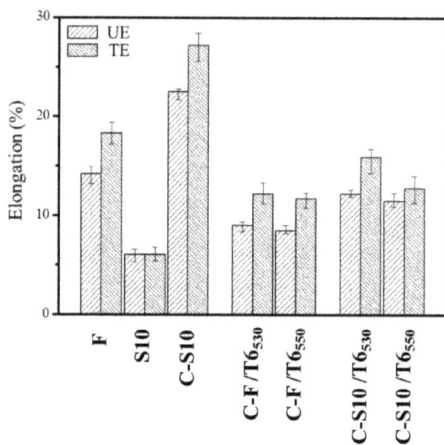

(b)

Figure 11. Mechanical properties of specimens: (**a**) tensile strength and (**b**) tensile elongation.

Figure 12 shows the microstructures of compressed TS-SIMA alloys after T6 heat treatment. It shows that the original compressed globular grains of compressed TS-SIMA alloys grew during solution treatment, as shown in Figure 12c,d. The microstructure of T6-heat-treated as-extruded alloys remained as fine recrystallized grains, as show in Figure 12e,f. The level of precipitation strengthening of compressed TS-SIMA alloys and as-extruded alloys should be similar because their compositions and T6 heat treatment conditions are the same. The slightly different mechanical properties are due to grain size according to Hall-Petch theory [14,15]. After T6 heat treatment, the strength of as-extruded alloys was higher than that of TS-SIMA forming alloys due to the former's fine grains. However, the high density of grain boundaries of T6-heat-treated as-extruded alloys slightly decreased elongation because dislocation slipping is restricted by grain boundaries. Therefore, the elongation of TS-SIMA forming alloys is higher than that of T6-heat-treated as-extruded alloys.

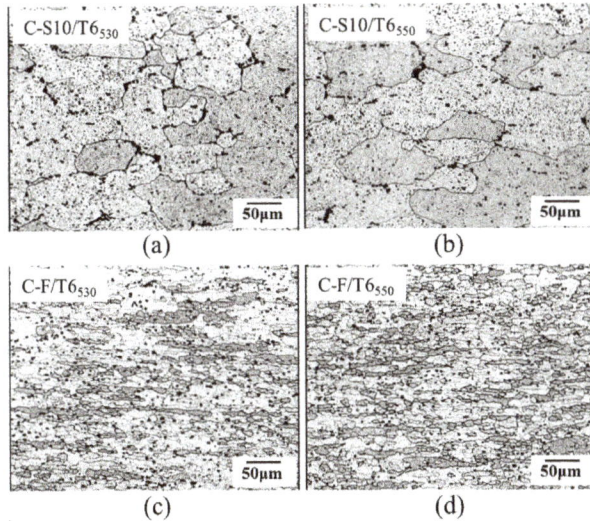

Figure 12. Morphologies of (**a**) C-S10/T6$_{530}$; (**b**) C-S10/T6$_{550}$; (**c**) C-F/T6$_{530}$; and (**d**) C-F/T6$_{550}$.

Figure 13 shows the elemental distribution of compressed TS-SIMA alloys after T6 heat treatment. It shows that all elements were distributed uniformly. This proves that T6 heat treatment made Cu, Mg, and Si solid-solute completely and precipitate.

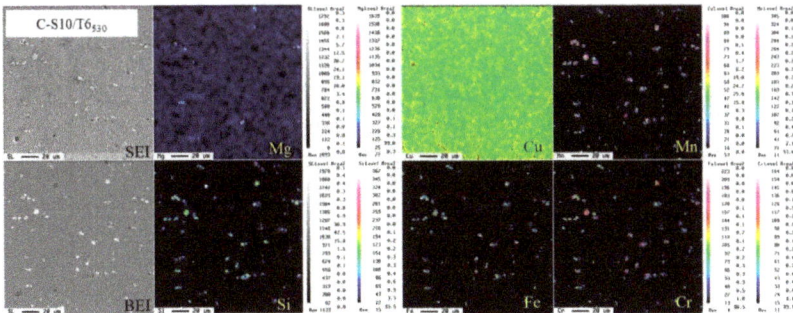

Figure 13. Elemental distribution of compressed TS-SIMA alloy after T6 heat treatment (C-S10/T6$_{530}$) obtained using EPMA.

The fracture mechanism of the above specimens can be interpreted from Figures 14 and 15. Figures 14a and 15a show the intergranular fracture characteristics of TS-SIMA alloys. The low-melting-point phases melted, penetrated, and solidified at globular grain boundaries, resulting in the grain boundary being more brittle and harder than the matrix. This led to stress concentration and the generation of cracks. The cracks initiated at grain boundaries and connected with each other, leading to intergranular fracture. In contrast, characteristic dimple fractures were found on the fracture surfaces of compressed TS-SIMA alloys and T6-heat-treated compressed TS-SIMA alloys, as shown in Figure 14c–f. Micro-void coalescence and ductile fracture caused these dimple fractures. The sub-surface morphologies of compressed TS-SIMA alloys and T6-heat-treated compressed TS-SIMA alloys are shown in Figure 15b–d. Intergranular fractures did not appear because the brittle and low-melting-point phases vanished after high-temperature compression. Therefore, good mechanical properties were obtained.

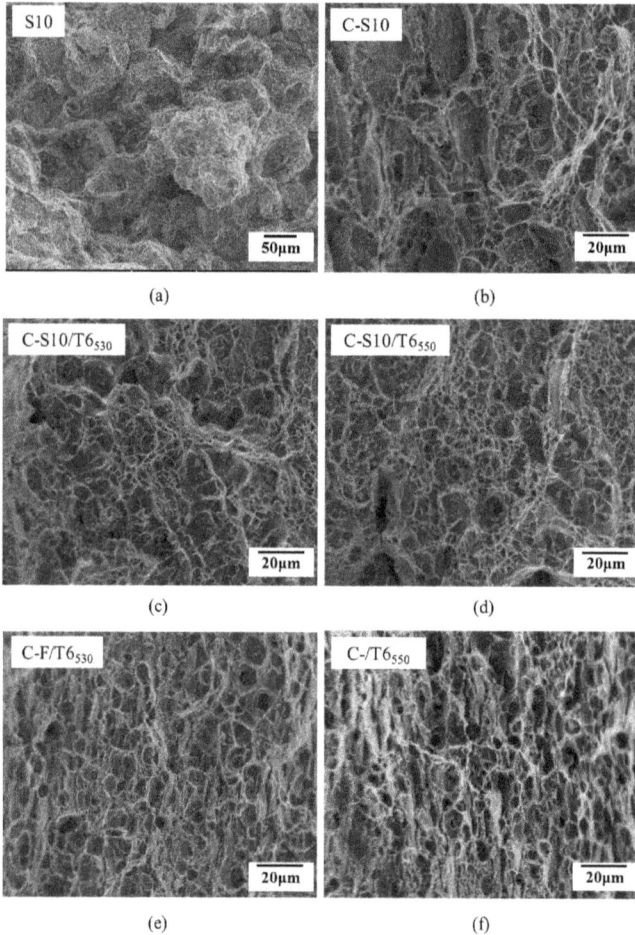

Figure 14. Fracture surfaces of (**a**) S10; (**b**) C-S10; (**c**) C-S10/T6$_{530}$; (**d**) C-S10/T6$_{550}$; (**e**) C-F/T6$_{530}$; and (**f**) C-F/T6$_{550}$.

Brittle and hard phases located at the grain boundaries of TS-SIMA alloys disappeared after high-temperature compression, improving elongation. Total elongation increased to about 30%. After T6 heat treatment, the tensile strength of TS-SIMA forming alloys reached about 430 MPa. The strength of such alloys is slightly less than that of T6-heat-treated as-extruded alloys (by about 20 MPa) and its elongation can be slightly higher than that. The mechanical properties of T6-heat-treated TS-SIMA forming alloys are sufficient for common applications.

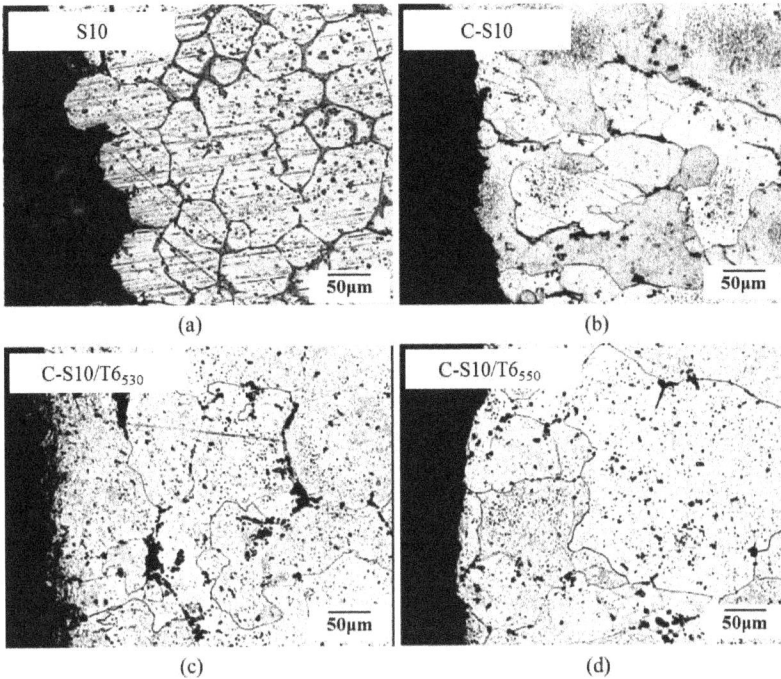

Figure 15. Sub-surfaces of (**a**) S10; (**b**) C-S10; (**c**) C-S10/T6$_{530}$; and (**d**) C-S10/T6$_{550}$.

4. Conclusions

(1) Globular grains were obtained using the TS-SIMA process. After a salt bath, the grains became globular and Cu, Mg, and Si were the major elements distributed on the globular grain boundaries. The globular grain boundaries were harder than the Al matrix.

(2) High-temperature compressive resistance can be reduced by the TS-SIMA process. With a 50% compression ratio, the TS-SIMA process decreased compression loading by about 35%. After high-temperature compression, Cu, Mg, and Si were no longer located at the compressed globular grain boundaries.

(3) High-temperature compression can improve the elongation of TS-SIMA alloys. The mechanical properties of TS-SIMA alloys can be enhanced by T6 heat treatment. The mechanical properties are sufficient for common applications.

Acknowledgments: The authors are grateful to the Instrument Center of National Cheng Kung University and the National Science Council of Taiwan (NSC MOST103-2221-E-006-056-MY2) for their financial support.

Author Contributions: Chia-Wei Lin, designed the experiments, performed the experiments, analyzed the data and wrote the paper; Fei-Yi Hung and Truan-Sheng Lui gave suggestions for improving experiments and analysis.

Conflicts of Interest: The authors declare no conflict of interest.

References

1. Hatch, J.E. *Aluminum: Properties and Physical Metallurgy*; ASM International: Materials Park, OH, USA, 1984; Volume 1, p. 50.
2. Zhen, L.; Fei, W.D.; Kang, S.B.; Kim, H.W. Precipitation behavior of Al-Mg-Si alloys with high silicon content. *J. Mater. Sci.* **1997**, *32*, 1895–1902.
3. Fan, Z. Semisolid metal processing. *Int. Mater. Rev.* **2002**, *47*, 49–85. [CrossRef]

4. Song, Y.B.; Park, K.T.; Hong, C.P. Recrystallization behavior of 7175 Al alloy during modified strain-induced melt-activated (SIMA) process. *Mater. Trans.* **2006**, *47*, 1250–1256. [CrossRef]

5. Tzimas, E.; Zavaliangos, A. A comparative characterization of near-equiaxed microstructures as produced by spray casting, magnetohydrodynamic casting and the stress induced, melt activated process. *Mater. Sci. Eng. A* **2000**, *289*, 217–227. [CrossRef]

6. Paes, M.; Zoqui, E.J. Semi-solid Behavior of New Al-Si-Mg Alloys for Thixoforming. *Mater. Sci. Eng. A* **2005**, *406*, 63–73. [CrossRef]

7. Parshizfard, E.; Shabestari, S.G. An investigation on the microstructural evolution and mechanical properties of A380 aluminum alloy during SIMA process. *J. Alloy. Compd.* **2011**, *509*, 9654–9658. [CrossRef]

8. Akhlaghi1, F.; Farhood, A.H.S. Characterization of globular microstructure in NMS processed aluminum A356 alloy: The role of casting size. *Adv. Mater. Res.* **2011**, *264–265*, 1868–1877. [CrossRef]

9. Tzimas, E.; Zavaliangos, A. Evolution of near-equiaxed microstructure in the semisolid state. *Mater. Sci. Eng. A* **2000**, *289*, 228–240. [CrossRef]

10. Emamy, M.; Razaghian, A.; Karshenas, M. The effect of strain-induced melt activation process on the microstructure and mechanical properties of Ti-refined A6070 Al alloy. *Mater. Des.* **2013**, *46*, 824–836. [CrossRef]

11. Lin, C.W.; Hung, F.Y.; Lui, T.S. High-temperature deformation resistance and forming behavior of two-step SIMA-processed 6066 alloy. *Mater. Sci. Eng. A* **2016**, *659*, 143–157. [CrossRef]

12. Lee, K.S.; Kim, S.; Lim, K.R.; Hing, S.H.; Kim, K.B.; Na, Y.S. Crystallization, High temperature defroemation behavior and solid-to-solid formability of a Ti-based bulk metallic glass within supercooled liquid region. *J. Alloy. Compd.* **2016**, *663*, 270–278. [CrossRef]

13. ASM International Alloy Phase Diagram and the Handbook Committees. *ASM Handbook*; ASM International: Materials Park, OH, USA, 1992; Volume 3, pp. 307–308.

14. Hall, E.O. The deformation and ageing of mild steel. *Proc. Phys. Soc.* **1951**, *64*, 747–753. [CrossRef]

15. Petch, N.J. The cleavage strength of polycrystals. *J. Iron Steel Int.* **1953**, *174*, 25–28.

metals

MDPI

Article

The Influence of Specimen Thickness on the Lüders Effect of a 5456 Al-Based Alloy: Experimental Observations

Yu-Long Cai, Su-Li Yang, Shi-Hua Fu and Qing-Chuan Zhang *

Chinese Academy of Science Key Laboratory of Mechanical Behavior and Design of Materials, University of Science and Technology of China, Hefei 230027, China; caiyl@mail.ustc.edu.cn (Y.-L.C.); yangsuli@mail.ustc.edu.cn (S.-L.Y.); fushihua@ustc.edu.cn (S.-H.F.)
* Correspondence: zhangqc@ustc.edu.cn; Tel.: +86-551-6360-1248

Academic Editor: Nong Gao
Received: 11 March 2016; Accepted: 17 May 2016; Published: 20 May 2016

Abstract: For the first time ever, a thickness dependence of the Lüders effect in an Al-based alloy is demonstrated. A three-dimensional digital image correlation method was used to gain insight into the Lüders band velocity and the Lüders strain. The results revealed that both the strain and velocity depend on the specimen thickness. The strain increases, whereas the velocity decreases, with decreasing specimen thickness. Moreover, the plot of the strain *vs.* the velocity concurs with the global deformation compatibility.

Keywords: Lüders effect; digital image correlation; thickness dependence; 5456 Al-based alloy

1. Introduction

Owing to their low density, high strength, and good formability, Al-based alloys are extensively used in the automotive industry [1]. These properties are advantageous from a manufacturing point of view. However, the heterogeneous deformation of Al-based alloys (even at room temperature) subjected to a high strain gradient represents a major drawback of these materials. This phenomenon leads to undesirable visible traces on the surface of the final products [2,3]. The induced heterogeneous deformation can be classified into two general categories: the Lüders effect and the Portevin-Le Chatelier (PLC) effect. These effects are governed by different microscopic mechanisms. The PLC effect is typically attributed to dynamic interactions between mobile dislocations and diffusing solutes (*i.e.*, dynamic strain ageing, DSA) [4–6]. On the other hand, the Lüders effect results from both the dislocation pinning/unpinning effect arising from Cottrell atmosphere constraints and the collective or self-organized dislocation multiplication and motion [7–11]. According to the Johnston theory [12,13], dislocation pinning by impurity atmospheres does not occur in materials with low densities of mobile dislocations. This seems to contradict the Cottrell theory [14], which describes locking or unlocking behaviors. However, these theories actually complement each other, as the low mobile dislocation density stems from the previously completed pinning process.

The Lüders effect, referred to as the yield point phenomenon, is characterized by a sharp yield point and a subsequent yield plateau. An ideal plastic plateau, due to the propagation of a localized plastic deformation, is manifested; the occurrence of this plateau is accompanied by a decrease in the yield stress. Studies focused on the features of the Lüders effect, e.g., the Lüders strain, the Lüders band velocity, or the morphology of the Lüders bands, have been conducted in recent years. These studies revealed that the Lüders strain is affected by the applied strain rate [8,15], solute concentration [15–18], test temperature [19], and the grain size [18,20,21]. For example, Johnson *et al.* [15], Winlock [16], Song *et al.* [17], and Tsuchida *et al.* [18] determined the influence of the carbon content on the magnitude of the Lüders strain in steel; in all cases, the strain exhibited a negative dependence on the carbon content. Jin *et al.* [20], Lloyd *et al.* [21], and Tsuchida *et al.* [18] found that the yield stress (in accordance with the Hall-Petch relationship) and the Lüders strain are both inversely proportional to the grain size. The digital image correlation (DIC) method and infrared thermography technique (IRT) have been used to perform kinematic and calorimetric analyses of Lüders bands. Nagarajan *et al.* [9,10] used DIC and IRT to determine the spatio-temporal evolution of the full-field strain and temperature contours; this yielded an improved understanding of band nucleation and propagation as well as the band growth mechanism. IRT has been used to determine spatial characteristics, such as the shape, orientation, and velocity [22] and morphological characteristics, such as the *X*-, *Y*- and *V*-shaped patterns [23] of Lüders bands. Based on the aforementioned experimental results, theoretical models, which takes both Cottrell atmosphere pinning and dislocation multiplication into consideration, have been used to describe the features of spatial coupling [13,24,25].

Since Yoshida *et al.* [7] reported a thickness dependence of the plastic behavior of 4–30 μm thick copper whiskers, a few studies have focused on this thickness dependence, on the millimeter scale, in polycrystals. In this work, we determine the influence of the thickness on the Lüders strain, Lüders band velocity, and the correlation between these features in a 5456 Al-based alloy.

2. Materials and Methods

The chemical composition (wt. %) of the 5456 Al-based alloy is shown in Table 1. Large plates of the alloy were subjected to an aging heat treatment (annealing at 673 K for 3 h followed by furnace-cooling to room temperature); dumbbell-shaped plate specimens were then cut from these plates. These 55 mm (gage length) × 20 mm (width) specimens had different thicknesses (t = 1, 2, and 3 mm). The stiffness of the testing machine influences the PLC effect and the Lüders plateau [26,27]. We used a hard machine, referred to as RGM-4050 (Reger Instrument Co., Ltd., Shenzhen, China), to determine the thickness dependence of the Lüders effect. Uniaxial tensile tests were performed at room temperature and constant strain rates ranging from 1.82×10^{-4} to 90.9×10^{-4} s^{-1}. The applied tensile load was recorded at a sampling rate of ~25 Hz.

Table 1. Chemical composition (wt. %) of the 5456 Al-based alloy.

Elements	Mg	Mn	Fe	Si	Zn	Ti	Cu	Cr	Al
Content	4.7–5.5	0.5–1.0	0.4	0.25	0.25	0.2	0.1	0.05–0.2	Balance

The three-dimensional DIC (3D-DIC) method is applied in the present study for surface strain mapping; this non-contact method is used to measure the full-field space coordinate, displacement, and strain distributions in a material. This method is precise, accurate, does not require special quakeproof equipment, and is therefore widely used in material testing [28,29]. 3D-DIC is adept at measuring strain localization. Optical methods, such as shadowgraphy [30], laser scanning extensometry [31], and digital speckle pattern interferometry [2,3] obtain indirect observations of the localized bands, and hence quantitative analysis is difficult. Direct images can be readily obtained using 3D-DIC. The accuracy of this method was determined through a coupling factor that takes into account the image noise, interpolation bias, and the calculation algorithm parameters [32]. Establishing the theoretical measurement resolution for high-gradient deformation conditions (e.g., within Lüders or the PLC band) is difficult. Specifically, the experiment value of the displacement and strain measurement error are ~0.01 pixels and 150 με, respectively [33]. In a previous study, we determined the effect of DIC parameters, such as the patch size, shape function, and the strain gradient, on the measurement error; based on the results, a moderate patch size was suggested for high-gradient inhomogeneous deformation [29].

In the present study, a self-developed 3D-DIC system (PMLAB DIC-3D; Nanjing PMLAB Sensor Tech Co., Ltd., Nanjing, China) was used to continuously capture deformed images (via synchronous image acquisition) at an image sampling rate of 3 fps. So-called sequence DIC and equal-interval DIC were both used in this work. In the case of sequence DIC, the correlation between a fixed reference image and the deformed image is determined; with equal-interval DIC, however, the correlation between frame n and frame $n + g$ is determined (g is the image interval; $g = 1$, time interval $\approx 1/3$ s in this work). Prior to the tensile tests, specimens were sprayed with a flat white lacquer and then oversprayed with random black spots. The following dimensions were used in the 3D-DIC system: array dimensions of each image = 2048 × 2048 pixels; calculation grid size = 7 pixels; patch size = 29 × 29 pixels; strain calculation window size = 15 × 15 points. A correspondence of ~16 pixel/mm was obtained between the real dimension and the acquired image. The right-handed coordinate was defined as follows: the transverse direction and the tensile direction were the X and Y axes, respectively.

3. Results

3.1. Lüders Strain

Figure 1 shows the engineering stress–strain curves, obtained at an applied strain rate of 9.09×10^{-4} s^{-1}, for specimens with different thicknesses. For clarity, we have depicted the curves separated vertically by a stress interval of 35 MPa. The inset indicates that the Lüders strain, *i.e.*, the length of the Lüders plateau formed during straining, decreases with increasing specimen thickness. A PLC effect is visible almost immediately after the Lüders effect. It consists of a serrated flow with numerous stress drops. These serrations are of similar characteristics for the three tests. These results indicate that the specimen thickness influences the Lüders effect. In contrast, the PLC effect seems not to be influenced by the specimen thickness.

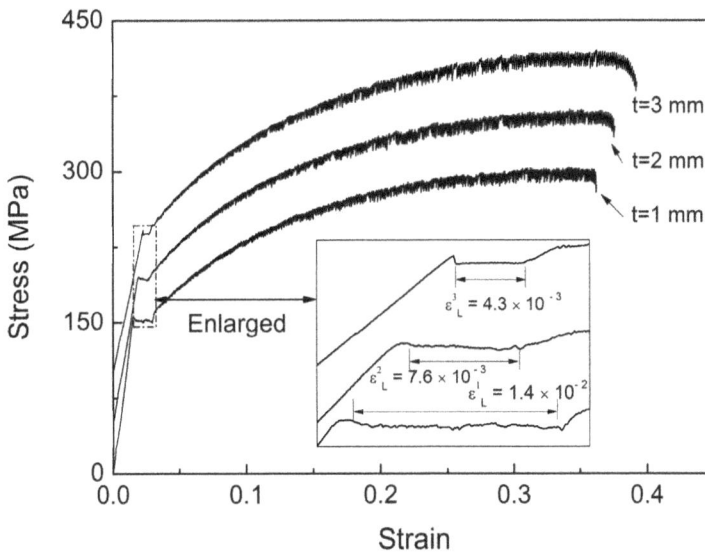

Figure 1. Engineering stress-strain curves obtained at 9.09×10^{-4} s^{-1} for specimens with different thicknesses. For clarity, these curves are separated vertically by a stress interval of 35 MPa. The inset provides a magnified view of the Lüders plateaus.

As previously stated, the specimen thickness has a pronounced effect on the Lüders strain. Therefore, specimens of different thicknesses were subjected to uniaxial tensile tests at various applied strain rates to determine the dominant factor (thickness or strain rate). Figure 2 shows the Lüders strain as a function of the specimen thickness. Data corresponding to the 2-mm-thick and 3-mm-thick specimens were obtained from six and four tests performed at strain rates ranging from 9.09×10^{-4} to 0.909×10^{-4} s^{-1} and 1.82×10^{-4} to 9.09×10^{-4} s^{-1}, respectively. In fact, previous studies [8,15] have revealed a power-law dependence of the Lüders strain on the strain rate. However, this dependence was not observed in the present study. The insets of Figure 2 show the Lüders strain of the 2- and 3-mm specimens as a function of the applied strain rate. The strain increases with increasing strain rate, albeit with some scatter, which to a certain extent contradicts previous studies. Compared to the applied strain rate, the specimen thickness generally has a more pronounced effect on the Lüders strain. Figure 2 confirms a negative relationship between the strain and specimen thickness.

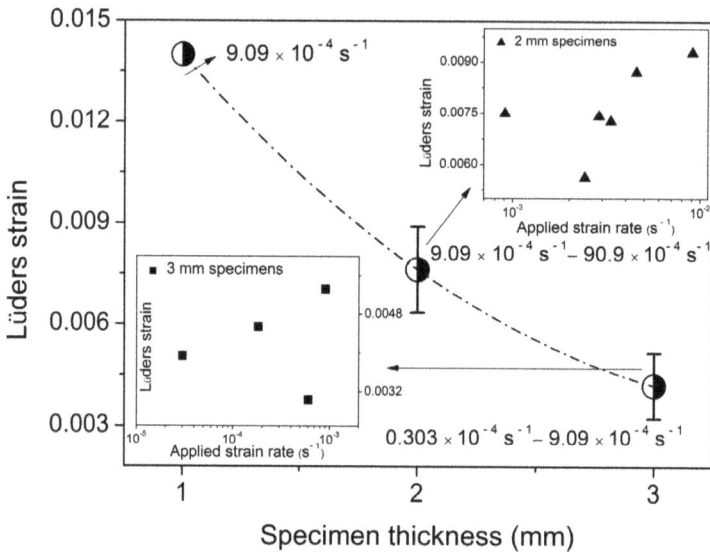

Figure 2. Dependence of the Lüders strain on the specimen thickness at various applied strain rates. The insets show the strain as a function of the strain rate at a given specimen thicknesses.

3.2. Lüders Band Velocity

As a form of plastic instability, the Lüders effect differs compared to the PLC effect mainly on the serration morphology and the strain-localized band propagation. Generally, the continuous sweep corresponds to the smooth plateau of the Lüders effect; nevertheless, the propagation of PLC bands can be continuous, hopping, and even random along the specimen with abrupt serrations. Here, to exhibit the difference in the straining process of these two types of plastic instabilities, we present the example of the 1-mm specimen. Figure 3a shows the illustration of the selected data for DIC calculations. Figure 3b,c respectively show the accumulated-strain mappings of the Lüders and PLC band. It can be seen that the Lüders band manifests as a continuous localized deformation band from one end of the specimen to the other. The PLC bands were characterized by discrete strip bands which belong to type B serrations. The DIC results revealed that the strain in the tensile direction of the PLC band is much higher than that of the Lüders band.

As previously stated, the Lüders strain is correlated with the specimen thickness. Usually, a yield plateau occurs during the propagation process of a Lüders band; hence, the dependence of the Lüders band velocity on the specimen thickness was subsequently explored. Figure 4 shows the strain mapping (both the equal-interval strain mapping (a) and the accumulated-strain mapping (b)) along the tensile direction of specimens with different thicknesses. The strain rate of the test was 9.09×10^{-4} s^{-1}. Figure 4a shows that localized deformation occurs within an inclined strip only, and then propagates from one end of the specimen to the other. The growing stair (or step) in the accumulated strain mappings (Figure 4b) reveals the propagation of the Lüders band. As the figure shows, the Lüders band velocity (in contrast to the Lüders strain) exhibits a positive dependence on the specimen thickness (see Figure 4c).

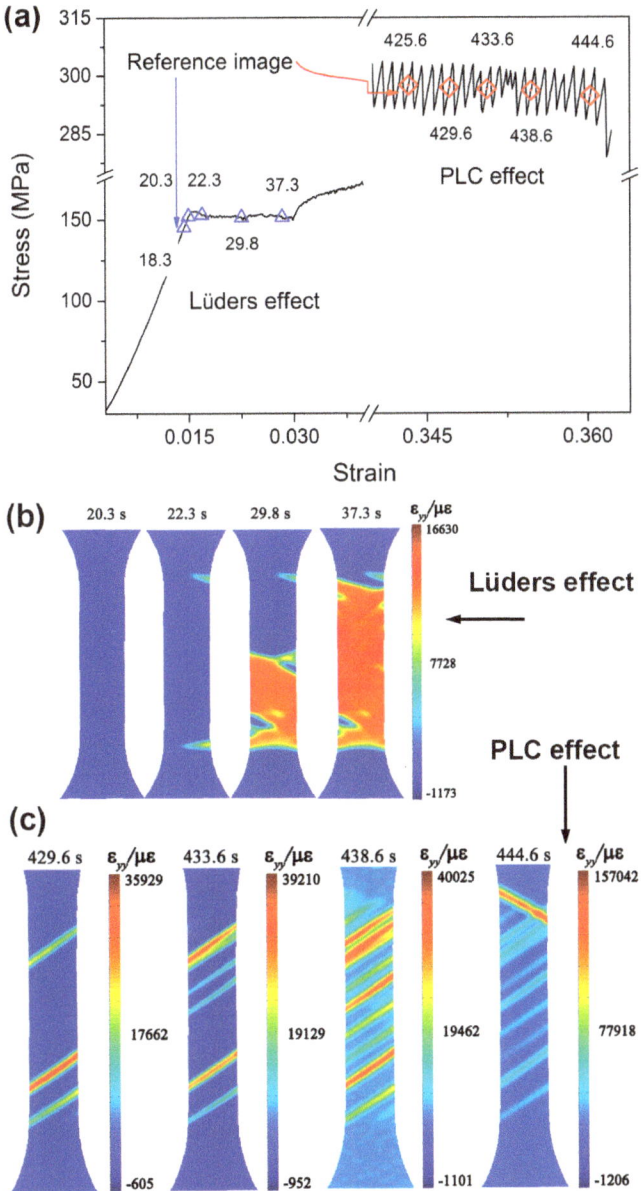

Figure 3. Test on a 1-mm-thick specimen at 9.09×10^{-4} s^{-1}. (**a**) Stress-strain curve that shows selected image data for digital image correlation (DIC) calculation; (**b**) Longitudinal strain maps revealing the Lüders effect; (**c**) Longitudinal strain maps revealing the Portevin-Le Chatelier (PLC) effect. Reference images are defined at ~18.3 and ~425.6 s, respectively. Sequence DIC method was used for this processing.

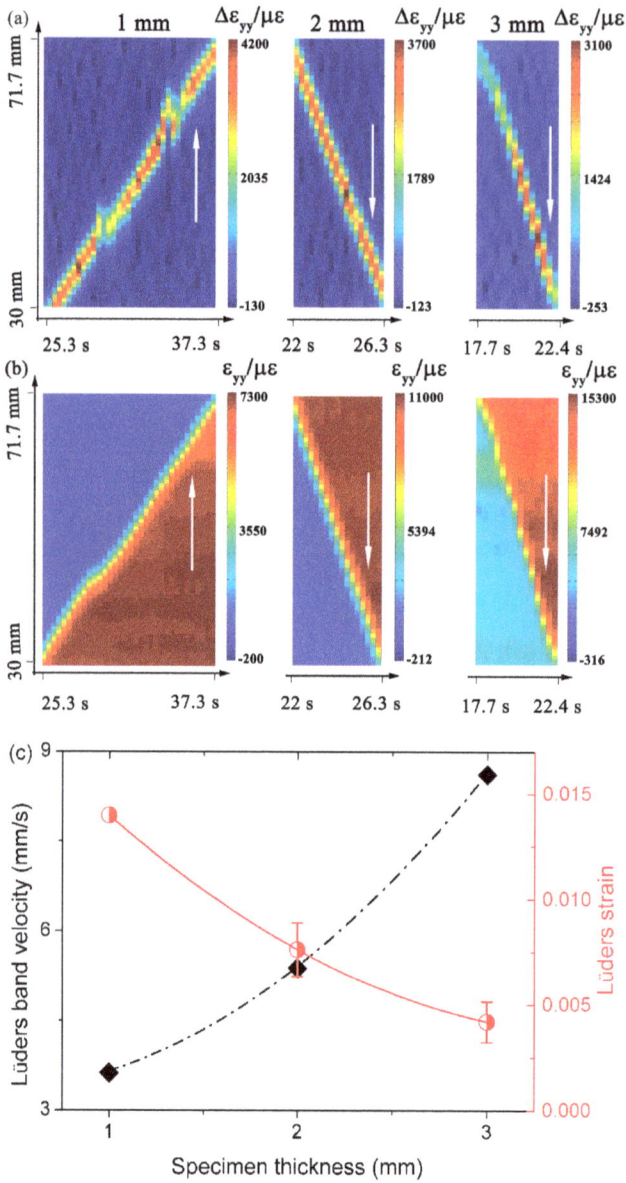

Figure 4. Test on three different thickness specimens at 9.09×10^{-4} s^{-1}. (**a**) Equal-interval strain mapping; (**b**) Accumulated-strain mapping in the tensile direction; (**c**) Dependence of the Lüders band velocity and the Lüders strain on the specimen thickness, at a fixed strain rate of 9.09×10^{-4} s^{-1}. The white arrows in (**a**) and (**b**) denote the propagation directions. The time interval in the equal-interval DIC calculation is approximately $1/3$ s.

3.3. Global Deformation Compatibility

A single Lüders band propagates at a constant velocity along the gage length of the specimen [15]. This propagation may be expressed mathematically as follows:

$$\varepsilon_L l_0 = V_{crosshead} \tag{1}$$

where ε_L, $\Delta t = l_0/V_{band}$, l_0, , $V_{crosshead}$, and V_{band} are the Lüders strain, the duration of the plateau, gage length, the velocities of the crosshead of the test machine, and the Lüders band, respectively. Equation (1) indicates that the Lüders plateau ends immediately after the band sweeps the entire specimen. Rewriting Equation (1) yields:

$$\varepsilon_L = V_{crosshead}/V_{band} \tag{2}$$

Equation (2) reveals that the Lüders strain, the applied strain rate, and the Lüders band velocity satisfy the global deformation compatibility requirement, in accordance with the theoretical predictions of Hall *et al.* [34] and Estrin *et al.* [35]. Furthermore, this relationship does not rely on the specimen thickness, whereas the Lüders strain only depends on the thickness. Therefore, at a given applied strain rate, the band velocity and strain exhibit opposite dependences on the specimen thickness; moreover, for a given thickness, the velocity is proportional to the applied strain rate.

The aforementioned results indicate that a plot of the Lüders strain *vs.* the Lüders band velocity can only be obtained by considering specimens of differing thicknesses. As Figure 5 shows, the relationship between the strain and the velocity concurs with the predicted tendency.

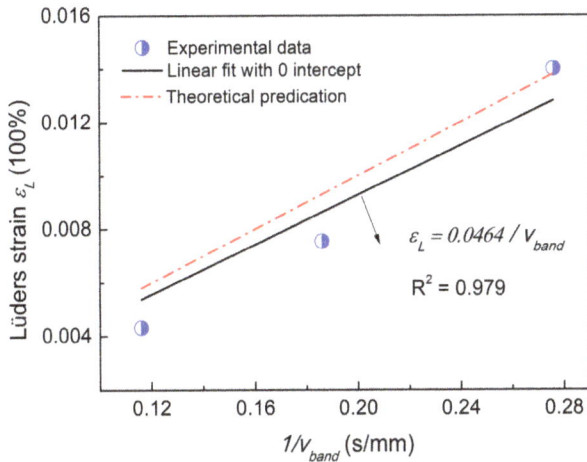

Figure 5. Relationship between the Lüders strain and the Lüders band velocity. The theoretical prediction can be expressed as: $\varepsilon_L = \frac{3 \text{ mm}}{60 \text{ s}} \times \frac{1}{V_{band}} = 0.05 \times \frac{1}{V_{band}}$.

3.4. Spatial Characteristics

Figure 6 shows the displacement and strain mappings, including the in-plane and out-of-plane displacement, of specimens with different thicknesses. A Lüders band is clearly visible in each specimen. We note that, in all cases, the localized-deformation region lies almost completely within a strip inclined at ~60° with respect to the tensile direction. However, the band inclination in Figure 6a is >60°, resulting possibly from the crossover of localized bands with different orientations. This is verified by Figure 7, where the crossover evolution of the localized band in a 1-mm-thick specimen is revealed. As the figure shows, growth of the partial band leads to changes in the band inclination.

The inclination of the band front (with respect to the tensile direction) increases, gradually leading, eventually, to a nearly flat band (the newly formed band in Figures 6a and 7). Hill [36] attributed the occurrence of necking to the plastic flow of anisotropic metals. Analysis (including the concept of von Mises plasticity) of the strain localization under plane stress uniaxial tension yields the classical value (~54.74°) of the band inclination in an isotropic sheet; this value is moderately lower than the experimentally determined value obtained in the present study. We have shown [37] that the deformation mode of the PLC localized region under uniaxial tensile tests in the sheet plane is simple shear, which demonstrates that the strain-localized region appears along the shear direction. Based on our research, the Lüders band and the PLC band are similar in macroscopic deformation, which probably implies the same deformation mode. Coer *et al.* [38] investigated the Lüders effect of an Al-Mg alloy during the simple shear tests; the literature gave direct observations of the localized deformation at the simple shear state. These two studies all reveal that the strain-localization and the simple shear state have some sort of inner link, which needs more effort to explore.

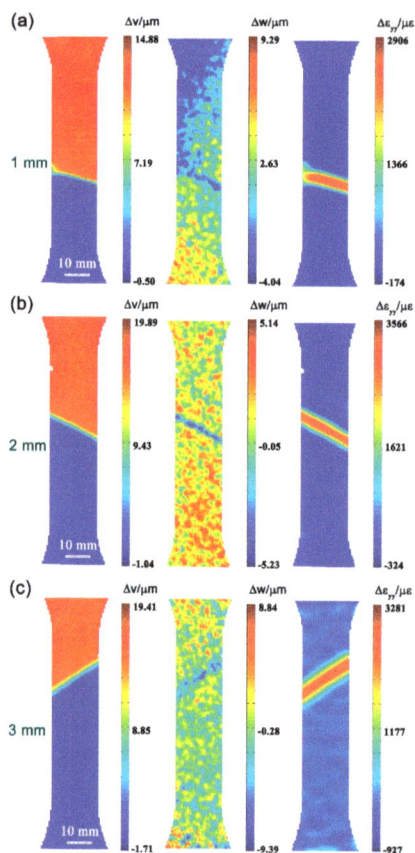

Figure 6. Test on three different thickness specimens at 9.09×10^{-4} s^{-1}. The Lüders band, for specimens of differing thicknesses, as determined via the equal-interval DIC method; 1-mm-, 2-mm-, and 3-mm-thick specimens are shown in (**a–c**), respectively. Incremental results are presented (*i.e.*, (**a–c**) show the correlation between frames 72 and 73, frames 80 and 81, and frames 118 and 119, respectively. Furthermore, u and ε_{yy} are the respective displacement and strain increment in the tensile direction, and w is the out-of-plane displacement increment.

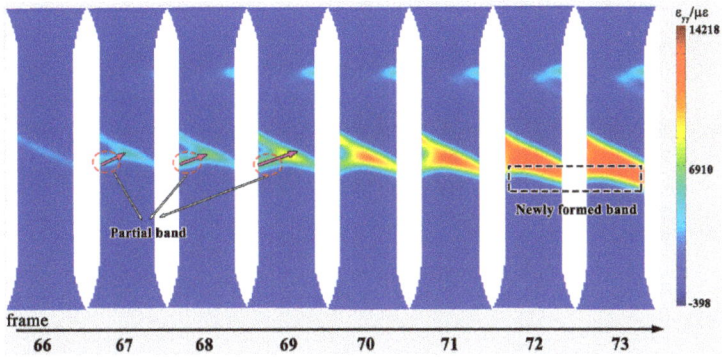

Figure 7. Crossover evolution of the Lüders band in a 1-mm-thick specimen at 9.09×10^{-4} s^{-1}. Sequence DIC method was used, and frame 65 was considered as the reference image. Changes in the band inclination are induced by growth of the partial band. This growth (frames 67–69) leads to a change in the band front direction.

Experimentally, the localized deformation in the out-of-plane displacement field is obtained due to the incompressibility during the plastic stage. In fact, the maximum out-of-plane displacement increases with increasing thickness at a maximum strain increment (~3000 µε), which is consistent with our previous study on PLC bands [39]. The in-plane displacement mappings (profiles on the left-hand side of Figure 6) show that the distance between the band front and band tail generally increases with increasing specimen thickness. This spatial characteristic was investigated in further detail.

Here we define the width of the Lüders band (w_{band}) as follows: the distance at the mid-height of the localized strain band of the strain increment (see Figure 8). The maximum strain increment (ε_{max}) of the Lüders band is defined as the maximum strain increment value of the center line (Line 0 in Figure 8a). Figure 8 shows the dependence of w_{band} and ε_{max} (obtained at a fixed image-sampling rate of 3 fps) on the specimen thickness, for a given strain rate of 9.09×10^{-4} s^{-1}. As the figure shows, ε_{max} and w_{band} remain approximately constant and increase, respectively, with increasing specimen thickness.

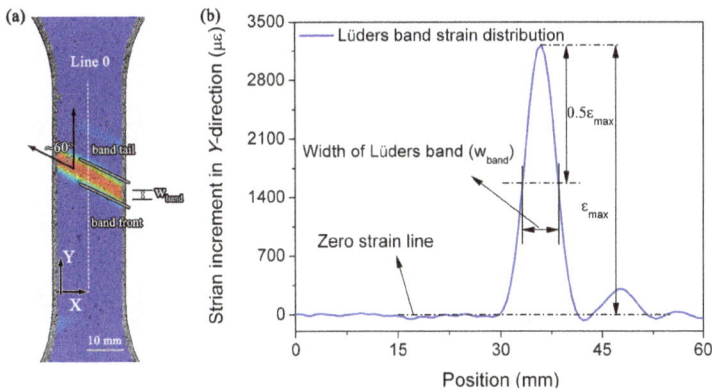

Figure 8. Test on a 2-mm-thick specimen at 9.09×10^{-4} s^{-1}. Schematic showing the calculation of the Lüders band width. The strain mapping in (**a**) was performed via equal-interval digital image correlation method (frame 80 and 81). The strain distribution along the *Y*-direction, (**b**), was determined for Line 0 in (**a**).

As previously mentioned, the Lüders band velocity increases with increasing specimen thickness. Consistent with the results shown in Figure 9, w_{band} increases owing to continuous propagation of the band. The band sweeps the gage length of the specimen (in general); hence, the accumulated strain varies with the equal-interval maximum strain increment and the equal-interval band width. However, the obtained experimental results reveal that the accumulated strain depends only on w_{band}. This strain increases with increasing specimen thickness, consistent with the results shown in Figure 4b.

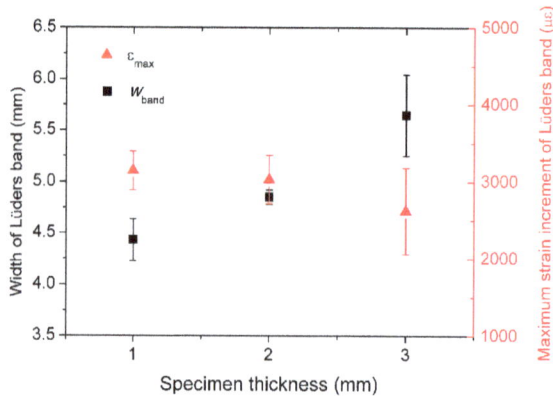

Figure 9. Dependence of the width and the maximum strain increment of the Lüders band on the specimen thickness, for a given strain rate of 9.09×10^{-4} s^{-1}.

4. Conclusions

We investigated the dependence of the Lüders effect on specimen thickness. Characteristics, such as the Lüders strain, Lüders band velocity, and the correlation between these features, were determined by using the 3D-DIC method. The major conclusions can be summarized as follows:

(1) This is the first report demonstrating the dependence of the Lüders strain on the specimen thickness of a 5456 Al-based alloy.
(2) The Lüders strain increases, but the Lüders band velocity decreases, with decreasing specimen thickness.
(3) Uniquely, the plot of the Lüders strain *vs.* the Lüders velocity was obtained via direct observation.

Acknowledgments: The authors gratefully acknowledge the financial support received from the National Natural Science Foundation of China (NSFC) under grant No. 11332010, 51271174, and 11372300.

Author Contributions: Yu-Long Cai performed the experiments, analyzed the data, and wrote the paper; Shi-Hua Fu, Su-Li Yang and Qing-Chuan Zhang discussed the results.

Conflicts of Interest: The authors declare no conflict of interests.

References

1. Fridlyander, I.N.; Sister, V.G.; Grushko, O.E.; Berstenev, V.V.; Sheveleva, L.M.; Ivanova, L.A. Aluminum alloys: Promising materials in the automotive industry. *Met. Sci. Heat Treat.* **2002**, *44*, 365–370. [CrossRef]
2. Zhang, Q.; Jiang, Z.; Jiang, H.; Chen, Z.; Wu, X. On the propagation and pulsation of Portevin-Le Chatelier deformation bands: An experimental study with digital speckle pattern metrology. *Int. J. Plast.* **2005**, *21*, 2150–2173. [CrossRef]
3. Jiang, H.; Zhang, Q.; Jiang, Z.; Wu, X. Experimental investigations on kinetics of Portevin-Le Chatelier effect in Al-4wt. %Cu alloys. *J. Alloys Compd.* **2007**, *428*, 151–156. [CrossRef]

4. Benallal, A.; Berstad, T.; Børvik, T.; Hopperstad, O.S.; Koutiri, I.; Nogueira de Codes, R. An experimental and numerical investigation of the behaviour of AA5083 aluminium alloy in presence of the Portevin-Le Chatelier effect. *Int. J. Plast.* **2008**, *24*, 1916–1945. [CrossRef]
5. Fu, S.; Cheng, T.; Zhang, Q.; Hu, Q.; Cao, P. Two mechanisms for the normal and inverse behaviors of the critical strain for the Portevin-Le Chatelier effect. *Acta Mater.* **2012**, *60*, 6650–6656. [CrossRef]
6. Cai, Y.; Tian, C.; Fu, S.; Han, G.; Cui, C.; Zhang, Q. Influence of γ′ precipitates on Portevin-Le Chatelier effect of Ni-based superalloys. *Mater. Sci. Eng. A* **2015**, *638*, 314–321. [CrossRef]
7. Yoshida, K.; Gotoh, Y.; Yamamoto, M. The thickness dependence of plastic behaviors of copper whiskers. *J. Phys. Soc. Jpn.* **1968**, *24*, 1099–1107. [CrossRef]
8. Sun, H.B.; Yoshida, F.; Ohmori, M.; Ma, X. Effect of strain rate on Lüders band propagating velocity and Lüders strain for annealed mild steel under uniaxial tension. *Mater. Lett.* **2003**, *57*, 4535–4539. [CrossRef]
9. Nagarajan, S.; Narayanaswamy, R.; Balasubramaniam, V. Advanced imaging for early prediction and characterization of zone of Lüders band nucleation associated with pre-yield microstrain. *Mater. Sci. Eng. A* **2013**, *561*, 203–211. [CrossRef]
10. Nagarajan, S.; Narayanaswamy, R.; Balasubramaniam, V. Study on local zones constituting to band growth associated with inhomogeneous plastic deformation. *Mater. Lett.* **2013**, *105*, 209–212. [CrossRef]
11. Van Rooyen, G.T. The stress and strain distribution in a propagating Lüders front accompanying the yield-point phenomenon in iron. *Mater. Sci. Eng.* **1968**, *3*, 105–117. [CrossRef]
12. Johnston, W.G.; Gilman, J.J. Dislocation velocities, dislocation densities, and plastic flow in lithium fluoride crystals. *J. Appl. Phys.* **1959**, *30*, 129–144. [CrossRef]
13. Hahn, G.T. A model for yielding with special reference to the yield-point phenomena of iron and related bcc metals. *Acta Metall.* **1962**, *10*, 727–738. [CrossRef]
14. Cottrell, A.H.; Bilby, B.A. Dislocation theory of yielding and strain ageing of iron. *Proc. Phys. Soc.* **1949**, *62*. [CrossRef]
15. Johnson, D.H.; Edwards, M.R.; Chard-Tuckey, P. Microstructural effects on the magnitude of Lüders strains in a low alloy steel. *Mater. Sci. Eng. A* **2015**, *625*, 36–45. [CrossRef]
16. Winlock, J. The influence of the rate of deformation on the tensile properties of some plain carbon sheet steels. *J. Metall.* **1953**, *5*, 797–803.
17. Song, R.; Ponge, D.; Raabe, D. Improvement of the work hardening rate of ultrafine grained steels through second phase particles. *Scr. Mater.* **2005**, *52*, 1075–1080. [CrossRef]
18. Tsuchida, N.; Tomota, Y.; Nagai, K.; Fukaura, K. A simple relationship between Lüders elongation and work-hardening rate at lower yield stress. *Scr. Mater.* **2006**, *54*, 57–60. [CrossRef]
19. Fisher, J.C.; Rogers, H.C. Propagation of Lüder's bands in steel wires. *Acta Metall.* **1956**, *4*, 180–185. [CrossRef]
20. Jin, H.; Lloyd, D.J. Effect of a duplex grain size on the tensile ductility of an ultra-fine grained Al-Mg alloy, AA5754, produced by asymmetric rolling and annealing. *Scr. Mater.* **2004**, *50*, 1319–1323. [CrossRef]
21. Lloyd, D.; Court, S.A.; Gatenby, K.M. Lüders elongation in Al-Mg alloy AA5182. *Mater. Sci. Technol.* **1997**, *13*, 660–666. [CrossRef]
22. Louche, H.; Chrysochoos, A. Thermal and dissipative effects accompanying Lüders band propagation. *Mater. Sci. Eng. A* **2001**, *307*, 15–22.
23. Delpueyo, D.; Balandraud, X.; Grediac, M. Calorimetric signature of the Portevin-Le Chatelier effect in an aluminum alloy from infrared thermography measurements and heat source reconstruction. *Mater. Sci. Eng. A* **2016**, *651*, 135–145. [CrossRef]
24. Mazière, M.; Forest, S. Strain gradient plasticity modeling and finite element simulation of Lüders band formation and propagation. *Continuum Mech. Thermodyn.* **2015**, *27*, 83–104. [CrossRef]
25. Wenman, M.R.; Chard-Tuckey, P.R. Modelling and experimental characterisation of the Lüders strain in complex loaded ferritic steel compact tension specimens. *Int. J. Plast.* **2010**, *26*, 1013–1028. [CrossRef]
26. Wang, H.D.; Berdin, C.; Mazière, M.; Forest, S.; Prioul, C.; Parrot, A.; Le-Delliou, P. Experimental and numerical study of dynamic strain ageing and its relation to ductile fracture of a C-Mn steel. *Mater. Sci. Eng. A* **2012**, *547*, 19–31. [CrossRef]
27. Nadai, A. *Theory of Flow and Fracture of Solids*; McGraw Hill: New York, NY, USA, 1950; Volume 1.
28. Gao, Y.; Cheng, T.; Su, Y.; Xu, X.; Zhang, Y.; Zhang, Q. High-efficiency and high-accuracy digital image correlation for three-dimensional measurement. *Opt. Laser Eng.* **2015**, *65*, 73–80. [CrossRef]

29. Xu, X.; Su, Y.; Cai, Y.; Cheng, T.; Zhang, Q. Effects of various shape functions and subset size in local deformation measurements using DIC. *Exp. Mech.* **2015**, *55*, 1575–1590. [CrossRef]
30. Chihab, K.; Estrin, Y.; Kubin, L.P.; Vergnol, J. The kinetics of the Portevin-Le Chatelier bands in an Al-5 at. % Mg alloy. *Scr. Metall.* **1987**, *21*, 203–208. [CrossRef]
31. Ziegenbein, A.; Hhner, P.; Neuhuser, H. Correlation of temporal instabilities and spatial localization during Portevin-Le Chatelier deformation of Cu-10 at. % Al and Cu-15 at. % Al. *Comput. Mater. Sci.* **2000**, *19*, 27–34. [CrossRef]
32. Su, Y.; Zhang, Q.; Gao, Z.; Xu, X.; Wu, X. Fourier-based interpolation bias prediction in digital image correlation. *Opt. Express* **2015**, *23*, 19242–19260. [CrossRef] [PubMed]
33. Sutton, M.A.; Turner, J.L.; Bruck, H.A.; Chae, T.A. Full-field representation of discretely sampled surface deformation for displacement and strain analysis. *Exp. Mech.* **1991**, *31*, 168–177. [CrossRef]
34. Hall, E.O. *Yield Point Phenomena in Metals and Alloys*, 1st ed.; Plenum Press (a Division of Plenum Publishing Corporation): New York, NY, USA, 1970; pp. 34–35.
35. Estrin, Y.; Kubin, L.P. *Continuum Models for Materials with Microstructures*; Wiley: New York, NY, USA, 1995; p. 395.
36. Hill, R. A theory of the yielding and plastic flow of anisotropic metals. *Proc. R. Soc. A Math. Phys. Eng. Sci.* **1948**, *193*, 281–297. [CrossRef]
37. Cai, Y.L.; Zhang, Q.C.; Yang, S.L.; Fu, S.H.; Wang, Y.H. Characterization of the deformation behaviors associated with the serrated flow of a 5456 Al-based alloy using two orthogonal digital image correlation systems. *Mater. Sci. Eng. A* **2016**, *664*, 155–164. [CrossRef]
38. Coer, J.; Manach, P.Y.; Laurent, H.; Oliveira, M.C.; Menezes, L.F. Piobert-Lüders plateau and Portevin-Le Chatelier effect in an Al-Mg alloy in simple shear. *Mech. Res. Commun.* **2013**, *48*, 1–7. [CrossRef]
39. Cai, Y.L.; Zhang, Q.C.; Yang, S.L.; Fu, S.H.; Wang, Y.H. Experimental study on three-dimensional deformation field of Portevin-Le Chatelier effect using digital image correlation. *Exp. Mech.* **2016**, 1–13. [CrossRef]

Article

Effects of Cryogenic Forging and Anodization on the Mechanical Properties and Corrosion Resistance of AA6066–T6 Aluminum Alloys

Teng-Shih Shih [1,*], Hwa-Sheng Yong [1,2] and Wen-Nong Hsu [1,2]

[1] Department of Mechanical Engineering, National Central University, Jhongli District, Taoyuan City 32001, Taiwan; nite-star@hotmail.com (H.-S.Y.); nong88@yam.com (W.-N.H.)

[2] Graduate Student, National Central University, Jhongli District, Taoyuan City 32001, Taiwan

*   Correspondence: t330001@cc.ncu.edu.tw; Tel.: +886-3-4267317; Fax: +886-3-4254501

Academic Editor: Nong Gao

Received: 11 December 2015; Accepted: 23 February 2016; Published: 3 March 2016

Abstract: In this study, AA6066 alloy samples were cryogenically forged after annealing and then subjected to solution and aging treatments. Compared with conventional 6066-T6 alloy samples, the cryoforged samples exhibited a 34% increase in elongation but sacrificed about 8%–12% in ultimate tensile strength (UTS) and yield stress (YS). Such difference was affected by the constituent phases that changed in the samples' matrix. Anodization and sealing did minor effect on tensile strength of the 6066-T6 samples with/without cryoforging but it decreased samples' elongation about 8%–10%. The anodized/sealed anodic aluminum oxide (AAO) film enhanced the corrosion resistance of the cryoforged samples.

Keywords: cryoforging; anodization; tensile properties; corrosion resistance

1. Introduction

Al–xMg–ySi alloys (6xxx series Al alloys) are commonly used as extruded shapes and forged for making bicycle parts. Their characteristics include ample formability, machinability, weldability, and corrosion resistance, as well as good strength and elongation after heat treatment. These alloys are also readily available on the market.

Aluminum possesses a high stacking fault energy and readily undergoes dynamic recovery during deformation. Plastic deformation at low temperatures, such as cryorolling, is beneficial for refining grains in an aluminum alloy matrix [1,2]. Chen studied equal-channel deformation of an Al–Mg alloy at cryogenic temperatures and found that high-density dislocations distorted grains to a refining grain size [3]. Lee *et al.* also found that cryorolling 5083 alloy could obtain 200 nm fine grains to increase the ultimate tensile strength (UTS) from 315 to 522 MPa [4]. For an Al–Mg–Si alloy, cryorolling with a 90% reduction could cause heavy plastic deformation to produce nanosized extra-fine grains [5–7].

Aluminum alloys that become deformed at cryogenic temperatures could suppress the dynamic recovery that occurs during plastic deformation, and could acquire fine grains featuring high-angle grain boundaries [8,9]. The interaction of high-density dislocations enhanced the precipitation capability for producing a high density of nanosized precipitates [10]. Yin *et al.* found nanograins that were less than 500 nm in size in 7075 alloy samples subjected to compression forging at cryogenic temperatures [11].

Krishina *et al.* [8,12] produced ultrafine-grained Al–4Zn–2Mg alloys by cryorolling and indicated that the driving force for precipitation could be enhanced by differential scanning calorimetry. As a result, the precipitates of the η phase became finer compared with conventional aging treatment. Sarma [13] found that cryorolling significantly changed aging behavior, leading to a reduction in the aging temperature from 190 to 125 °C and in aging time from 24 to 8 h for treating 2219 alloy (Al–Cu alloy).

Jayaganthan [14] used X-ray analyses to study the aging behavior of 6061 alloy subjected to solution treatment then cryorolling. They found that increasing true strain in cryorolling tended to enhance the dissolution of alloys in the alloy matrix and promoted the driving force for precipitation. The UTS was improved from 300 to 365 MPa and elongation was raised from 11% to 13%. Cryorolling apparently reduced the size of intermetallic compounds contained in the matrix of 7075–T73 alloys. As a result, the corrosion resistance of anodized and sealed 7075–T73 alloy could be significantly enhanced [15].

Anodization and sealing improves the corrosion resistance of Al alloys by forming amorphous alumina and hydrate alumina in the anodized film. This process has been widely used in industry. Forging is a common process used for making bicycle and automobile parts. Determining the influence of cryoforging on the corrosion resistance of Al alloys with or without anodization should provide more values for designing and using Al alloys.

Copper was added as an alloying element in a 6xxx series alloy to enhance mechanical strength by precipitation hardening after heat treatment. For example, 6066 Al alloy contains some high-strength Cu (0.8–1.4 Mg, 0.9–1.8 Si, 0.6–1.1 Mn, and 0.7–1.2 Cu) to obtain a UTS of 393 MPa and a yield stress (YS) of 359 MPa after a T6 treatment. This study introduced cryoforging to further improve the toughness of 6066–T6 alloys and their corrosion resistance. The effects of anodization and sealing on the tensile properties and corrosion resistance of AA6066–T6 with and without cryoforging were also evaluated.

2. Experimental Procedures

2.1. Materials

As-extruded 6066 alloy bars, ∅42 × 100 mm in size, were supplied by Tzan Wei Aluminum Co., Ltd. (Tainan, Taiwan). The chemical compositions of the alloys (in wt. %) were 1.38 Si, 0.15 Fe, 1.18 Cu, 1.00 Mn, 1.09 Mg, 0.16 Cr, 0.04 Zn, and 0.02 Ti.

After annealing at 688 K for 120 min, the samples were divided into two groups. The first group was subjected to a solution treatment (803 K for 120 min) and artificial aging (450 K for 480 min); these samples were coded as T6 samples. The second group was subjected first to cryogenic forging, achieving a 40% reduced thickness (from 28 to 16.8 mm), then immersed in liquid nitrogen again, rotated by 90°, and subjected to a second round of cryogenic forging. A 500-ton hydraulic press equipped with one open die set was used to conduct compression forging. After forging, the second group of samples was subjected to the solution to get CFT4 sample and followed by artificial aging treatments: 450 K for 480 min for CFT6a samples and 540 min for CFT6b samples.

2.2. Tensile and Fatigue Tests

The specimens used for testing tensile and rotating bending fatigue strengths were machined from heat-treated samples according to the ASTM B557 [16] (gage diameter: 6 mm) and JIS Z2274 [17] (gage diameter: 8 mm) specifications as shown in Figure 1a,b. The machined test bars were polished by a series of abrasive papers (2000 grit) and an alumina slurry to achieve a surface roughness of less than 0.1 μm (*Ra*).

2.3. X-ray Tests

X-ray diffraction (XRD) measurements were performed by using a NANO-Viewer Advance (Rigaku, Tokyo, Japan) equipped with a Cu target to identify the precipitates (or second-phase particles) formed in the matrix of different samples, including the T6, CFT4, and CFT6a samples. The cryoforged samples were solution-treated to get a CFT4 sample and followed by aging treatment to acquire a CFT6a sample. A power of 30 KV and current of 10 mA were used in this study. The sample sizes were 10 × 10 × 1 mm.

(a)

Nominal Diameter	Dimensions,mm
G-gage length	30±0.06
D-Diameter	6.00±0.10
R-Radius of fillet	6
A-Length of reduced section	36

(b)

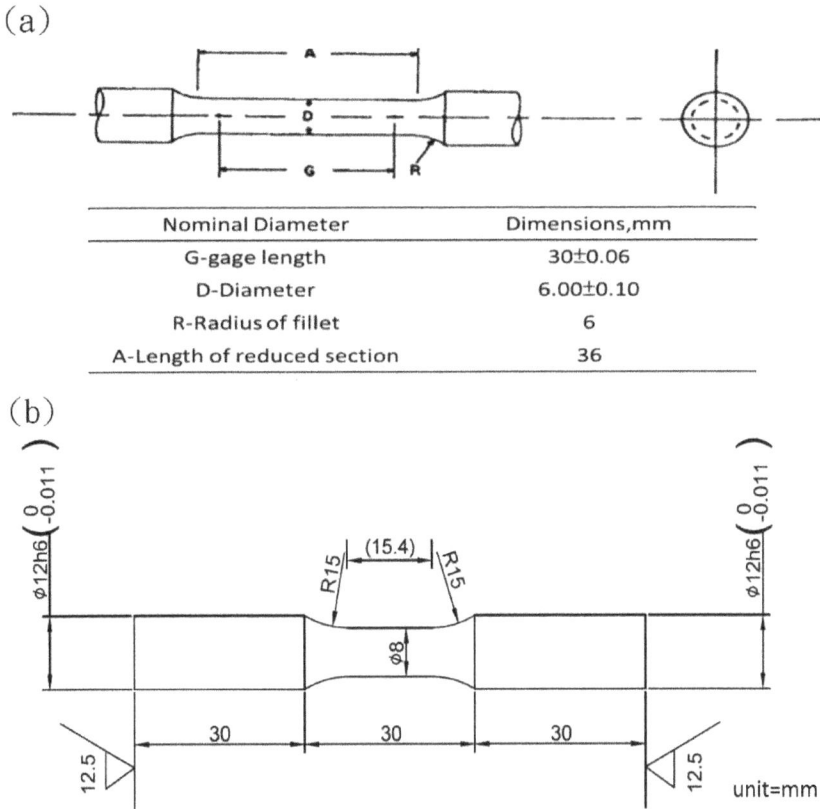

Figure 1. The dimensions of specimens used for (**a**) tensile test (gage diameter: 6 mm); and (**b**) rotating bending test (gage diameter: 8 mm).

2.4. Anodization Process

Before anodization, all samples were polished to a surface roughness of approximately $Ra \leqslant 0.1$ µm and then dipped into methanol and ultrasonically vibrated. The specimens were initially degreased by immersion in an alkaline solution (5 mass % NaOH) at 60 °C for 30 s and were then rinsed with water for 1–2 min. For the pickling process, specimens were submerged in an aqueous solution of HNO_3 (30 vol. %) for 90 s at room temperature and then rinsed with water for 1–2 min. The anodization was conducted at 15 mA·cm^{-2} at 15 °C for 900 s in a 15 mass % sulfuric acid solution. The anodized samples were sealed in hot water at 95 °C for 1200 s [18]. After anodization, scanning electron microscopy was conducted and determined that the anodic film was 12–14 µm thick. The anodized samples were sealed in hot water at 368 K for 20 min.

3. Results and Discussion

3.1. Microstructure Observation and Tensile Properties

Table 1 shows the measured mechanical properties of different samples. The CFT6b samples, which aged for 60 min longer than did the CFT6a samples, had increased strength but reduced elongation. The CFT6a samples exhibited a similar strength but superior elongation to the CFT6b samples, and they were adopted in this study for evaluation of their corrosion resistance and fatigue strength. In addition, the CFT6a samples obtained a matrix with a uniform hardness (HV minimum deviation: 1.7).

Table 1. Mechanical properties of T6, CFT6a and CFT6b samples.

Sample	Tensile Strengths			Vickers Hardness (HV)
	UTS (MPa)	YS (MPa)	Elongation (%)	
T6	460 (0.5)	438 (1.4)	10.6 (1.7)	133.6 (2.9)
CFT6a	431 (6.6)	394 (9.2)	14.2 (0.2)	136.6 (1.7)
CFT6b	435 (5.4)	404 (10.1)	13.4 (0.9)	137.1 (2.9)

Note: the deviations are listed in parentheses.

Figure 2a,b show the second-phase (SP) and/or intermetallic compounds (IMC) located at the cross-sections in the transverse direction of the T6 and CFT6a samples. Different sizes of particles were counted and are listed in Table 2. The T6 samples exhibited more coarse SP/IMC particles than the CFT6a samples did as revealed by arrows in Figure 2. During the cryogenic forge, a shear stress was generated to act on the samples' matrix, breaking down the particles. As a result, the particles became finer and dissolution was enhanced during the solution treatment. The total SP/IMC particle count decreased from 864 to 734 counts/mm^2.

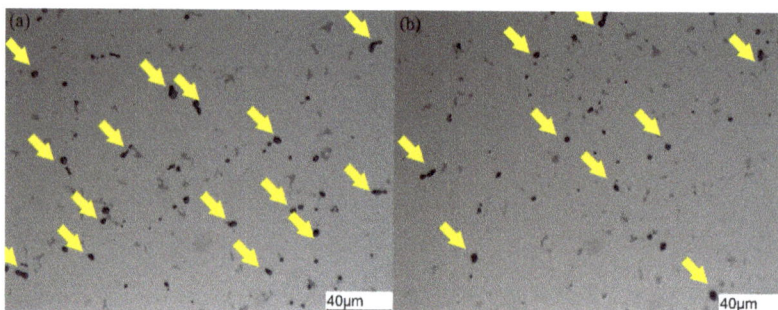

Figure 2. Optical micrographs show second-phase particles located in the matrix of (**a**) T6; (**b**) CFT6a sample; non-etched cross-section in transverse direction.

Table 2. The SP and/or IMC particles measured from T6 and CFT6a samples; maximum particle size and count population were included.

Sample	Second Phase (Coarse Precipitates) Counts/mm^2				Max. Diameter of Particle, μm
	1–10	11–20	>20	Total Count	
T6	838	25	0	864	17
CFT6a	719	15	0	734	17

Electron backscattered diffraction (SUPRA ULTRA 55 field emission scanning electron microscope, ZEISS, Jena, Germany) was performed to measure the misorientation angles of grain boundaries in the longitudinal direction of the tensile test bar samples, as illustrated in Figure 3a,b for the CFT6a and T6 samples, respectively. The CFT6a sample had a high fraction of high-angle grain boundaries (HAGBs). The matrix of the T6 sample had grains featuring mainly low-angle grain boundaries, as shown in Figure 3b. The cyclic loaded sample after being subjected to 250 MPa and fractured at 2.56×10^5 life cycles is shown in Figure 3c, revealing further increasing HAGBs compared with those in Figure 3a. This increase could have been affected by dynamic recrystallization during cyclic loading.

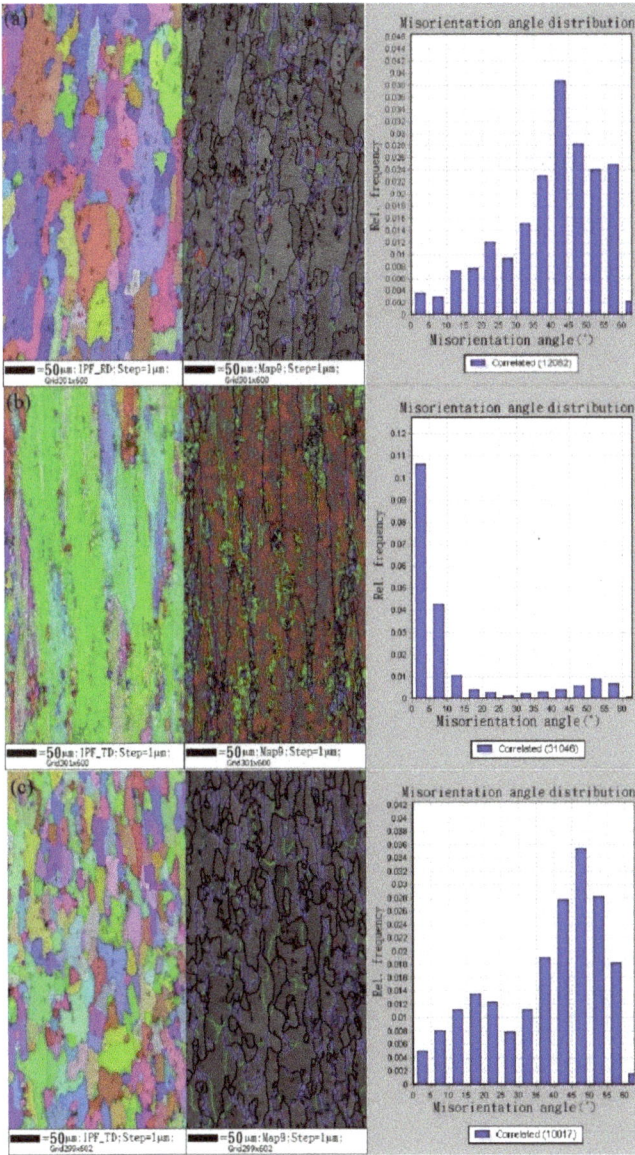

Figure 3. Inverse pole figure maps obtained from EBSD data and measured misorientation angle of grain boundaries from (**a**) CFT6a tensile tested sample; (**b**) T6 tensile tested sample; and (**c**) CFT6a sample after subjected to 250 MPa and fractured at 2.56×10^5 life cycles.

The T6 sample comprised α-Al, *Q*-(AlCuMgSi), Al(Mn,Fe)Si, and some Mg_2Si phases, as shown in Figure 4. After solution treatment, the cryoforged sample was tested and indicated that SP/IMC particles were highly dissolved in its matrix, as confirmed by the CFT4 sample in Figure 4. After artificial aging, the intensity of the *Q* phase was notably decreased, presenting complex phases of Mg_2Si, Q-(AlCuMgSi), and Al(Mn,Fe)Si, as well as some Cu_9Al_4 and CuMgSi phases; see CFT6a sample. Kim *et al.* annealed copper wire and an aluminum pad at 150 to 300 °C and found a Cu_9Al_4 phase with an X-ray spectra peak at 43.9° [19].

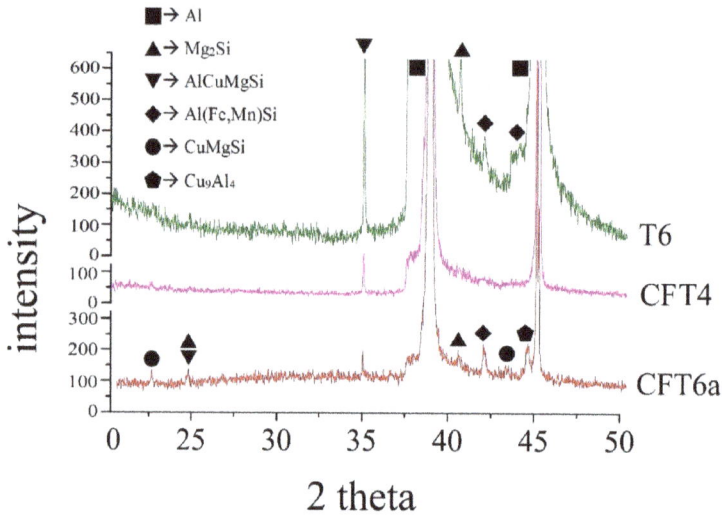

Figure 4. XRD patterns of 6066 alloy samples; including solution and aging T6 sample; cryoforged samples after solution treatment CFT4 and aging CFT6a, respectively.

The diffusivity of alloying elements in the aluminum matrix is in the order of $D_{Cu/Al}$ (4.44×10^{-5} m^2/s), $D_{Mg/Al}$ (1.49×10^{-5} m^2/s), and $D_{Si/Al}$ (1.38×10^{-5} m^2/s) in face-centered cubic Al [20]. Cu atoms tend to segregate around the metastable phase in the Al–Mg–Si alloy, move to grain boundaries during the solution treatment, and diffuse to Mg–Si nanoparticles located at or near grain boundaries to finally form type-C precipitates [21] and the Q phase [22]. As a result, in this study, the T6 sample mainly contained the Q phase and Mg_2Si precipitates after aging. The present XRD spectra did not detect the S-Al_2CuMg phase from the T6 sample. Neither could Vieira find the S-Al_2CuMg phase in his study, in which two age-hardening heat treatments of Al–10Si–4.5Cu–2Mg were completed [23].

The CFT6a sample increased the HAGBs in its matrix to provide more potent sites for accommodating Cu atoms after the solution treatment. During quench or aging in room temperature, Cu diffused to tie up with Al forming Cu_9Al_4 phase. During artificial aging, a Mg–Si cluster formed *in situ*; this either led to the formation of CuMgSi precipitates through movement of the Cu atoms, or the Cu–Mg clusters formed first and subsequently incubated the CuMgSi precipitate in the matrix. Figure 5a,b shows the transmission electron microscopy photos of the T6 and CFT6a samples, respectively. For a given solution and aging conditions, the CFT6a sample achieved finer precipitates (less than 100 nm) than the T6 sample did. The main constituted phases likely included Al(Mn,Fe)Si in block and plate shapes, and some fine particles (less than 100 nm) likely being Cu_9Al_4 and CuMgSi precipitates or Cu atoms surrounding fine Mg_2Si phases, as shown in Figure 5b. The α-Al(Mn,Fe)Si dispersoids could have a block-shaped or plate-shaped morphology in the size of 50–200 nm, as reported by Li *et al.* [24].

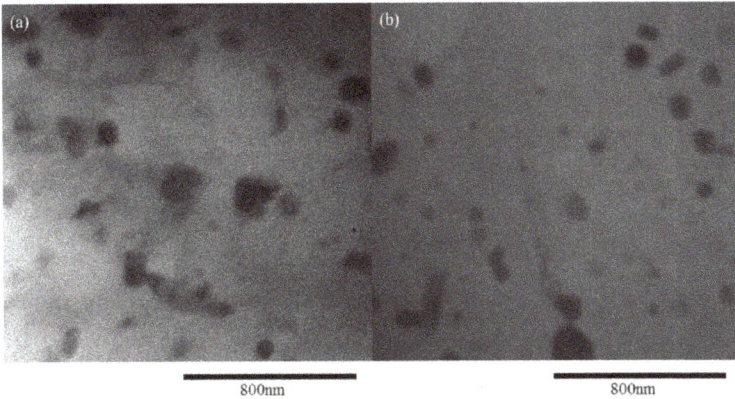

Figure 5. TEM photo show the plate-shape and lump-shape precipitate in the matrix of (**a**) T6 sample, and (**b**) CFT6a sample.

An electron probe X-ray microanalyzer (JXA-8200, JEOL USA, Inc., Peabody, MA, USA) was used to obtain images of Mg, Si, Mn, and Cu mappings from the matrix of the CFT6a sample. The white aggregates in Figure 6a are the Al(Mn,Fe)Si and *Q* phase, and the black lumps are Mg_2Si particles. The arrows indicate the locations of *Q*-phase precipitates. The CuMgSi and Cu_9Al_4 precipitates are nanoscale and were difficult to detect by mapping. In Figure 6b, the white particles shown in the matrix of the T6 samples are Al(Mn,Fe)Si and *Q*-AlCuMgSi phases. An Al(MnFe)Si particle attached with Cu atoms could be distinguished and are indicated by an arrow. The black spots are the Mg_2Si phase.

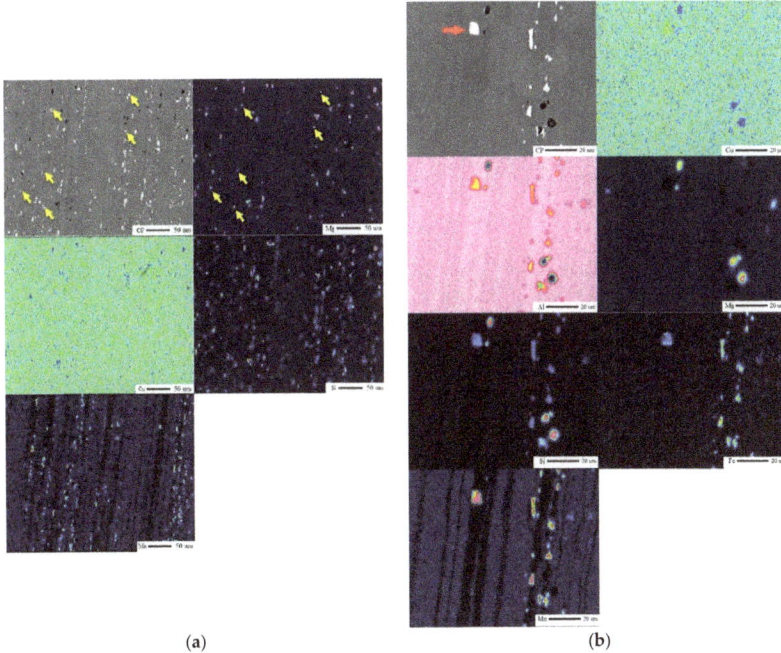

(**a**)   (**b**)

Figure 6. SEM photo and alloying elements mappings obtained from (**a**) T6 sample and (**b**) CFT6a sample.

The Q phase, Al(Mn,Fe)Si, and Mg_2Si (larger than 80 nm) contained in the aluminum alloy sample are non-shearable particles in the α-Al matrix [25]. Therefore, the T6 samples gained high strength. By contrast, the CFT6a samples contained Al(Mn,Fe)Si, Mg_2Si, fine CuMgSi, and Cu_9Al_4 precipitates. The fine Mg_2Si or CuMgSi and Cu_9Al_4 precipitates are shearable. Figure 7 illustrates the dislocations (marked as 1-1 and 2-2) intersected with two fine precipitates. In addition, the SP/IMC particle counts (Figure 2) were lower in the matrix of the CFT6a samples. As a result, the CFT6a sample had decreased numbers of barrier sites for tangling dislocations to reduce strength but enhance elongation.

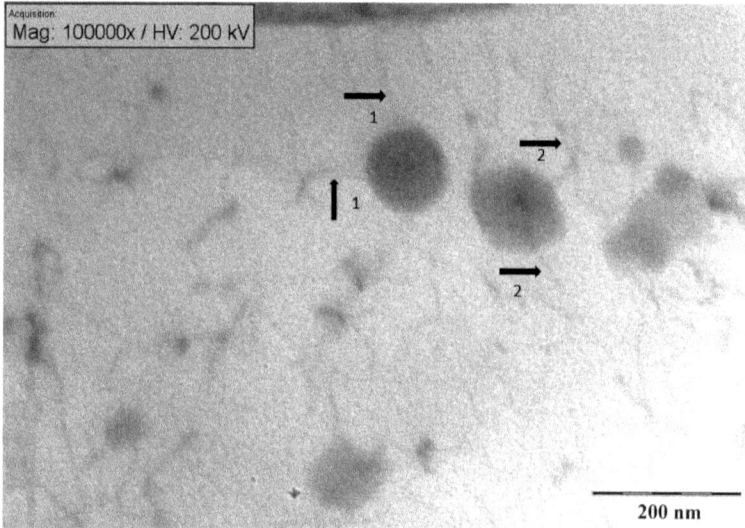

Figure 7. TEM photo shows the intersection of dislocation 1-1 and 2-2 with two fine precipitates in the matrix of CFT6a sample, respectively.

Table 3 compares the surface roughness of T6 and CFT6a samples before and after anodization/sealing. Compared with the samples without anodization, these anodized and sealed samples showed an approximately 8%–10% reduction in elongation. Such a decrease is likely caused by the dissolution of SP/IMC at the film/matrix interface leading to degrade surface roughness and elongation. Before anodization, the surface roughness of the T6 and CFT6a sample was approximately 0.07 (0.04) μm; after anodization and sealing, the surface roughness became 0.16 (0.03) and 0.13 (0.06) μm.

Table 3. Measured surface roughness of different samples before and after anodization.

Sample	Surface Roughness (μm)
T6; CFT6a	0.07 (0.04)
T6-A	0.16 (0.03)
CFT6a-A	0.13 (0.06)

Note: the deviations are listed in parentheses.

3.2. Fatigue and Corrosion Tests

Applying the cryoforge before the solution treatment reduced the SP/IMC particle counts and transformed the Q phase into nanoscale CuMgSi and Cu_9Al_4 precipitates, which reduced UTS and YS but increased elongation. Figure 8 shows that both the T6 and CFT6a samples obtained fatigue strength of 180 MPa at 1×10^7 life cycles. As shown in Figure 3a,c, we found that cyclic loading functioned to shift the peak of relative frequency from the misorientation angle of 40°–45° to 45°–50°, and increase some grain boundaries at angles of 10°–20°. During cyclic loading, shear stress drove part of the dislocations to conduct polygonization and/or reorganization.

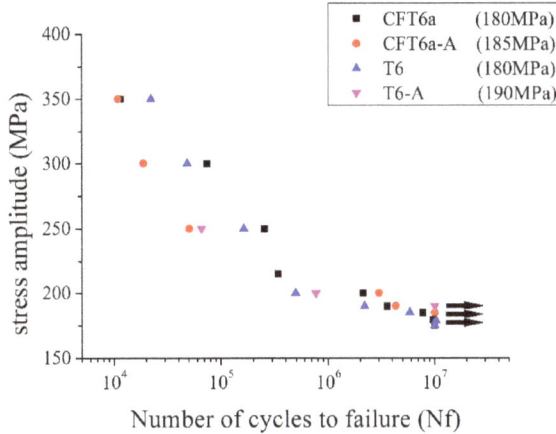

Figure 8. S–N curves for different samples; including T6 and CFT6a with/without anodization/sealed treatment; "A" represented anodization/sealed sample.

Figure 9a,b show the fractured surface of the T6 and CFT6a samples, both of which were subjected to 185 MPa and which performed 5.8×10^6 and 7.7×10^6 life cycles, respectively. The CFT6a sample exhibited narrower striation spacing than did the T6 samples and achieved a longer fatigue life. The T6 sample obtained fracture steps, as revealed in Figure 9a, which can be attributed to the Q phase located at or near grain boundaries that served to strengthen the grain boundaries [22,26]. Increasing fine precipitates in the matrix of the CFT6a sample likely more effectively to consume the crack propagation energy and thus slightly enhanced the life cycles.

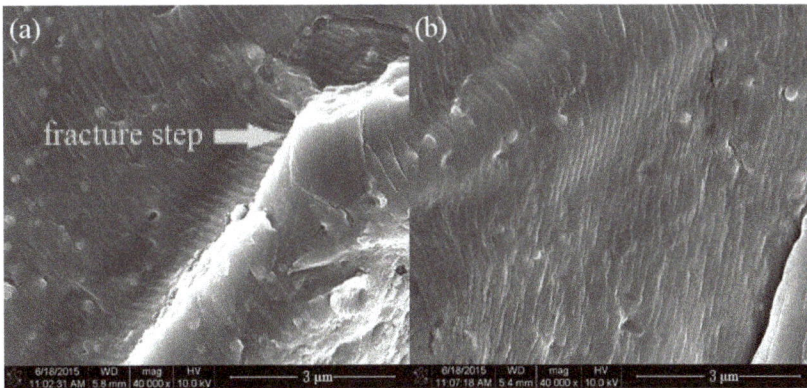

Figure 9. Fracture surface prepared from (**a**) T6; (**b**) CFT6a samples; both subjected to 185 MPa and fractured at 5.8×10^6 and 7.7×10^6 life cycles, respectively.

The SP/IMC particles were small to be less than 17 µm. Anodization did not yield the deteriorated effect of reducing fatigue strength. The two anodized and sealed samples (T6-A and CFT6a-A) showed similar fatigue strength (185–190 MPa) at 1×10^7 life cycles as did those without anodization and sealing (180 MPa).

The polarization curves for all the samples with and without sealed anodic aluminum oxide (AAO) films are shown in Figure 10. The corrosion behavior of the samples was affected by the constituent phases of each. The T6 samples contained AlCuMgSi, Al(FeMn)Si, and Mg_2Si phases, whereas the CFT6a samples acquired mainly Al(FeMn)Si and Mg_2Si phases as well as some CuMgSi and Cu_9Al_4 precipitates. Mg_2Si is anodic relative to the aluminum matrix, but Al(FeMn)Si and AlCuMgSi are cathodic. The Cu–Al precipitates are more vulnerable to attack during the immersion test, compared with the Q phase [23]. The CFT6a samples exhibited slightly inferior E_{corr} (−1.0 V) and I_{corr} (4.7×10^{-6} A/cm^2) than the T6 sample did (−0.93 V and 2.6×10^{-6} A/cm^2, respectively).

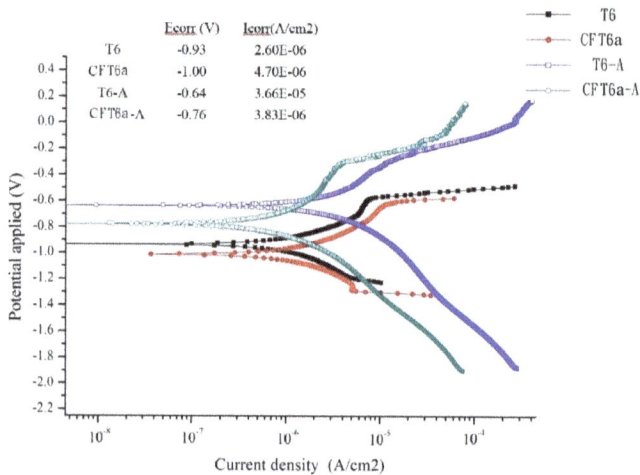

	E_{corr} (V)	I_{corr} (A/cm2)
T6	-0.93	2.60E-06
CFT6a	-1.00	4.70E-06
T6-A	-0.64	3.66E-05
CFT6a-A	-0.76	3.83E-06

Figure 10. Polarization curves obtained from immersion tests of T6 and CFT6a samples with/without anodization/sealing treatment.

After anodization and sealing, the anodized films on the aluminum alloy samples contained mainly amorphous alumina and a few hydrated alumina [27]. These AAO films could significantly enhance the corrosion resistance of the aluminum alloys [28]. Therefore, the anodized films deposited on the T6 and CFT6 samples improved E_{corr} from (−0.9 to −1.0 V) to (−0.64 to −0.76 V).

Affected by a greater number of coarse SP/IMC (Table 2) and/or a higher particle population, the anodized and sealed film on the T6 sample was entrapped with particles along with air pockets, as shown in Figure 11a. By contrast, the film on the CFT6a sample is relatively sound with few trapped particles; see Figure 11b. The trapped particles could function as corrosion channels to accelerate chloride attacks. Therefore, the anodized and sealed T6 sample obtained a higher current density (I_{corr} of 3.6×10^{-5} A/cm^2) than did the anodized and sealed CFT6a sample (I_{corr} of 3.8×10^{-6} A/cm^2).

The dissolution of SP/IMC at the film–matrix interface drove aluminum and magnesium ions to move toward electrolytes leaving the Si particles remaining in the film, as shown in Figure 11c [27]. Consequently, the anodized and sealed T6 sample obtained inferior corrosion current density to the bare T6 sample (3.6×10^{-5} A/cm^2 *vs.* 2.6×10^{-6} A/cm^2). Decreasing SP/IMC particle size and counts also reduced the size and counts of the Si particles that remained in the anodized film to undergo a minor change in corrosion current density (4.7×10^{-6} *vs.* 3.86×10^{-6} A/cm^2) in the bare CFT6a and anodized/sealed CFT6a samples.

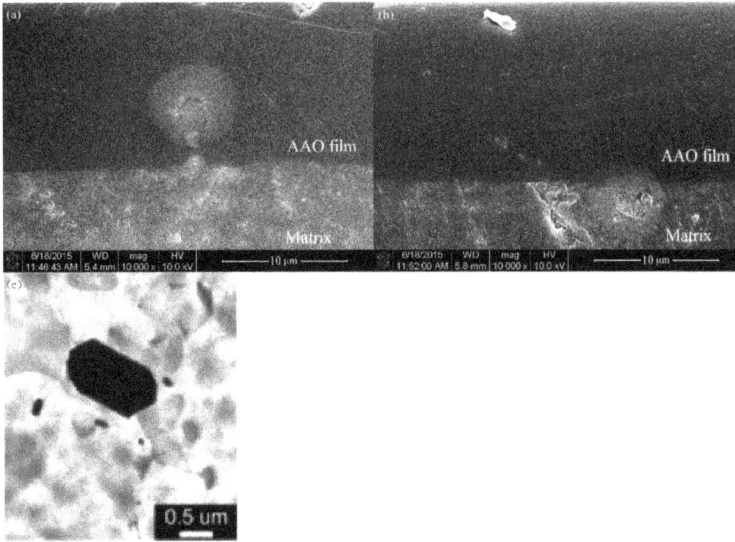

Figure 11. SEM photos show the cross-section of deposited anodized/sealed films on (**a**) T6 sample; (**b**) CFT6a sample; and (**c**) anodized/sealed 6063-T6 samples [27].

4. Conclusions

Adding cryoforging to the process changed the microstructure and mechanical properties of the AA6066–T6 alloy samples; specifically, it decreased SP/IMC particle counts and reduced the Q phase but increased fine CuMgSi and Cu_9Al_4 precipitates in the matrix of the CFT6a sample. As a result, the tensile property of elongation was increased by approximately 34% comparing with T6 sample. The anodized and sealed AAO film on the CFT6a sample improved its corrosion resistance but decreased its elongation by approximately 10% (from 14.2% to 12.7%).

Acknowledgments: We gratefully acknowledge the financial support from the Ministry of Science and Technology of the Republic of China (MOST 103-2221-E-008-026-MY2). Many thanks also to National Central University for providing the SEM and TEM tests and to National Sun-Yat Sen University for the EBSD analysis.

Author Contributions: Hwa-Sheng Yong and Wen-Nong Hsu ran experiments of this study and did OM, SEM, TEM observation and EBSD analyses. Main contribution also included experimental data collection. All authors provided equal contribution.

Conflicts of Interest: The authors declare no conflict of interest.

References

1. Wang, Y.M.; Ma, E.; Chen, M.W. Enhanced tensile ductility and toughness in nanostructured Cu. *Appl. Phys. Lett.* **2002**, *80*, 2395–2397. [CrossRef]
2. Rangaraju, N.; Raghuram, T.; Krishna, B.V.; Rao, K.P.; Venugopal, P. Effect of cryo-rolling and annealing on microstructure and properties of commercially pure aluminium. *Mater. Sci. Eng. A Struct.* **2005**, *398*, 246–251. [CrossRef]
3. Chen, Y.J.; Roven, H.J.; Gireesh, S.S.; Skaret, P.C.; Hjelen, J. Quantitative study of grain refinement in Al–Mg alloy processed by equal channel angular pressing at cryogenic temperature. *Mater. Lett.* **2011**, *65*, 3472–3475. [CrossRef]
4. Lee, Y.B.; Shin, D.H.; Park, K.T.; Nam, W.J. Effect of annealing temperature on microstructures and mechanical properties of a 5083 Al alloy deformed at cryogenic temperature. *Scr. Mater.* **2004**, *51*, 355–359. [CrossRef]
5. Panigrahi, S.K.; Jayagathan, R. A study on the mechanical properties of cryorolled Al–Mg–Si alloy. *Mater. Sci. Eng. A Struct.* **2008**, *480*, 299–305. [CrossRef]

6. Panigrahi, S.K.; Jayagathan, R.; Chawla, V. Effect of cryorolling on microstructure of Al–Mg–Si alloy. *Mater. Lett.* **2008**, *62*, 2626–2639. [CrossRef]

7. Panigrahi, S.K.; Jayagathan, R. Development of ultrafine-grained Al 6063 alloy by cryorolling with the optimized initial heat treatment conditions. *Mater. Des.* **2011**, *32*, 2172–2180. [CrossRef]

8. Krishna, K.G.; Sivaprasad, K.; Venkateswarlu, K.; Kumar, K.C.H. Microstructural evolution and aging behavior of cryorolled Al–4Zn–2Mg alloy. *Mater. Sci. Eng. A Struct.* **2012**, *535*, 129–135. [CrossRef]

9. Das, P.; Jayaganthan, R.; Singh, I.V. Tensile and impact-toughness behavior of cryorolled Al 7075 alloy. *Mater. Des.* **2011**, *32*, 1298–1305. [CrossRef]

10. Panigrahi, S.K.; Jayaganthan, R. Effect of ageing on microstructure and mechanical properties of bulk, cryorolled, and room temperature rolled Al 7075 alloy. *J. Alloy. Compd.* **2011**, *509*, 9609–9016. [CrossRef]

11. Yin, J.; Lu, J.; Ma, H.; Zhang, P. Nanostructural formation of fine grained aluminum alloy by severe plastic deformation at cryogenic temperature. *J. Mater. Sci.* **2004**, *39*, 2851–2854. [CrossRef]

12. Krishna, K.G.; Sivaprasad, K.; Narayanana, T.S.N.S.; Kumar, K.C. Localized corrosion of an ultrafine grained Al–4Zn–2Mg alloy produced by cryorolling. *Corros. Sci.* **2012**, *60*, 82–89. [CrossRef]

13. Shanmugasundaram, T.; Murty, B.S.; Sarma, V.S. Development of ultrafine grained high strength Al–Cu alloy by cryorolling. *Scr. Mater.* **2006**, *54*, 2013–2017. [CrossRef]

14. Rao, P.N.; Jayaganthan, R. Effects of warm rolling and ageing after cryogenic rolling on mechanical properties and microstructure of Al 6061 alloy. *Mater. Des.* **2012**, *39*, 226–233.

15. Huang, Y.S.; Shih, T.S.; Chou, J.H. Electrochemical behavior of anodized AA7075-T73 alloys affected by matrix structures. *Appl. Surf. Sci.* **2013**, *283*, 249–257. [CrossRef]

16. *ASTM B557-15, Standard Test Methods for Tension Testing Wrought and Cast Aluminum- and Magnesium-Alloy Products*; ASTM International: West Conshohocken, PA, USA, 2015. [CrossRef]

17. *JIS Z2274, Method of Rotating Bending Fatigue Testing of Metals*; Japanese Standards Association: Tokyo, Japan, 1978.

18. Shih, T.S.; Lee, T.H.; Jhou, Y.J. The effects of anodization treatment on the microstructure and fatigue behavior of 7075-T73 aluminum alloy. *Mater. Trans.* **2014**, *55*, 1280–1285. [CrossRef]

19. Kim, H.J.; Lee, J.Y.; Paik, K.W.; Koh, K.W.; Won, J.; Choe, S.; Lee, J.; Moon, J.T.; Park, Y.J. Effects of Cu/Al intermetallic compound (IMC) on copper wire and aluminum pad bondability. *IEEE Trans. Pack. Technol.* **2003**, *26*, 367–374.

20. Du, Y.; Chang, Y.A.; Huang, B.; Gong, W.; Jin, Z.; Xu, H.; Yuan, Z.; Liu, Y.; He, Y.; Xie, F.Y. Diffusion coefficients of some solutes in fcc and liquid Al: Critical evaluation and correlation. *Mater. Sci. Eng. A Struct.* **2003**, *363*, 140–151. [CrossRef]

21. Matsuda, K.; Teguri, D.; Sato, T.; Uetani, Y.; Ikeno, S. Cu Segregation around Metastable Phase in Al–Mg–Si Alloy with Cu. *Mater. Trans.* **2007**, *48*, 967–974. [CrossRef]

22. Svenningsen, G.; Larsen, M.H.; Walmsley, J.C.; Nordlien, J.H.; Nisancioglu, K. Effect of artificial aging on intergranular corrosion of extruded AlMgSi alloy with small Cu content. *Corros. Sci.* **2006**, *48*, 1528–1543. [CrossRef]

23. Vieira, A.C.; Pinto, A.M.; Rocha, L.A.; Mischler, S. Effect of Al_2Cu precipitates size and mass transport on the polarisation behaviour of age-hardened Al–Si–Cu–Mg alloys in 0.05 M NaCl. *Electrochim. Acta* **2011**, *56*, 3821–3828. [CrossRef]

24. Li, Y.J.; Muggerud, A.M.F.; Olsen, A.; Furu, T. Precipitation of partially coherent α-Al(Mn,Fe)Si dispersoids and their strengthening effect in AA 3003 alloy. *Acta Mater.* **2012**, *60*, 1004–1014. [CrossRef]

25. Cabibbo, M. Microstructure strengthening mechanisms in different equal channel angular pressed aluminum alloys. *Mater. Sci. Eng. A Struct.* **2013**, *560*, 413–432. [CrossRef]

26. Zhaia, T.; Jianga, X.P.; Lia, J.X.; Garratt, M.; Bray, G.H. The grain boundary geometry for optimum resistance to growth of short fatigue cracks in high strength Al-alloys. *Int. J. Fatigue* **2005**, *27*, 1202–1209. [CrossRef]

27. Huang, Y.S.; Shih, T.S.; Wu, C.E. Electrochemical behavior of anodized AA6063-T6 alloys affected by matrix structures. *Appl. Surf. Sci.* **2013**, *264*, 410–418. [CrossRef]

28. Shih, T.S.; Chiu, Y.W. Corrosion resistance and high-cycle fatigue strength of anodized/sealed AA7050 and AA7075 alloys. *Appl. Surf. Sci.* **2015**, *351*, 997–1003. [CrossRef]

metals

MDPI

Article

Temperature Effects on the Tensile Properties of Precipitation-Hardened Al-Mg-Cu-Si Alloys

J.B. Ferguson [1], Hugo F. Lopez [1], Kyu Cho [2] and Chang-Soo Kim [1,*]

[1]   Materials Science and Engineering Department, University of Wisconsin-Milwaukee, Milwaukee, WI 53211, USA; jbf2@uwm.edu (J.B.F.); hlopez@uwm.edu (H.F.L.)
[2]   U.S. Army Research Laboratory, Weapons and Materials Research Directorate, Aberdeen Proving Ground, MD 21005, USA; kyu.c.cho2.civ@mail.mil
*   Correspondence: kimcs@uwm.edu; Tel.: +1-414-229-3085

Academic Editor: Nong Gao
Received: 5 January 2016; Accepted: 15 February 2016; Published: 23 February 2016

Abstract: Because the mechanical performance of precipitation-hardened alloys can be significantly altered with temperature changes, understanding and predicting the effects of temperatures on various mechanical properties for these alloys are important. In the present work, an analytical model has been developed to predict the elastic modulus, the yield stress, the failure stress, and the failure strain taking into consideration the effect of temperatures for precipitation-hardenable Al-Mg-Cu-Si Alloys (Al-A319 alloys). In addition, other important mechanical properties of Al-A319 alloys including the strain hardening exponent, the strength coefficient, and the ductility parameter can be estimated using the current model. It is demonstrated that the prediction results based on the proposed model are in good agreement with those obtained experimentally in Al-A319 alloys in the as-cast condition and after W and T7 heat treatments.

Keywords: precipitation-hardened alloys; Al alloys; tensile properties; temperature effects; property modeling

1. Introduction

Precipitation-hardened alloys or age-hardened alloys based on Al, Mg, Ni, and/or Ti are widely used in structural applications due to the enhanced strength that can be achieved by heat treating [1–3]. The mechanical properties of these types of alloys are deeply influenced by (i) the thermodynamic stability of the precipitates in the alloys; and (ii) the concentration of impurity elements that influence which precipitates ultimately evolve in the alloy microstructure. Because differing heat treatment procedures are employed to control the quantity, size, and composition of the intermetallics that form in the alloys, it is possible to develop different combinations of room temperature strength and ductility. However, these alloys are frequently used in applications well above room temperature, where the effect of heat treatment/precipitation condition is not well-described in general. Al alloy A319 (Al-A319) is a good example of this type of alloy. Al-A319 cast alloys have been increasingly used in the manufacture of engine blocks due to a combination of good abrasive properties and mechanical strength [1–3]. The microstructural constituents present in this alloy are typically complex multiphases comprising eutectic (acicular) Si as well as numerous intermetallic phases. Al-A319 alloys nominally contain several percents of Si. Other elements such as Cu and Mg are typically incorporated in Al-A319 alloys to improve the room and high temperature strength [4,5].

Because yield strength, failure stress, strain-to-failure, and strain hardening coefficient that can be varied depending on different service temperatures are key factors to determine the performance of precipitation-hardened alloys, it would be advantageous to have a prediction model to describe the temperature-dependent behavior of these types of alloys. However, despite the long history of developing these precipitation-hardened alloys in the metallurgy field, there are no tools to estimate the temperature dependence of the mechanical properties of the alloys. In the present study, therefore, an analytical model involving an empirical fitting approach has been developed to predict the dependence of temperature on strength, ductility, and strain hardening coefficient of precipitation-hardened alloys. We previously developed an analytical model to predict the tensile and compressive properties of bimodal metallic materials at room temperature [6]. The model showed how the properties of true yield stress, true failure stress, strain hardening exponent, strain-to-failure, and a newly defined ductility parameter are interrelated. The interrelatedness of these parameters need not be limited to only bimodal materials. In this work, an extended version of the model is proposed to account for the effect of temperature on Young's modulus, true yield stress, true failure stress, and strain-to-failure for precipitation-hardened alloys. We will show that, from these basic parameters, all other mechanical parameters of interest including strength coefficient, ductility parameter (which will be described later), and strain hardening exponent can be calculated. The prediction results of these combined models were compared to experimental data from tensile tests of Al-A319 alloys in the as-cast, W (solutionized for 4.5 h at 450 °C), and T7 (W treatment + age-hardened for 4.67 h at 230 °C) heat treatment conditions. Specimens with W and T7 heat treatment conditions were selected in this work because solutionization (*i.e.*, W condition) and subsequent age-hardening (*i.e.*, T7 condition) are commonly applied to Al-A319 alloys to reduce the residual stress accumulated during the casting process and to improve homogeneity in the microstructure of materials. Because of the relatively short time intervals between the W heat treatment and the tensile testing, the effect of natural aging was not considered in the current work.

2. Analytical Model Development

2.1. Temperature-Dependent Model

Based on the evaluation of experimental data, the empirical relationships of Equations (1) to (5) are proposed to account for the temperature-dependent properties of Young's modulus (Y), true yield stress (σ_{t-y}), true failure stress ($\sigma_{t-\text{fail}}$), and strain-to-failure ($\varepsilon_{t-\text{fail}}$).

Previous research has shown that the Young's modulus of pure Al decreases linearly with temperature from some initial maximum value, Y_0 [7]. However, in alloy systems, the behavior is more complicated and often shows a more marked non-linear decrease. We assume that the behavior of the Young's modulus can be adequately described by a transitional function such as the logistic function to account for such non-linearity. This is based on an assumption that the Young's modulus of Al alloys would not be much influenced by the temperature changes at extreme temperatures, which is true near the melting temperature and at very low temperature. The logistic function requires the use of a low temperature Young's modulus value, Y_{LT}, as well as a high temperature value, Y_{HT}. The subscripts LT and HT denote low temperature and high temperature, respectively. It would be reasonable to consider these LT and HT are the temperatures that are applicable as the lower and upper bound of the logistic function. Therefore, we set the initial maximum value of modulus, Y_0, as the Y_{LT}. Based on this, in this work, Equation (1) is used to describe the Young's modulus (Y) as a function of temperature (T) using suitable choices of empirical constants.

$$Y = Y_o - \frac{\Delta Y}{\Delta T}T - \frac{Y_{LT} - Y_{HT}}{1 + e^{(T^*-T)/\Psi}} = Y_{LT} - \frac{\Delta Y}{\Delta T}T - \frac{Y_{LT} - Y_{HT}}{1 + e^{(T^*-T)/\Psi}} \tag{1}$$

In Equation (1), T^* is the alloy softening temperature at which Young's modulus with temperature shows the highest rate of change (*i.e.*, the highest $\frac{dY}{dT}$) [8]. Here, Ψ represents another experimentally determined transition parameter to govern the rapidity with which the transition takes place. As will be demonstrated in the "Results and Discussion" section, the linear dependence term ($\frac{\Delta Y}{\Delta T}T$) in

Equation (1) does not considerably affect the variations of the Young's modulus with temperature changes for the Al-A319 systems considered in this work.

The temperature dependence of true yield stress (σ_{t-y}) is determined by adjusting the low temperature true yield stress ($\sigma_{t-y_{LT}}$) for the loss of stiffness that accompanies the increase in temperature. We approximate that this adjustment would be proportional to the elastic portion of the true yield strain, i.e., $\varepsilon_{t-y_{LT}} = \sigma_{t-y_{LT}}/Y_{LT}$, and the elastic modulus difference, i.e., $(Y_{LY} - Y)$, as given in Equation (2), where C_y is an empirically determined proportionality constant. This implies that the decrease in the $\sigma_{t-y_{LT}}$ could be linearly approximated by the change in the Young modulus with temperature, $(Y_{LY} - Y)$. Here, note that we do not claim that the loss of true yield strength is the same as the product of the elastic portion of the true yield strain and the elastic modulus difference. Instead, we attempted to describe the true yield stress behaviors using the simplest approach.

$$\sigma_{t-y} = \sigma_{t-y_{LT}} - C_y \left(\frac{\sigma_{t-y_{LT}}}{Y_{LT}} \right) (Y_{LT} - Y) = \sigma_{t-y_{LT}} \left[1 - C_y \left(\frac{Y_{LT} - Y}{Y_{LT}} \right) \right] \tag{2}$$

Now, the temperature dependence of the true failure stress ($\sigma_{t-\text{fail}}$) can be determined in the same manner, where the adjustment is likewise proportional to the elastic portion of the true failure strain, $\varepsilon_{t-\text{fail}_{LT}} = \sigma_{t-\text{fail}_{LT}}/Y_{LT}$, and the elastic modulus difference, $(Y_{LY} - Y)$.

$$\sigma_{t-\text{fail}} = \sigma_{t-\text{fail}_{LT}} - C_{\text{fail}} \left(\frac{\sigma_{t-\text{fail}_{LT}}}{Y_{LT}} \right) (Y_{LT} - Y) = \sigma_{t-\text{fail}_{LT}} \left[1 - C_{\text{fail}} \left(\frac{Y_{LT} - Y}{Y_{LT}} \right) \right] \tag{3}$$

In Equation (3), C_{fail} is a proportionality constant that can be empirically determined. Note that C_y and C_{fail} values would be varied by the material systems of interest. In the case of Al-A319 for all heat treatment conditions, it was found that $C_y = C_{\text{fail}}$, which indicates that the decreasing rates of σ_{t-y} and $\sigma_{t-\text{fail}}$ are same from their LT values ($C_y = C_{\text{fail}}$).

For the representation of the true failure strain ($\varepsilon_{t-\text{fail}}$), it is proposed that $\varepsilon_{t-\text{fail}}$ can be described by the Arrhenius-like Equation (4) in which true failure strain ($\varepsilon_{t-\text{fail}}$) increases as temperature increases up to the solidus temperature of the metal (T_{solidus}).

$$\ln \frac{\varepsilon_{t-\text{fail}}}{\varepsilon_{t-\text{fail}}^*} = \frac{-Q_{\text{fail}}}{R(T - T_{\text{solidus}})} \tag{4}$$

$$\varepsilon_{t-\text{fail}} = \varepsilon_{t-\text{fail}}^* \exp \left(\frac{-Q_{\text{fail}}}{R(T - T_{\text{solidus}})} \right) \tag{5}$$

where Q_{fail} is the activation energy change for failure and R is the gas constant, respectively. In Equations (4) and (5), mathematically, $\varepsilon_{t-\text{fail}}^*$ corresponds to the failure strain of alloys at infinitely low temperature (i.e., $T \approx -\infty$). Physically, it would be reasonable to approximate as $\varepsilon_{t-\text{fail}}^* = Y_{LT}/\sigma_{t-\text{fail}}$ at LY. The parameter Q_{fail} can be determined using linear regression analysis. Here, $\ln \varepsilon_{t-\text{fail}}^*$ and Q_{fail} can be graphically interpreted as the y-intercept value and the slope of $\ln \varepsilon_{t-\text{fail}}$ vs. $\dfrac{-1}{R(T - T_{\text{solidus}})}$ plot, respectively.

2.2. Interrelated Mechanical Properties

There are advanced models to describe the evolution of dislocation densities upon processing and the subsequent hardening behavior of alloys. These models include the Kocks-Mecking model [9–11], the Alflow model [12], and Myhr et al.'s model [13]. Although these sophisticated models may include the effects of stacking fault energy, grain size, impurity element content, precipitation particle sizes/contents, etc., on the hardening behavior of alloys, we elected to use a simpler Hollomon equation ($\sigma_t = K\varepsilon_t^n$, where σ_t, ε_t, K, and n are true stress, true strain, strength coefficient, and strain hardening exponent, respectively) to represent the hardening behavior of Al-A319 alloys. When the work hardening phenomena of the material can be described by the Hollomon relationship, then Equations (6) and (7) can be used to determine n and K [6].

$$n = \frac{\ln\left(\sigma_{t-fail}/\sigma_{t-y}\right)}{\ln\left(\varepsilon_{t-fail}/\varepsilon_{t-y}\right)} \tag{6}$$

$$K = \frac{\sigma_{t-y}}{\varepsilon_{t-y}^n} \tag{7}$$

The ductility parameter (A) is defined as the area under the ($\ln\sigma_t$ *vs.* $\ln\varepsilon_t$) curve between the yield stress and maximum (*i.e.*, failure) stress [6]. This parameter is considered to govern the ductility of the material and is defined by Equation (8).

$$A = \frac{\ln\left(\sigma_{t-y}\sigma_{t-fail}\right)}{2}\ln\left(\varepsilon_{t-fail}/\varepsilon_{t-y}\right) \tag{8}$$

3. Experimental Section

To compare the results based on the analytical models developed in the previous section, Al-A319 alloy was supplied in the form of sectioned chilled blocks of 100 mm × 15 mm × 15 mm having a dendrite arm spacing (DAS) ranging from 20 to 50 µm. In this alloy, Si modification and grain refinement were achieved by employing an Al-10% Sr master alloy, and a commercial Ti-B (5% Ti-1% B) alloy. Final composition of the alloy was Al-8.6 Si-3.8 Cu-0.36 Mg-0.5 Fe-0.3 Mn-0.012 Sr-0.05 Cr-0.023 Ni-0.015 Pb. From the chilled as-cast blocks, square bars were cut and heat-treated resulting in the following material conditions: (a) as-cast, (b) W—solutionized (4.5 h at 450 °C) and (c) T7—age-hardened (W treatment + 4.67 h at 230 °C). Figure 1 shows the micrographs for (a) as-cast and (b) W specimens obtained using optical microscope (Olympus model Mx50, Olympus Corp., Center Valley, PA, USA). In the micrographs, light and dark contrasts generally represent the α-Al and modified Si particles in Al-A319 alloys, respectively. As can be seen in Figure 1, after W heat treatments, the general microstructures did not exhibit much difference from the microstructure of original as-cast specimen. Micrographs from T7 conditions also displayed similar microstructures to those from the as-cast and W conditions.

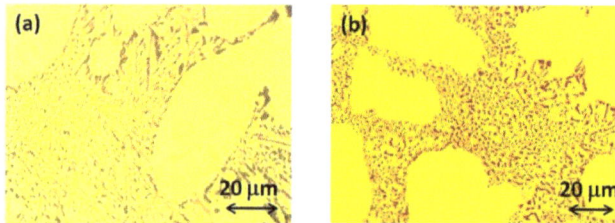

Figure 1. Optical micrographs from the (**a**) as-cast; and (**b**) W specimens. Light and dark contrasts generally represent the α-Al phase and the modified Si particles in the Al-A319 alloys, respectively.

Tensile specimens of the as-cast and heat treated bars were then machined into samples 9 mm in diameter and 80 mm in length according to the ASTM standards E21-921998 and B557-02. Tensile testing was carried out on an MTS 810 machine (MTS Systems Corp., Eden Prairie, MN, USA) at a strain rate of 10^{-4} s^{-1}. The tensile testing machine was instrumented with an ambient chamber to maintain the testing temperatures within ±2 °C. Tensile testing was carried out at −90, −60, −30, 0, 150, 180, 210, 240, 270, 320, 370, and 400 °C. Prior to tensile testing, the specimens were heated to the desired temperature and held for approximately 10 min to achieve temperature stability. Up to four samples were tested at each test temperature to obtain statistically reliable tensile testing results.

An extensometer (axial extensometer Epsilon 3542, Epsilon Technology Corp., Jackson, WY, USA) was placed on the gage length and total elongation values were measured to the points of fracture. Figure 2 shows the examples of stress-strain curves at −90, 0, 150, 270, and 400 °C measured using these (a) as-cast, (b) W, and (c) T7 specimens, respectively. Experimental Young's modulus (Y), true yield stress (σ_{t-y}), true failure stress ($\sigma_{t-\mathrm{fail}}$), true failure strain ($\varepsilon_{t-\mathrm{fail}}$), strength coefficient (K), ductility parameter (A), and strain hardening exponent (n) data were extracted based on these tensile testing results.

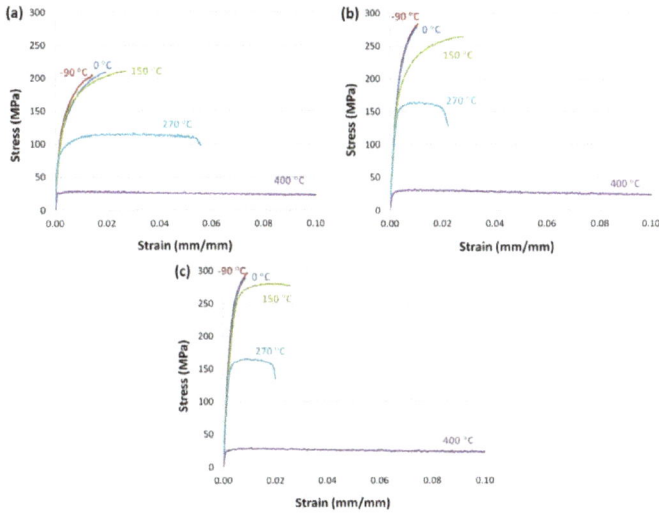

Figure 2. Examples of tensile stress-strain curves at various temperatures (−90, 0, 150, 270, and 400 °C) from (**a**) as-cast; (**b**) W; and (**c**) T7 specimens.

4. Results and Discussion

Using the property prediction model developed above and the experimentally/empirically determined values of parameters contained in Table 1, Figures 3–7 present the predicted property values in comparison to the results obtained from the tensile tests of Al-A319 alloys. In Table 1, the parameters (Ψ, C_y, C_{fail}, and Q_{fail}) in the shaded boxes were empirically-determined and all other parameters were experimentally determined.

Table 1. Experimental and empirical model parameters for Al-A319 alloys in various heat treatment conditions.

Property	Parameters	As-Cast	W	T7
Y (MPa)	$\Delta Y/\Delta T$ (MPa·K^{-1})		0	
	$Y_0 = Y_{LT}$ (MPa)		74,000	
	Y_{HT} (MPa)		30,000	
	T^* (°C)		300	
	Ψ (°C)		55	
σ_{t-y} (MPa)	$\sigma_{t-y_{LT}}$ (MPa)	141.0	240.0	262.0
	C_y		1.6	
$\sigma_{t-\mathrm{fail}}$ (MPa)	$\sigma_{t-\mathrm{fail}_{LT}}$ (MPa)	211.0	279.0	292.0
	C_{fail}		1.6	
$\varepsilon_{t-\mathrm{fail}}$ (%)	$\varepsilon_{t-\mathrm{fail}}^{*}$ (%)	0.2851	0.3770	0.3946
	Q_{fail} (J·mol^{-1})	5269.6	3906.4	3932.1
	T_{solidus} (°C)		515	

It is important to note that the Al-A319 chemical composition is such that many different types of intermetallics are evolved during solidification, some of which can be dissolved by a solutionizing heat treatment and others (especially Fe-containing species) that cannot [8,14]. Rincon *et al.* [8,14] have shown that these brittle intermetallics can be large and significantly degrade tensile properties due to their tendency to initiate and propagate cracks. In Al-A319 in particular, and precipitation-hardenable alloys in general, the as-cast condition is likely to have a variety of intermetallics that are non-uniformly distributed throughout the material. This is due to differences in cooling rates in various parts of the casting, which causes differing degrees of concentration gradients during the solidification. Differences in cooling rate will also result in variability in grain size and grain morphology. Both of these factors will affect the mechanical properties. Therefore, it is to be expected that the material properties of the as-cast metal will possess the most variability. Material in the W condition would be expected to show less variability due to the homogenizing effect of the solutionizing treatment, where some intermetallics are dissolved, and differences in grain sizes are to a certain extent evened out. The material in the T7 condition would be expected to show the least variability in mechanical properties, because (i) it has undergone the solutionizing heat treatment to form more homogenous microstructure; and (ii) it derives its strength primarily from the Orowan strengthening mechanism [15], which depends on concentration, size, and distribution of nanoscopic precipitates—making it relatively insensitive to grain size effects on the yield stress of alloys.

Figure 3 shows the temperature-dependence of Young's modulus for Al-A319 alloys in various heat treatment conditions. In the figure, the symbols and dashed curve represent experimental data and the prediction of the model using Equation (1), respectively. It should be noted that in the case of the Young's modulus of Al-A319, the scatter in the data does not allow for the adequate determination of the slight linear decrease in Young's modulus ($\Delta Y/\Delta T = 0$). Therefore, to include a more marked non-linearity in Equation (1), $\Delta Y/\Delta T$ was set to 0 for all heat treatment conditions. Also, because the modulus did not decrease with decreasing temperature below room temperatures, we selected Y_0 (=Y_{LT}) as the nominal modulus of Al-319, *i.e.*, 74 GPa. The Young's modulus at HT was assumed as the Y_{HT} at $T_{solidus}$ and was asymptotically extrapolated based on the experimental measurement data (*i.e.*, $Y_{HT} \approx 30$ GPa). The alloy softening temperature (T^*) was set as 300 °C for all of the Al-A319 alloy systems tested in the present study [8]. The transition parameter (Ψ) of the logistic function is the only empirical parameter to describe the temperature-dependent behaviors of the Young's modulus. As shown in Figure 3, the experimental measurements exhibit considerable scatter in the Young's modulus. In particular, the Young's moduli of the as-cast sample show large variations, and the measured values sometimes display a decreasing trend with decreasing temperatures, which can be explained by the inhomogeneous non-uniform distributions of intermetallics as addressed in the previous paragraph. Though there is considerable scatter in the modulus data, Figure 3 shows that Equation (1) with the empirically determined transition parameter (Ψ) value of 55 °C provides an adequate description of the temperature-dependent behavior of the material. Due to the scatter, it was not feasible to determine if Young's modulus depends significantly on the heat treatment conditions. A significant difference in the modulus would only be expected if (i) the various intermetallic or dispersed phases in the different heat treated materials are substantially more or less stiff than the metallic phase; and (ii) the intermetallics are present in sufficient concentrations that they have a non-negligible effect. Examination of microstructures with different heat treatment conditions did not show such difference. Further, from Figure 3, it is clear that any obvious trend is not observed among the as-cast, W, and, T7 samples. In addition, because the Young's modulus used in Equations (2) and (3) to predict the yield (σ_{t-y}) and failures stresses (σ_{t-fail}) will demonstrate below that these predictions are seemingly accurate, the assumption that Young's modulus is roughly independent of the heat treatment conditions seems justified as a first approximation, at least in the case of Al-A319 alloys considered in the present work.

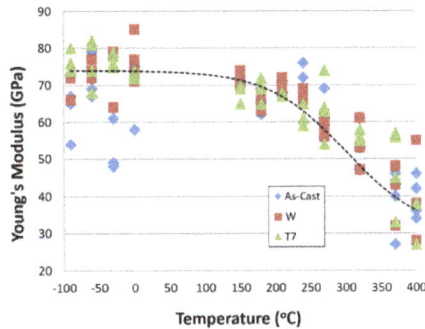

Figure 3. Temperature-dependence of Young's modulus for Al-A319 in various heat treatment conditions. Symbols represent experimental data and the dashed curve is the prediction of the model, respectively.

In Figure 4, using Equations (2) and (3), the temperature dependence of (a) true yield stress (σ_{t-y}) and (b) true failure stress ($\sigma_{t-\text{fail}}$) for Al-A319 alloys in various heat treatment conditions is plotted. Again, symbols represent experimental data and the curves are the prediction of the model, respectively. The $\sigma_{t-y_{LT}}$ and $\sigma_{t-\text{fail}_{LT}}$ values were determined by the averages of σ_{t-y} and $\sigma_{t-\text{fail}}$ in the temperature ranges of -90 °C and 30 °C, because it was observed that these properties are not affected by the temperature below the room temperature for Al-A319 alloys [8]. In determining the experimental σ_{t-y} values, we used the 0.2% offset method from the measured stress-strain curves. By the regression analysis, we identified that the proportionality constant C_y and C_{fail} values of 1.6 produce the best-fit to the experimental observations. This implies that the loss of σ_{t-y} and $\sigma_{t-\text{fail}}$ from LT due to the temperature increase can be successfully expressed by the multiples of elastic true strain of LT and the loss conversion constants for both cases are 1.6. Also, this validates that the approximation of Equations (2) and (3) could be applied to the estimation of true yield stress and true failure strain in Al-A319 systems considered in this work. From the figure, it is seen that the analytical prediction shows good agreements with the experimental observations. As expected, Figure 4a clearly shows that true yield stress (σ_{t-y}) depends significantly on the heat treatment conditions. Figure 4b indicates that in general the as-cast material fails at much lower stresses compared with the W and T7 materials. Low failure stress is likely the result of higher concentrations of brittle intermetallics in grain boundary regions of the as-cast material. It is, however, notable that, at a given temperature, there is little difference between the failure stresses ($\sigma_{t-\text{fail}}$) of the W and T7 conditions, though age-hardening (*i.e.*, T7) induces slightly higher yield stress (σ_{t-y}). It is known from Al-5083 bimodal materials that the grain size has a significant effect on the failure stress [6], but it would appear that in homogenized Al-A319 (*i.e.*, W and T7 samples), the failure stress is influenced by alloy composition (*i.e.*, chemical composition) rather than the concentration, size, and distribution of precipitates. This indicates it is likely that grain size and chemical composition primarily control failure stress in homogenized Al-A319. An analysis of failure stress data from Talamantes-Silva *et al.* [16] for Al-206 in W and T7 heat treat conditions over a range of grain sizes has shown that there is also little difference in true failure stress between W and T7 conditions if the samples have the same grain size. It is, however, thought that further investigation would be necessary to determine if this is the case for alloys other than Al-A319 and Al-206. At temperatures of 210 °C and 270 °C, the behavior of the as-cast material significantly departs from its general behavior and takes on the attributes of the T7 material. However, at a temperature of 240 °C, the behavior exhibits its predicted behavior. The W material also seems to take on the attributes of the T7 material in the temperature range of 210 °C to 240 °C. The abrupt changes in behavior are likely due to the dissolution and/or precipitation of various intermetallic phases. At temperatures of 300 °C (*i.e.*, alloy softening temperature) and above, there appears to be little difference between the materials.

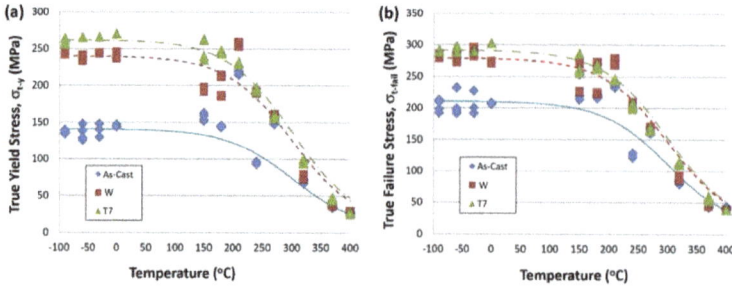

Figure 4. Temperature-dependence of (**a**) true yield stress (σ_{t-y}); and (**b**) true failure stress ($\sigma_{t-\text{fail}}$) for Al-A319 alloys in various heat treatment conditions. Symbols represent experimental data and the curves are the prediction of the model, respectively.

Figure 5 shows the Arrhenius-like dependence of true failure strain ($\varepsilon_{t-\text{fail}}$) for Al-A319 alloys using the expression given in Equations (4) and (5) in (a) as-cast, (b) W, and (c) T7 heat treatment conditions. Again, the symbols represent experimental data and the lines are the prediction of the model, respectively. $\varepsilon^*_{t-\text{fail}}$ values were calculated using $\varepsilon^*_{t-\text{fail}} = Y_{LT}/\sigma_{t-\text{fail}}$ for each heat treatment condition. After obtaining the y-intercept values ($\varepsilon^*_{t-\text{fail}}$), the linear regression was used to predict the slopes shown in Figure 5. R-squared values (R^2) of the linear regression are also provided in the figure. The data in the figure evidently exhibit a higher true failure strain ($\varepsilon_{t-\text{fail}}$) with higher temperature (*i.e.*, higher $\dfrac{-1}{R(T - T_{solidus})}$). From the figures, it is clearly seen that the predictions from the proposed Equation (4) provide a reasonably accurate estimate of the true failure strain ($\varepsilon_{t-\text{fail}}$), although there are instances in the as-cast and W data where there is some departure from the general trend. The slopes of the prediction lines in Figure 5 can be used to understand the temperature effects on the changes of the true failure strain ($\varepsilon_{t-\text{fail}}$). From the figure, it is seen that the true failure strain ($\varepsilon_{t-\text{fail}}$) of the as-cast sample shows the highest failure activation energy (*i.e.*, high Q_{fail} value in Equations (4) and (5)), while the heat treated samples (W and T7) show a lower failure activation energy (*i.e.*, smaller Q_{fail} value in Equations (4) and (5)). This can be explained from the different microstructural homogeneity of as-cast and heat treatment samples. As-cast materials possess a large variation in intermetallic phase size, shape, and composition, and therefore, possess many inhomogeneous microstructural features that could lead to failure. Hence, the as-cast samples likely show more temperature-sensitive behavior for the strain failure. On the other hand, W and T7 materials are more refined and homogenous and thus the types/degrees of defects that can cause failure are less in these solutionized and age-hardened materials, which will exhibit less sensitivity to the temperature changes with regards to the temperature changes.

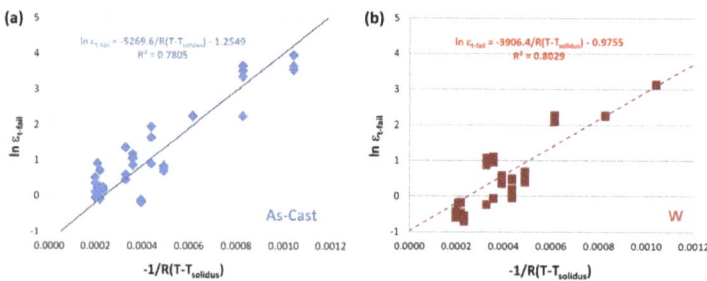

Figure 5. *Cont.*

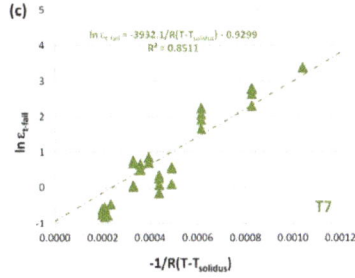

Figure 5. Arrhenius-like dependence of true failure strain ($\varepsilon_{t-\text{fail}}$) for Al-A319 alloys in (**a**) as-cast; (**b**) W; and (**c**) T7 heat treatment conditions. Symbols represent experimental data and the lines are the prediction of the model, respectively.

In addition to yield stress (σ_{t-y}), failure stress ($\sigma_{t-\text{fail}}$), and failure-to-strain ($\varepsilon_{t-\text{fail}}$) of alloys, in Figure 6, the predictions of other mechanical parameters of interest including the strain hardening exponent (n), strength coefficient (K), and ductility parameter (A) are compared with values calculated based on the Holloman equation ($\sigma_t = K\varepsilon_t^n$) with the experimental data. Figure 6a,b shows that there is good agreement between the predicted and experimental values of strength coefficient (K) and ductility parameter (A) with the same some deviations at 210 °C and 240 °C for the as-cast condition and the W condition from 240 °C to 320 °C. From these figures, it is seen that the strength coefficient (K) increases and the ductility parameter (A) decreases with W/T7 heat treatments, respectively, which indicates that the heat treatment will increase the strength coefficient (K) (~48%) as well as the true yield stress (σ_{t-y}) and the true failure stress ($\sigma_{t-\text{fail}}$) as shown in Figure 4, and it will decrease the ductility of alloys. Further, it is observed that the general trend of these two parameters (*i.e.*, K and A) for the W and T7 samples are similar, although the K and A parameters of W condition are slightly lower and higher, respectively, compared with those from T7 specimen. On the other hand, Figure 6c shows that there is considerable scatter in the experimental data for the strain hardening exponent (n) and the agreement is relatively poor between the experimental values and the predicted trend, especially at temperatures above 180 °C, which indicates that either the growth or dissolution of precipitates can abruptly affect the mechanical properties.

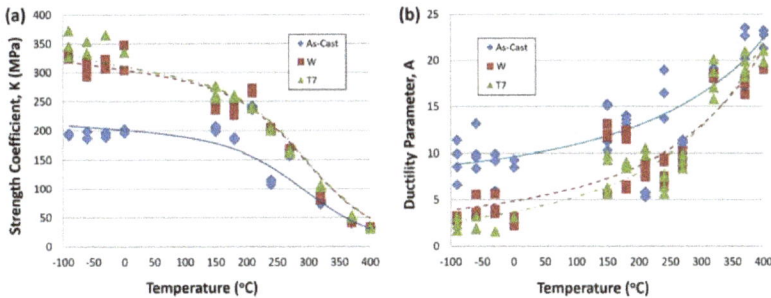

Figure 6. *Cont.*

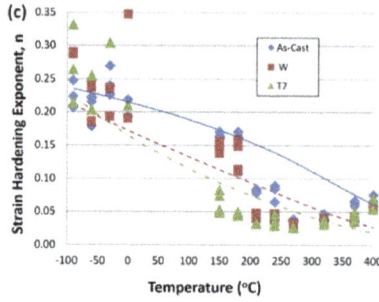

Figure 6. Temperature-dependence of (**a**) strength coefficient (*K*); (**b**) ductility parameter (*A*); and (**c**) strain hardening exponent (*n*), for Al-A319 alloys in various heat treatment conditions. Symbols represent experimental data and the curves are the prediction of the model, respectively.

In Figure 7, the relationships between the ductility parameter (*A*) and the strain hardening exponent (*n*) are displayed for the (a) as-cast, (b) W, and (c) T7 specimens. The symbols and curve lines represent the experimental and prediction data, respectively. From the figures, it is clear that, in general, the *A* and *n* parameters exhibit an inverse relationship; that is, *A* decreases and *n* increases as the temperature decreases. Further, the experimental data in Figure 7 show that there is a complicated temperature-dependent behavior of *n* and *A*, which can account for the relative inconsistency between experimental and predicted values in Figure 6c, and again shows that either the growth or dissolution of precipitates can abruptly affect the mechanical properties. Although there are some deviations from the experimental measurement data, the analytical predictions show reasonable averaged behavior over a wide range of temperatures for as-cast and heat treated specimens.

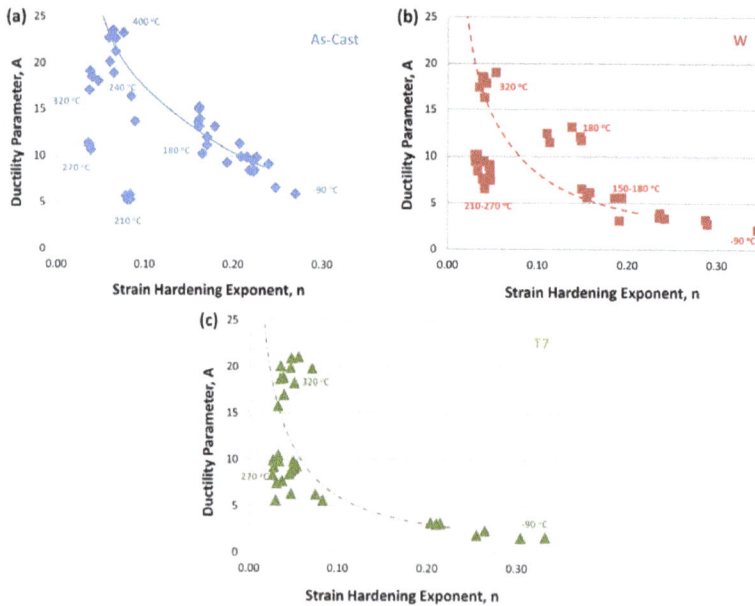

Figure 7. Temperature-dependent relationships of ductility parameter (*A*) and strain hardening exponent (*n*) for Al-A319 alloys in (**a**) as-cast; (**b**) W; and (**c**) T7 heat treatment conditions. Symbols represent experimental data, and solid and dashed lines represent predictions of the model, respectively.

Across Figures 3–7 we have shown that the developed analytical model can predict various mechanical properties such as Young's modulus (Y), true yield stress (σ_{t-y}), true failure stress ($\sigma_{t-\text{fail}}$), true failure strain ($\varepsilon_{t-\text{fail}}$), strength coefficient (K), ductility parameter (A), and strain hardening exponent (n) of Al-A319 alloys at different temperatures. Although we have only demonstrated that the model can be applied to the Al-A319 material systems at various temperatures with different heat treatment conditions in the current study, it is generally expected that the presented analytical model approach can be applied to predict the mechanical behaviors of precipitation-hardened alloys where the stress-strain curves are described by the Hollomon equation under various temperatures.

5. Conclusions

In this work, an analytical model using some empirical parameters was proposed to describe the temperature dependence of Young's modulus (Y), true yield stress (σ_{t-y}), true failure stress ($\sigma_{t-\text{fail}}$), and true failure strain ($\varepsilon_{t-\text{fail}}$) of precipitation-hardened alloys majorly strengthened by the Orowan strengthening mechanism. Strain hardening exponent (n) and strength coefficient (K) of the Hollomon relation are derivable from these properties, as is the ductility parameter (A). The analytical model is majorly based on the experimentally measured data at a reference temperature (*i.e.*, LT in our work) and three empirical parameters (Ψ, $C_y = C_{\text{fail}}$, and Q_{fail}). The predictions of the model have been found to provide consistent descriptions with experimental observations for the temperature-dependent tensile mechanical behavior of Al-A319, except in the cases where high temperature results in either dissolution or precipitation of intermetallic phases. In general, it was demonstrated that the model adequately describes the true yield stress (σ_{t-y}), the true failure stress ($\sigma_{t-\text{fail}}$), and the true failure strain ($\varepsilon_{t-\text{fail}}$) of the Al-A319 material. It is expected that the developed model can be applied to predict the mechanical behaviors of precipitation-hardened alloys strengthened by the Orowan mechanism under various temperatures. Other than the reliability of the prediction of the developed model, the following lists the finding attained through this work.

- The temperature-dependent behavior of Young's modulus (Y) of Al-A319 alloys does not significantly depend on heat treatment conditions of the material. Such temperature-dependent behavior of modulus can be described by adopting the logistic function.
- The loss of stiffness that accompanies with temperature increase can be proportional to the elastic portion of the true yield strain ($\varepsilon_{t-y_{LT}} = \sigma_{t-y_{LT}}/Y_{LT}$) at the reference temperature (LT) and the elastic modulus difference ($Y_{LT} - Y$) from the reference temperature (LT). In homogenized materials (*i.e.*, W and T7 heat treatment conditions in Al-A319), the failure stress ($\sigma_{t-\text{fail}}$) is relatively insensitive to heat treatment condition—meaning that for materials of the same grain size, the overall chemical composition of the alloy determines the failure stress.
- The tensile properties of strength coefficient (K) and ductility parameter (A) as calculated from the predicted Young's modulus (Y), true yield stress (σ_{t-y}), true failure stress ($\sigma_{t-\text{fail}}$) and true failure strain ($\varepsilon_{t-\text{fail}}$) agree with the values estimated from the experimental data.

Acknowledgments: This materials is based upon work supported by the U.S. Army Research laboratory under Cooperative Agreement No. W911NF-15-2-0005. The views, opinions, and conclusions made in this document are those of authors and should not be interpreted the official policies, either expressed or implied, of Army Research Laboratory or the U.S. Government. The U.S. Government is authorized to reproduce and distribute reprints for Government purposes notwithstanding any copyright notation herein.

Author Contributions: J.B. Ferguson developed the model and wrote the manuscript. H.F. Lopez contributed to the experimental testing and K. Cho validated the model and revised the manuscript. C.-S. Kim supervised the progress of the overall work and wrote the manuscript.

Conflicts of Interest: The authors declare no conflicts of interest.

References

1. Han, Y.Y.; Samuel, A.M.; Doty, H.W.; Valtierra, S.; Samuel, F.H. Optimizing the tensile properties of Al-Si-Cu-Mg 319-type alloys: Role of solution heat treatment. *Mater. Design.* **2014**, *58*, 426–438. [CrossRef]

2. Mohamed, A.M.A.; Samuel, F.H.; Kahtani, S.A. Influence of Mg and solution heat treatment on the occurrence of incipient melting in Al-Si-Cu-Mg cast alloys. *Mater. Sci. Eng. A* **2012**, *543*, 22–34. [CrossRef]

3. Sjolander, E.; Seifeddine, S. The heat treatment of Al-Si-Cu-Mg casting alloys. *J. Mater. Process. Technol.* **2010**, *210*, 1249–1259. [CrossRef]

4. Caceres, C.H.; Djurdjevic, M.B.; Stockwell, T.J.; Sokolowski, J.H. The effect of Cu content on the level of microporosity in Al-Si-Cu-Mg casting alloys. *Scripta Mater.* **1999**, *40*, 631–637. [CrossRef]

5. Li, Z.; Samuel, A.M.; Samuel, F.H.; Ravindran, C.; Valtierra, S.; Doty, H.W. Parameters controlling the performance of AA319-type alloys: Part I. tensile properties. *Mater. Sci. Eng. A* **2004**, *367*, 96–110. [CrossRef]

6. Ferguson, J.B.; Tabandeh-Khorshid, M.; Rohatgi, P.K.; Cho, K.; Kim, C.-S. Predicting tensile and compressive mechanical properties of biomodal nano-aluminum alloys. *Scripta Mater.* **2014**, *72–73*, 13–16. [CrossRef]

7. Gerlich, D.; Fisher, E.S. The high temperature elastic moduli of aluminum. *J. Phys. Chem. Solids* **1969**, *30*, 1197–1205. [CrossRef]

8. Rincon, E.; Lopez, H.F.; Cisneros, M.M.; Mancha, H. Temperature effects on the tensile properties of cast and heat treated aluminum alloy A319. *Mater. Sci. Eng. A* **2009**, *519*, 128–140. [CrossRef]

9. Bouaziz, O. Revised storage and dynamic recovery of dislocation density evolution law: Toward a generalized Kocks-Mecking model of strain-hardening. *Adv. Eng. Mater.* **2012**, *14*, 759–761. [CrossRef]

10. Cao, X.; Jahazi, M.; Al-Kazzaz, H.; Medraj, M. Modeling of work-hardening behavior for laser welded magnesium alloy. *Int. J. Mater. Res.* **2008**, *99*, 216–221. [CrossRef]

11. Zolotorevsky, N.Y.; Solonin, A.; Churyumov, A.Y.; Zolotorevsky, V. Study of work hardening of quenched and naturally aged Al-Mg and Al-Cu alloys. *Mater. Sci. Eng. A* **2009**, *502*, 111–117. [CrossRef]

12. Nes, E.; Marthinsen, K. Modeling the evolution in microstructure and properties during plastic deformation of f.c.c.-metals and alloys-an approach towards a unified model. *Mater. Sci. Eng. A* **2002**, *322*, 176–193. [CrossRef]

13. Myhr, O.R.; Grong, O.; Schafer, C. An extended age-hardening model for Al-Mg-Si alloys incorporating the room-temperature storage and cold deformation process stages. *Metall. Mater. Trans. A* **2015**, *46A*, 6018–6039. [CrossRef]

14. Rincon, E.; Lopez, H.F.; Cisneros, M.M.; Mancha, H.; Cisneros, M.A. Effect of temperature on the tensile properties of an as-cast aluminum alloy A319. *Mater. Sci. Eng. A* **2007**, *452–453*, 682–687. [CrossRef]

15. Dieter, G. *Mechanical Metallurgy*, 3rd ed.; McGraw-Hill Book Co.: New York, NY, USA, 1986.

16. Talamantes-Silva, M.; Rodríguez, A.; Talamantes-Silva, J.; Valtierra, S.; Colás, R. Effect of solidification rate and heat treating on the microstructure and tensile behavior of an aluminum-copper alloy. *Metall. Mater. Trans. B* **2008**, *39*, 911–919. [CrossRef]

metals

Article

Phase Evolution and Mechanical Behavior of the Semi-Solid SIMA Processed 7075 Aluminum Alloy

Behzad Binesh and Mehrdad Aghaie-Khafri *

Faculty of Materials Science and Engineering, K. N. Toosi University of Technology, Tehran 1999143344, Iran; b.binesh@dena.kntu.ac.ir
* Correspondence: maghaei@kntu.ac.ir; Tel.: +98-912-308-8389; Fax: +98-21-886-74748

Academic Editor: Nong Gao
Received: 12 December 2015; Accepted: 6 February 2016; Published: 23 February 2016

Abstract: Microstructural and mechanical behaviors of semi-solid 7075 aluminum alloy were investigated during semi-solid processing. The strain induced melt activation (SIMA) process consisted of applying uniaxial compression strain at ambient temperature and subsequent semi-solid treatment at 600–620 °C for 5–35 min. Microstructures were characterized by scanning electron microscope (SEM), energy dispersive spectroscopy (EDS), and X-ray diffraction (XRD). During the isothermal heating, intermetallic precipitates were gradually dissolved through the phase transformations of α-Al + η (MgZn$_2$) → liquid phase (L) and then α-Al + Al$_2$CuMg (S) + Mg$_2$Si → liquid phase (L). However, Fe-rich precipitates appeared mainly as square particles at the grain boundaries at low heating temperatures. Cu and Si were enriched at the grain boundaries during the isothermal treatment while a significant depletion of Mg was also observed at the grain boundaries. The mechanical behavior of different SIMA processed samples in the semi-solid state were investigated by means of hot compression tests. The results indicated that the SIMA processed sample with near equiaxed microstructure exhibits the highest flow resistance during thixoforming which significantly decreases in the case of samples with globular microstructures. This was justified based on the governing deformation mechanisms for different thixoformed microstructures.

Keywords: 7075 Al alloy; SIMA; phase transformation; semi-solid; thixoforming

1. Introduction

Semi-solid metal processing is a single step fabrication method for production of near net shape metallic parts [1–3]. Advantages of semi-solid processing when it is compared with the conventional casting, forging, and powder metallurgy processes are (i) reduction of the porosity, macrosegregation, and surface cracks; (ii) high dimensional accuracy of the produced parts; (iii) extending of the forming die life due to decreasing of the casting temperature and required forming force; and (iv) the uniform filling of the die and reduction of the manufacturing stages and costs [4,5].

The main goal in the semi-solid forming processes is to achieve a thixotropic microstructure consisting of fine and spherical solid grains which are uniformly distributed in the liquid matrix [6,7]. Various routes have been developed for production of thixotropic aluminum, magnesium, lead, and also ferrous alloys including mechanical and magneto-hydrodynamic (MHD) stirring processes, the thermo-mechanical strain induced melt activation (SIMA) process and the spray casting process [7–9]. The SIMA process which is based on the mechanical work was introduced by Young *et al.* [10] in the 1980s. This process has received great attention in recent years due to its simplicity, low equipment costs, and its capability to process many engineering alloys [8]. The two main stages of the SIMA process are cold or warm plastic deformation and then partial remelting in the semi-solid range [11,12].

Many studies have been focused on the semi-solid microstructure of cast aluminum alloys due to superior semi-solid formability [12,13]. However, the cast products show poorer mechanical properties than wrought alloys e.g., 2xxx, 6xxx, and especially 7xxx series. This restricts their applicability in the high strength applications such as aerospace and automotive. On the other hand, control of the semi-solid heat treatment in the wrought alloys is relatively difficult due to their wide solidification range and high sensitivity of solid fraction to temperature variations (steep slope of the solid fraction *versus* temperature). Furthermore, the wrought alloys are severely susceptible to formation of hot tearing during the forming process [14]. Consequently, investigation on the semi-solid microstructure of wrought aluminum alloys to achieve an appropriate thixotropic microstructure is vital. Yong *et al.* [15] prepared an appropriate 7075 semi-solid feedstock by 50% swaging of the specimens at room temperature and then heat treating at 590 °C. They also investigated the properties of the thixoforged part. Multi step heat treatment in thixoforming processes was investigated by Chayong *et al.* [16,17]. Atkinson *et al.* [18] studied the recrystallization of 7075 Al alloy during the recrystallization and partial melting (RAP) process. Vaneetveld *et al.* [19] investigated the effects of heating parameters on the recrystallization in the RAP process and suggested using the high solid fraction to prevent hot tearing in the thixoforging process. Bolouri *et al.* [20] studied the effect of cold work on the semi-solid microstructure by compression straining of the samples up to 40% and Mohammadi *et al.* [21] investigated the effects of the isothermal treatment on the semi-solid microstructure of extruded samples.

Previous studies on the semi-solid processing of 7075 aluminum alloy were mainly restricted to investigations in the field of thixotropic feedstock preparation criteria and producing some thixoformed parts. However, detailed information of the phase transformations during partial remelting and the correlation between the initial microstructure and rheological behavior as well as their final mechanical properties needs to be further investigated. There is a lack of thorough research on this topic in the literature. Consequently, the present work aims to investigate the phase evolution and kinetics of partial remelting during isothermal treatment in the mushy zone and the semi-solid deformation behavior of AA7075 alloy.

2. Experimental Procedure

2.1. Materials and Thermal Analysis

Commercial extruded round bar of 7075-T6 aluminum alloy was used as the starting material, the chemical composition of which is shown in Table 1.

Table 1. Chemical composition of wrought 7075 Al alloy in wt. %.

Element	Al	Mn	Fe	Cr	Cu	Mg	Zn	Si
wt. %	Bal.	0.28	0.28	0.13	1.58	2.41	5.31	0.14

In order to determine the solidus and liquidus temperatures and the solid volume fraction *versus* the temperature curve of 7075 alloy, thermal analysis was carried out using a Metler Star SW 10 differential scanning calorimeter (DSC) apparatus, Metler, Giessen, Germany. Samples of the material (30 mg) were put into the alumina pan and then heated to 700 °C at 10 °C/min under nitrogen atmosphere. The solid volume fraction-temperature relationship was determined using the obtained DSC curve.

2.2. SIMA Process

The SIMA process was carried out to obtain a thixotropic microstructure in 7075 Al alloy. Cylindrical samples with a diameter of 30 mm and height of 35 mm were machined from the starting material. The samples were stress relieved at 460 °C for 1 h, and then air cooled. Compression of cylindrical samples, with a compression ratio of 20%, was carried out at room temperature using a high strain rate (~15 s^{-1}) hammer apparatus. The compressed specimens were machined to samples with a diameter of 20 mm and height of 15 mm to ensure uniform strain in the core section. Isothermal treatment of samples was carried out at three different temperatures of 600, 610, and 620 °C with an accuracy of ± 1 °C for 5–35 min in a resistance furnace. Heating cycles were interrupted at predetermined interval times and then samples were quenched to study the microstructure.

2.3. Compression Tests

The thixoforming process consisting of hot compression experiments at the semi-solid temperature range was carried out on the specimens of different initial microstructures. The cylindrical specimens with 11 mm in height and 7 mm in diameter, according to ASTM E209 standard, were prepared from the SIMA processed samples. The isothermal hot compression tests were carried out at temperatures of 600 and 610 °C and strain rate of 0.3 s^{-1} using a Gotech-Al7000 model universal testing machine, Gotech, Taichung, Taiwan, equipped with an electrical resistance furnace. The specimens were first preheated to the preset temperature and soaked for 5 min to homogenize the samples. The specimens were then compressed up to a strain of 0.6 followed by rapid quenching in water. A high accuracy load cell was used to record the true stress values, and the true strain values were computed using the displacement data. A very thin mica plate was used to reduce the friction effect and to prevent the adhesion of the specimen on the die. The specimens were sectioned parallel to the deformation direction to study the microstructural changes.

2.4. Metallography and Microstructural Characterization

The microstructure of the quenched samples was examined using the standard metallography method. The surface perpendicular to the direction of the compression was ground with standard SiC abrasive papers and polished with 0.25 µm diamond paste. Samples were etched using modified Keller etchant solution (3 mL HF, 2 mL HCl; 20 mL HNO$_3$; 175 mL H$_2$O). Microstructural investigations were carried out using a Neophot 32 optical microscope, Norfab, Trondheim, Norway, and Vega©Tescan, Tescan, Libušina, Czech Republic, and Mira3Tescan field emission scanning electron microscope, Tescan, Libušina, Czech Republic, equipped with EDS detector (Tescan, Libušina, Czech Republic). The linear intercept method was used to determine the average grain size. A series of straight lines with a specified length was considered on the optical micrographs for each sample and the average grain size was determined by division of the line length (*L*) by the number of intercepted grains (*N*). The shape factor of the solid grains was measured by means of Clemex professional edition image analyzer software (version 5.0, Clemex Technologies Inc., Longueuil, QC, Canada) and using Equation (1) [22]:

$$F = \frac{\sum_{N=1}^{N} 4\pi A/P^2}{N} \tag{1}$$

where *A* and *P* are area and perimeter of the solid grains, respectively, and *N* is the number of grains. For each sample, measurements were taken from the whole sectioned area including 300–400 solid particles per sample.

X-ray diffraction (XRD) analysis was used to investigate phase evolution in the semi-solid treated samples. XRD experiment was carried out using a Philips X'pert model apparatus, Panalytical, Eindhoven, The Netherlands, with CuKα target and wave length of 0.15406 nm.

3. Results and Discussion

3.1. Microstructure and Thermal Analysis of the Starting 7075 Alloy

Figure 1 shows the scanning electron microscope (SEM) micrographs of the as-received 7075 aluminum alloy. It can be observed (Figure 1a,b) that the microstructure of the alloy consists of directed α-Al grains and some precipitate particles in the direction of the initial extrusion process. It is worth noting that Zn, Cu, and Mg are the main alloying elements in 7075 alloy which play an important role in the formation of precipitates. In the 7xxx series of aluminum alloys, when the Zn:Mg ratios are between 1:2 and 1:3, $MgZn_2$ (η) precipitates are produced at ageing temperatures below 200 °C and are the main strengthening factor in 7xxx alloys [18]. The dimensions of these precipitates are only a few tens of nanometer which can only be revealed by TEM technique. Considering Figure 1b–d, EDS analysis results of the coarse precipitate particles, as summarized in Table 2, suggested them to be (I) Al_2CuMg intermetallic phase (which is usually denoted as S-phase) with relatively smooth corners (C_1 and C_2 in Figure 1b,d), (II) $Al_6(Cu,Fe)$ intermetallic phase which is also known as $Al_{23}Fe_4Cu$ or $\alpha(FeCu)$ with polygonal (B_1 in Figure 1b) or irregular (B_2 in Figure 1c) shapes, (III) Al_7Cu_2Fe (β) phase with irregular shapes which have been elongated along the extrusion direction (A in Figure 1b) and (IV) dark Mg_2Si precipitates with irregular shapes (D in Figure 1c). Considering the EDS analysis results, Fe-rich intermetallic phases ($Al_6(Cu,Fe)$ and Al_7Cu_2Fe) also contain small amounts of manganese which can be substituted for iron as a result of their close atomic radii.

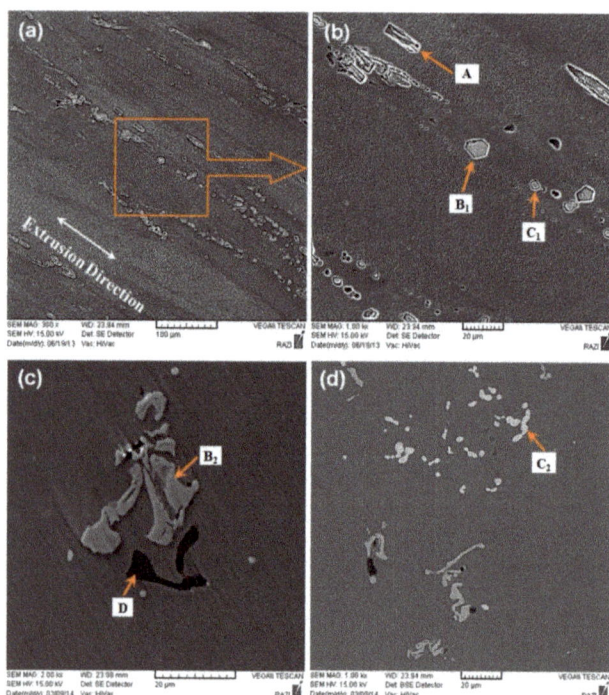

Figure 1. (**a**) Scanning electron microscope (SEM) secondary electron image; (**b**) SEM secondary electron image of selected area in (**a**); (**c**) SEM secondary electron image; (**d**) SEM backscattered electron image of as-received 7075 sample.

Figure 2a shows the DSC curve of the 7075 aluminum sample. The solidus and liquidus temperatures based on the DSC curve were 486 and 649 °C, respectively. The characteristic values of the areas under the DSC peaks calculated for mass unity are also included in Figure 2a. The liquid or solid fraction *versus* the temperature curve was obtained by integration of the DSC curve (Figure 2b). The results were used to determine the isothermal heat treatment temperatures in the SIMA process. It is worth noting that the liquid or solid volume fraction at a known temperature based on Figure 2b is not exactly the same as its values appearing in the micrographs. This is due to insufficient cooling rate of the semi-solid samples after isothermal treatment. In other words, some of the liquid fraction that transforms to the solid during quenching is considered as solid fraction and therefore the liquid volume fraction observed in the micrographs is always lower than that determined by Figure 2b.

Table 2. Scanning electron microscope (SEM)/energy dispersive spectroscopy (EDS) analysis results of selected constituent particles highlighted in Figure 1.

Content of Elements	Constituent Particles (at. %)					
	A	B_1	B_2	C_1	C_2	D
Al	61.35	78.39	82.23	66.72	57.59	-
Zn	0.31	0.20	1.19	0.17	0.05	-
Cu	20.00	5.30	2.67	17.16	18.64	-
Mg	0.03	0.02	0.81	13.83	23.38	69.42
Fe	13.71	11.61	10.64	0.52	0.02	-
Si	1.79	0.13	1.09	0.47	0.03	30.58
Mn	2.27	3.56	1.33	0.52	0.18	-
Cr	0.55	0.79	0.04	0.59	0.10	-
Phase shape	Irregular	polygonal	Irregular	Spherical	Irregular with round corners	Dark-Irregular
Closest phase	Al_7Cu_2Fe	$Al_6(Cu,Fe)$	$Al_6(Cu,Fe)$	Al_2CuMg	Al_2CuMg	Mg_2Si

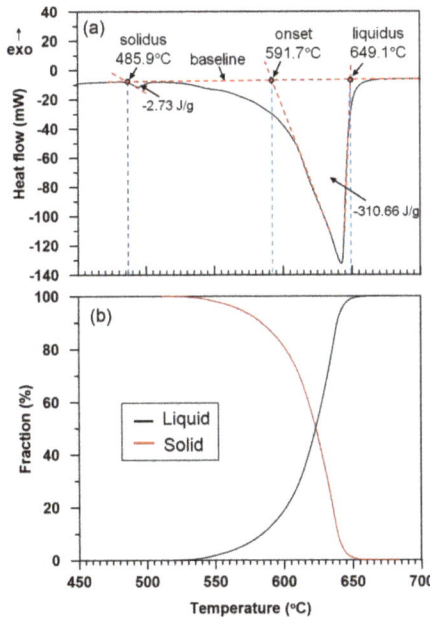

Figure 2. (**a**) Differential scanning calorimeter (DSC) curve of 7075 Al alloy at a heating rate of 10 °C/min; (**b**) solid and liquid volume fraction *versus* temperature derived from the DSC curve.

3.2. Microstructural Evolution during Partial Remelting of the Compressed 7075 Alloy

Microstructures of 7075 specimens after 20% compression and isothermal heat treating at the semi-solid temperature of 600 °C for various times during the SIMA process are shown in Figure 3. The initial microstructure consisted of plastically deformed grains that gradually transforms to a globular microstructure of spherical and equiaxed grains after heating and partial remelting at 600 °C. Micrographs of Figure 3a,b show that following 5 min heating of the sample at 600 °C, no evidence of recrystallization was observed. Following the increase of the heating time up to 10 min, a near equiaxed microstructure consisting of elongated and polygonal solid grains with what appears to be quenched liquid at the grain boundaries was formed (Figures 3c and 4a). In some zones, the size of these grains is quite small which implies a recrystallization phenomenon. Moreover, the formation of the liquid phase is expected to be as a result of partial remelting during isothermal holding at 600 °C. The results of research conducted by Atkinson *et al.* [18] revealed that liquids are first formed at the recrystallized grain boundaries of RAP processed 7075 alloy because these are areas with higher solute concentrations. However, the liquid phase has not penetrated thoroughly into the grain boundaries which results in incomplete grain boundary wetting, as shown in Figure 4a. The mechanism of low melting point phase formation in the semi-solid microstructure is extensively discussed in Section 3.4. With further extension of the heating time up to 15 min, the whole microstructure was transformed into individual polygonal grains due to separation and increase of the liquid phase (Figure 3d). The grain size significantly increased compared with that of the grains which had just recrystallized and separated, shown in Figure 3c. This indicates that the separation and coarsening were in competition.

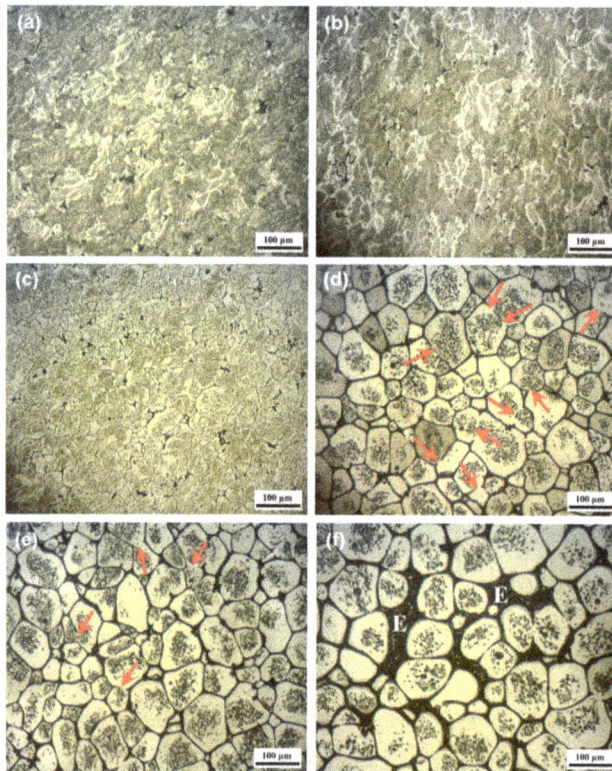

Figure 3. Optical micrographs of samples compressed 20% and isothermally heat treated at 600 °C for (**a**) 0 min; (**b**) 5 min; (**c**) 10 min; (**d**) 15 min; (**e**) 25 min; (**f**) 35 min.

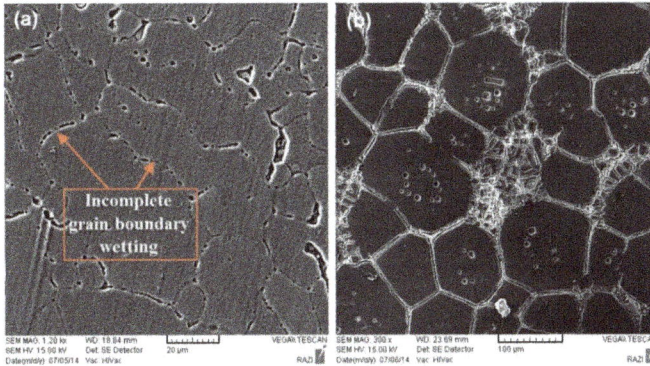

Figure 4. SEM images of 20% compressed 7075 alloy heat treated at 600 °C for (**a**) 10 min; (**b**) 35 min.

Comparing the primary grain size of the deformed sample and the resulting polygonal particle size, it can be suggested that the recrystallization and separation dominated during the period of 5 min to 10 min or at a longer time and then the coarsening became the dominant process. During the coarsening stage, it was found that the liquid amount is relatively small and the neighboring grains are only separated by a thin liquid layer (Figure 3c,d).

During the isothermal heating period of 15 min to 25 min, the interconnection between the solid grains reduced and the grains gradually became spherical (Figure 3e). The spheroidization mainly occurred during this stage. During the period of 25 min to 35 min, the liquid phase obviously increased (comparing Figure 3e,f) which finally led to intergranular formation of thick eutectic regions following 35 min heating of the sample (indicated by the E letter in Figure 3f). This is evidenced with the eutectic phase becoming the predominant feature at grain boundaries after semi-solid soaking (Figure 4b). It can be expected that the coarsening of spherical grains will occur with raising of the heating time over 35 min.

In addition, some intragranular islands which are claimed to be liquid pools just before quenching, were observed within the solid grains. Therefore, the semi-solid microstructure consisted of α-Al spherical solid grains, intergranular eutectic liquid film and entrapped liquid droplets within the solid grains. Entrapped liquid droplets were produced in two different ways. The first was as a consequence of alloying elements segregation within the solid particles during the partial remelting. In this case, some fine liquid droplets entrap within the solid grains. The second one was due to coalescence of solid grains with complex geometrical shape during the heating stage to decrease the solid-liquid phase interfacial energy. In this condition, larger droplets of the liquid phase, in comparison with the former, were created within the grains [23]. As is observed in Figure 3, with longer holding times some of the fine liquid droplets as a result of coalescence phenomena or atomic diffusion joined to the larger liquid droplets. Subsequently, these new liquid droplets become more spherical to reduce the solid/liquid interfacial energy. Finally, with further increase of the holding time, a lesser number of the entrapped liquid droplets was produced owing to diffusion into the liquid phase matrix that surrounded the solid grains.

3.3. Kinetics of the Microstructural Evolution during Partial Remelting

Considering the obtained results in the present study, the kinetics of the microstructural evolution of 7075 samples during the SIMA process can be divided into four steps. (I) recovery, recrystallization and structural separation, (II) coarsening of polygonal solid grains, (III) spheroidization of polygonal solid grains and (IV) coarsening of spherical grains. Recovery, recrystallization, and structural separation mainly occur in the early stages of the isothermal heating operation. During recovery and recrystallization, vacancies are joined to each other and dislocations are rearranged as low energy structures by climbing or cross slipping and subgrain boundaries are created. In this step, grains of high dislocation density are replaced by new subgrains with less dislocation density. Moreover, because the holding temperature is higher than the eutectic line, partial remelting also occurred [20,23]. It has been found that when the angle between subgrains is greater than 20°, the surface energy of the grain boundaries is greater than two fold of the solid/liquid interfacial energy ($\gamma_{gb} > 2\gamma_{sl}$). If these grain boundaries contact with liquid phase, the grain boundaries are replaced by a thin liquid film. Considering the *vice versa* case *i.e.* $\gamma_{gb} < 2\gamma_{sl}$, low energy grain boundaries will be formed which are not wetted by the liquid phase. If two solid grains reach together through such a low energy boundary, they can be joined together and stimulate the coalescence of solid particles [24].

With rising of the isothermal heating time, the liquid volume fraction becomes greater and isolated polygonal solid grains are created. By continuing the heating process, these grains grow and coarsen. The growth and coarsening of the solid grains in the SIMA process are controlled by two mechanisms of coalescence and Ostwald ripening. During the growing and coarsening of the microstructure, diffusion of the solid material from regions with high curvature to low curvature points occurs based on Equation (2) [12]. This subsequently provides the required driving force for spheroidization of solid grains.

$$\Delta T_r = -\frac{2\sigma T_m V_s k}{\Delta H} \tag{2}$$

where $\Delta T_r = T_m - T$ represents the difference to the equilibrium melting point, T_m is the equilibrium transformation temperature, k is the mean surface curvature of the solid particles, σ is the surface tension and ΔH (a negative value) is the molar change in the enthalpies of the solid and liquid.

As long as the grains are not spherical during semi-solid treatment, the spheroidization process leads to change of the particles shape from polygonal to spherical where the numbers per unit volume remain constant. Following this stage, smaller grains which have a lower melting point according to Equation (2), are melted in favor of the larger grains and the numbers per unit volume are reduced. The microstructure of samples that have been subjected to partial remelting in the SIMA process, are evolved by one of the mentioned mechanisms depending on the sphericity of solid grains and the amount of the liquid volume fraction [25].

Ostwald ripening is a diffusion controlled mechanism and acts as the dominant mechanism at high liquid volume fractions *i.e.* high heating temperatures and long holding times. This mechanism is less effective in particle growth but has a great influence on the spheroidization of solid grains. In contrast, the coalescence mechanism which needs shorter holding times and a small amount of liquid fractions, is more effective in grain growth and has a minor effect on the spheroidization process [25–27].

When the liquid fraction is low (early stages of heating operation), the solid grains are readily in contact with one another, and coalescence is the dominant mechanism in the coarsening of the microstructure. This can be verified by considering the microscopic images shown in Figure 3d–f. The arrows on the micrographs show that some grain boundaries are removed due to coalescence of the solid grains. Therefore, it can be concluded that the growing and coarsening of grains in the sample heated at 600 °C for 15 min mainly occurs through the coalescence mechanism due to the relatively low liquid fraction (Figure 3d). With further increase of the holding time (Figure 3e,f), the liquid fraction increased and Ostwald ripening mechanism was more effective in the coarsening process. When the majority of the solid grains gained uniform surface curvature through these two mechanisms, the growth of spherical grains commences. Therefore, spheroidization and further growth of solid grains by the two mentioned competing mechanisms are expected in the SIMA process [25].

3.4. Phase Evolution and Alloying Element Distribution during Partial Remelting of Compressed 7075 Alloy

A close examination of the micrographs in Figure 3 reveals that the volume fractions of precipitates in the microstructure reduce with the rise of the isothermal holding time. This is clearly confirmed by increasing the brightness of the recrystallized solid grains. XRD results shown in Figure 5 also confirm this phenomenon. The X-ray diffraction pattern of the 20% compressed sample prior to the isothermal treatment, indicated that the microstructure consists of α-Al phase and η, S, $Al_6(Cu,Fe)$, Al_7Cu_2Fe, and Mg_2Si intermetallic phases, which is similar to the as-received sample (Figure 1). According to the X-ray diffraction patterns of samples heat treated at 600 °C for 5 to 35 min, it is clear that the intensity of η, S, and Mg_2Si peaks decrease with the increase of the holding time. The diffraction peaks of the 20% compressed sample are shown by symbols and the corresponding peaks in each case are specified by arrows for the other samples in Figure 5. The diffraction peaks of η, S, and Mg_2Si phases almost disappeared after heating for 25 min. However, some diffraction peaks of Fe-rich intermetallic phases (Al_7Cu_2Fe and $Al_6(Cu, Fe)$) show little change with prolonging of the holding time. Considering the sample heated for 35 min, the presence of Al_7Cu_2Fe and $Al_6(Cu, Fe)$ phases was distinguishable. However, no evidence of η, S, and Mg_2Si phases was observed. This illustrates that Fe-rich secondary particles cannot be dissolved into the matrix for the semi-solid treatment conditions used in the present study.

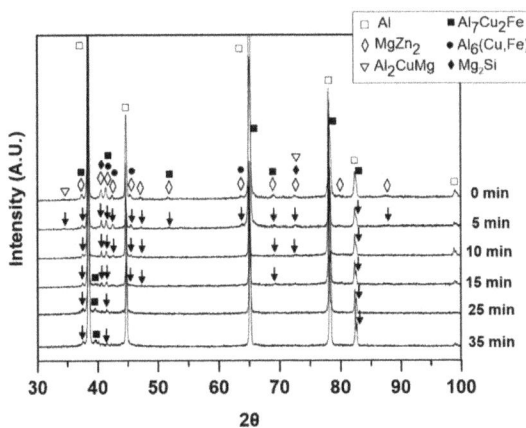

Figure 5. X-ray diffraction (XRD) patterns of 20% compressed 7075 alloys, heat treated at 600 °C for different times and then water quenched.

Previous work reported that the melting temperature of η-phase in 7xxx alloys is 475–480 °C [26,27]. This is fairly well correlated with the first peak in the DSC curve of the starting material with an enthalpy of 2.7 J/g in Figure 1. The solution temperatures of S and Mg_2Si phases are 490–501 °C [28,29] and 478–525 °C [30], respectively, depending on the Mg content of the alloy. Considering the XRD results (Figure 5) and solution temperature of intermetallic precipitates, it can be suggested that the dissolving of intermetallic particles commences with melting of η precipitates after heating for 10 min through the reaction of α-Al + η → L. Following the increase of the heating time, dissolving of S and Mg_2Si intermetallic particles also occurs through the phase transformation of α-Al + S + Mg_2Si → L. These precipitates are fully dissolved after isothermal heating for 25 min.

The Fe-rich intermetallic precipitates have higher solution temperatures and are usually stable over 600 °C [31]. Microstructural observations revealed that the Fe-rich intermetallic particles, shown in Figure 6, precipitate at grain boundaries during partial remelting. Precipitation of these particles at the grain boundaries has a great influence on pinning of the grain boundaries movement. Therefore, it may cause the retardation of the particle growth and coarsening during the isothermal treatment. This is discussed in more detail in Section 3.5.

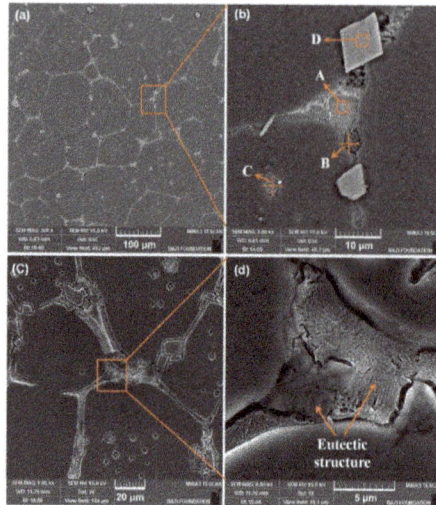

Figure 6. SEM electron images of the strained samples heat treated at (**a**) 600 °C for 25 min; (**b**) 600 °C for 25 min in higher magnification; (**c**) 610 °C for 25 min; (**d**) 610 °C for 25 min in higher magnification.

It can be concluded from the above discussion that the distribution of alloying elements in the microstructure is significantly changed during the isothermal treatment and solution of the constituent particles. Segregation and distribution of the alloying elements has a great influence on the microstructural evolution during semi-solid treatment. Moreover, this affects the mechanical properties of the subsequent thixoformed parts [32–34]. Figure 6 shows the SEM images of the compressed samples heat treated at 600 °C (Figure 6a,b) and 610 °C (Figure 6c,d) for 25 min. The EDS analysis results of the marked areas (A to D) in Figure 6b are shown in Figure 7. Intense concentration of Cu at the grain boundaries (point A) was observed. In addition, Si intensely segregated at other regions of the grain boundaries (point B). The entrapped liquid droplets within the solid grains (point C) also showed high concentration of Cu. Although the segregation of Cu at the grain boundaries and within the liquid droplets has been reported in the literature [35,36], the segregation of Si is an unreported phenomenon in the semi-solid microstructure of 7075 Al alloy. Solid grains are depleted from Cu as a consequence of the segregation of Cu at the grain boundaries. This results in a rise in the solidus temperature in these regions and a decline of the temperature at the grain boundaries. Finally, chemical composition approaches Al-Cu eutectic composition which facilitates grain boundary melting. The segregation of Si at other regions of the grain boundaries derives grain boundary chemistry closer to Al-Si eutectic composition. Therefore, it can be concluded that the formation of low melting phases at grain boundaries is mainly influenced by the high content of Cu and Si.

Grain boundary regions with low solidus temperature remelt during the isothermal heating. Thus, a eutectic liquid film surrounds the solid grains. Melting of solid grains occurs on further increasing the heating temperature. However, it should be noted that the segregation of Cu and Si at the grain boundaries improves on raising the heating temperature due to more intense atomic diffusion. The EDS analysis of the precipitate particles at the grain boundaries (point D) in Figure 6 (Figure 7d) demonstrates Fe-rich intermetallic phases. This is also confirmed by XRD patterns in Figure 5. As discussed earlier, these particles cannot be dissolved at 600 °C and appear mainly as square particles at the grain boundaries in the semi-solid microstructure. The results of the microstructural investigations in this research showed that the amount of these particles reduce on increasing the isothermal heating temperature to 620 °C and isothermal heating for longer times (>15 min).

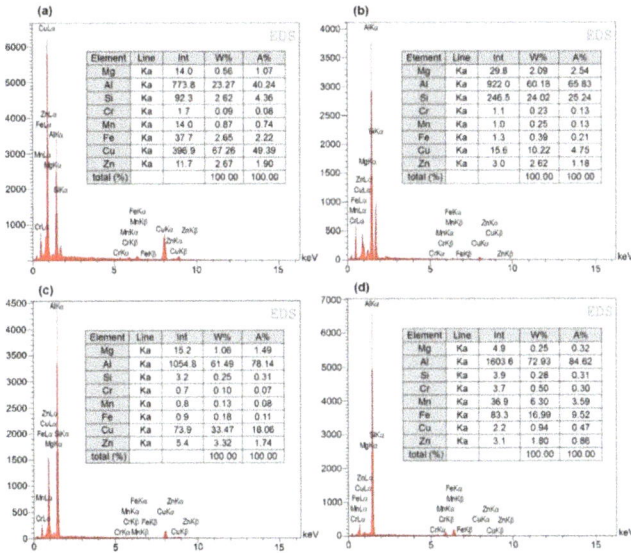

(a)

Element	Line	Int	W%	A%
Mg	Ka	14.0	0.56	1.07
Al	Ka	773.8	23.37	40.24
Si	Ka	92.3	2.62	4.36
Cr	Ka	1.7	0.09	0.08
Mn	Ka	14.0	0.87	0.74
Fe	Ka	37.7	2.65	2.22
Cu	Ka	396.9	67.26	49.39
Zn	Ka	11.7	2.67	1.90
total (%)			100.00	100.00

(b)

Element	Line	Int	W%	A%
Mg	Ka	29.8	2.09	2.54
Al	Ka	922.0	60.18	65.83
Si	Ka	246.5	24.02	25.24
Cr	Ka	1.1	0.23	0.13
Mn	Ka	1.0	0.25	0.13
Fe	Ka	1.3	0.39	0.21
Cu	Ka	15.6	10.22	4.75
Zn	Ka	3.0	2.62	1.18
total (%)			100.00	100.00

(c)

Element	Line	Int	W%	A%
Mg	Ka	15.2	1.06	1.49
Al	Ka	1054.8	61.49	78.14
Si	Ka	3.2	0.25	0.31
Cr	Ka	0.7	0.10	0.07
Mn	Ka	0.8	0.13	0.08
Fe	Ka	0.9	0.18	0.11
Cu	Ka	73.9	33.47	18.06
Zn	Ka	5.4	3.32	1.74
total (%)			100.00	100.00

(d)

Element	Line	Int	W%	A%
Mg	Ka	4.9	0.25	0.32
Al	Ka	1603.6	72.93	84.62
Si	Ka	3.9	0.28	0.31
Cr	Ka	3.7	0.50	0.30
Mn	Ka	36.9	6.30	3.59
Fe	Ka	83.3	16.99	9.52
Cu	Ka	2.2	0.94	0.47
Zn	Ka	3.1	1.80	0.86
total (%)			100.00	100.00

Figure 7. SEM/EDS analysis of various points in Figure 6: (**a**) A; (**b**) B; (**c**) C; (**d**) D.

Figure 8 shows the EDS mapping of the main alloying elements of the strained samples heat treated at 600 °C for 25 min. Considerable segregation of Cu and Si at the grain boundaries was also confirmed by the EDS mapping analysis of Cu and Si (Figure 8c,d). In addition, an appreciable depletion of Mg was observed at the grain boundaries (Figure 8e). However, as it can be observed from the EDS analysis results in Figure 7 and the EDS mapping of Mg in Figure 8e, the Al-Si eutectics have a higher content of Mg compared with the Al-Cu eutectics. There was no appreciable change in Zn content at various point of the microstructure (Figure 8f). This is in contrast to the result of researches conducted by Bolouri et al. [35] and Shim et al. [37] who reported the segregation of Zn at the grain boundaries of a SIMA processed 7075 alloy and semi-solid Al-Zn-Mg alloy prepared by the cooling plate method, respectively. Atkinson et al. [18] also indicated that Zn and Mg are enriched at the grain boundaries of RAP processed 7075 alloy. Moreover, as it can be observed in Figure 8g,h, the grain boundary precipitates contain a high amount of Fe and Mn. As the EDS analysis revealed (Figure 7c,d), these precipitates are attributed to Fe-rich intermetallic particles. An overview of the alloying elements distribution in the semi-solid microstructure of 7075 Al alloy can be observed in Figure 8i.

Figure 9a–e shows SEM images of compressed samples heat treated at 600 °C for 5–35 min. Variations of the main alloying elements of 7075 alloy (i.e. Cu, Zn, and Mg) at the grain boundaries and grain centers in these samples are also demonstrated in Figure 9f. The EDS analysis results of the grain centers and grain boundaries (Figure 9f), showed that the Cu concentration reduces within the solid grains on increasing the isothermal holding time at 600 °C. In other words, Cu segregation rises at grain boundaries with increasing holding time. Mg concentration declined at the grain boundaries on raising the holding time and no significant change was observed in Zn concentration at various parts of the microstructure.

Considering that (i) Zn, Mg, Si, and Cu-bearing intermetallic precipitates (η, S and Mg_2Si) are dissolved gradually on prolonging the holding time according to XRD patterns of the SIMA processed samples (Figure 5), (ii) Zn and Mg have higher solubility in Al compared with Cu and Si according to the Al-Zn, Al-Mg, Al-Cu and Al-Si binary phase diagrams [38], and (iii) no distinct peak for these elements appeared in the XRD patterns (Figure 5), it can be concluded that Zn and Mg elements are easily dissolved in α-Al, while Cu and Si significantly segregate at the grain boundaries. Since,

Zn has a smaller atomic radius relative to aluminum (0.133 nm compare to 0.143 nm) [39], it can be concluded that dissolving Zn in the aluminum crystal lattice leads to reduction of the lattice parameter of aluminum. However, dissolving Mg in Al, as a result of its greater atomic radius, can result in increase of the lattice parameter of aluminum. The shifting of the Al main peaks in XRD patterns (Figure 5) to higher angles, as shown in Figure 10a for (111) reflection of the aluminum matrix, revealed that the lattice parameter of Al decreases on prolonging the heating time (Figure 10b) based on Bragg's equation ($n\lambda = 2d\sin\theta$). It seems that dissolving Zn in Al is more effective than Mg. This may be as a result of the higher Zn content of 7075 alloy. According to Figure 10b, the lattice parameter is gently reduced with prolonging holding time to 15 min and then with further extension of the holding time to 25 min, the lattice parameter precipitously declines. Finally, further increasing of the heating time results in a gentle reduction of the lattice parameter. These observations clearly correlate with the XRD results in Figure 5. According to the XRD patterns, great amounts of η precipitates dissolve during the isothermal heating of the sample for 25 min. This results in dissolving of a higher amount of Zn atoms in the aluminum lattice and therefore leads to a significant decline of the lattice parameter.

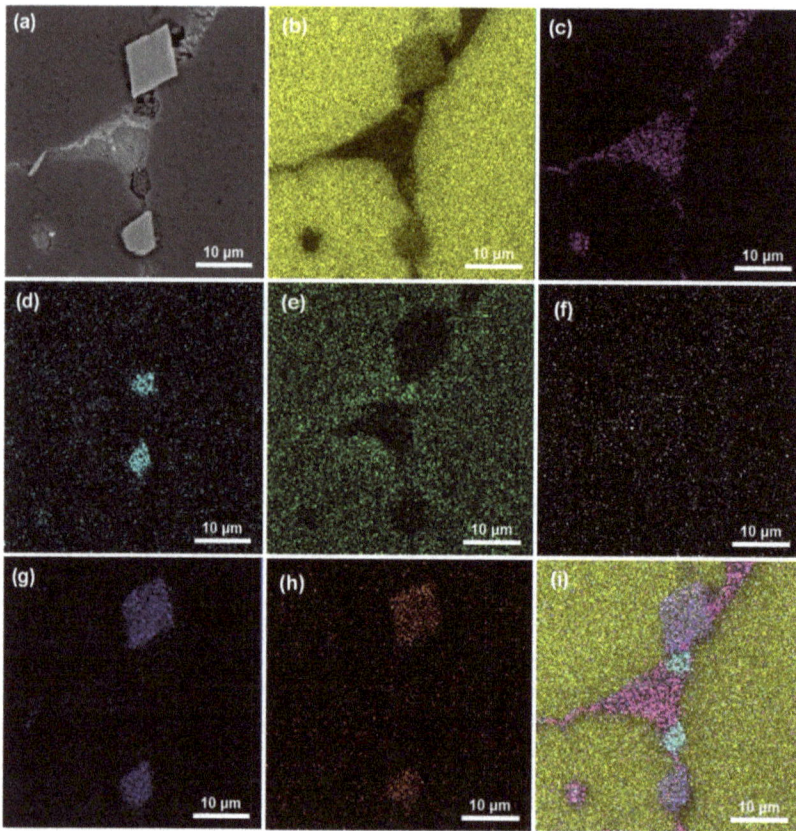

Figure 8. (**a**) SEM image; (**b**) Al; (**c**) Cu; (**d**) Si; (**e**) Mg; (**f**) Zn; (**g**) Fe; (**h**) Mn; (**i**) Al Fe Si Mn Cu X-ray image analysis of a compressed sample isothermally heat treated at 600 °C for 25 min.

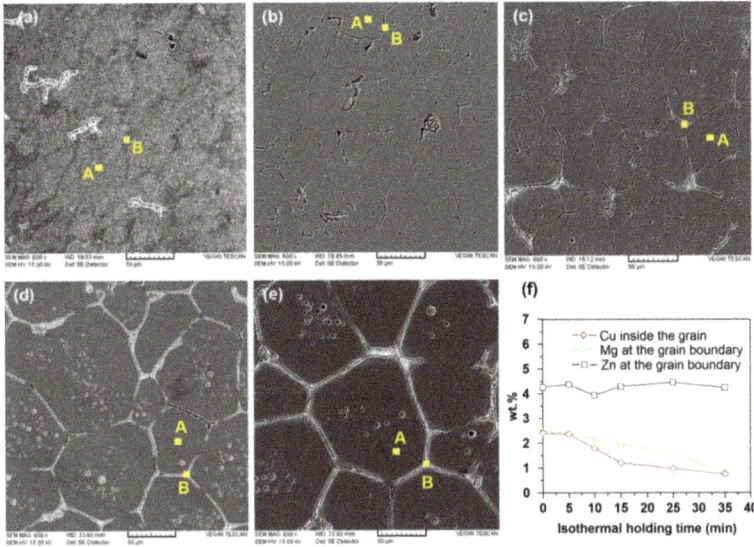

Figure 9. SEM images of sample compressed 20% , heat treated at 600 °C for (**a**) 5 min; (**b**) 10 min; (**c**) 15 min; (**d**) 25 min; (**e**) 35 min and then water quenched; (**f**) EDS analysis results of the grain centers (A point) and grain boundaries (B point).

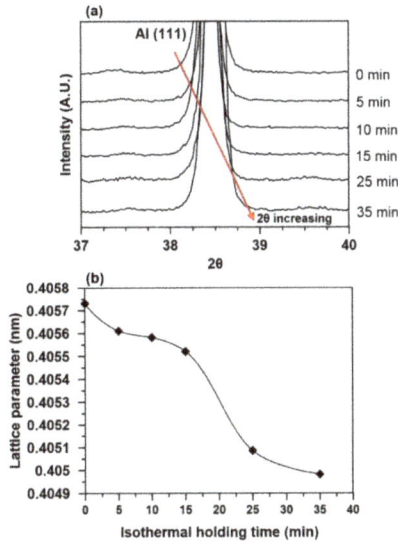

Figure 10. (**a**) XRD peaks corresponding to the Al(111) plane in XRD patterns of Figure 5; (**b**) lattice parameter variations of aluminum in 20% compressed samples heated at 600 °C for different times.

3.5. *Effects of Isothermal Holding Temperature and Time on the Semi-Solid Microstructure*

Figure 11 shows the quenched microstructures of 7075 samples which compressed 20% and heated at 600, 610, and 620 °C for different times. A close examination of the micrographs in Figure 11 reveals that the liquid fraction and the sphericity of solid grains become greater on raising the isothermal holding temperature and time. Another fact that can be observed in the micrographs is the reduction of the number of entrapped liquid droplets within the grains and increase of their size on extending the heating temperature and time.

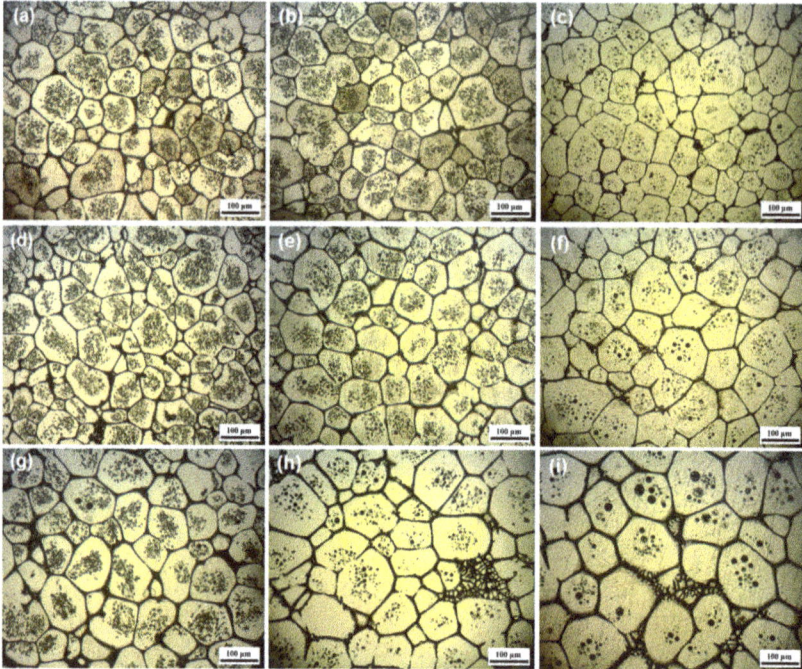

Figure 11. Optical micrographs of 7075 samples with 20% compression ratio heat treated at (**a**) 600 °C/15 min; (**b**) 610 °C/15 min; (**c**) 620 °C/15 min; (**d**) 600 °C/25 min; (**e**) 610 °C/25 min; (**f**) 620 °C/25 min; (**g**) 600 °C/35 min; (**h**) 610 °C/35 min; (**i**) 620 °C/35 min.

Variations of the average grain size and shape factor of strained samples heated at different temperatures and times are shown in Figure 12. As expected, the average grain size and shape factor become greater with increase of the holding temperature and time. However, it is worth noting that no significant variation of the average grain size of samples heated for 15 min on raising the holding temperatures from 600 to 620 °C is observed. Disintegration of solid particles owing to liquid phase penetration between the solid grains occurred during the primary heating time. This is due to the fact that a large number of solid grains with the same crystallographic orientations make contact with each other. It seems that the mechanism of particle disintegration is in equilibrium with the coalescence mechanism at longer heating times up to 15 min. Therefore, the solid grain sizes show no appreciable variation. In contrast, with further extension of holding times at various heating temperatures, the liquid film thickness between the solid grains and the liquid fraction becomes greater. Consequently, the Ostwald ripening mechanism can be considered as a dominant mechanism in the solid grain coarsening process. Figure 3e,f shows separated spherical grains with large entrapped liquid droplets which is a fact that supports the discussion above.

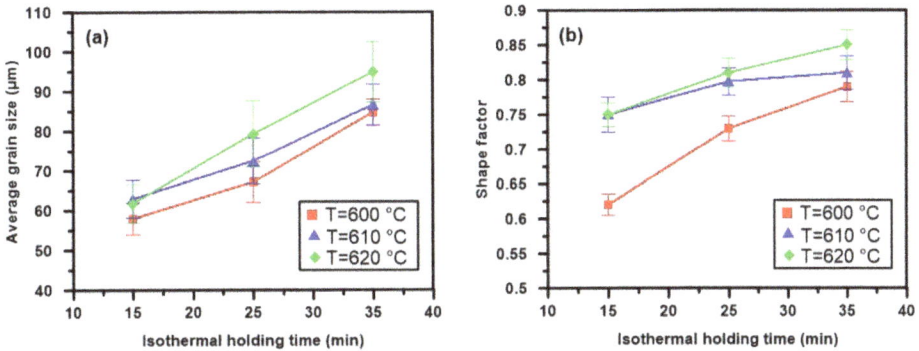

Figure 12. Variations of (**a**) the average grain size; (**b**) shape factor *versus* isothermal holding time of compressed samples heated at 600, 610, and 620 °C.

Results of research conducted by Vaneetveld *et al.* [40] showed that the semi-solid feedstock with high solid fraction ($f_s \approx 0.84$) shows better thixotropic behavior during semi-solid forming and also significantly prevents the formation of some defects such as porosity and shrinkage during the forming process. According to the solid fraction *versus* temperature curve in Figure 2b, the solid fraction is relatively low at 620 °C ($f_s \approx 0.55$). Thus, heating samples at such a temperature leads to undesirable coarsening of the solid grains. In addition, the solid fraction is more sensitive to temperature variations at 620 °C, when compared to the lower temperatures in which control of the semi-solid forming parameters is difficult. Consequently, the semi-solid microstructure obtained by isothermal heating at 620 °C is inappropriate for thixotropic applications. However, high solid fractions were observed following heating at temperatures of 600 and 610 °C. The solid volume fractions at 600 and 610 °C were 0.8 and 0.7, respectively (Figure 2b). Average grain size smaller than 75 μm and shape factor greater than 0.7 were obtained for strained samples heated at 600 and 610 °C for 25 min. Therefore, the isothermal heating temperature range of 600 to 610 °C and holding time of 25 min can be considered the optimum semi-solid treatment condition.

The coarsening kinetics can be expressed by the Lifshitz-Slyozov-Wagner (LSW) relationship [41,42]:

$$D^n - D_0^n = kt \tag{3}$$

where D and D_0 are the final and initial grain sizes, respectively, t is the isothermal holding time, k is the coarsening rate constant and n is the power exponent. It has been found that n is 3 for volume diffusion controlled systems in the semi-solid state [12].

In the present research, the coarsening rate constant (k) was calculated by fitting a power relationship to the experimental results. Figure 13 shows the cube of grain size variations *versus* isothermal holding time for compressed samples heated at 600, 610, and 620 °C, where D_0 is the average grain size when the holding time is 15 min. The regression coefficients of the fitted equations are close to 1. Thus, it can be concluded that the coarsening kinetics of solid particles during the isothermal heating of deformed 7075 samples at the semi-solid temperature range are fairly well correlated with the LSW equation. The values of the coarsening rate constant k for the samples are summarized in Table 3. The main characteristics of the variation of k values when the temperature was decreased from 620 to 600 °C are as follows; (i) the k value significantly reduced on decreasing the temperature from 620 to 610 °C; (ii) this was followed by a slight increase at 600 °C.

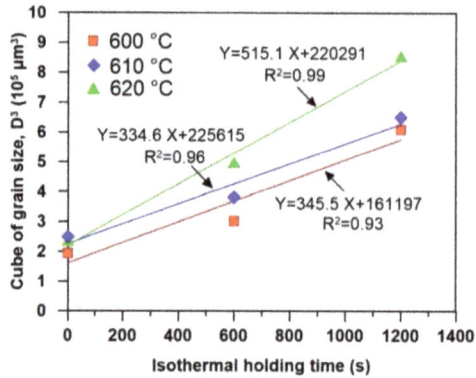

Figure 13. Variations of the cube of average globule size *versus* isothermal holding time of compressed samples at heating temperatures of 600–620 °C where R^2 is the regression coefficient.

Table 3. Values of coarsening rate constants of compressed samples at various isothermal holding temperatures.

Compression ratio (%)	Coarsening rate constant, k ($\mu m^3 \cdot s^{-1}$)		
	600 °C	610 °C	620 °C
20	345	334	515

According to the Doherty theorem [32], the coarsening rate accelerates with rising solid fraction (f_s) for f_s higher than 0.6. However, the research conducted by Manson-Whitton *et al.* [41] showed opposite results for higher solid fractions ($f_s \geqslant 0.7$) in the case of spray formed Al-4% Cu. Therefore, they proposed a modified model considering the solid-solid contacts effect during coarsening and a transition solid fraction ($f_s \approx 0.7$). For the solid fractions greater than the transition value, the k value decreased with an increase in the solid fraction. However, for the solid fraction lower than the transition value, the k value increased on increasing the solid fraction. Considering the holding temperatures that correspond to the calculated solid fractions (Figure 2b), for solid fractions greater than 0.7 (at 610 °C) and the transition value of 0.8 (at 600 °C), the coarsening process of solid grains is similar to the model proposed by Manson-Whitton *et al.* [41]. On the other hand, an unexpected increase in k values is observed for solid fractions lower than 0.7. This can be attributed to (i) further increase of the atomic diffusion and more effective Ostwald ripening mechanism and (ii) a less retarding effect of the precipitate particles (Fe-rich intermetallic particles) for a holding temperature up to 620 °C. These particles bring about convoluting grain boundaries as is marked by arrows in Figure 14, suggesting pinning and retardation of the grain boundary liquid film migration during the solid grains coarsening. The effect of the presence of intermetallic precipitates at the grain boundaries on the coarsening rate of aluminum alloys has also been reported by Manson-Whitton *et al.* [41] and de Freitas *et al.* [43]. The results obtained in the present research indicate that Fe-rich precipitates mainly dissolve or become smaller than the thickness of the liquid film on raising the heating temperature up to 620 °C. Thus, the movement of grain boundaries can easily occur and leads to greater values of the coarsening rate constant.

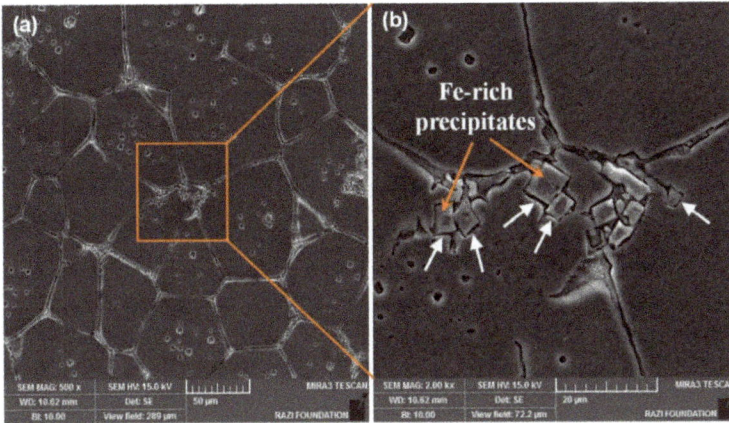

Figure 14. (**a**) SEM image of a strained sample heat treated at 600 °C for 15 min; (**b**) SEM image of selected area in (a).

3.6. Semi-Solid Deformation Behavior

Figure 15 shows a typical load-displacement curve obtained from the hot compression experiment. As it can be observed, the load rises to a maximum (L_{max}) and then decreases to a minimum value (L_{min}). The thixotropic flow behavior consisted of a transient regime and a steady state regime, as shown in Figure 15. Semi-solid alloys with an equiaxed microstructure are considered as deformable semi-cohesive granular solids saturated with liquid [44]. The solid grains are partially interconnected by unwetted grain boundaries and liquid fills the interstitial spaces. Large deformation results in the destruction of some solid grain boundary bonds to allow grains to move freely and rearrange through sliding and rolling. The cohesion at the grain level reflects the strength of unwetted grain boundaries and results in a flow resistance. With further deformation breakdown the solid agglomerates, the load (stress) decreases to a steady state level.

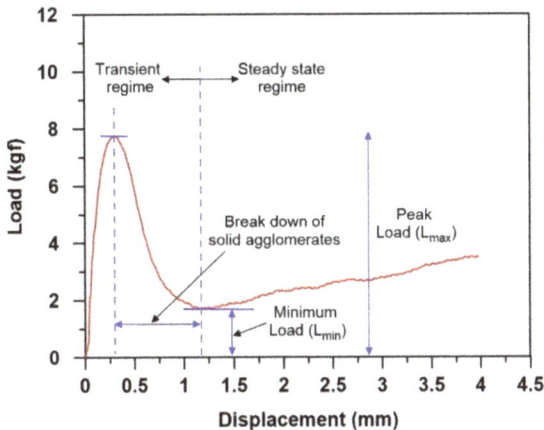

Figure 15. Typical load *versus* displacement curve of the hot compressed semi-solid 7075 sample (isothermally treated at 600 °C for 15 min) at 600 °C.

Tzimas and Zavaliangos [44] suggested that the flow resistance of semi-solid alloys at high solid volume fraction is controlled by four mechanisms: (i) elastoplastic deformation at grain contacts, (ii) destruction of cohesive bonds between solid grains, (iii) resistance to the flow of liquid relative to the solid, and (iv) resistance to grain rearrangement. Mechanism (i) is dominant in the rising part of the flow curve and mechanism (ii) results in the breakdown of solid particles and a significant drop in the load value. Mechanisms (iii) and (iv) are dominant in the second stage, and their relative contribution to the overall resistance to flow depends on the volume fraction of solid.

The stress-strain curves obtained from the SIMA processed 7075 samples with different initial microstructures after hot compression at 600 °C (for the SIMA samples isothermally treated at 600 °C for 10–35 min) and 610 °C (for the SIMA sample isothermally treated at 610 °C for 15 min) are shown in Figure 16. The stress exhibits a maximum value at a relatively low strain in all samples. The strain at the peak stress varies between 0.01 and 0.04. The peak stress then drops to a steady state value, typically at a strain of 0.1–0.18. The stress at the steady state shows no significant change with strain. Figure 17 shows the variation of the peak and steady state stresses for the SIMA processed samples with different initial microstructures. As is evident from Figures 16 and 17 the sample semi-solid treated at 600 °C for 10 min exhibits the highest peak and steady state stresses. On increasing the isothermal holding time to 25 min at 600 °C, which results in a coarser and more spherical solid grains in the initial microstructure (see Figure 12), both the peak and steady state stresses decreased significantly. The peak stress showed a slight increase while a considerable increase in the steady state stress was observed with further increase of the holding time to 35 min in the SIMA process. Moreover, the increase of the deformation temperature had a great influence on the stress-strain curve and resulted in a decrease of thixotropic strength, as is observed for the sample semi-solid treated at 610 °C for 15 min.

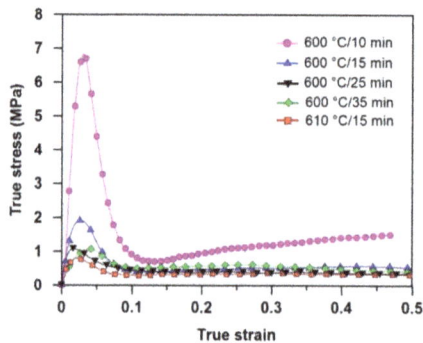

Figure 16. Stress-strain curves of strain induced melt activation (SIMA) processed samples with different initial microstructures.

The descending trend of the thixotropic strength (decreasing of the peak stress) with increase of the isothermal holding time in the SIMA process can be attributed to the geometry of the solid grains. As was previously described, the deformation of samples with equiaxed microstructure mainly occurs by grain rearrangement through sliding and rolling. However, the microstructures consisting of non-equiaxed and elongated grains are deformed by plastic deformation due to which the relative movement among solid grains is constrained. Figure 18 shows the hot compressed microstructure of SIMA processed samples with near equiaxed and globular initial microstructure. The microstructure of various regions (1 to 3 in Figure 18a) of the hot compressed SIMA sample processed at 600 °C for 10 min (see Figure 3c) revealed that this sample cannot be deformed by grain rearrangement due to the significant geometric interference of the solid grains which results in plastic deformation of the grains (Figure 18b–d). As is observed in Figure 18b–d, the solid grains are severely deformed and elongated in the direction of compression which results in a textured microstructure. This leads to a much higher resistance to flow compared to samples with a completely spheroidized microstructure. On prolonging the isothermal heating time, the sphericity of solid grains increases and the interconnection decreases (Figure 3) which results in easier movement and sliding during deformation and reduction of flow resistance. The high flow stress of the SIMA sample heated for 35 min can be ascribed to the coarsened initial microstructure with an average grain size of 85 μm. According to Clarke [45], an increase in the particle size leads to an increase in the apparent viscosity of the semi-solid slurry. Therefore, a sample with coarsened initial microstructure possesses a more viscous flow and requires higher loads for thixoforming.

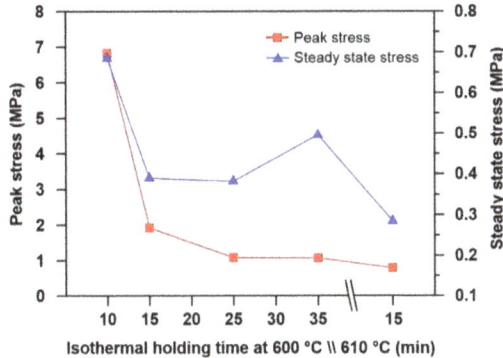

Figure 17. Peak and steady state stresses of the SIMA processed samples with different initial microstructures.

Figure 18e,f shows the microstructure of the center region (region 2 in Figure 18a) of the hot compressed semi-solid samples prepared by heating at 600 °C for 25 and 35 min, respectively. After compression, no evidence of the plastic deformation of the solid grains was observed in the microstructure of the samples. There was no difference in the microstructure of the center regions and edge regions, which indicated that the solid and the liquid flowed together during the compression. However, it is worth mentioning that the liquid fraction in the edge regions was slightly high, which was due to the liquid phase being squeezed out during the deformation. The microstructural observations also revealed that the solid grains do not experience plastic deformation in the SIMA samples prepared by heating at 600 and 610 °C for 15 min. According to Chen and Tsao [46], the plastic deformation of solid particles (PDS), sliding between the solid particles (SS), liquid flow (LF), and the flow of liquid incorporating solid particles (FLS) are four dominant mechanisms controlling deformation of alloys in the semi-solid state. The two former mechanisms are active when the solid particles are in contact with each other and the other two mechanisms are dominant when the solid particles are surrounded by the liquid phase. Therefore, it can be deduced that the sample with near equiaxed initial microstructure deform through the mechanism of PDS. In the case of globular initial microstructures, the SS and FLS mechanisms are dominant, however, the effectiveness of the SS mechanism may dwindle and the FLS mechanism is reinforced on prolonging the isothermal holding temperature and time due to the decrease of the interconnection between the solid grains.

The close examination of the hot compressed microstructure of different samples (Figure 18) also reveals that the microstructure of the SIMA processed sample with near equiaxed and elongated initial grains contains a higher amount of porosity, especially in regions 2 and 3, (shown by arrows in Figure 18c,d) compared to the samples with globular microstructure. In the sample with near equiaxed initial microstructure, the low amount of liquid content leads to higher viscosity, which may prevent the homogenous flow of the liquid and solid particles during deformation. Therefore, this sample shows a high amount of porosity in the microstructure. In addition, as shown by dotted ovals in Figure 18c, recrystallization occurs in some parts of the center region in the sample with an incomplete spheroidized initial microstructure during hot compression. It is believed that this recrystallization may occur dynamically as a result of higher induced strain in the center regions compared to the edge regions during hot compression.

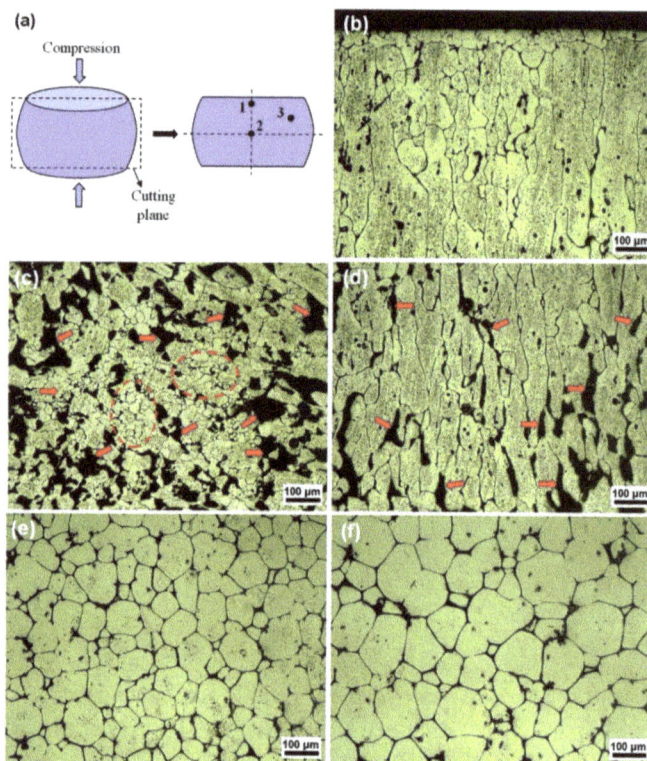

Figure 18. (a) Schematic of a hot compressed SIMA sample showing compression direction; Optical micrographs of (b) region 1; (c) region 2; (d) region 3 of SIMA sample processed at 600 °C for 10 min; region 2 of SIMA sample processed at 600 °C for (e) 25 min; (f) 35 min after hot compression at 600 °C.

4. Conclusions

The SIMA process consisting of applying uniaxial compression strain at ambient temperature and subsequent semi-solid treatment in the range of 600–620 °C for 5–35 min was investigated. It has been found that using the cold compression process following by heating in the semi-solid range results in some phase evolution that is suitable for obtaining a semi-solid microstructure for subsequent thixoforming. The following results can be drawn from the analysis:

The results showed that the partial remelting kinetics during the SIMA process consist of four steps including recrystallization and structural separation, coarsening of polygonal solid particles, spheroidization of polygonal particles, and coarsening of spherical particles. The growth and coarsening of the solid particles in the SIMA process are controlled by two mechanisms of coalescence (for low liquid fractions) and Ostwald ripening (for high liquid fractions).

The XRD results of the compressed samples heated at 600 °C for different times showed that $MgZn_2$ (η), Al_2CuMg (S), and Mg_2Si precipitates are dissolved gradually during isothermal heating through the phase transformations of α-Al + η → L and then α-Al + S + Mg_2Si → L. However, Fe-rich precipitates aggregate as square particles at grain boundaries due to their higher melting points.

An intense segregation of Si and Cu was observed at the grain boundaries in the semi-solid microstructure which results in shifting of the grain boundary composition toward the Al-Si and Al-Cu eutectics. In contrast, a significant depletion of Mg was observed at the grain boundaries and Zn distribution showed no appreciable change during isothermal treatment.

Microstructural observations indicated that the isothermal heating temperature range of 600–610 °C for 25 min can be considered an optimum condition for the SIMA process. Coarsening kinetics of the solid particles is fairly well correlated to the LSW theory during isothermal heating. Despite the higher liquid fraction of samples heated at 610 °C, the coarsening rate at 610 °C was lower than 600 °C.

Samples with near equiaxed initial microstructure containing elongated and polygonal solid grains and samples with globular initial microstructure with average grain size of 61 μm show the greatest and the lowest flow resistance during the thixoforming process at 600 °C, respectively. The flow resistance decreases on raising the deformation temperature.

Author Contributions: Behzad Binesh performed the experiments and wrote the paper under Mehrdad Aghaie-Khafri's guidance, and contributed to all activities.

Conflicts of Interest: The authors declare no conflict of interest.

References

1. Kirkwood, D.H.; Suery, M.; Kapranos, P.; Atkinson, H.V. *Semi-Solid Processing of Alloys*; Springer: London, UK, 2009; pp. 109–112.
2. Flemings, M.C. Behavior of metal alloys in the semisolid state. *Metall. Trans. A* **1991**, *22*, 957–981. [CrossRef]
3. Jayaraj, J.; Fleury, E.; Kim, K.-B.; Lee, J.-C. Globulization mechanism of the primary Al of Al-15Cu alloy during slurry preparation for rheoforming. *Met. Mater. Int.* **2005**, *11*, 257–262. [CrossRef]
4. Zhang, Q.Q.; Cao, Z.Y.; Zhang, Y.F.; Su, G.H.; Liu, Y.B. Effect of compression ratio on the microstructure evolution of semisolid AZ91D alloy. *J. Mater. Process. Technol.* **2007**, *184*, 195–200. [CrossRef]
5. Saklakoglu, N.; Saklakoglu, I.E.; Tanoglu, M.; Oztas, O.; Cubukcuoglu, O. Mechanical properties and microstructural evaluation of AA5013 aluminum alloy treated in the semi-solid state by SIMA process. *J. Mater. Process. Technol.* **2004**, *148*, 103–107. [CrossRef]
6. Chen, T.J.; Hao, Y.; Sun, J. Microstructural evolution of previously deformed ZA27 alloy during partial remelting. *Mater. Sci. Eng. A* **2002**, *337*, 73–81. [CrossRef]
7. Tzimas, E.; Zavaliangos, A. A comparative characterization of near-equiaxed microstructures as produced by spray casting, magnetohydrodynamic casting and the stress induced melt activated process. *Mater. Sci. Eng. A* **2000**, *289*, 217–227. [CrossRef]
8. Lin, H.Q.; Wang, J.G.; Wang, H.Y.; Jiang, Q.C. Effect of predeformation on the globular grains in AZ91D alloy during strain induced melt activation (SIMA) process. *J. Alloys Compd.* **2007**, *431*, 141–147. [CrossRef]
9. Kang, B.K.; Hong, C.P.; Choi, B.H.; Jang, Y.S.; Sohn, I. Microstructural evolution in semisolid forging of A356 alloy. *Met. Mater. Int.* **2015**, *21*, 153–158. [CrossRef]
10. Young, K.P.; Kyonka, C.P.; Kamado, S. Fine Grained Metal Composition. U.S. Patent 4415374, 30 March 1983.
11. Jiang, J.; Wang, Y.; Atkinson, H.V. Microstructural coarsening of 7005 aluminum alloy semisolid billets with high solid fraction. *Mater. Charact.* **2014**, *90*, 52–61. [CrossRef]
12. Yan, G.; Zhao, S.H.; Ma, S.H.; Shou, H. Microstructural evolution of A356.2 alloy prepared by the SIMA process. *Mater. Charact.* **2012**, *69*, 45–51. [CrossRef]
13. Jung, H.K.; Seo, P.K.; Kang, C.G. Microstructural characteristics and mechanical properties of hypo-eutectic and hyper-eutectic Al-Si alloys in the semi-solid forming process. *J. Mater. Process. Technol.* **2001**, *113*, 568–573. [CrossRef]
14. Curle, U.A.; Govender, G. Semi-solid rheocasting of grain refined aluminum alloy 7075. *Trans. Nonferrous Met. Soc. China* **2010**, *20*, s832–s836. [CrossRef]
15. Yong, L.S.; Hwan, L.J.; Seon, L.Y. Characterization of Al 7075 alloys after cold working and heating in the semi-solid temperature range. *J. Mater. Process. Technol.* **2001**, *111*, 42–47. [CrossRef]

16. Chayong, S.; Atkinson, H.V.; Kapranos, P. Thixoforming 7075 aluminum alloys. *Mater. Sci. Eng. A* **2005**, *390*, 3–12. [CrossRef]
17. Chayong, S.; Atkinson, H.V.; Kapranos, P. Multistep induction heating regimes for thixoforming 7075 aluminum alloy. *Mater. Sci. Technol.* **2004**, *20*, 490–496. [CrossRef]
18. Atkinson, H.V.; Burke, K.; Vaneetveld, G. Recrystallisation in the semi-solid state in 7075 aluminum alloy. *Mater. Sci. Eng. A* **2008**, *490*, 266–276. [CrossRef]
19. Vaneetveld, G.; Rassili, A.; Atkinson, H.V. Influence of parameters during induction heating cycle of 7075 aluminum alloys with RAP process. *Solid State Phenom.* **2008**, *141–143*, 719–724. [CrossRef]
20. Bolouri, A.; Shahmiri, M.; Kang, C.G. Study on the effects of the compression ratio and mushy zone heating on the thixotropic microstructure of AA 7075 aluminum alloy via SIMA process. *J. Alloys Compd.* **2011**, *509*, 402–408. [CrossRef]
21. Mohammadi, H.; Ketabchi, M.; Kalaki, A. Microstructure evolution of semi-solid 7075 aluminum alloy during reheating process. *J. Mater. Eng. Perform.* **2011**, *20*, 1256–1263. [CrossRef]
22. Seo, P.K.; Kang, C.G. The Effect of raw material fabrication process on microstructural characteristics in reheating process for semi-solid forming. *J. Mater. Process. Technol.* **2005**, *162–163*, 402–409. [CrossRef]
23. Zhang, L.; Liu, Y.B.; Cao, Z.Y.; Zhang, Y.F.; Zhang, Q.Q. Effects of isothermal process parameters on the microstructure of semisolid AZ91D alloy produced by SIMA. *J. Mater. Process. Technol.* **2009**, *209*, 792–797. [CrossRef]
24. Doherty, R.D.; Lee, H.-I.; Feest, E.A. Microstructure of stir-cast metals. *Mater. Sci. Eng.* **1984**, *65*, 181–189. [CrossRef]
25. Tzimas, E.; Zavaliangos, A. Evolution of near-equiaxed microstructure in the semisolid state. *Mater. Sci. Eng. A* **2000**, *289*, 228–240. [CrossRef]
26. Wang, T.; Yin, Z.-M.; Sun, Q. Effect of homogenization treatment on microstructure and hot workability of high strength 7B04 aluminum alloy. *Trans. Nonferrous Met. Soc. China* **2007**, *17*, 335–339. [CrossRef]
27. Fan, X.-G.; Jiang, D.-M.; Meng, Q.-C.; Zhang, B.-Y.; Wang, T. Evolution of eutectic structures in Al-Zn-Mg-Cu alloys during heat treatment. *Trans. Nonferrous Met. Soc. China* **2006**, *16*, 577–581. [CrossRef]
28. Lim, S.T.; Lee, Y.Y.; Eun, I.S. Microstructural evolution during ingot preheat in 7xxx aluminum alloys for thick semiproduct applications. *Mater. Sci. Forum* **2006**, *519–521*, 549–554. [CrossRef]
29. Li, N.-K.; Cui, J.-Z. Microstructural evolution of high strength 7B04 ingot during homogenization treatment. *Trans. Nonferrous Met. Soc. China* **2008**, *18*, 769–773. [CrossRef]
30. Jiang, L.-T.; Wu, G.-H.; Yang, W.-S.; Zhao, Y.-G.; Liu, S.-S. Effect of heat treatment on microstructure and dimensional stability of ZL114A aluminum alloy. *Trans. Nonferrous Met. Soc. China* **2010**, *20*, 2124–2128. [CrossRef]
31. Boettinger, W.J.; Kattner, U.R.; Moon, K.W.; Perepezko, J.H. DTA and heat-flux DSC measurements of alloy melting and freezing. In *Methods for Phase Diagram Determination*, 1st ed.; Zhao, J.C., Ed.; Elsevier: Ames, IA, USA, 2007; pp. 194–200.
32. Annavarapu, S.; Doherty, R.D. Inhibited coarsening of solid-liquid microstructures in spray casting at high volume fractions of solid. *Acta Metall. Mater.* **1995**, *43*, 3207–3230. [CrossRef]
33. Liu, D.; Atkinson, H.V.; Kapranos, P.; Jones, H. Effect of heat treatment on properties of thixoformed high performance 2014 and 201 aluminum alloys. *J. Mater. Sci.* **2004**, *39*, 99–105. [CrossRef]
34. Cavaliere, P.; Cerri, E.; Leo, P. Effect of heat treatments on mechanical properties and fracture behavior of a thixocast A356 aluminum alloy. *J. Mater. Sci.* **2004**, *39*, 1653–1658. [CrossRef]
35. Bolouri, A.; Shahmiri, M.; Kang, C.G. Coarsening of equiaxed microstructure in the semisolid state of aluminum 7075 alloy through SIMA processing. *J. Mater. Sci.* **2012**, *47*, 3544–3553. [CrossRef]
36. Jiang, H.; Li, M. Microscopic observation of cold-deformed Al-4Cu-Mg alloy samples after semi-solid heat treatments. *Mater. Charact.* **2005**, *54*, 451–457. [CrossRef]
37. Shim, S.-Y.; Kim, D.-H.; Chio, S.-H.; Kim, Y.-H.; Bai, H.; Lim, S.-G. The analysis of variance on process parameters affecting the microstructures of semi-solid Al-Zn-Mg alloy billet by cooling plate method. *Met. Mater. Int.* **2010**, *16*, 1009–1017. [CrossRef]
38. Baker, H. *ASM Handbook, Alloy Phase Diagrams*; ASM International: Materials Park, OH, USA, 2004; Volume 3, pp. 295–331.
39. Shackelford, J.F.; Alexander, W. *Materials Science and Engineering Handbook*, 3rd ed.; CRC Press LLC: Boca Raton, FL, USA, 2001.

40. Vaneetveld, G.; Rassili, A.; Pierret, J.C.; Beckers, J.L. Conception of tooling adapted to thixoforging of high solid fraction hot-crack-sensitive aluminum alloys. *Trans. Nonferrous Met. Soc. China* **2010**, *20*, 1712–1718. [CrossRef]

41. Manson-Whitton, E.D.; Stone, I.C.; Jones, J.R.; Grant, P.S.; Cantor, B. Isothermal grain coarsening of spray formed alloys in the semi-solid state. *Acta Mater.* **2002**, *50*, 2517–2535. [CrossRef]

42. Li, P.; Chen, T.; Zhang, S.; Guan, R. Research on semisolid microstructural evolution of 2024 aluminum alloy prepared by powder thixoforming. *Metals* **2015**, *5*, 547–564. [CrossRef]

43. De Freitas, E.R.; Ferracini, E., Jr.; Ferrante, M. Microstructure and rheology of an AA2024 aluminum alloy in the semi-solid state, and mechanical properties of a back-extruded part. *J. Mater. Process. Technol.* **2004**, *146*, 241–249. [CrossRef]

44. Tzimas, E.; Zavaliangos, A. Mechanical behavior of alloys with equiaxed microstructure in the semisolid state at high solid content. *Acta Mater.* **1999**, *47*, 517–528. [CrossRef]

45. Clarke, B. Rheology of coarse settling suspensions. *Trans. Inst. Chem. Eng. Chem. Eng.* **1967**, *45*, 251–256.

46. Chen, C.P.; Tsao, C.-Y.A. Semi-solid deformation of non-dendritic structures-phenomenological behavior. *Acta Mater.* **1997**, *45*, 1955–1968. [CrossRef]

metals

MDPI

Article

Guideline for Forming Stiffened Panels by Using the Electromagnetic Forces

Jinqiang Tan, Mei Zhan * and Shuai Liu

State Key Laboratory of Solidification Processing, School of Materials Science & Engineering,
Northwestern Polytechnical University, Xi'an 710072, China; jayson2005@126.com (J.T.);
liushuai_npu@126.com (S.L.)
* Correspondence: zhanmei@nwpu.edu.cn; Tel.: +86-29-8846-0212 (ext. 805); Fax: +86-29-8849-5632

Academic Editor: Nong Gao
Received: 9 July 2016; Accepted: 28 October 2016; Published: 7 November 2016

Abstract: Electromagnetic forming (EMF), as a high-speed forming technology by applying the electromagnetic forces to manufacture sheet or tube metal parts, has many potential advantages, such as contact-free and resistance to buckling and springback. In this study, EMF is applied to form several panels with stiffened ribs. The distributions and variations of the electromagnetic force, the velocity and the forming height during the EMF process of the bi-directional panel with gird ribs are obtained by numerical simulations, and are analyzed via the comparison to those with the flat panel (non-stiffened) and two uni-directional panels (only with X-direction or Y-direction ribs). It is found that the electromagnetic body force loads simultaneously in the ribs and the webs, and the deformation of the panels is mainly driven by the force in the ribs. The distribution of force in the grid-rib panel can be found as the superposition of the two uni-directional stiffened panels. The velocity distribution for the grid-rib panel is primarily affected by the X-directional ribs, then the Y-directional ribs, and the variation of the velocity are influenced by the force distribution primarily and secondly the inertial effect. Mutual influence of deformation exists between the region undergoing deformation and the deformed or underformed free ends. It is useful to improve forming uniformity via a second discharge at the same position. Comparison between EMF and the brake forming with a stiffened panel shows that the former has more advantages in reducing the defects of springback and buckling.

Keywords: electromagnetic forming (EMF); stiffened panel; numerical simulation

---

1. Introduction

The development of the modern fuselage structure in aerospace industry makes it necessary to pursue possible methods to form the parts with a desired contour. The stiffened panels (integrally stiffened structure) have become one of the important parts of modern aircrafts, benefiting from their high strength, high structural efficiency and low weight. The stiffened panels forming technology, is therefore one of the key technologies in aerospace industry. However, the improved structural stiffness of the stiffened panels, due to the stiffened ribs, increases the forming difficulty.

The traditional forming processes for the stiffened panels (panels after machining the ribs with desired arrangement form from plate) include creep age forming, shot peen forming, roll forming, brake forming, etc. [1]. Creep age forming (CAF), which appeared in the 1980s, is accomplished by combining creep forming and age hardening simultaneously. This process is widely applied in aircraft manufacture [2]. Eberl et al. verified the feasibility of CAF in forming the commercial aircraft stiffened plates via practical experiments [3]. One of the key problems during CAF is the springback because the plastic strain level is very low and the elastic strain maintains a high level after forming [4]. The comparison of springback with different plates, including flat, beam stiffened, waffle and isogrid plates, was studied through experiment and numerical simulation, which showed that the springback of the four plates ranged from 12.2% to 15.7% [5]. Unavoidable and unpredictable springback, even about 70% [6], which makes it difficult for the accurate design of the forming tools to compensate for the elastic strain. For the shot peen forming, the advantages lies in no die use and fatigue property improvement because of the existence of the residual compressive stress in the formed surface of the parts, which make the shot peen forming one of the preferred forming methods [7]. However, the surface roughness after shot peen forming is poor and thus subsequent correction is necessary for the panels with complicated structure [8]. Meanwhile, due to the limited forming ability, it is unsuitable or difficult to form the stiffened panels with a complex contour or curvature, as well as high stiffened ribs panels, with shot peen forming [9]. Roll forming, also being called roll bending based on the continuous local plastic deformation, is a method with low cost tools and low time consuming, thus can be adaptable to different contours. However, there are certain limits in the forming of the stiffened panels, due to the occurrence of mark-off, support material needed in the pockets of the panels sometimes and simple contour only [10]. The basic principle of the brake forming (e.g., air bending) is three point bending. The workpiece undergoes discontinuous local plastic deformation under the press of a punch. As a traditional forming method, brake forming is widely used in the formation of aircraft stiffened panels owing to several advantages, such as the low tool cost, strong applicability to various part shapes and compound stiffened styles. For example, Yan et al. utilized the incremental-press bending method to form a stiffened panel with grid ribs successfully after designing suitable forming path with the help of a back-propagation neural network response surface method [11]. However, because the forces apply directly on the ribs by the punch and then the deformation of the web is driven by the ribs, the defects, such as springback and fracture, are the main disadvantages with this technology. Studies on the defects during the brake forming process can be found with respect to buckling [12] and springback [13]. In a word, all these forming limitations and forming equality issues lying in the traditional process for forming the stiffened panels make the exploration of innovative forming approaches urgent to meet the developing requirements in aircraft industry.

Electromagnetic forming (EMF) is a high-speed forming process that utilizes the electromagnetic forces, produced under the effects of the eddy current induced by the coil magnetic field and the magnetic fields stemming from the coil and the eddy current itself, to manufacture parts. According to Daehn [14], the advantages of this process can be deemed as reduction of die fabricate cost and production cycle, the improvement of the materials mechanical properties and the decrease of springback and residual stress. In addition, because of the induced eddy currents flowing through the entire panel, the web of the stiffened panel is also affected by the electromagnetic field, that is, there are electromagnetic body forces loading in the ribs and the web at the same time. Different from the traditional stiffened panel forming methods where the forces load on the surface of the web or the ribs only, EMF produces the body force on the part, which with will be useful to increase the forming ability. These unique advantages make EMF become a potential method for forming the aircraft stiffened panels of aluminum alloys, which are with good electrical conductivity. However, most of researches on EMF, according to Psyk et al. [15], focus on relatively smaller and simpler parts (non-stiffened) that can be formed through one to several discharges with a coil which is usually fixed at a given location. The fixed coil makes it difficult to apply EMF directly to form large-scale components, such as the aircraft stiffened panels.

An electromagnetic incremental forming (EMIF) technology was developed by Cui et al. [16], and EMIF was validated to be effective to solve the difficulty in the large-scale component forming. In the EMIF process, a small coil is utilized, which moves along some certain 2D/3D paths with accompanying discharging for many times, to produce the large-scale and complicated-shape parts. By comparing with the traditional EMF process, some new parameters, such as coil overlap rate, discharge pass etc., are introduced with EMIF. Kamal and Daehn [17] reported the large clearance between the coil and the workpiece can reduce the efficiency of EM induction, and the multi-discharge cannot increase the forming depth significantly while can improve the quality of the workpiece.

Zhao et al. [18] simulated the EMIF process of tubes, and analyzed the influence of coil path and coil overlap rate on the forming results. Their results showed the forming uniformity increased with higher overlap rate. Cui et al. [19] produced a circular plate successfully by using a six-turn coil with the discharging energy no more than 6 kJ, and contrastively studied the effect of several discharging paths and discharging parameters to verify the feasibility of the EMIF.

Because the EMF process is a complicated process involving the coupled effects of magnetic field and deformation field, the finite element method (FEM) becomes an effective tool for the processing study. In a typical EMF simulation, the control of the air distortion is one of the key technologies because the deformation of the workpiece will conversely lead to severe deformation of the air, resulting in the distortion of the air elements. Through a 2D FE simulation for electromagnetic sheet forming process, Fenton and Daehn [20] claimed that the introduction of the Arbitrary Lagrange Euler (ALE) method can effectively control the air distortion and then a more accurate result was obtained. Ma et al. [21] utilized the ALE method to simulate EMF process of a flat panel (non-stiffened), where the distortion of the air mesh was controlled availably. In the simulation of sheet process with EMIF, Cui et al. [19] used the morphing and remeshing technology to describe the air movement caused by the workpiece deformation, and pointed that the remeshing method is more suitable than the morphing method for the overlap region simulation during the EMIF. In addition, the coupling strategy adopted is another key technology. Generally, multi-physics problems can be solved by two methods: direct coupling and indirect coupling. The direct coupling method reflects the most accurate physical mechanism, however, the calculation efficiency of this method is low. Thus, it is practical to adopt the latter coupling method. Oliveira and Worswick [22] proposed a "loose" coupling strategy to simulate a 3D free-bulging of aluminum alloy sheet, and their FE model is verified by comparing the final geometry and strain distributions of the workpiece between the simulated and experimental results. Yu et al. [23] mentioned a sequential coupling method considering the mesh morphing technology to simulate the EMF process with tubes. Their results supported the sequential model with higher accuracy than the results by using loose coupling method. Employing the sequential coupling method to realize the iterative coupling between magnetic field and structural field, Cui et al. [24] analyzed electromagnetic sheet bulging process and obtained the change regulation of the magnetic forces.

For the stiffened panels, the existence and deformation of the ribs increase the forming difficulty. Additionally, the distribution of the ribs will make the panel forming process significantly different from the traditional sheet EMF process, which can be attributed to the distinct distribution of the magnetic forces. Therefore, it is necessary to investigate the EMF process of the stiffened panel. In this study, the contents on materials study, experimental methods and numerical simulation preparations are presented in Section 2. The FE model established for the stiffened panels during the EMIF process is given in Section 3. Finally, the distribution of electromagnetic force, the evolvement rule of velocity and height of the stiffened panels with different rib arrangements, as well as the comparison of forming quality of the stiffened panels obtained by the brake forming and EMIF are discussed in Section 4. Conclusions are presented in Section 5.

2. Material and Methods

2.1. Material and Blank

The 2A12-T4 aluminum alloy is used in this study [25]. With EMF, the strain rate effect must be considered [26], and so the stress–strain data at various strain rates via quasi-static tests and Hopkinson bar tests are provided here. The quasi-static experiments were measured by SANS® CMT5205 (Shenzhen, China) electronic universal testing machine at room temperature under a fixed strain rate of 10^{-3} s^{-1}. The specimens used were standard tablet specimen referring to ISO 6892-1:2009 [27] with a gauge length of 50 mm. The dynamic experiments, in which the cylindrical specimens with the sizes of Φ5 mm $\times$ 5 mm were used, were conducted and measured by a self-designed split Hopkinson pressure bar (SHPB) machine at room temperature under different strain rates of 1100 s^{-1}, 2800 s^{-1},

3900 s^{-1} and 5000 s^{-1}, respectively. The original material was a rolled sheet, and all the test specimens were machined to have the coincident axis with the rolling direction. All the test experiments were conducted repeatedly at least three times under each condition, especially for the dynamic experiments, to ensure repeatable and consistent. The results of the quasi-static and dynamic tests are shown in Figure 1. Remarkable strain rate sensibility can be found for the alloy, and the larger strain rate leads to the higher yield stress and strength limit while stress softening at a large strain.

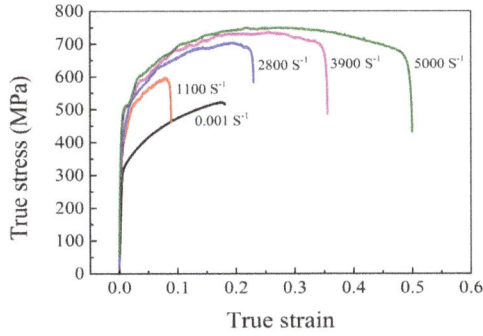

Figure 1. True stress–strain of 2A12-T4 alloy at various strain rates.

The stiffened panel is usual composed of a web and several ribs. In this study, a grid-rib panel blank is adopted, as shown in Figure 2. The size of the main deformation region is 150 mm × 100 mm. The height of the web is 2 mm and the heights of all the ribs are 4 mm. In comparison, three decomposition panels from the grid-rib panel (GP), that is, panel with transversal ribs only (called X-rib panel, XP), panel with longitudinal ribs only (Y-rib panel, YP) and panel without any ribs (flat panel, FP) are studied, where the sizes of these three panels are consistent with the grid-rib panel.

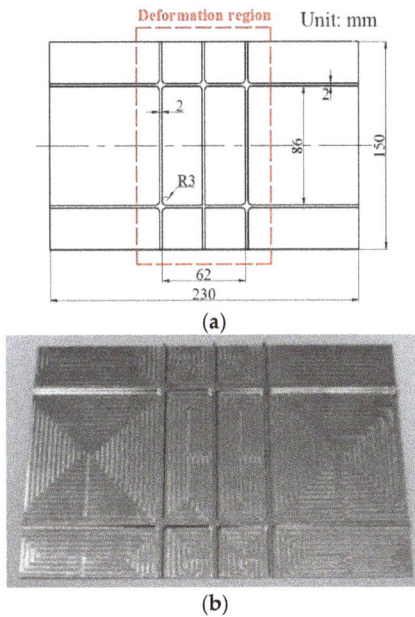

(a)

(b)

Figure 2. The grid-rib panel: sizes (**a**); and the experimental blank sample (**b**).

2.2. Forming Set-Up

The experimental set-up is shown in Figure 3, which includes two blank holder blocks with guide grooves and two coil support blocks with arc grooves and guide grooves simultaneously, which are connected by bolts. The incremental distances were accurately and manually controlled using a Vernier caliper. The initial gap between the bottom surface of the copper wire and the top surface of the ribs of the panel blank was 5 mm.

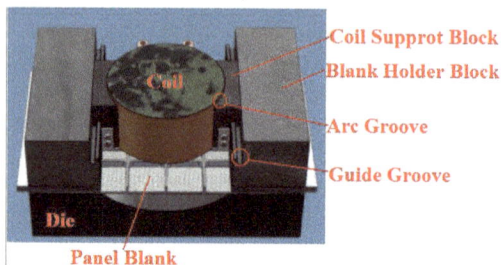

Figure 3. Forming equipment during electromagnetic incremental forming (EMIF) of stiffened panels.

2.2.1. Die

The single curvature die, 230 mm × 150 mm × 50 mm in length, width and height, respectively, is used, as shown in Figure 4. The curvature radius is 80 mm and the transition fillet radius is 10 mm.

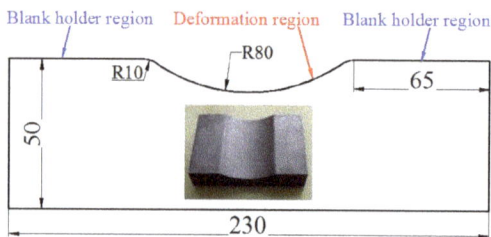

Figure 4. The single curvature die.

2.2.2. Coil

A flat circular spiral coil is selected in this study due to its reliable performance and large load endurance capacity and the simplified form is shown in Figure 5. The distance between two adjacent turns of the coil is designed as a gradual decrease from the inside to the outside (Table 1), which makes more epoxy resin in the coil to improve the support force.

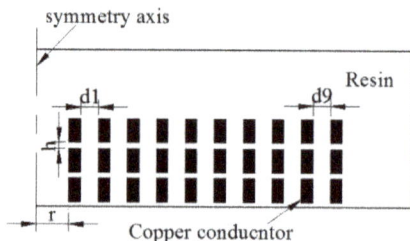

Figure 5. Structure schematic diagram of the flat spiral coil.

Table 1. Size parameters of a three-layer coil with variable turn gaps.

Parameters	Value	Parameters	Value	Parameters	Value
Rectangular section	2 mm × 4 mm	Layer gap h	0.4 mm	Turn gap d_5	0.95 mm
Total turns N	30	Turn gap d_1	1.8 mm	Turn gap d_6	0.95 mm
Layers	3	Turn gap d_2	1.8 mm	Turn gap d_7	0.7 mm
Inner radius R	7 mm	Turn gap d_3	1.7 mm	Turn gap d_8	0.7 mm
Height H	12.8 mm	Turn gap d_4	1.2 mm	Turn gap d_9	0.7 mm

2.3. Forming Stations and Coil Paths

For the sake of describing forming stations (coil position) and paths, a Cartesian coordinate system is constructed, which is oriented from the center of the panel blank, as shown in Figure 6, and the forming stations are designed along the X axis, where station B is in the center of the panel, while station A and station C are symmetrical about station B. The distance between two forming stations is 40 mm. Two coil support blocks, which are installed on two arc grooves, are designed to move the coil accurately, as shown in Figure 3.

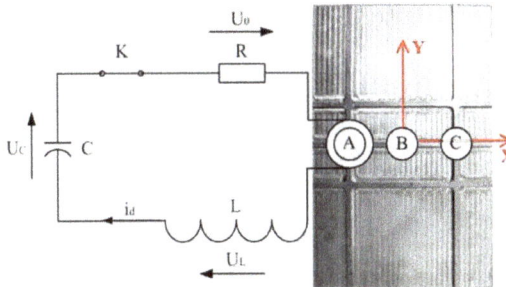

Figure 6. Defined coordinate system of the workpiece.

To analyze the forming rule of the stiffened panels in the EMIF process, the forming processes of four panels (Table 2), are studied using the established FE model. As seen from Table 2, the ratio of rib height to web thickness is relatively small, as a result, a capacitor with a capacity of 80 µF is adopted. As seen from Table 3, in the first station there are twice discharges with different voltages to study the forming rule under the same station and the multi-discharging conditions.

Table 2. Original sizes of the four panels blanks.

Blank Form	Rib Height (mm)	Rib Width (mm)	Web Thickness (mm)
Flat panel (FP)	0	0	2
X-rib panel (XP)	4	2	2
Y-rib panel (YP)	4	2	2
Grid-rib panel (GP)	4	2	2

Table 3. System parameters and forming conditions during electromagnetic incremental forming for stiffened panels (EMIF-SP).

Materials & Boundary Condition	Parameter	Value
Air Coil (copper)	Relative permeability	1
	Relative permeability	1
	Resistance	20 mΩ
	Inductance	15 μH
	Reference resistivity	1.7×10^{-8} Ω·m
	Density	8.9×10^3 kg/m^3
	Elastic modulus	90 GPa
	Poisson's ratio	0.33
Panel (2A12-T4 Al alloy)	Relative permeability	1
	Reference resistivity	3×10^{-8} Ω·m
	Density	2.77×10^3 kg/m^3
	Elastic modulus	69 GPa
	Poisson's ratio	0.31
	Yield strength	284 MPa
	Ultimate strength	495 MPa
Die/Blank holder block (42CrMo4)	Relative permeability	1
	Density	7.85×10^3 kg/m^3
	Elastic modulus	206 GPa
	Poisson's ratio	0.3
Circuit line	Line resistance	25 mΩ
	Line inductance	6.5 μH
Contact set (Blank holder block-Panel)	Static friction factor	0.17
	Dynamic friction factor	0.15
Contact set (Die-Panel)	Static friction factor	0.17
	Dynamic friction factor	0.15

3. Numerical Simulation

3.1. Establishment of the FE Model

The EMIF process of the stiffened panels (EMIF-SP) is composed of four basic stages: generation of the induced eddy current and the magnetic field surround the workpiece, plastic formation, shift of the coil station and unloading springback. Mutual effects and multi-factors occur at every stage in the forming process. Therefore, it is necessary to build a whole-process model that includes these stages to accurately analyze the mechanisms and governing principles of EMIF-SP. By using the ANSYS V8.1 software (Pittsburgh, PA, USA), a bilaterally coupled model of EMIF-SP has been established in this study to analyze the EMIF process. In the model, an electromagnetic model, a structural model and a springback model are included. The detailed calculation flowchart is shown in Figure 7. The electromagnetic model is established to simulate the magnetic field existing surround the workpiece, which is resulted by the coil magnetic field when an impulse current flow through the coil. By dividing the coil current into *n* increments and then loading each increment into the electromagnetic model in a step-by-step manner, the electromagnetic force can be obtained. Then, by loading the electromagnetic force into the workpiece in the structural model, the plastic forming process of the stiffened panel can be simulated. When the deformation completes at every increment, the geometry of the workpiece and air must be updated in the electromagnetic model for the next-increment magnetic field calculation. This cycle is repeated until the total current loading is finished. The stiffened panel continues to deform under inertial effect until the velocity of each point reaches zero. For the next station discharge, the coil must be moved to the next forming station. The electromagnetic force calculation and deformation analysis are applied repeatedly on each new station using the same method until all forming stations have been loaded. Finally, the holder is removed, and the workpiece springback due to unloading can be simulated using the springback model.

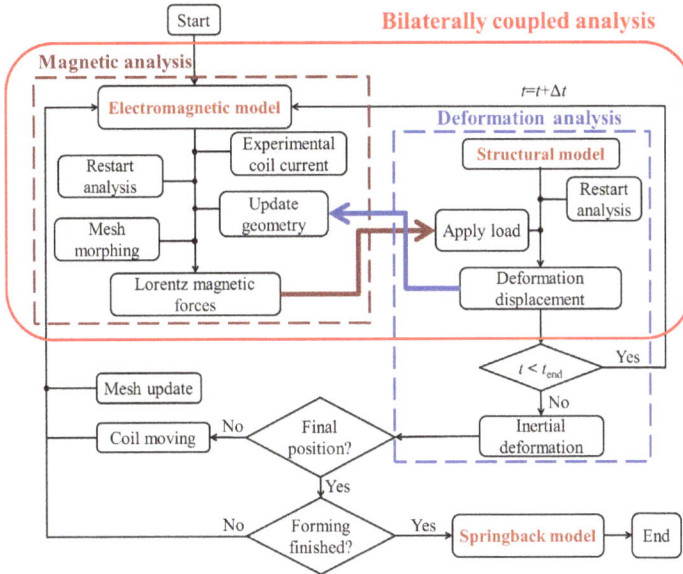

Figure 7. Flowchart of FE (finite element) model of EMIF for stiffened panels.

The electromagnetic circuit model is based on the EMF system in Cao et al. [28], where an additional crowbar circuit is used to effectively reduce the temperature rise of the coil.

Considering the symmetry of deformation process and boundary conditions about X axis, a half multi-physics coupling FE model of the entire EMIF process for the stiffened panels is established according to the aforementioned four stages, shown as Figure 8.

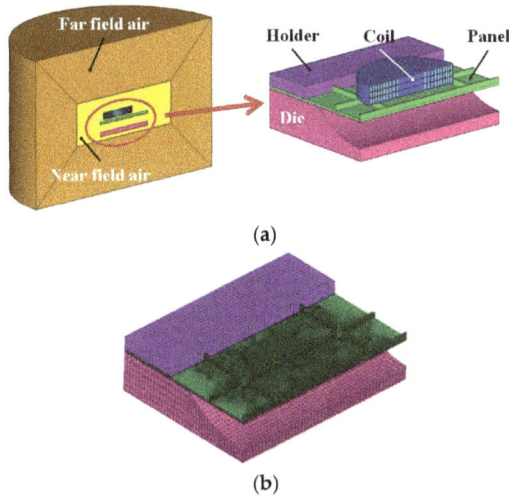

(a)

(b)

Figure 8. *Cont.*

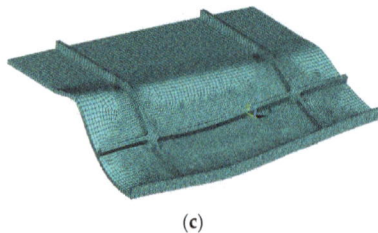

(c)

Figure 8. The multi-physics coupling model of the grid-rib panel during EMIF: (**a**) electromagnetic field model; (**b**) forming model; and (**c**) springback model.

The main system parameters during the EMIF-SP process are listed in Table 3.

The panel and the coil are surrounded by air in the electromagnetic model, thus large distortion of the air meshes occurs because there is large plastic deformation or location change in the coil, which will interrupt the analysis. To solve this problem, the tetrahedral elements are adopted for the near-field air because the update of the air via remeshing operation could be easily achieved to guarantee excellent meshes with greater changes in the coil or the panel for the purpose of successful computation. Such changes in the meshes during the simulating process are shown in Figure 9, where the yellow part is the one-layer meshes closed to the panel. It can also be seen that the air meshes can be adjusted according to the panel geometry or the coil station.

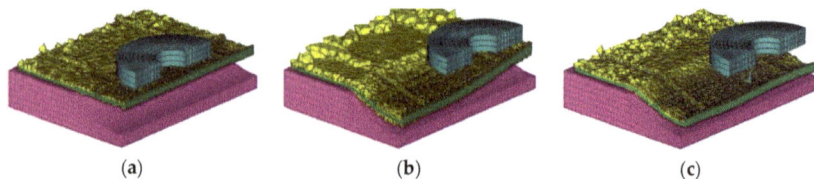

(a)   (b)   (c)

Figure 9. Air meshes at different moments: (**a**) initial; (**b**) deformation after the first station; and (**c**) deformation after the second station.

The ALE algorithm and smoothing treatment are employed for the air elements. To improve computational efficiency, the singe point integral algorithm is used, though it is prone to increase the hourglass energy in explicit dynamic analysis, which can be fixed through choosing the hourglass control type to be viscous form. The hourglass energy does not exceed 10% of the internal energy in this study, as shown in Figure 10.

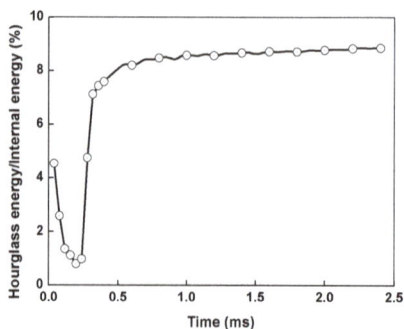

Figure 10. Ratio of hourglass energy to total energy.

3.2. Verification of the FE Model

To verify the established model, experiments and simulations for single discharge at station A and then C (Figure 6) are conducted. Parameters used in the experiments are as follows:

(1)   The capacitance is 160 μF.
(2)   For station A: X coordinate is −28 mm and the voltage is 7 kV. For station C, X coordinate is 28 mm and the voltage is 9.8 kV.

The system parameters in Table 3 are adopted in the experiments.

For the simulation, the same parameters in the experiments are adopted. The current data in the simulations are obtained from the experimental measurement. The simulation approach is based on the bilaterally coupled model established in Section 3.1.

In order to compare the experimental and simulated results, several paths are defined, as shown in Figure 11. Figure 11a shows the paths from the top view of the panel and Figure 11b is from the bottom view of the panel.

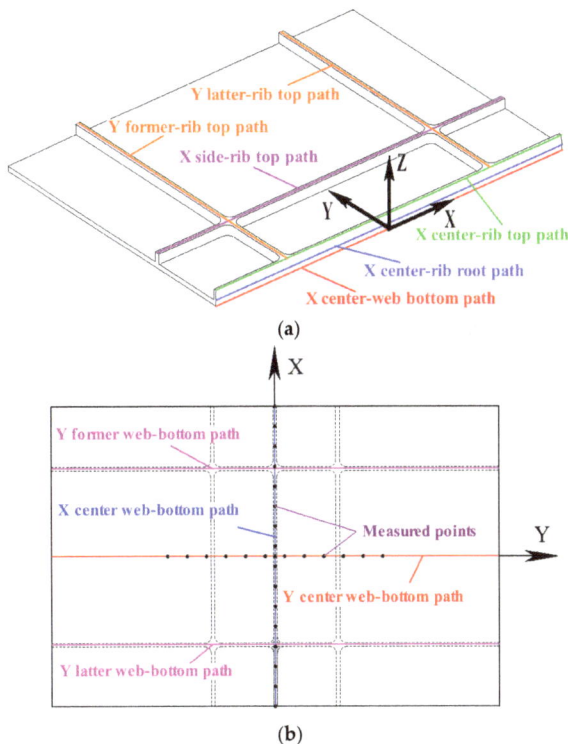

(a)

(b)

Figure 11. The defined paths of the grid-rib panel: (**a**) from the top view of the panel, 1/2 model shown only; and (**b**) from the bottom view of the panel.

Figure 12 shows the comparison between the experimental and simulation results after discharging at station A and station C. Comparisons of the formed profile and the forming height distribution along the X center web-bottom path (Figure 11) shows good agreement in the formed profile between experiments and simulations. After the first discharge at station A (Figure 12a), the workpiece touches the die at the free end close to station A while the deformation of the part far from the coil are insufficient; the forming height decreases with the increase of X coordinate.

The workpiece acquires a higher forming height after the second discharge at station C (Figure 12b); the forming height increases slightly and then come to steady with the increase of X coordinate. Comparison between experimental and simulated results shows that the established FE model for the entire EMIF process of the stiffened panels is reliable.

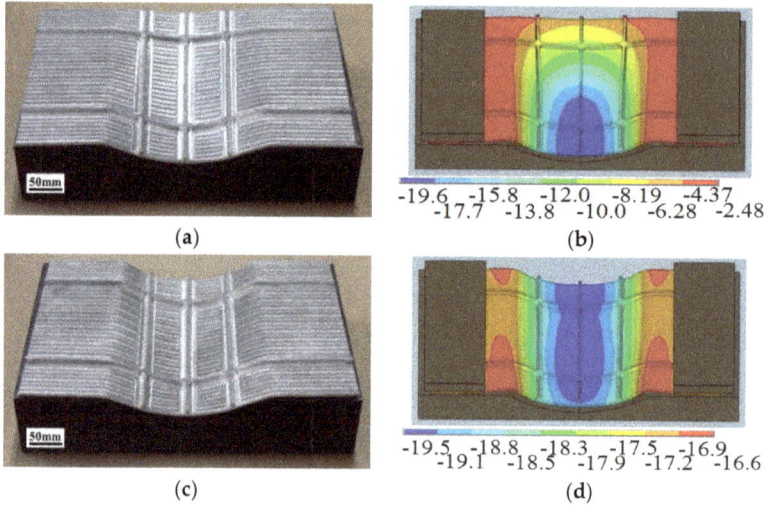

Figure 12. Profile comparisons between the experimental and simulated results: (**a,b**) experimental and simulated results after station A, respectively; and (**c,d**) experimental and simulated results after station C, respectively. The units for (**b**) and (**d**) are both millimeters.

To quantify the comparison, Figure 13 shows the forming depths of the panel along the X center web-bottom path and the Y center web-bottom path after discharges at the first and then second stations. The experimental data at 16 and 15 equal-interval points are measured. Figure 13a,b shows the comparison of the simulated results with the experimental data after discharging at the first station (station A) along the two paths. It can be observed from Figure 13a,b that the simulation and experimental results coincide well with a maximum error of only 6.5% along X direction path and 9.2% along Y direction path, respectively. Figure 13c,d shows the comparison after discharging at the second station (station B). It can be found that the maximum error is 5.4% and 7.2% along X and Y direction paths, respectively. This comparison indicates that the model established in this study is reliable.

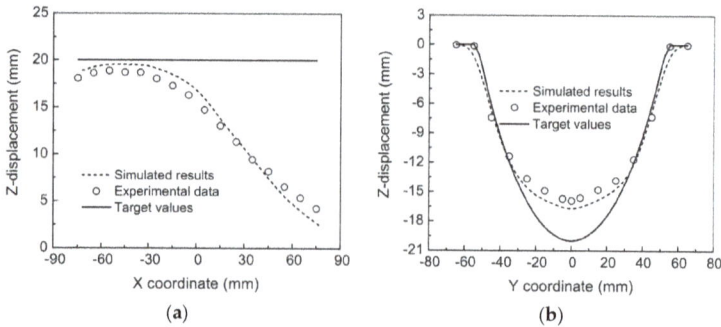

Figure 13. *Cont.*

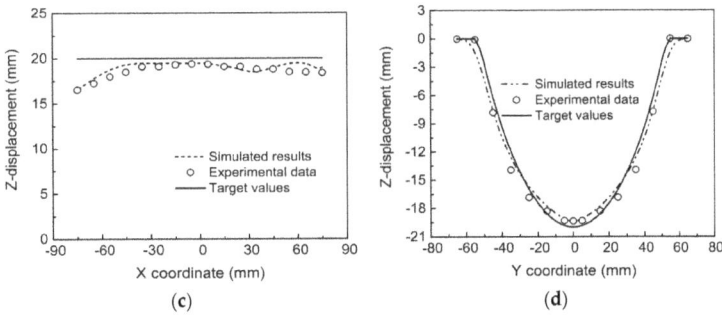

(c)   (d)

Figure 13. Comparison of Z-displacement between experimental and simulated results along X center web-bottom path (a,c) and Y center web-bottom path (b,d): (a,b) after the first station; and (c,d) after the second station.

In addition, the thickness distribution in the web between experimental and simulated results after discharging at the second station is quantified here, which is shown in Figure 14. It can be found that along the X center web-bottom path (Figure 14a), there are local thinning both in the center ($X = 0$) and the positive free end ($X = 75$ mm); local thickening occurs mainly in the negative free end ($X = -75$ mm). The maximal relative error between the simulated and experimental results is 2.1%. From Figure 14b, along the Y center web-bottom path, there are obvious thinning phenomena along almost the entire path, except for the center position, and the maximal relative error is 3.1%. The comparisons of the web thickness between the simulated and experimental results verify again the reliability of the established model.

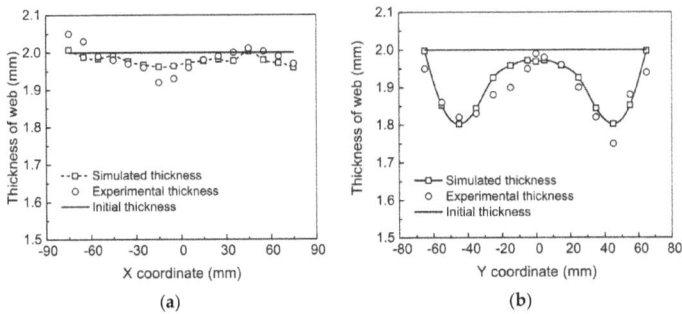

(a)   (b)

Figure 14. Comparison of web thickness between experimental and simulated results along: X center web-bottom path (a); and Y center web-bottom path (b).

3.3. Determination of the Discharge Time and Deformation Time

The discharge current of the coil in the experiment is taken as the input in the simulation. The measured coil current is shown in Figure 15, and the detailed description of the current characteristic is presented by Cao et al. [28]. In order to improve the calculation efficiency of the model, especially the EM model, the current–time curve in the simulation is also idealized, according to the research of Oliveira et al. [22]. Only a part of the actual current data is adopted in the EM field, and the rest during the inertial deformation are set to zero in the structural field. In this study, the discharge time is set to 0.45 ms. It can be seen from Figure 15 that the change trend of the coil generally characterizes a single sine shape, which is with a large increase rate before the peak value and then decayed gradually and at last comes to zero at the end of the forming. The current takes only 75 µs to reach the peak, which is useful to obtain a high strain rate in a short time.

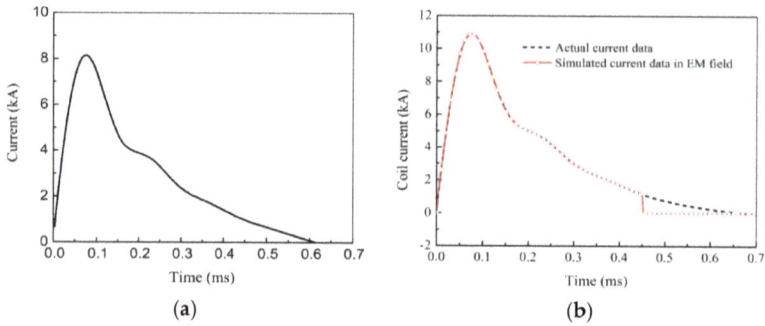

Figure 15. Experimental and simulated coil current data ((**a**) U = 9.9 kV, C = 80 µF; and (**b**) U = 12 kV, C = 80 µF).

Due to the high-speed forming of EMF, the inertial effect of the workpiece is the major factor to keep the deformation proceeding when the action of the electromagnetic force disappears or can be neglected. It is necessary to determine the total deformation time, which consists of the discharge time and the inertial deformation time. Figure 16 demonstrates the evolution of the displacement and velocity with time under different voltages and the capacitance of each voltage maintains 320 µF, which is sampled at the center node at the top of the rib along the X center rib-top path. It can be seen from Figure 16 that the displacement mainly produces before the half of the discharge time; the effect of the voltage on the displacement time is close to each other, and the displacement changes no more than 1 mm after t = 0.75 ms under the voltage of 4.5 kV and t = 0.85 ms under the other two voltages; similar rules can also be observed with the velocity. The fluctuation of the displacement in period after the calculated total deformation time is mainly caused by the explicit algorithm adopted in the dynamic analysis in this study, and thus it should be neglected. In the simulation, three times discharges take place in a sequential but discontinuous mode, and the time of each discharge lasts 0.75 ms, 0.85 ms and 0.85 ms, respectively. That is to say, 0–0.75 ms is the first discharge period, 0.75–1.60 ms the second period and 1.60–2.45 ms the third period.

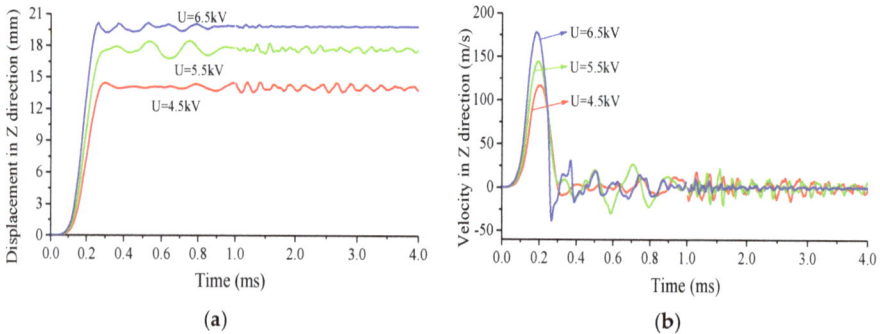

Figure 16. Change of central node's: displacement (**a**); and velocity (**b**) at different U (C = 320 µF).

4. Results and Discussion

In order to study the influence of ribs on the formation of the stiffened panels, four types of panels are simulated here and the results are analyzed contrastively. The simulation conditions are listed in Table 4. The current data in Figure 15 are used as the input in the simulation.

Table 4. Basic simulation conditions for EMF.

Discharge Capacity (µF)	X Coordinate at the 1st Pass (mm)	Voltage at the 1st Pass (kV)	X Coordinate at the 2nd Pass (mm)	Voltage at the 2nd Pass (kV)	X Coordinate at the 3rd Pass (mm)	Voltage at the 3rd Pass (kV)
80	−20	9.9	−20	12	+20	12

4.1. Electromagnetic Force Distribution

Figure 17 shows the peak electromagnetic forces loading in the four panels during the first discharge. In this section, the distribution of electromagnetic forces for the grid-rib is analyzed via comparison with flat panel, X-rib panel and Y-rib panel. For three stiffened panels, the force distributions include results in the entire panel and the web only. As shown in Figure 17a, the force distribution on FP exhibits several "rings"; the maximal value locates in the region corresponding to the 2/3 coil radius; the forces in the center of the "rings", corresponding to the 1/6 of the coil radius, are near zero. From Figure 17b, different from that on FP, the force on XP mainly concentrates on the ribs, and the maximal value is approximately three times larger than that on the web, which can be attributed to the smaller distance between the rib and the coil and the larger one between the web and the coil. The force distribution on the web is discontinuous near the side of the X-direction ribs (as defined in Figure 11), where the maximal value concentrates in the region close to the symmetry plane, also corresponding to the 2/3 coil radius. From Figure 17c, force distribution in YP shows that the force also concentrates in the Y-direction former ribs (Figure 11). There is also discontinuous distribution of the force in the web and the large force locates in the region between the former-rib and the free end, instead of the symmetry plane like the case of XP. Because the bending direction is identical to the ribs direction of YP, the force concentrates in the former rib and the web near the former rib, which benefits for the reduction of the forming difficulty of YP. For GP in Figure 17d, the comparison of the force in the ribs of XP and YP shows that the large force concentrates in both of the two directions ribs (X direction side-rib and Y direction former-rib) with GP, which can be treated as a superposition effect of X and Y directions ribs. However, the comparison of the maximal value in the ribs shows that the force in GP are larger than that in the other two stiffened panels, which may be caused by the closed structure with grid ribs availing to reduce the leak of magnetic flux in comparison to the open structure of XP and YP. The distribution of the force in the web is more discontinuous with GP than with the other two panels, while the maximal value locates at the same position with YP. Note that the maximal value in the web of GP is smaller than that of the other three panels due to the fact that the induced eddy current mainly concentrates in the closed structure of the ribs of GP.

It can also be found that the force distributes simultaneously in the ribs and web of the stiffened panels; the maximal forces in the ribs of each stiffened panel are much larger than that in the webs, indicating that the deformation of the stiffened panels is mainly driven by the deformation of the ribs. Considering the electromagnetic force distributing in the entire volume of the conductive materials, Liu et al. [29] showed that the electromagnetic force plays a positive role in improving the formability of sheets. It is the same for promoting the formability of the stiffened panels, which distinguishes from the traditional forming processes of the stiffened panels, such as brake forming process [11], where the force acts on the top surface of the ribs, creep age forming process [2] where the force acts on the bottom surface of the web. In addition, the surface defects of the parts, e.g., mark-off and scratch in brake forming [1], can also be avoided because there is no mechanical contact between the forming tool (the coil) and the workpiece in the EMF process.

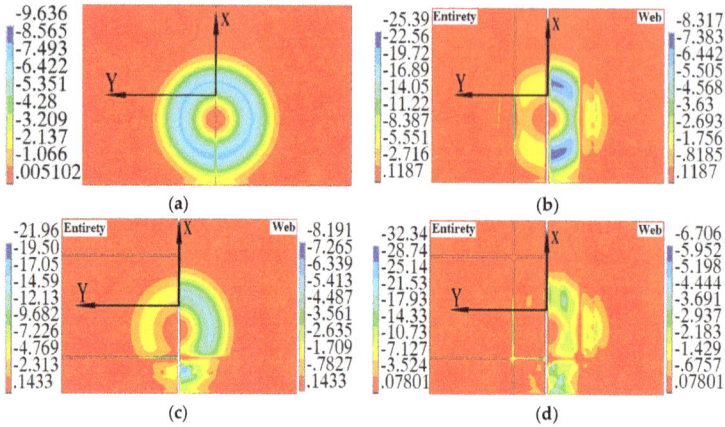

Figure 17. Electromagnetic force (N) distribution in four panels in the first discharge during EMIF: (**a**) Flat panel (FP); (**b**) X-rib panel (XP); (**c**) Y-rib panel (YP); and (**d**) Grid-rib panel (GP).

4.2. Velocity Distribution

In the sheet forming with a die, it is very important to obtain the distribution of sheet velocity for the sake of controlling the deformation of the sheet [30], especially for the high-speed forming process, e.g., electromagnetic forming. By considering that the largest velocity of the grid-rib panel appears in the discharging forming stage for each pass, the Z-velocity distributions of aforementioned four panels during the first discharging forming stage in the EMIF process are studied with the velocity variation characteristics being concerned, which is shown in Figure 18.

From Figure 18a, at the beginning of the discharge (t = 0.05 ms), the velocity exhibits a ring-shape distribution, where the center of the ring corresponds to the center of coil. It can be explained as that the maximal magnetic force in the flat panel corresponds to the half of the coil radius and almost zero at the coil center (Figure 17a). The velocity ring becomes an elliptical ring (t = 0.10 ms) and gradually narrows into an oval at t = 0.15 ms, 1/3 of single discharge time. At t = 0.25 ms, the elliptical region reduces further, the large velocity also appears at the free end of the workpiece adjacent to the coil (the negative free end area, NEA). The maximal velocity reaches 130.5 m/s at t = 0.15 ms in the electromagnetic forces loading stage. Note that there is reverse velocity distribution (along the positive direction of Z axis) in local areas, which firstly distributes around the velocity ring (t = 0.10 ms), and then appears at the free end area of the workpiece far from the coil (the positive free end area, PEA) over time. The maximal value of the reverse velocity occurs in the middle stage of discharge (t = 0.25 ms).

Due to the influence of the X direction ribs, the velocity distribution of the X-rib panel, as shown in Figure 17b, is significantly different from the flat panel (Figure 18a). At the beginning (t = 0.05 ms), the large velocity region in the X-rib panel distributes like two crescents and disconnected in the center-rib. This crescent-shape velocity distribution can be explained as that the initial velocity distribution is greatly related to the distribution of the electromagnetic force in the X-rib panel at the beginning of the forming process, and thus the both concentrate in the same region (Figure 17b) at t = 0.05 ms. The inhomogeneous velocity distribution will lead to that the velocity near the center rib lags behind the large velocity region (i.e., the two side ribs). Then, the high velocity concentration area in the two crescents turns into two small elliptical regions at t = 0.10 ms, locating near the X-direction side ribs. At t = 0.25 ms, the large velocity region extends from the elliptical region to the NEA with a peak shape. The largest velocity also appears at t = 0.15 ms, reaching up to 111.6 m/s, which is smaller than the flat panel at the same time. It can be explained as that the increase of structural stiffness caused by the ribs and the velocity lag of the center rib increases the forming difficulty of the

X-rib panel, though the largest electromagnetic force on the X-rib panel is about three times larger than that on the flat panel. In addition, Figure 18b shows a reverse velocity distribution, which mainly distributes in PEA at $t = 0.15$ ms.

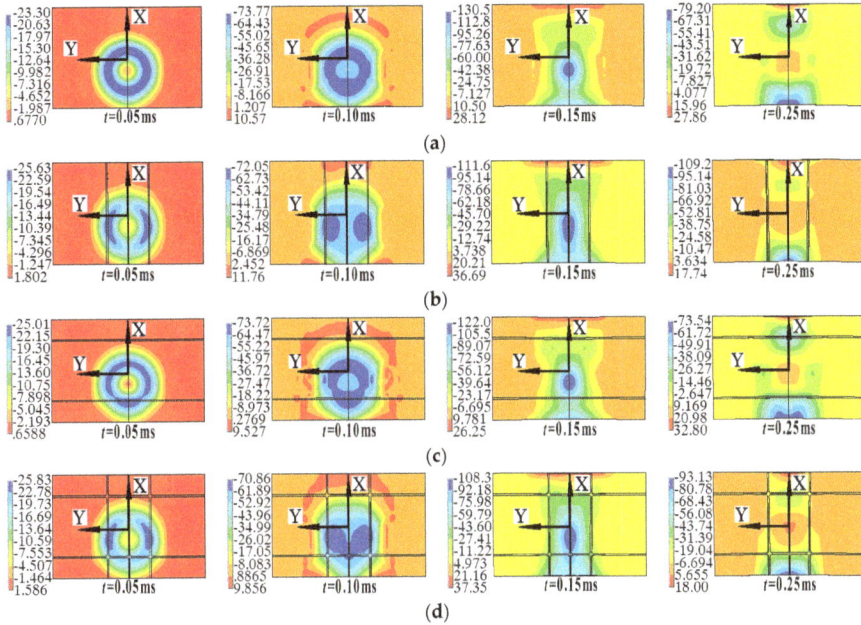

Figure 18. Z-velocity (m/s) distribution of: flat panel (**a**); X-rib panel (**b**); Y-rib panel (**c**); and grid-rib panel (**d**) during EMIF.

Figure 18c shows that the velocity distribution and the largest velocity in the Y-rib panel are similar to those in the flat panel (Figure 18a) at the same time. The difference between the two panels lies in that there is the large velocity in the Y-rib panel former-rib (as defined in Figure 11). The largest velocity of the Y-rib panel is 122 m/s at $t = 0.15$ ms, which is slightly lower than that of the flat panel at the same time. The reverse velocity distribution ($t = 0.10$–0.25 ms) in the Y-rib panel still mainly distributes in PEA, and the largest reverse velocity is 38.9 m/s at $t = 0.35$ ms. Figure 18c indicates that the influence of the Y-direction ribs on the velocity is weaker than that of the X-direction ribs, because the distance between two adjacent X-direction ribs is less than that between two adjacent Y-direction ribs and thus bend deformation of the ribs may occur along the Y direction.

The velocity distribution of the grid-rib panel, as shown in Figure 18d, is similar to that of the X-rib panel (Figure 18b), indicating the velocity distribution in the gird-rib panels is affected by the X-direction ribs and then the Y-direction ribs. Under the influence of the Y-direction ribs, in the initial stage of discharge ($t = 0.05$ ms), the large crescent-shape velocity distribution in the grid-rib panel cuts off by the Y-direction ribs, and then the crescent-shape velocity distribution changes to elliptic distribution area ($t = 0.10$ ms). The reverse velocity distribution of grid-rib panel is consistent with that of the X-rib panel in Figure 18b.

Figure 18 shows that, at $t = 0.25$ ms, the maximal velocity reduces by comparing with the one at $t = 0.15$ ms for all the panels. The decrement value for XP is the smallest, the largest for FP and YP and the middle for GP, which indicates the ribs parallel to the bending direction (Y direction) and the flat panel increase or promote the velocity attenuation, while ribs perpendicular to the bending direction prevent the velocity attenuation. For the grid-rib panel, it can be seen as the superposition effect of the X-direction ribs and the Y-direction ribs.

In summary, as shown in Figure 18, the large velocity distribution of the four panels evolves from the large magnetic force region at the beginning of the process, then to the coil center region, and finally to NEA velocity distribution; the existence of ribs affects the velocity distribution, and the influence of X-direction ribs shows significantly higher than that of the Y direction ribs; there is reverse velocity distribution for all of the panels during the EMIF process. The velocity exceeds 100 m/s for all the panels, which is far greater than that of the traditional forming process, for example the brake forming. The higher velocity with EMF is conducive to reduce the springback [14].

4.3. Forming Height

Forming heights of the four panels along two X direction paths (X center web-bottom path and X side web-bottom path in Figure 11, shorten as Xc path and Xs path, respectively) at different times are shown in Figure 19. It can be seen from Figure 19a that, along the Xc path, the forming height curves for FP and YP almost coincide with each other in the early stage of the first discharge ($t = 0.15$ ms) due to the similar distribution of velocity in Figure 18a,c. The forming height distributes in a bimodal shape, where the peaks correspond to the velocity "ring" at the same time in Figures 18a and 18c, respectively, and the trough between the two peaks locates correspondingly to the coil center. The height curves for XP and GP are almost coincident, distributing in a unimodal shape, where the peak locates in the region corresponding to the coil center. The forming heights of FP and YP are larger than those of the rest two panels along the Xc path. Along the Xs path, the height curves of the four panels exhibit a unimodal distribution, where the position of the peak approximates correspondingly to the coil center. The forming heights at this time are influenced by the distribution of velocity.

From Figure 19b, along both the X direction paths, the forming height of the four panels at the end of the first pass ($t = 0.75$ ms) reduces parabolically with the X coordinate; by comparing to the early stage of discharge, the forming height increases obviously as the result of the inertial effect because the velocity reduces after $t = 0.15$ ms (Figure 18); the forming height along the Xc path remains larger than that along the Xs path. Figure 19b shows that the forming height in NEA after the first discharging at station A are larger than the other regions for the four panels due to the inertial effect. It can also be found that the forming height for FP and YP, XP and GP has the identical values in pairs along the two paths at the beginning of discharging stage, but with different distributions at the end of the inertial deformation stage. The final forming height for the four panels follows the sequence of XP, FP, GP and YP (from the largest to the smallest) as the result of the different velocity attenuation effect of ribs as shown in Figure 18.

From Figure 19c, in the second discharge at the same station ($t = 0.90$ ms), the forming height of the four panels along both of the two X direction paths continues to increase. Along the Xc path, the forming height of the four panels reduce parabolically with X coordinate, and the largest height appears in NEA. Along the Xs path, the forming height of the Y-rib panel decreases parabolically with X coordinate, but for the other three panels it increases slightly and then decreases. The forming height reduces and changes to the reverse direction (along the positive direction of Z axis) in the region far from the coil (PEA). The height differentiations at the end of the first station discharge (Figure 19b) diminish after the second discharge at the same station.

From Figure 19d, along both the two X direction paths, the forming heights of the four panels at the end of the second pass ($t = 1.60$ ms) reduce parabolically with the increase of the X coordinate. Along the Xc path, the order of the forming height in NEA from the largest to smallest is similar to that in Figure 19b at the same place; the forming height for the grid-rib panel in PEA is 0.56 mm, and there are reverse deformation for the other three panels. Along the Xs path in NEA, the maximal forming height appears with FP (10.95 mm); there is also are reverse deformation for the other three panels. The shape changes of the height curves in Figure 19d show that the uniformity improves after the second discharge in comparison to the on at the end of the first discharge in Figure 19b.

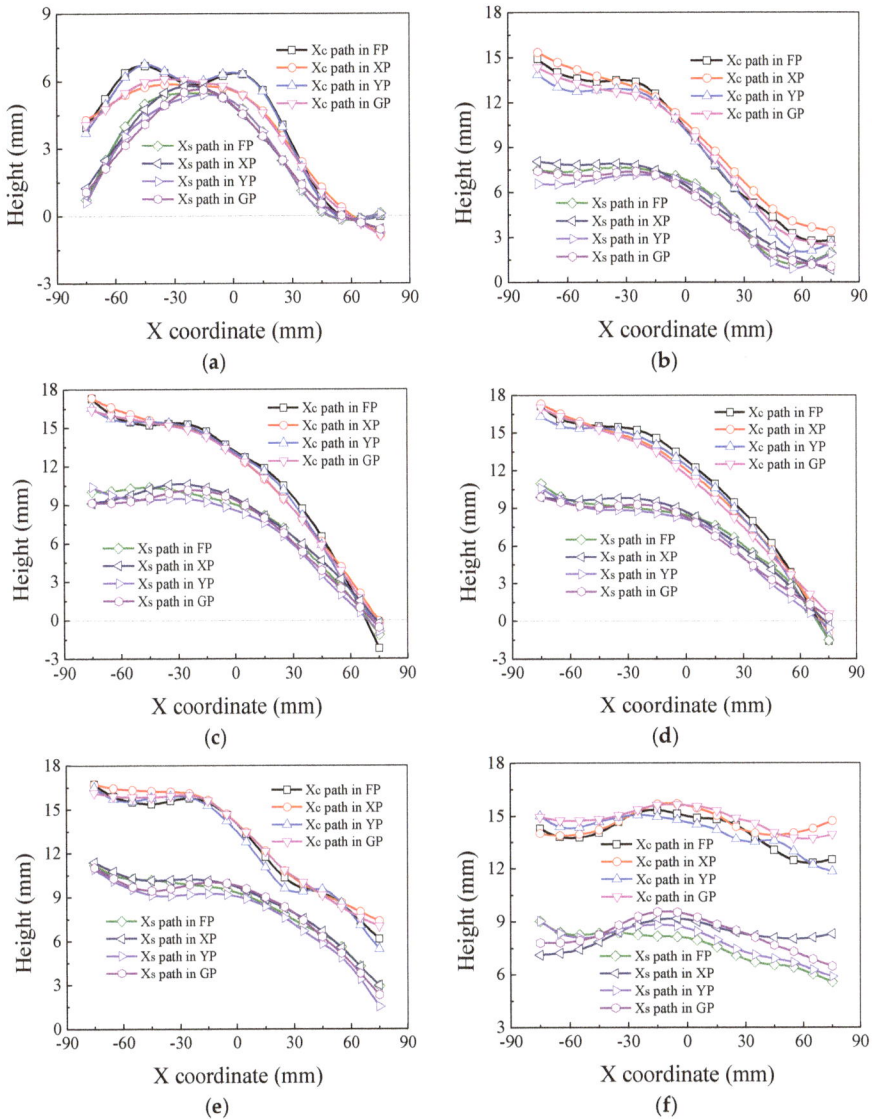

Figure 19. Z direction height distribution along X direction paths of four panels during EMIF: (**a**) *t* = 0.15 ms; (**b**) *t* = 0.75 ms; (**c**) *t* = 0.90 ms; (**d**) *t* = 1.60 ms; (**e**) *t* = 1.75 ms; and (**f**) *t* = 2.45 ms. Xc and Xs represent X center web-bottom path and X side web-bottom path, respectively.

From Figure 19e, along both the two X direction paths, the forming height of the four panels in the early stage of the discharge at the second station (*t* = 1.75 ms) changes slightly in comparison to the one at *t* = 1.60 ms, and reduces also parabolically with the increase of X coordinate, however, the reduction extent is less than that in Figure 19c,d. There is no reverse deformation.

It can be seen from Figure 19f that, the height curves for the four panels distribute in the style of the letter "W" at the end of the second station (*t* = 2.45 ms), where the position of the peak locates in the geometric center of the panels. Along both of the two X direction paths, the forming height increases mainly in the region under the coil in the new station. The forming height along the X_c path reduces in the region far from the coil (NEA), due to the effect of the main deformation region. Comparison of the forming heights at $t = 1.75$ ms and $t = 2.45$ ms in NEA shows that there is also reverse deformation in the third discharge stage. In addition, the different distributions occur again at the end of the second station forming stage.

From Figure 19, there are reverse deformation regions, mainly locating in the free end region and resulting from the influence of deforming region on the deformed or undeformed regions. It is obvious that the reverse deformation reduces the forming height, and thus is disadvantage to improve the fittability of the workpiece to the die. In addition, there are less electromagnetic forces in the free ends during the entire forming process, which indicates that the deformation of these regions is mainly driven by the inertial effect of the panels.

Figure 20 shows the forming height of four panels along two Y direction paths (Y former web-bottom path and Y latter web-bottom path, shorten as Yf path and Yl path respectively) at different times during the EMIF process. It can be seen from Figure 20 that the distribution curves of forming height for the four panels along both of the two Y direction paths exhibit a parabolically declining law. From Figure 20a, in the early stage of the first discharge ($t = 0.15$ ms), the forming height along the Yf path is larger than that the one along the Yl path.

From Figure 20b, at the end of the first pass forming ($t = 0.75$ ms), the increasing rates of the forming height of the four panels are larger than that at $t = 0.15$ ms in Figure 20a. Along the two Y direction paths, the forming height increases about twice from $t = 0.15$ ms to $t = 0.75$ ms in NEA, which can be attributed to the inertial effect on the deformation region.

From Figure 20c, the forming height of the four panels in the early stage of the second discharge ($t = 0.90$ ms) continues to increase in comparison to the first pass forming. The large forming height mainly occurs in the rib center along the two paths.

At the end of the second pass forming ($t = 1.60$ ms), as shown in Figure 20d, along the Yf path, except for the flat panel, the forming height for the rest three panels decreases in comparison to the early stage of the second discharge in Figure 20c, which indicates that there is reverse deformations for these three panels. Along the Yl path, the forming height reduces in comparison to the height in the second discharge, which also indicates there is reverse deformation.

From Figure 20e, in the early stage of the discharge at the second station ($t = 1.75$ ms), along the Yf path, except for the flat panel, the forming height for the rest three panels increases in comparison to the second pass forming in Figure 20d, while the forming height of the flat panel decreases, which indicates that there is reverse deformations for the flat panel. Along the Yl path, the forming height in the rib center increased for all of the four panels in comparison to the second pass forming.

From Figure 20f, at the end of the second station forming ($t = 2.45$ ms), the forming height curves are very close to each other along the two different Y direction paths. Along the Yf path, the forming height of the four panels decreases in comparison to the early stage of the discharge at the current station in Figure 20e, which indicates that there is reverse deformations for all the panels. Along the Yl path, the forming height in the rib center increases significantly in comparison to the height in the early stage of discharge at the present station.

Metals **2016**, *6*, 267

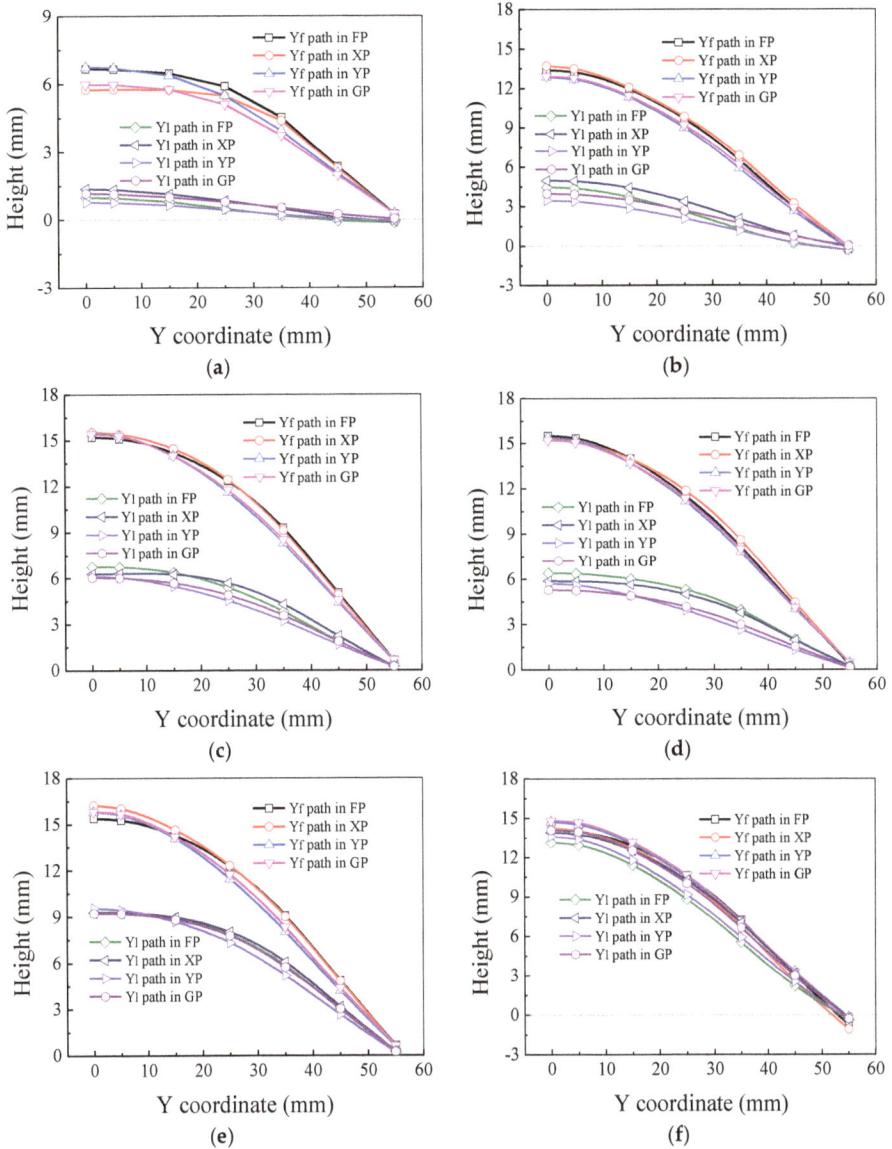

Figure 20. *Z* direction height distribution along *Y* direction paths of four panels during EMIF:
(**a**) *t* = 0.15 ms; (**b**) *t* = 0.75 ms; (**c**) *t* = 0.90 ms; (**d**) *t* = 1.60 ms; (**e**) *t* = 1.75 ms; and (**f**) *t* = 2.45 ms.
Yf and Yl represent *Y* former web-bottom path and *Y* latter web-bottom path, respectively.

4.4. Comparison with Brake Forming

As mentioned in Section 1, the brake forming is widely used in the aircraft industry due to its high applicability. In this section, a comparison of the simulated results between one-pass brake forming (BF) and one-discharge electromagnetic forming (EMF) is conducted. The sizes of the grid-rib panel in Figure 2 are used, except for the height of the ribs. In this section, the height of the ribs is set to 10 mm in order to observe the instability more easily. In the EMF model, the input parameters, including geometric sizes of the coil, die and blank holder block, electromagnetic parameters, etc., are identical to those in Table 3 and Section 4; the coordinate of the initial station at the coil center is (0,0); the data in Figure 15 are used as the input current for the EMF model. In the BF model, the sizes of the die and blank holder block are similar to those in the one-discharge EMF process. The axis of the punch is perpendicular to the width direction of the grid-rib panel, as shown in Figure 21. From Figure 22, the Z direction displacements in the two forming processes reach the same value (11.4 mm) at the end of EMF.

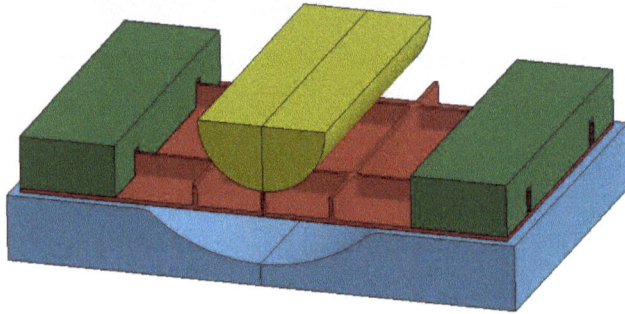

Figure 21. Geometric model of the brake forming (BF) process.

As shown in Figure 22a, the tensile and compressive normal stresses along the Y direction with BF are larger than those with EMF, especially for the compressive stress. The large compressive stress concentrates at the cross position of the center rib along the X-direction and Y-direction ribs, which means buckling is more inclined to occur with BF than with EMF. However, large tensile or compressive stress with EMF mainly concentrates in the web under the X-direction ribs. The buckling of the web can be prevented or weakened by the ribs. The distribution of the equivalent elastic strain in Figure 22b shows that the maximal elastic strain with BF is larger than that with EMF, indicating a smaller springback during the EMF process. Moreover, the elastic strain mainly concentrates in the ribs for the BF, more easily leading to springback than the EMF where the elastic strain concentrates in the web. Figure 22c shows that plastic strain mainly occurs at the cross position of the X and Y-direction ribs with BF; with EMF, the maximal plastic strain mainly concentrates in the transition fillet area, however, the plastic strain level of the cross ribs in the concerned deformation regions is obviously lower than that with BF.

Due to the characteristics of the distribution of stress and strain with BF as discussed above, buckling and springback are the main defects that limit the application of BF to form the complex parts, e.g., the stiffened panels with high ribs. Comparison of the distributions of stress and strain with the brake forming to those with the electromagnetic forming in Figure 22 implies that the advantages of smaller elastic strain in the web and plastic strain in the cross ribs make the electromagnetic forming technology competent for forming the stiffened panels.

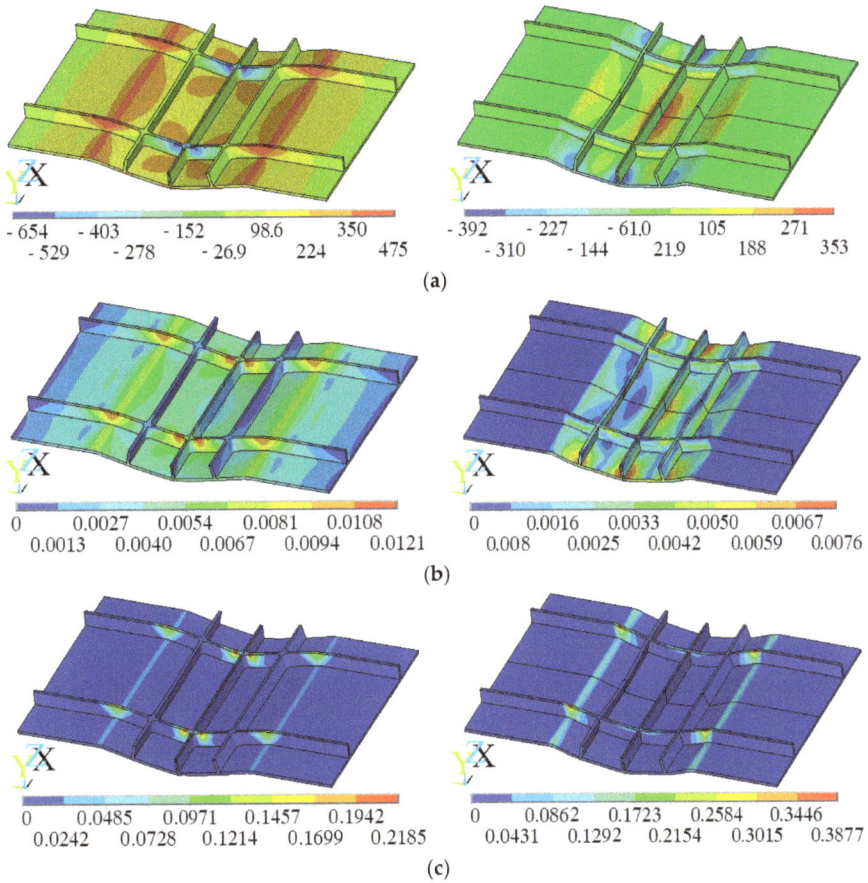

Figure 22. Comparison of simulated results between BF (left) and EMF (right): (**a**) σ_y (MPa); (**b**) equivalent elastic strain; and (**c**) equivalent plastic strain.

5. Conclusions

The forming rules of the bi-directional stiffened panels with grid ribs during the EMIF process are studied by FE simulations aided by experiments. The distributions of the electromagnetic force, the forming velocity and the height are analyzed by the comparison among the grid-rib panel, flat panel and two uni-directional stiffened panels (the panel with the X-direction ribs, the panel with the Y-direction ribs). The main conclusions are as follows:

(1) Different from traditional stiffened panels forming process, there are electromagnetic body forces loading in the ribs and webs simultaneously, which increases the forming ability of the stiffened panels. The large force of GP concentrates in both the two direction ribs, indicating the superposition effect of the two direction ribs.

(2) The velocity distribution of the grid-rib panel is mainly affected by the X-direction ribs, next is the Y-direction ribs; the ribs along the bending direction and the flat panel promote the velocity attenuation while the ribs perpendicular to the bending direction prevent the velocity attenuation. For the grid-rib panel, it can be seen as the superposition of the X-direction ribs and the Y-direction ribs.

(3)　The decrease of the forming height at the free ends of the four panels at the end of each pass indicates that the mutual influence exists between the region undergoing deformation and the deformed or undeformed regions; the reverse velocity distribution exists for all the panels during the EMIF process, which mainly locates in the two free ends and thus is harmful for improving the forming height. Therefore, the reverse velocity should be avoided.

(4)　The comparison of the simulation results with the brake forming to those with the electromagnetic forming shows that EMF has a small elastic deformation in the web and a smaller plastic deformation in the ribs in the deformation region, which helps prevent or reduce the defects, such as springback and buckling that commonly happen in the traditional brake forming process.

Acknowledgments: This work is supported by the National Science Fund for Distinguished Young Scholars of China (51625505), the Key Program Project of the Joint Fund of Astronomy and National Science Foundation of China (Project U1537203) and the National Key Basic Research Program of China (973 Program, Grant No. 2011CB012804). The authors would also express thanks to the Wuhan National High Magnetic Field Center of Huazhong University of Science and Technology, for the help in the forming experiment. The author also wishes to thank Hongwei Li for his kindly help in improving this article.

Author Contributions: Jinqiang Tan performed the experiments, simulation and wrote this paper under the guidance of Mei Zhan; Shuai Liu assisted in performing experiments and analyzing simulation results.

Conflicts of Interest: The authors declare no conflict of interest.

References

1.　Munroe, J.; Wilkins, K.; Gruber, M. *Integral Airframe Structures (IAS)—Validated Feasibility Study of Integrally Stiffened Metallic Fuselage Panels for Reducing Manufacturing Costs*; Boeing Commercial Airplane Group: Seattle, WA, USA, 2000.

2.　Zhan, L.H.; Lin, J.G.; Dean, T.A. A review of the development of creep age forming: Experimentation, modeling and applications. *Int. J. Mach. Tools Manuf.* **2011**, *51*, 1–17. [CrossRef]

3.　Erbel, F.; Gardiner, S.; Campanile, G.; Surdon, G.; Venmans, M.; Prangnell, P. Ageformable panels for commercial aircraft. *J. Aerosp. Eng.* **2008**, *222*, 873–886.

4.　Ho, K.C.; Lin, J.G.; Dean, T.A. Modelling of springback in creep forming thick aluminum sheet. *Int. J. Plast.* **2004**, *20*, 733–754. [CrossRef]

5.　Lam, A.C.L.; Shi, Z.S.; Yang, H.L.; Wan, L.; Davies, C.M.; Lin, J.G.; Zhou, S.J. Creep-age forming AA2219 plates with different stiffener designs and pre-form age conditions: Experimental and finite element studies. *J. Mater. Proc. Technol.* **2015**, *219*, 155–163. [CrossRef]

6.　Ribeiro, F.C.; Marinho, E.P.; Inforzato, D.J.; Costa, P.R.; Batalha, G.F. Creep age forming: A short review of fundaments and applications. *J. Achiev. Mater. Manuf. Eng.* **2010**, *43*, 353–361.

7.　Yamada, Y.; Takahashi, T.; Ikeda, M.; Sugimoto, S.; Ohta, T. Development of shot peening for wing integral skin for continental business jets. *Mitsubishi Heavy Ind. Tech. Rev.* **2002**, *39*, 57–61.

8.　Miao, H.Y.; Demers, D.; Larose, S.; Perron, C.; Levesque, M. Experimental study of shot peening and stress peen forming. *J. Mater. Proc. Technol.* **2010**, *210*, 2089–2102. [CrossRef]

9.　Wang, X.F.; Guo, X.L.; Chen, G.N.; Silvanus, J. Remark of integral panel forming. *Mod. Manuf. Technol. Equip.* **2008**, *3*, 1–4.

10.　Lai, S.B.; Chen, T.X.; Yu, D.Y. Dynamic explicit analysis method for roll bending forming of integrally stiffened panel with rubber filler. *Spacecr. Eng.* **2012**, *3*, 41–47.

11.　Yan, Y.; Wan, M.; Wang, H.B.; Huang, L. Design and optimization of press bend forming path for producing aircraft integral panels with compound curvatures. *Chin. J. Aeronaut.* **2010**, *23*, 274–282.

12.　Yu, Y.; Wang, H.B.; Min, M. Prediction of stiffener buckling in press bend forming of integral panels. *Trans. Nonferr. Met. Soc. China* **2011**, *21*, 2459–2465.

13.　Yu, Y.; Wang, H.B.; Min, M. FEM modeling for press bend forming of doubly curved integrally stiffened aircraft panel. *Trans. Nonferr. Met. Soc. China* **2012**, *22*, s39–s47.

14.　Daehn, G.S. High Velocity Metal Forming. In *ASM Handbook, Volume 14B, Metalworking: Sheet Forming*; ASM International: Columbus, OH, USA, 2006; pp. 405–418.

15.　Psyk, V.; Risch, D.; Kinsey, B.L.; Tekkaya, A.E.; Kleiner, M. Electromagnetic forming—A review. *J. Mater. Proc. Technol.* **2011**, *211*, 787–829. [CrossRef]

16. Cui, X.H.; Mo, J.H.; Li, J.J.; Zhao, J.; Xiao, S.J. Produce a large aluminum alloy sheet metal using electromagnetic-incremental (EM-IF) forming method: Experiment and Numerical simulation. In Proceedings of the 5th International Conference on High Speed Forming, Dortmund, Germany, 24–26 April 2012; pp. 59–70.

17. Kamal, M.; Daehn, G.S. A uniform pressure electromagnetic actuator for forming flat sheets. *J. Manuf. Sci. Eng.* **2007**, *129*, 369–379. [CrossRef]

18. Zhao, J.; Mo, J.H.; Cui, X.H.; Qiu, L. Research on numerical simulation and forming uniformity of electromagnetic incremental tube bulging. *J. Plast. Eng.* **2012**, *19*, 92–99. (In Chinese)

19. Cui, X.H.; Mo, J.H.; Li, J.J.; Zhao, J.; Zhu, Y.; Huang, L.; Li, Z.W.; Zhong, K. Electromagnetic incremental forming (EMIF): A novel aluminum alloy sheet and tube forming technology. *J. Mater. Proc. Technol.* **2014**, *214*, 409–427. [CrossRef]

20. Fenton, G.K.; Daehn, G.S. Modeling of electromagnetically formed sheet metal. *J. Mater. Proc. Technol.* **1998**, *75*, 6–16. [CrossRef]

21. Ma, S.G.; Zhan, M.; Tan, J.Q.; Zhang, P.P. 3D sequentially coupled model to simulate electromagnetic sheet forming. *J. Plast. Eng.* **2012**, *19*, 54–58. (In Chinese)

22. Oliveira, D.A.; Worswick, M. Electromagnetic forming of aluminium alloy sheet. *J. Phys. IV France* **2003**, *110*, 293–298. [CrossRef]

23. Yu, H.P.; Li, C.F.; Deng, J.H. Sequential coupling simulation for electromagnetic mechanical tube compression by finite element analysis. *J. Mater. Proc. Technol.* **2009**, *209*, 707–713.

24. Cui, X.H.; Mo, J.H.; Xiao, S.J.; Du, E.H. Numerical simulation of electromagnetic sheet bulging based on FEM. *Int. J. Adv. Manuf. Technol.* **2011**, *57*, 127–134. [CrossRef]

25. ISO 209:2007. *Aluminium and Aluminum Alloy—Chemical Composition*; ISO: Geneva, Switzerland, 2007.

26. Tan, J.Q.; Zhan, M.; Liu, S.; Huang, T.; Guo, J.; Yang, H. A modified Johnson-cook model for tensile flow behaviors of 7050-T7451 aluminum alloy at high strain rates. *Mater. Sci. Eng. A* **2015**, *631*, 214–219. [CrossRef]

27. ISO 6892-1:2009. *Metallic Materials-Tensile Testing. Part 1: Method of Test at Room Temperature*; ISO: Geneva, Switzerland, 2009.

28. Cao, Q.L.; Han, X.T.; Lai, Z.P.; Xiong, Q.; Zhang, X.; Chen, Q.; Xiao, H.X.; Li, L. Analysis and reduction of coil temperature rise in electromagnetic forming. *J. Mater. Proc. Technol.* **2015**, *225*, 185–194. [CrossRef]

29. Liu, D.H.; Zhou, W.H.; Li, C.F. Influence of body force effect of the pulsed magnetic forces of the dynamic forming limits of AA5052 sheets. *J. Plast Eng.* **2013**, *6*, 62–67.

30. Neugebauer, R.; Bouzakis, K.D.; Denkena, B.; Klocke, F.; Sterzing, A.; Tekkaya, A.E.; Wertheim, R. Velocity effects in metal forming and machining processes. *CIRP Ann. Manuf. Technol.* **2011**, *2*, 627–650. [CrossRef]

metals

MDPI

Article

The Hot Deformation Activation Energy of 7050 Aluminum Alloy under Three Different Deformation Modes

Deli Sang [1,2], Ruidong Fu [1,2,*] and Yijun Li [1,2]

1   State Key Laboratory of Metastable Materials Science and Technology, Yanshan University, Qinhuangdao 066004, China; S13930385984@126.com (D.L.S.); liyijun1987@126.com (Y.J.L.)
2   College of Materials Science and Engineering, Yanshan University, Qinhuangdao 066004, China
*   Correspondence: rdfu@ysu.edu.cn; Tel.: +86-335-807-4792; Fax: +86-335-807-4545

Academic Editor: Nong Gao
Received: 21 January 2016; Accepted: 23 February 2016; Published: 29 February 2016

Abstract: In this study, the hot deformation activation energy values of 7050-T7451 aluminum alloy, calculated with two different methods under three deformation modes, were compared. The results showed that the hot deformation activation energy values obtained with the classical constitutive equation are nearly equivalent under the hot tensile, compression, and shear-compression deformation modes. Average values exhibited an obvious increase when calculated with the modified constitutive equation because it can reflect the variation of activation energy with deformation conditions such as deformation temperature, strain rate and strain state. Moreover, the values under tensile and compression deformation modes were nearly the same regardless of the calculation method. The higher average value under the shear-compression deformation mode with modified equation indicates that the strain state has a significant effect on the hot deformation activation energy. In addition, when the activation energy was investigated for various deformation conditions, the effect of the strain state on the activation energy was more significant. Under a certain condition, the activation energy was the same for the three deformation modes.

Keywords: activation energy; hot deformation; aluminum alloy; deformation mode

1. Introduction

Actual production processes of metals and alloys are always accompanied by various hot deformation processes such as forging, rolling, extrusion, and welding. Therefore, the hot deformation behavior of metals and alloys has been an important scientific issue. As an indication of the degree of difficulty during hot deformation processes, hot deformation activation energy is typically used to estimate the hot workability and to optimize the hot working process of metals and alloys [1–3]. For example, a typical application of activation energy is that it is used to derive the constitutive equation, which generally consists of the response of flow stress, strain rate, and deformation temperature. Among the various constitutive models and equations available, the hyperbolic sine law proposed by Sellars and McTegart [4] is the most applicable for a wide range of stresses and has been extensively used to study the hot deformation behavior of various metals and alloys. Moreover, activation energy is always treated as a constant in the constitutive equation regardless of the applied hot deformation conditions such as temperature, strain rate, and load state.

However, the hot deformation activation energy is correlated with not only material composition and heat treatment conditions but also deformation conditions and deformation mode. For instance, Reyes-Calderón *et al.* [5] found that the presence of microalloying elements such as Nb, V, and Ti increased the activation energy of the nonmicroalloyed material from 366 kJ·mol^{-1} to 446 kJ·mol^{-1} of V-microalloyed twinning induced plasticity steel. For aluminum alloys, the hot deformation activation energy is affected significantly by the initial state of the material because the solute atoms diffuse to the dislocation cores and are pinned at the dislocations [6–9]. Recently, Zhang *et al.* [10] found that the hot deformation activation energy of Ti-15-3 titanium alloy increases with increasing temperature and decreasing strain rate. Shi *et al.* [11] found that the hot deformation activation energy of an AA7150 aluminum alloy is not constant and decreases with increasing deformation temperature and strain rate. Therefore, a modified constitutive equation was proposed, wherein the term for activation energy is treated as a variant involving temperature and strain rate.

Recently, considerable work has been performed on the strain state effects. Although the research has primarily focused on aluminum alloys under cold working conditions, variations in strain state have a significant effect on the deformation microstructure, particularly the development of dislocation structures [12,13]. Therefore, as a reflection of the energy barrier to dislocation motion for metals and alloys during hot deformation, the hot deformation activation energy may be affected by the deformation modes for different strain states. However, it has not yet been verified.

In this paper, hot tensile (HT), hot compression (HC), and hot shear-compression (HSC) tests of a 7050-T7451 aluminum alloy have been performed at different temperatures and strain rates. The hot deformation activation energy under these three deformation modes was calculated using the classical constitutive equation and the modified equation. The effect of deformation conditions, modes, and calculation methods on activation energy was discussed.

2. Material and Methods

The chemical composition (wt. %) of the investigated 7050-T7451 aluminum alloy is shown in Table 1. The alloy was obtained in an as-rolled plate with a gauge thickness of 22 mm. The hot deformation specimens were prepared by using wire-electrode cutting. The HT and HC specimens are 100 mm and 12 mm in length and 6 mm and 8 mm in diameter, respectively. The HSC test is same to the classical HC test. However, because of the special design of the HSC specimen [14] schematically presented in Figure 1, that shear and compression strain are induced simultaneously by axial force in the specimen. The three hot deformation tests were performed on a Gleeble 3500 thermo-mechanical simulator (Dynamic Systems Inc. (DSI), Poestenkill, NY, USA) at axial strain rates and deformation temperatures ranging from 0.01 s^{-1} to 10 s^{-1} and from 523 K to 723 K, respectively. During the tests, all the specimens were heated to the selected deformation temperature at a heating rate of 20 K·s^{-1} and held at the deformation temperature for 30 s prior to deformation. The thermal cycle with no solution treatment and short dwell time is on account of modeling the process such as friction welding process [15]. The HT, HC, and HSC specimens were deformed to snap break, height reductions of 50%, and a total axial displacement of 4.5 mm, respectively. Thereafter, all the specimens were immediately water quenched.

Table 1. Chemical composition (wt. %) of the 7050 aluminum alloy.

Chemical Composition (wt. %)									
Si	Fe	Cu	Mn	Mg	Cr	Zn	Ti	Zr	Al
0.12	0.15	2.0~2.6	0.1	1.9~2.6	0.04	5.7~6.7	0.06	0.08~0.15	Bal.

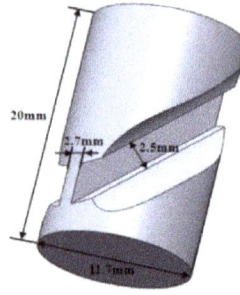

Figure 1. Schematic presentation of the shear-compression specimen.

3. Results

3.1. Flow Stress during the Three Hot Deformation Processes

The flow stress *vs.* strain curves during the three hot deformation processes are shown in Figure 2. As can be seen, in all the cases, the flow stress increases rapidly in the initial stage of deformation then either decreases to a certain degree or remains constant after the peak stress. Moreover, the flow stress increases with an increase in strain rate and decrease in deformation temperature. Flow stress is a function of dislocation density [16]. During the hot deformation process, the dislocation density is affected by the primary microstructure, temperature, and strain rate. If these parameters are constant, the variation of flow stress is related to the microstructure evolution. Before the peak stress is reached, the dislocations multiply drastically, and the work hardening process prevails over the softening process. Therefore, the flow stress increases rapidly. After the initial deformation stage, the work hardening and softening processes induced by dynamic recovery (DRV) and dynamic recrystallization (DRX) compete. Moreover, the flow stress still increases, but the rate of increase decreases continuously. After the peak value, the softening process becomes dominant. Initially, the peak stresses were determined from the flowing stress *vs.* strain curves under different deformation conditions and were used to calculate the activation energy. The choice is on account of two aspects: In aluminum alloys, the peak stress is generally used and refers to a dynamic equilibrium between the work hardening and dynamic softening processes [17,18]. All the deformation processes can be assumed to be homogeneous up to the peak stress excluding the effects of strain state induced by bulging or necking.

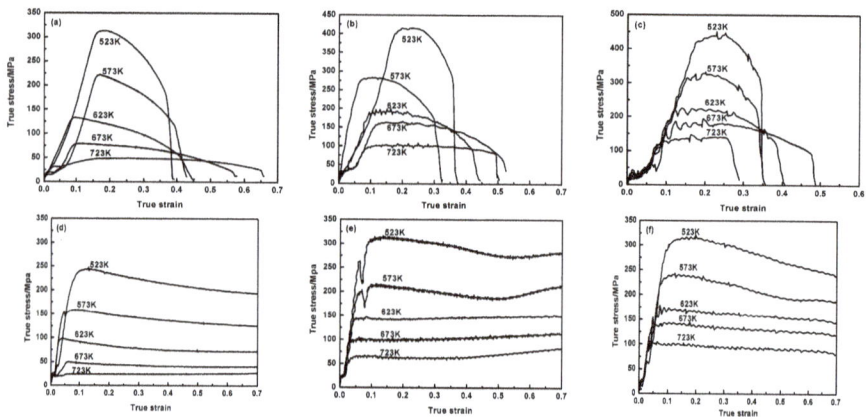

Figure 2. *Cont.*

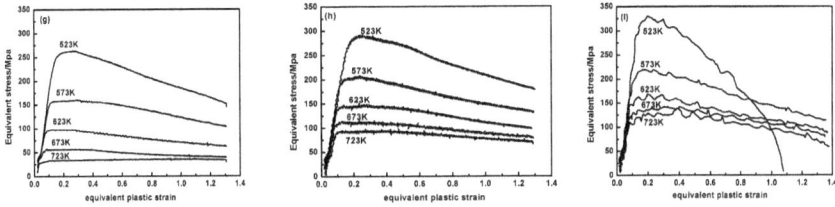

Figure 2. Flow stress *vs.* strain curves for 7050 aluminum alloy at hot tensile (HT) mode: (**a**) 0.01 s^{-1}; (**b**) 1 s^{-1}; (**c**) 10 s^{-1}; HC mode: (**d**) 0.01 s^{-1}; (**e**) 1 s^{-1}; (**f**) 10 s^{-1}; and HCS mode: (**g**) 0.01 s^{-1}; (**h**) 1 s^{-1}; (**i**) 10 s^{-1}.

3.2. Calculation of the Hot Deformation Activation Energy

The hot deformation activation energy parameter provides important information about the deformation mechanism associated with microstructural evolution, especially the dislocation movement, DRV, DRX, and the movement of grain boundaries. If the activation energy is treated as a constant, according to the process of established classical constitutive relations, we can calculate the activation energy using the hyperbolic sine constitutive law [4]:

$$\dot{\varepsilon} = A \left[\sinh(\alpha\sigma)\right]^n \exp\left(-Q/RT\right) \tag{1}$$

where α, n and A are material constants, and R is the universal gas constant (8.314 J·mol^{-1}·K); T is the deformation temperature (K); and Q is the hot deformation activation energy (kJ·mol^{-1}). Moreover, σ can either be the peak stress or steady flow stress (MPa). In this study, the peak stress σ_p was used.

We obtained approximately equal values of activation energy with the classical equation (Q_1 shown in Table 2) under the three hot deformation modes. In the classical constitutive equation, n and A are regarded as material constants, and it defines α as $\alpha = \beta / n_1$ [4], where β and n_1 are the slopes of the plots of $\ln \dot{\varepsilon}-\sigma_p$ and $\ln \dot{\varepsilon}-\ln(\sigma_p)$, respectively. Furthermore, the constitutive equation involves the assumption that during the hot deformation process, any microstructure transformation mechanism is absent [17]. In addition, the material components and heat treatment conditions are identical. Consequently, the activation energy may be assumed to be a constant independent of the deformation conditions and modes on the basis of the approximately equal values. On the other hand, the activation energy (Q_1) calculated with Equation (1) can be comprehended to be, in fact, an average value.

Table 2. Values of material constants and activation energy Q_1 for the 7050 aluminum alloy.

Deformation Mode	A	$\alpha = \beta/n_1$ (MPa^{-1})	n	S	Q_1 (kJ·mol^{-1})
HT	1.27×10^{20}	0.0051	8.05	3.62	241.38
HC	2.31×10^{19}	0.0066	7.34	3.88	236.69
HSC	1.67×10^{20}	0.0054	8.59	3.36	239.93

However, ensuring the microstructure constant during hot deformation is difficult because the microstructures involve various complex interactions among the dislocations, solutes or precipitates, grain boundaries, or periodic friction of the lattice. These processes are strongly affected by strain, strain rate, and deformation temperature. Therefore, the hot deformation activation energy will vary with these deformation conditions. From Equation (2), in which the activation energy is temperature and strain rate dependent [11], the activation energy, Q_{2v}, under different deformation conditions were calculated. The results are listed in Table 3. As can be seen, Q_{2v} decreases with an increase in strain rate and temperature, and the deformation modes significantly affect it.

$$\dot{\varepsilon} = A(T, \dot{\varepsilon}) \left[\sinh(\alpha \sigma p)\right]^{n(T)} \exp\left(-\frac{Q(T, \dot{\varepsilon})}{RT}\right) \qquad (2)$$

Table 3. Values of activation energy Q_{2v} for the 7050 aluminum alloy under three deformation modes.

Deformation Temperature (K)	Strain Rate (s^{-1})								
	HT			HC			HSC		
	0.01	1	10	0.01	1	10	0.01	1	10
523	391.3	323.0	304.5	517.5	379.4	326.0	654.9	398.6	374.3
573	349.4	288.3	271.8	429.0	314.7	270.0	597.2	385.7	341.3
623	307.4	253.5	239.0	340.5	250.0	214.0	398.4	242.5	227.7
673	265.5	218.8	206.3	252.0	185.3	158.1	251.7	153.2	143.8
723	223.6	184.0	173.5	163.5	120.6	102.1	176.8	107.6	101.1
Average (Q_2)		266.7			268.2			314.3	

4. Discussion

Generally, the hot deformation activation energy can qualitatively reflect the energy barrier to dislocation motion for metals and alloys during hot deformation. A higher value of activation energy signifies the existence of higher dragging forces to the movement of the dislocation in hot deformation. The activation energy Q_1 under the three deformation modes are greater than those for the homogenized and aged 7xxx aluminum alloys [7–9]. This deviation is mainly due to the difference in the microstructure state of the alloy in experiments that the specimens were heated rapidly to the testing temperature and held for 30 s before loading. In such a short dwell time, the precipitates may not be completely dissolved and the dragging forces to the movement of the dislocation still exist in the subsequent deformation. Therefore, the activation energy exhibits a higher value.

A comparison of Q_1 and Q_2 values for the three deformation modes is shown in Figure 3. The values of Q_1, which can be considered as the average values under the different deformation conditions, are approximately equal under the three deformation modes. Moreover, the deformation modes do not affect the thermal activation processes, which lead to confusion in the current understanding the nature of hot deformation activation energy. In comparison, an obvious increase can be found in Q_2 values for the three deformation modes. Moreover, Q_2 of the HT mode is nearly identical to the value of the HC mode, but is lower than the value of the HSC mode (see in Table 3). Notably, the values of the activation energies Q_1 and Q_2 are approximate equivalent under both the HT and HC modes irrespective of the calculation method employed. This behavior can be attributed to the similar loading path for the HT and HC deformation modes when the stress reaches the peak value. Tensile or compressive loading can activate a larger number of slip systems when compared with torsion or shear loading [12]. In HT and HC deformations, a large number of slip systems are activated. Nevertheless, the increase in activation energy Q_2 resulting from the calculated methods should be paid more attention, particularly to the significant increase of activation energy Q_2 under the HSC deformation mode. When compared with HT and HC deformations, relatively fewer slip systems can be activated with the complex strain state in the slot position of the HSC specimen. Complicated interactions of different slip systems and intricate dislocation structures induce a much higher obstruction resulting in the higher activation energy in HSC. These results indicate that the modified equation is more suitable to calculate the activation energy for the hot deformation under a complex strain state. Moreover, the hot deformation activation energy is affected by deformation conditions as well as the strain state.

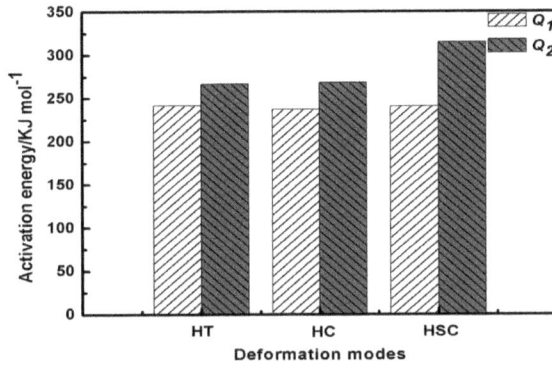

Figure 3. Comparison between the average values of the activation energy Q_1 and Q_2 for the three deformation modes.

The effect of the strain state on the activation energy can be seen in Figure 4. It shows the variation of the activation energy Q_{2v} as a function of temperature for the three deformation modes. As can be seen, Q_{2v} is not constant but decreases with increasing temperature. Furthermore, the relationship is perfectly linear under HT and HC deformations. This phenomenon may be attributed to the thermodynamic mechanism of dislocation movement, which is a thermally activated process [19]. It is related to the critically resolved shear stress [20]. When the temperature increases, rearrangement and annihilation of dislocations occur to improve DRV, which reduces the resistance of dislocation motion and leads to a reduced activation energy. Specifically, the variation under HSC deformation is similar to that under the other modes, but the relationship is not linear. This difference is due to the complex strain state and increased external shear stress induced by HSC deformation.

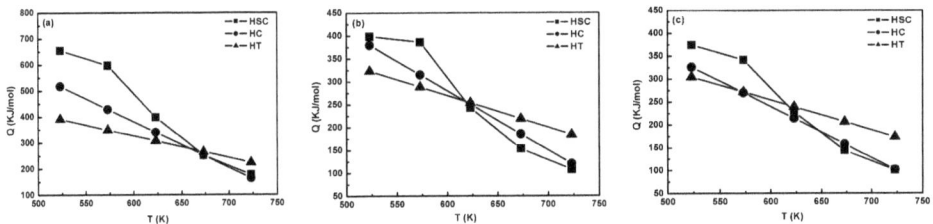

Figure 4. Relationship of hot deformation activation energy *vs.* deformation temperature for 7050 aluminum alloy under three deformation modes: (**a**) 0.01 s^{-1}; (**b**) 1 s^{-1}; (**c**) 10 s^{-1}.

An interesting phenomenon in Figure 4 is the intersection of the curves, which indicates that the activation energy Q_{2v} is approximately same at some deformation conditions for these deformation modes. In addition, the intersection varies with increasing strain rate and decreasing temperature, which may be due to precipitates, dislocation patterns, stress state, or other factors affecting the development of dynamic microstructure. However, the exact cause of the phenomenon is unclear and must be studied in the future.

5. Conclusions

1. By comparing the hot deformation activation energy values calculated using the classical and modified equations, the modified equation is found to be more suitable to calculate the hot deformation activation energy under the complex strain state.

2. With the modified constitutive equation, the hot deformation activation energy is greater under the HSC deformation mode than under the HT and HC deformation modes. It is attributed to the fewer and different slip systems and intricate dislocation structures promoted in HSC deformation. Therefore, the significantly higher obstruction induced by their complicated interactions results in the higher activation energy in HSC deformation case.

3. The effect of the strain state on the hot deformation activation energy can also be seen from the variation of the activation energy with the deformation conditions under the three deformation modes. It indicates that the complex strain state facilitates the increase activation energy under the same deformation conditions.

Acknowledgments: The authors thank the China Friction Stir Welding Center, Beijing Friction Stir Welding Technology Limited Company for financial support.

Author Contributions: Deli Sang performed research, analyzed the data and wrote the paper; Yijun Li helped in the experimental part; Ruidong Fu assisted in the data analysis and revised manuscript.

Conflicts of Interest: The authors declare no conflict of interest.

Abbreviations

HT	Hot Tensile
HC	Hot Compression
HSC	Hot Shear-Compression
DRV	Dynamic Recovery
DRX	Dynamic Recrystallization

References

1. Deschamps, A.; Brechet, Y. Influence of quench and heating rates on the ageing response of an Al-Zn-Mg-(Zr) alloy. *Mater. Sci. Eng. A* **1998**, *251*, 200–207. [CrossRef]
2. McQueen, H.J.; Spigarelli, S.; Kassner, M.; Evangelista, E. *Hot Deformation and Processing of Aluminum Alloys*; CRC Press: Boca Raton, FL, USA, 2011.
3. Zhang, M.J.; Li, F.G.; Wang, S.Y.; Liu, C.Y. Characterization of hot deformation behavior of a P/M nickel-base superalloy using processing map and activation energy. *Mater. Sci. Eng. A* **2010**, *527*, 6771–6779.
4. Sellars, C.M.; McTegart, W.J. On the mechanism of hot deformation. *Mem. Sci. Rev. Met.* **1966**, *63*, 731–746. [CrossRef]
5. Reyes-Calderón, F.; Mejía, I.; Cabrera, J.M. Hot deformation activation energy (QHW) of austenitic Fe-22Mn-1.5Al-1.5Si-0.4C TWIP steels microalloyed with Nb, V, and Ti. *Mater. Sci. Eng. A* **2013**, *562*, 46–52. [CrossRef]
6. Rokni, M.R.; Zarei-Hanzaki, A.; Roostaei, A.A.; Abolhasani, A. Constitutive base analysis of a 7075 aluminum alloy during hot compression testing. *Mater. Des.* **2011**, *32*, 4955–4960. [CrossRef]
7. Jin, N.P.; Zhang, H.; Han, Y.; Wu, W.; Chen, J. Hot deformation behavior of 7150 aluminum alloy during compression at elevated temperature. *Mater. Charact.* **2009**, *60*, 530–536. [CrossRef]
8. Sheppard, T.; Jackson, A. Constitutive equations for use in prediction of flow stress during extrusion of aluminum alloys. *Mater. Sci. Technol.* **1997**, *13*, 203–209. [CrossRef]
9. Cerri, E.; Evangelista, E. Comparative hot workability of 7012 and 7075 alloys after different pretreatments. *Mater. Sci. Eng. A* **1995**, *197*, 181–198. [CrossRef]
10. Zhang, J.Q.; Di, H.S.; Wang, H.T.; Mao, K.; Ma, T.J.; Cao, Y. Hot deformation behavior of Ti-15-3 titanium alloy: A study using processing maps, activation energy map, and Zener-Hollomon parameter map. *J. Mater. Sci.* **2012**, *47*, 4000–4011. [CrossRef]
11. Shi, C.J.; Mao, W.M.; Chen, X.-G. Evolution of activation energy during hot deformation of AA7150 aluminum alloy. *Mater. Sci. Eng. A* **2013**, *571*, 83–91. [CrossRef]
12. Miller, M.P.; McDowell, D.L. Modeling large strain multiaxial effects in FCC polycrystals. *Int. J. Plast.* **1996**, *12*, 875–902. [CrossRef]
13. Davenport, S.B.; Higginson, R.L. Strain path effects under hot working: An introduction. *Mater. Pro. Technol.* **2000**, *98*, 267–291. [CrossRef]

14. Rittel, D.; Lee, S.; Ravichandran, G. A shear-compression specimen for large strain testing. *Exper. Mech.* **2002**, *42*, 58–64. [CrossRef]
15. Masaki, K.; Sato, Y.S.; Maeda, M.; Kokawa, H. Experimental simulation of recrystallized microstructure in friction stir welded Al alloy using a plane-strain compression test. *Scr. Mater.* **2008**, *58*, 355–360. [CrossRef]
16. Dieter, G.E. *Mechanical Metallurgy*, 3rd ed.; McGraw-Hill: New York, NY, USA, 1987.
17. McQueen, H.J.; Ryan, N.D. Constitutive analysis in hot working. *Mater. Sci. Eng. A* **2002**, *322*, 43–63. [CrossRef]
18. Medina, S.F.; Hernandez, C.A. General expression of the Zener-Hollomon parameter as a function of the chemical composition of low alloy and microalloyed steels. *Acta Metar.* **1996**, *44*, 137–148. [CrossRef]
19. Caillard, D.; Martin, J.L. *Thermally Activated Mechanisms in Crystal Plasticity*; Pergamon Press: Oxford, UK, 2003.
20. Honeycombe, R.W.K. *The Plastic Deformation of Metals*, 2nd ed.; Edward Arnold Ltd.: Baltimore, MD, USA, 1984.

metals   MDPI

Article

Dry Machining Aeronautical Aluminum Alloy AA2024-T351: Analysis of Cutting Forces, Chip Segmentation and Built-Up Edge Formation

Badis Haddag [1,*], Samir Atlati [1], Mohammed Nouari [1] and Abdelhadi Moufki [2]

[1]   LEMTA CNRS-UMR 7563, Lorraine University, Mines Albi, Mines Nancy, GIP-InSIC, 27 rue d'Hellieule, 88100 Saint-Dié-des-Vosges, France; samir.atlati@univ-lorraine.fr (S.A.); mohammed.nouari@univ-lorraine.fr (M.N.)

[2]   LEM3 CNRS-UMR 7239, Lorraine University, Ile du Saulcy, 57045 Metz, France; abdelhadi.moufki@univ-lorraine.fr

*   Correspondence: badis.haddag@univ-lorraine.fr; Tel.: +33-032-942-1821

Academic Editor: Nong Gao

Received: 10 June 2016; Accepted: 11 August 2016; Published: 24 August 2016

Abstract: In this paper, machining aeronautical aluminum alloy AA2024-T351 in dry conditions was investigated. Cutting forces, chip segmentation, and built-up edge formation were analyzed. Machining tests revealed that the chip formation process depends on cutting conditions and tool geometry. So continuous and segmented chips are generated. Under some cutting conditions, built-up edge formation occurs. A predictive machining theory, based on a finite elements method (FEM), was applied to reproduce and explain these phenomena. Thermomechanical behaviors of the work material and the tool-work material interface were considered. Results of the proposed modelling were compared to experimental data for a wide range of cutting speed. It was shown that the feed force is well reproduced by the ALE-FE (arbitrary lagrangian-eulerian finite element) formulation and highly underestimated by the lagrangian finite element (LAG-FE) one. While, the periodic localized shear band, leading to a chip segmentation, is well reproduced with the Lagrangian FE formulation. It was found that the chip segmentation can be correlated to the cutting force evolution using the defined chip segmentation intensity parameter. For the built-up edge (BUE) phenomenon, it was shown that it depends on the contact/friction at the tool-chip interface, and this is possible to simulate by making the friction coefficient time-dependent.

Keywords: aluminum alloy AA2024-T351; dry machining; cutting/feed forces; chip segmentation; built-up edge; FE modeling

1. Introduction

Aluminum alloy AA2024-T351 has been used for over 30 years in the aeronautic and aerospace industries, especially for its good resistance to fatigue [1]. Also, this metal alloy has a low density in comparison to steels. Often, components made of aluminum alloy AA2024-T351 need machining, using various processes (turning, drilling, milling, etc.), to obtain required shape of the component and also to satisfied the high exigency on the surface quality (low roughness). When machining this alloy, considered as ductile material, several phenomena occur, depending on cutting conditions (cutting speed, feed rate, dry/wet cutting, etc.).

The first one is the chips segmentation which should promotes chips fragmentation, hence ease evacuation of chips. However, this alloy is difficult to fragment, due to its relative low ductility, in comparison with titanium alloy Ti-6Al-4V for example, also used widely in aeronautics. Mechanisms of chip formation have been widely studied by many researchers [2–7]. Komanduri and Brown [2] classified chips on four types, according to their morphology (wavy, discontinuous, segmented, and catastrophic shear chip), and gave the definition of each one. Globally, the origin of each type depends on cutting conditions, cutting tool geometry, cutting angles, and the machined material (thermo-mechanical characteristics: Soft or hard materials). To understand mechanisms of chip formation, micrographic analysis was conducted in several research works (e.g., [3,4], [2,5]). For example, Bayoumi and Xie [3] analyzed metallurgical aspects of the chip formation when cutting the usual titanium alloy Ti-6Al-4V.

The second one is the BUE formation, which may alter the surface quality, since the adhered work material can pass under the tool flank face and acts on the newly generated surface. Also, the BUE edge affects the chips morphology since the BUE changes the tool rake angle. The literature review shows that the analysis of the BUE formation during the cutting process was a subject of several theoretical and experimental studies. For experimental aspects, Ernst et al. [8] reported that the BUE can be often formed under high friction conditions at the tool-chip interface and its morphology is significantly influenced by the state of stress around the tool cutting edge. Shaw et al. [9] attribute the BUE formation to the temperature gradient across the chip and to the brittle behavior of the workpiece material. In addition, Philip [10] concluded that the BUE formation is the result of seizure and sub-layer flow at the tool-chip interface. The strain hardening of the work material promotes the formation of a stagnant build-up at the cutting edge.

In this paper, to analyze the machining process of AA2024–T351 (cutting forces, chip segmentation, and BUE formation), experimental and modelling studies were developed. In the experimental study, cutting speed, feed rate, and tool rake angle were varied to highlight there effect on the cutting process. The modelling study was performed to give more insight on revealed phenomena. This also allows simulating other cutting conditions not performed experimentally. Friction coefficient was varied in order to analyze the mechanism of BUE formation.

2. Experimental and Modelling Aspects

2.1. Experimental Aspects

To analyze the chip segmentation phenomenon, the experimental study previously done by one of the authors in [11] has been investigated. Orthogonal cutting tests, in dry conditions, were performed on a planer machine at low cutting speed to observe with CCD (charge-coupled device) camera the cutting process (chip formation), and on a CNC (computer numeric control) lathe to allow variation of the cutting speed in a large range, as shown in Figure 1a,b. Uncoated carbide inserts, of type K4, have been used as cutting tools with two rake angles (0° and 15°), see Figure 1c. The value of the clearance angle is kept constant (7°) for all tools and all cutting conditions. Each tool has a cutting edge radius of 0.01 mm. The depth of cut in both planing and turning processes is 4 mm for all cutting conditions.

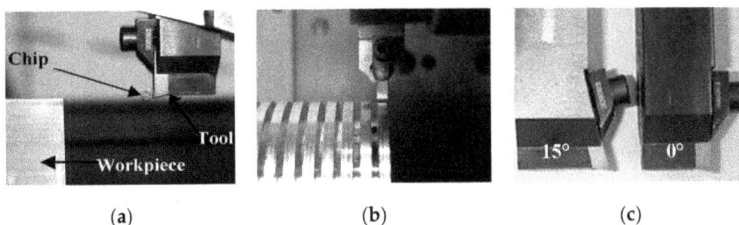

(a)   (b)   (c)

Figure 1. Experimental setup of orthogonal cutting tests: (**a**) planar machining; (**b**) turning; and (**c**) cutting tools with two rake angles (0° and 15°) [11].

The inserts material is a WC-Co cemented tungsten carbide with cobalt as binder phase. The chemical analysis on a polished surface inside the tool gives a composition with 6 wt. % of cobalt and no mixed carbides such as TiC, TaC, and NbC have been detected in the microstructure. The cobalt binder is uniformly distributed with WC grains (Figure 2b). The workpiece material is the usual aeronautical aluminum alloy AA2024-T351 (Figure 2a). Its chemical composition is given in Table 1.

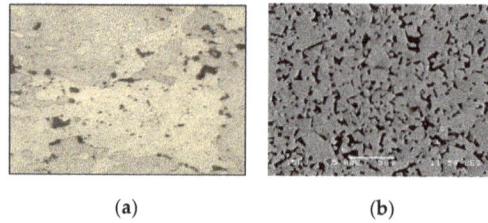

(a)   (b)

Figure 2. Micrographs of a polished surfaces of (a) aluminum alloy AA2024-T351 [6] and (b) cemented carbide tool WC-Co [11].

Table 1. Chemical composition (wt. %) of the AA2024-T351 [11].

Al	Cr	Cu	Fe	Mg	Mn	Si	Ti	Zn
Balanced	Max. 0.1	3.8–4.9	Max. 0.5	1.2–1.8	0.3–0.9	Max. 0.5	Max. 0.15	Max. 0.25

Using a high speed camera and dynamometer table, instantaneous images of the cutting process and cutting forces were obtained. As shown in Figure 3, for a low feed (0.1 mm) a continuous chip was observed, while for a large feed (0.3 mm) segmented chip was obtained with the two tools (0° and 15°), but the chip segmentation is different. Average cutting forces (evaluated in stabilized range or steady state of cutting) and contact lengths (evaluated from instantaneous images of the cutting process) are reported in Table 2. The analysis of the results shows that there is a strong correlation between cutting conditions (cutting speed and feed), cutting parameters (rake angle), and the chip morphology.

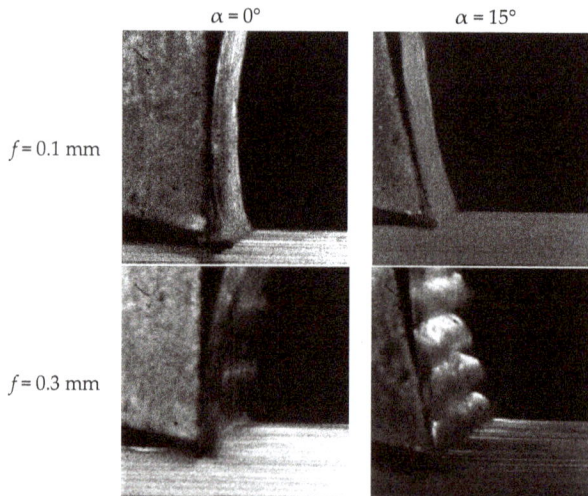

Figure 3. Example of experimental chips morphology obtained with cutting speed = 60 m/min.

The BUE occurs at particular cutting conditions when machining ductile metals, like aluminum alloy AA2024-T351. The effect of uncut chip thickness on the formation of the BUE was examined on an instrumented planar machine, where the tool rake angle was varied. Different sequences of the cutting process were recorded by the CCD camera and then analyzed. From instantaneous images of Figure 3, it can be observed that the BUE occurs for the low rake angle (i.e., 0°) and uncut chip thicknesses of 0.1 mm. Figure 4 highlights the BUE formation in the vicinity of the tool tip. Indeed, a low rake angle makes the work material flow difficult. This promotes the sticking contact at the tool tip, which results in accumulation of the work material in this zone and hence the BUE formation.

Table 2. Average cutting force, feed force and contact length for different cutting speeds [12].

Cutting Speed (m/min)	Cutting Force (N)	Feed Force (N)	Contact Length (mm)
80	510	375	0.30
95	475	300	0.28
160	450	280	0.26
195	430	265	0.25
320	410	240	0.25
390	405	235	0.23
500	400	220	0.20

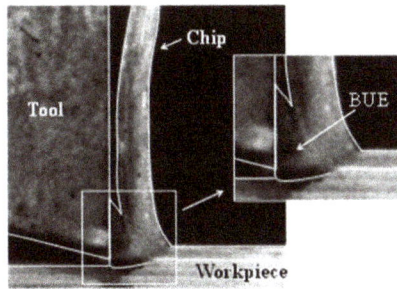

Figure 4. BUE (built-up edge) formation for cutting speed = 60 m/min, feed = 0.1 mm and rake angle = 0°.

Additional experimental tests have been done under orthogonal turning to bring out the effect of cutting speed, with cutting tool angle of 0° and feed rate fixed to 0.1 mm/rev. The depth of cut is the same as in planing process (i.e., 4 mm). Results are reported in Table 2 and Figure 5. The apparent friction coefficient, reported in Figure 5, is determined as the ratio of feed force by cutting force in the case of considered rake angle (i.e., 0°).

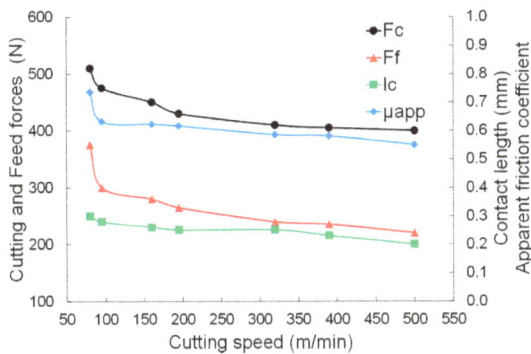

Figure 5. Experimental cutting force (*F*c), feed force (*F*f), contact length (*l*c), and apparent friction coefficient (μapp).

2.2. Modelling Aspects

In order to reproduce and then analyze observed cutting phenomena (cutting forces evolution, chips segmentation, and BUE formation), a predictive modelling theory, based on FEM, was developed. Two formulations (2D LAG-FE and 2D ALE-FE) were developed in Abaqus/Explicit FE code [13] to represent orthogonal cutting tests. The 2D LAG-FE was developed in order to reproduce the chips segmentation phenomenon. While the 2D ALE-FE was developed in order to reproduce the BUE phenomenon. Figure 6 illustrates the two FE models.

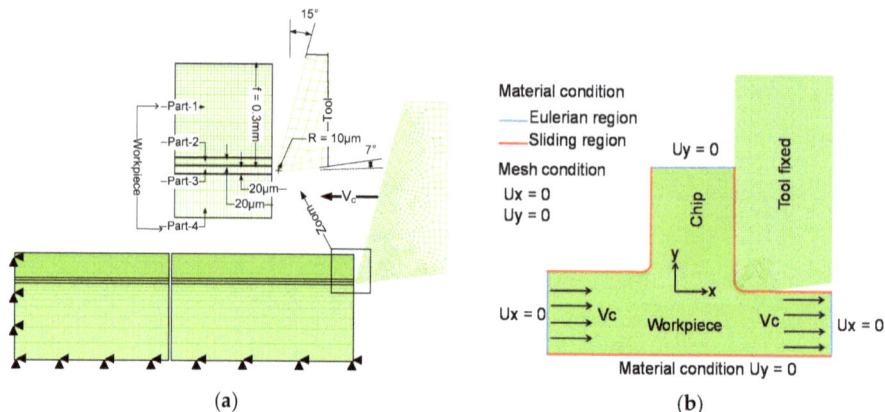

(a)   (b)

Figure 6. Illustration of (**a**) 2D LAG-FE (lagrangian finite element) and (**b**) ALE-FE (arbitrary lagrangian-eulerian finite element) models of the orthogonal cutting.

Physical properties of the workpiece and tool materials are given in Table 3.

Table 3. Mechanical and thermal properties of work material and tool [6].

Physical Parameter	Workmaterial (AA2024-T351)	Tool (WC-Co)
Density, ϱ (kg/m^3)	2700	11,900
Elastic modulus, E (GPa)	73	534
Poisson's ratio, ν	0.33	0.22
Specific heat, Cp (J/kg/°C)	$Cp = 0.557T + 877.6$	400
Thermal conductivity, λ (W/m/C)	$25 \leq T \leq 300 : \lambda = 0.247T + 114.4$ $300 \leq T \leq Tm : \lambda = 0.125T + 226$	50
Thermal expansion, α (μm·m/°C)	$\alpha = 8.9 \times 10^{-3}T + 22.2$	-
T_m (°C)	520	-
T_0 (°C)	25	25

To represent the of the workpiece material behavior during machining, a Johnson-Cook thermo-viscoplastic-damage model has been adopted. The flow stress is given as:

$$\bar{\sigma} = \left[A + B\left(\bar{\varepsilon}^p\right)^n\right]\left[1 + C\ln\left(\frac{\dot{\bar{\varepsilon}}^p}{\dot{\bar{\varepsilon}}_0}\right)\right]\left[1 - \left(\frac{T - T_0}{T_m - T_0}\right)^m\right]$$ (1)

The damage behavior is described by a damage initiation criterion and a damage evolution law up to fracture. The damage initiation criterion is given by:

$$w_d = \int \frac{d\bar{\varepsilon}^p}{\bar{\varepsilon}_d^p} \quad \text{with} \quad 0 \le w_d \le 1$$

$$\bar{\varepsilon}_d^p = \underbrace{\left[d_1 + d_2 e^{(d_3 \frac{p}{\bar{\sigma}})} \right]}_{\text{Stress triaxiality}} \underbrace{\left[1 + d_4 \ln \dot{\bar{\varepsilon}}^* \right]}_{\text{Viscosity}} \underbrace{\left[1 - d_5 T^* \right]}_{\text{Temperature}} \tag{2}$$

The damage evolution can be expressed by the following relationships:

$$d = \begin{cases} \dfrac{\bar{u}^p}{\bar{u}_f} = \dfrac{L\bar{\varepsilon}^p}{\bar{u}_f} = \dfrac{2G_f L\bar{\varepsilon}^p}{\bar{\sigma}} & \text{linear evolution} \\[2ex] 1 - \exp\left(-\int\limits_0^{\bar{u}^p} \dfrac{\bar{\sigma}}{G_f} d\bar{u}^p \right) & \text{exponential evolution} \end{cases} \tag{3}$$

The true stress tensor is defined as:

$$\sigma = (1 - d)\, \tilde{\sigma} \tag{4}$$

where $\tilde{\sigma}$ is the effective stress, representing a stress state that would exist in the material if no damage occurs. The behavior parameters of the workpiece material are given in Table 4.

Table 4. Johnson-Cook viscoplastic-damage parameters of AA2024-T351 [6].

Viscoplastic Parameters				
A (Mpa)	B (Mpa)	n	C	m
352	440	0.42	0.0083	1

Damage Parameters						
d_1	d_2	d_3	d_4	d_5	K_C^I (MPa·m$^{1/2}$)	K_C^{II} (MPa·m$^{1/2}$)
0.13	0.13	1.5	0.011	0	37	26

As the mechanical behavior is affected by temperature, the mechanical plastic work generates heat flux which results in temperature rise. The heat flux due to this phenomenon is described the following relation:

$$\dot{q}_p = \eta_p \bar{\sigma} : \dot{\bar{\varepsilon}}^p \tag{5}$$

The contact behavior at the tool-workpiece interface is defined by the relationship between the normal friction stress σ_n and the shear friction stress τ_f, as follows:

$$\tau_f = \min\left(\mu \sigma_n, \tau_{\max} \right) \tag{6}$$

The friction at the contact interface may generate a heat flux which is evaluated by the following relation:

$$\dot{q}_f = f_f \eta_f \tau_f V_s \tag{7}$$

The mechanical plastic work of the chip may affect heat exchange at the tool-workpiece interface. To take account of this energy in the heating of the tool a heat conduction flux, $\dot{q}_c$, is introduced, so the heat balance at the interface can be written as follows:

$$\dot{q}_{\to tool} = f_f \dot{q}_f + \dot{q}_c$$

$$\dot{q}_{\to workpiece} = \left(1 - f_f \right) \dot{q}_f - \dot{q}_c \tag{8}$$

$$\text{with } \dot{q}_c = h(T_{int-w} - T_{int-t})$$

The tool-workpiece interface parameters depend on the adopted FE model (LAG-FE or ALE-FE). Particularly, the coefficient of friction (COF) is adjusted to better fit cutting forces.

3. Results and Discussion

3.1. Cutting Forces Analysis

As illustrated in Figure 7, both FE models effectively predict cutting force and contact length (upper values with ALE-FE model and lower values with LAG-FE model). However, LAG-FE model highly underestimates feed force, due to the FE deletion in a thin layer defined between the chip and workpiece. Indeed, this induces a loss of contact at the flank face which, in turn, induces a loss of contact pressure acting in the feed direction. The apparent friction coefficient is globally well estimated by the ALE-FE model, except for the lower cutting speed (80 m/min). While since LAG-FE model fails to predict feed force, this impacts the apparent friction coefficient (recall that it is defined as a ratio of feed force by cutting force in the case of rake angle of 0°). Comparison between LAG-FE model, ALE-FE model and experimental data is reported in Tables 5 and 6.

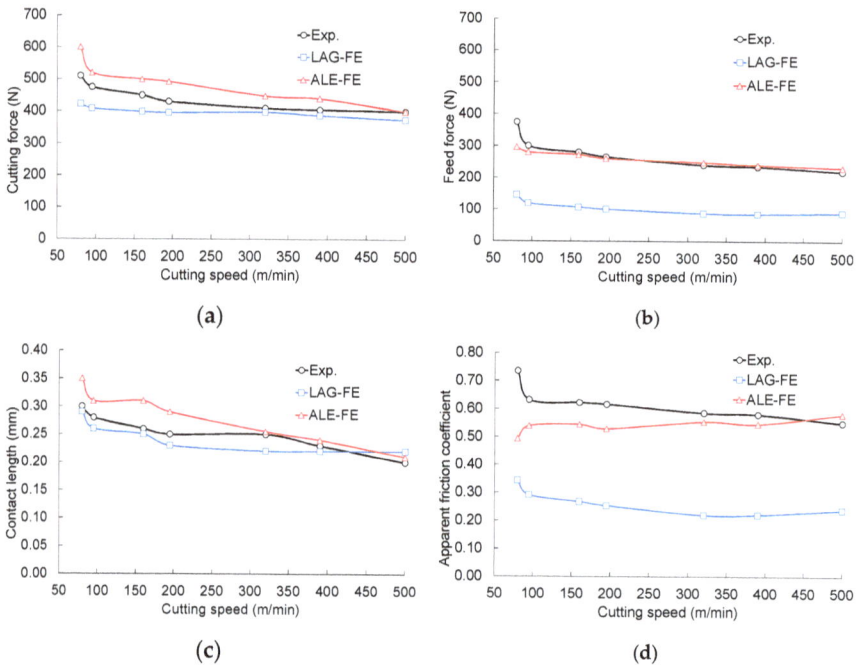

Figure 7. Experimental vs. numerical modelling (ALE-FE and LAG-FE) of (**a**) cutting force; (**b**) feed force; (**c**) contact length, and (**d**) apparent friction coefficient.

Table 5. Comparison between experimental data and LAG-FE results.

Vc (m/min)	Fc (N)			Ff (N)			Lc (mm)			μapp		
-	Exp.	Num.	Err. (%)	Exp.	Num.	Err. (%)	Exp.	Num.	Err. (%)	Exp.	Num.	Err. (%)
80	510	422	17	375	145	61	0.30	0.29	3	0.74	0.34	53
95	475	408	14	300	118	61	0.28	0.26	7	0.63	0.29	54
160	450	398	12	280	106	62	0.26	0.25	4	0.62	0.27	57
195	430	395	8	265	100	62	0.25	0.23	8	0.62	0.25	59
320	410	397	3	240	87	64	0.25	0.22	12	0.59	0.22	62
390	405	387	4	235	86	64	0.23	0.22	4	0.58	0.22	62
500	400	374	7	220	89	60	0.20	0.22	10	0.55	0.24	57

Table 6. Comparison between experimental data and ALE-FE results.

Vc (m/min)	Fc (N)			Ff (N)			Lc (mm)			μapp		
-	Exp.	Num.	Err. (%)	Exp.	Num.	Err. (%)	Exp.	Num.	Err. (%)	Exp.	Num.	Err. (%)
80	510	600	18	375	296	21	0.30	0.35	17	0.74	0.49	33
95	475	520	9	300	280	7	0.28	0.31	11	0.63	0.54	15
160	450	500	11	280	272	3	0.26	0.31	19	0.62	0.54	13
195	430	492	14	265	260	2	0.25	0.29	16	0.62	0.53	14
320	410	448	9	240	248	3	0.25	0.26	2	0.59	0.55	5
390	405	440	9	235	240	2	0.23	0.24	4	0.58	0.55	6
500	400	400	0	220	232	5	0.20	0.21	5	0.55	0.58	5

At first view, the ALE-FE gives better results, since it effectively predicts both cutting force and feed force. However, this is not sufficient, since the capability of each model should be analyzed regrading other phenomena, like reproducing continuous or segmented chips and BUE formation. This is what is developed in the two following sections.

3.2. Chip Morphology-Segmented vs. Continuous Chip

To analyze finely the chip morphology, particularly chip segmentation phenomenon, orthogonal cutting tests performed on a planer machine (see Figures 1a and 3) were firstly simulated with LAG-FE model. The model is able to reproduce non-continuous chip, like chip segmentation. This should highlight the effect of cutting conditions (cutting speed, feed, and tool-rake angle) on the chip morphology, especially on the chip segmentation phenomenon.

The chip morphology often carries the signature of correct behavior chosen in simulations. From Figure 8, it can be seen that simulated chips are in good agreement with what it is observed by CCD camera. For the small feed (0.1 mm) continuous shape is obtained regardless of rake angle. For the large feed (0.3 mm) segmented chips are well reproduced for both rake angles.

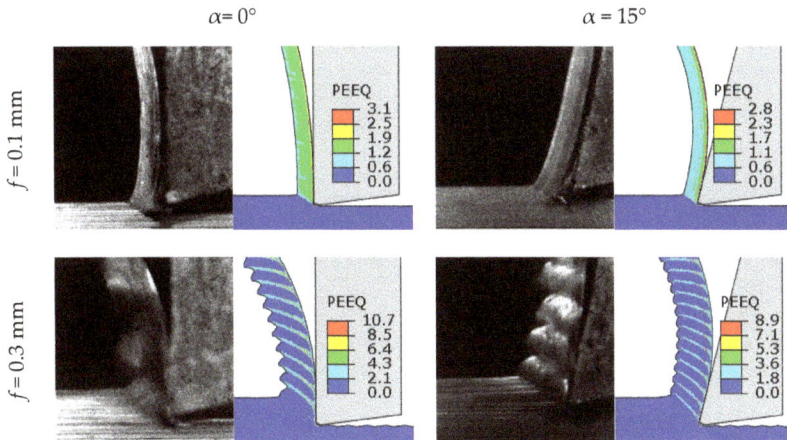

Figure 8. Experimental vs. numerical chips morphology for two feeds and two rake angles.

To quantify the chip morphology some parameters are introduced as follows:

$$f_s = \frac{V_c}{l_s}$$
$$SIR_g = \frac{\Delta h}{f} = \frac{h_{max} - h_{min}}{f}$$
$$SIR_l = \frac{\bar{\varepsilon}^p_{in}}{\bar{\varepsilon}^p_{out}}$$

$$(9)$$

where f_s and l_s are the chip segmentation frequency and the distance between two successive segments. Note that these are only indicators of the apparition of successive shear localization bands in the chip, but it does not give information about the intensity of the segmentation phenomenon. So for this purpose SIR_g and SIR_l parameters are defined. SIR_g is the global segmentation intensity ratio and SIR_l is the local segmentation intensity ratio. h_{max} and h_{min} are the maximum and minimum chip thickness, respectively. $\bar{\varepsilon}_{in}^p$ and $\bar{\varepsilon}_{out}^p$ are the plastic equivalent strain in and out of the shear band, respectively. Geometric parameters for the assessment of defined chip morphology parameters are illustrated in Figure 9.

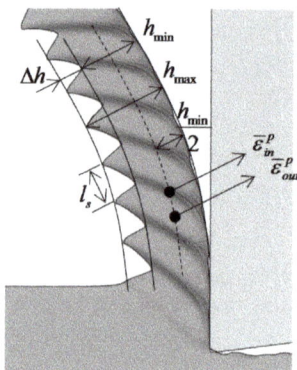

Figure 9. Geometric parameters for the assessment of chip morphology parameters.

Figure 10 shows the evolution of classical chip segmentation parameters f_s and l_s with cutting speed. f_s Increases quasi-linearly with cutting speed for each tool rake angle, while l_s is quasi-insensitive to the cutting speed. However, effect of the rake angle is not negligible and influences both f_s and l_s Indeed, increasing the rake angle decreases the chip segmentation length, so the slope of f_s curve as function of cutting speed increases when the rake angle increases. The consequence of this is that the gap $\Delta f_s = f_s(15°) - f_s(0°)$ increases when cutting speed increases. Globally, f_s and l_s give an indication on the number of shear bands within the same chip length, but these two quantities are not sufficient to quantify the chip segmentation, since they do not give an indication on the intensity of the phenomenon.

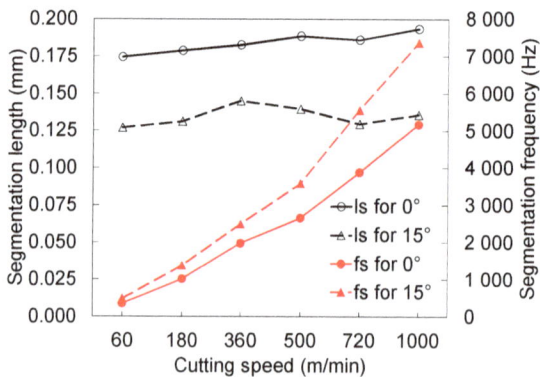

Figure 10. Segmentation length vs. segmentation frequency as function of cutting speed, for fixed feed (0.3 mm) and two rake angles (0° and 15°).

Hence, to quantify the intensity of chip segmentation phenomenon, introduced SIR_g and SIR_l parameters were assessed. As shown in Figure 11, these two parameters increase with cutting speed and tend to stagnate at high cutting speed. This confirms that increasing cutting speed promotes the chip segmentation, as was often observed in experimental tests. Hence, these parameters are adequate to quantify the intensity of chip segmentation phenomenon. In addition, Figure 11 brings out the close link between chip segmentation intensity and average cutting force. For each rake angle, as cutting speed increases, average cutting force decreases and SIR_g and SIR_l parameters increase. This confirms that the chip segmentation phenomenon is at the origin of the cutting force reduction when machining aluminum alloy AA2024-T351.

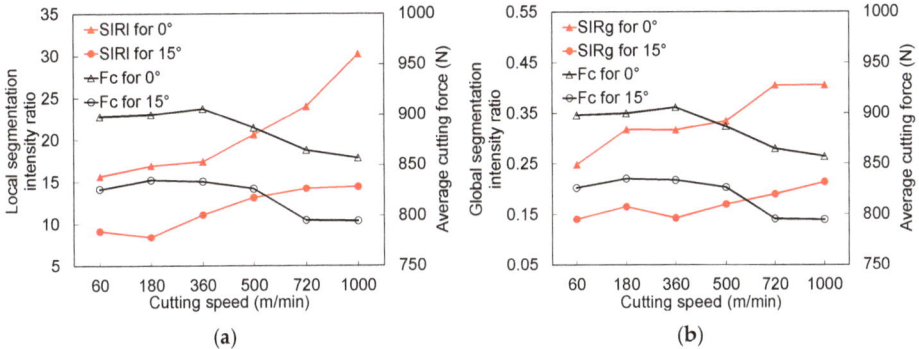

Figure 11. Correlation between average cutting force and segmentation intensity ratio: (a) SIR_l and (b) SIR_g.

3.3. Built-Up Edge Formation-Time-Dependent Friction

The BUE phenomenon can be considered as the consequence of a gradual increase of the friction at the tool-chip interface up to reach a critical level leading to a complete or partial sticking contact at the tool-chip interface. The proposed idea to analyze the process of BUE formation consists to vary the local coefficient of friction (COF), defined as the ratio of frictional stress by contact pressure (see Equation (6)), during the chip flow on the rake face. This is physically an admitted assumption, since the BUE corresponds to the adhesion of the work material at the rake face close to the tool tip. Between the time where the cutting process is stable and the time where the BUE is formed, the friction evolves from a certain value to a higher one that induces the adhesion of the work material on the rake face.

Different possibilities can be proposed for the evolution of the friction at the contact interface. Here two cases were considered, corresponding, respectively, to an abrupt change and a gradual evolution of COF. The effect of friction change on BUE formation was investigated by examining particular cutting forces and the work material flow velocity on rake face. The ALE-FE model was adopted to simulate the cutting process with varying COF. The simulated cutting case is the reported one in Figure 4, where the BUE was observed by the CCD camera.

One possible mechanism of BUE formation is an abrupt change of the friction at the tool–chip interface. Two successive steps are then defined. In the first one (step 1), the cutting process was simulated with COF equal 0.2. In the followed step (step 2), the simulation is continued with COF equal 0.4 or 0.6. The impact of this abrupt change of COF on BUE formation is highlighted in Figure 12, through the work material flow velocity at the secondary shear zone. It can be deduced from Figure 12 that the work material flow velocity is affected by the increase of COF. The change in work material flow is more pronounced for a higher COF in the second step (i.e., COF = 0.6). The sticking zone becomes large as the friction is higher. For the low friction, (COF = 0.2 in step 1) a negligible sticking zone can be noticed and, consequently, no BUE can form. Therefore, it can be deduced that a change in the nature of the tool-work material contact have a direct effect on the BUE formation.

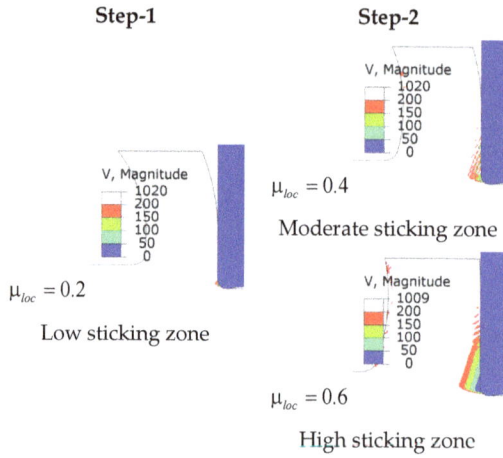

Figure 12. Effect of COF (coefficient of friction) on the material flow velocity at the tool-chip interface.

In addition, the abrupt change of the COF also has a significant effect on cutting and feed forces, as shown in Figure 13. So increasing COF increases cutting and feed forces. The chip flows with more difficultly at the rake face, resulting in the increase and fluctuation of cutting forces. Note that feed force is more affected by the friction change (see from Figure 13b), since it is in the direction of the friction stress on the rake face (rake angle = $0°$).

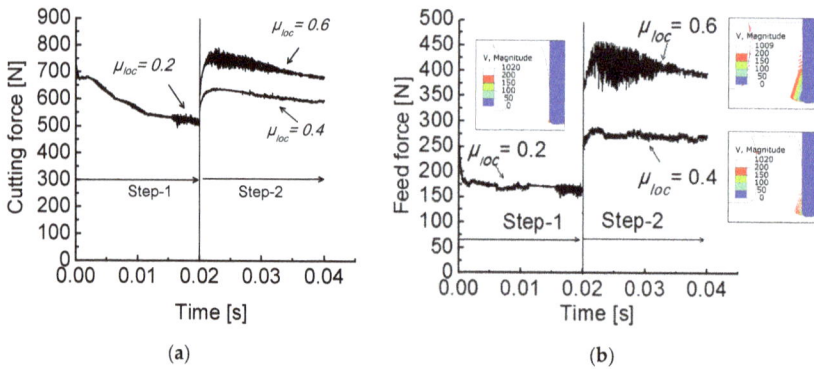

Figure 13. Effect of an abrupt change of COF on (a) the cutting force and (b) the feed force.

The second possible mechanism of BUE formation corresponds to a gradual evolution of the friction at the tool-work material interface during chip formation. This is represented by the increase of COF gradually (increment of 0.1 is taken) from 0.2 to 1. This assumes that BUE is formed when the friction increases gradually to a higher level that leads to a complete adhesion of the work material at the rake face (sticking contact). According to Figure 14, the process of BUE formation can be viewed from the evolution of work material flow velocity with COF. The gradual increase of the friction increases the amount of the sticking zone. This later represents the work material having low flow velocity at the tool-chip interface. So the variation of the sticking zone with the friction can be correlated with the BUE formation.

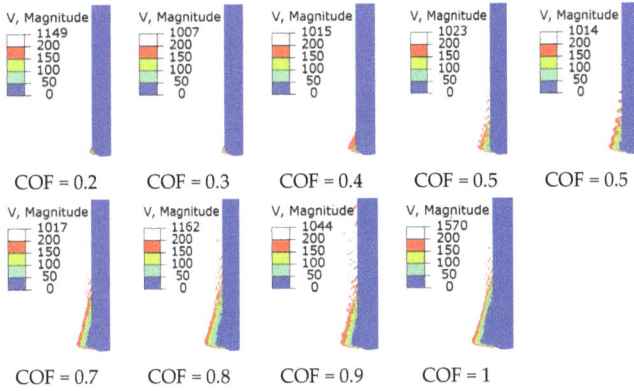

Figure 14. Effect of a gradual evolution of COF on the work material flow velocity at the tool-chip interface (increase of the sticking zone).

The gradual increase of the friction also affects cutting forces, as shown in Figure 15, with more effect on feed force. The apparent friction coefficient follows the same trend as feed force. Globally, the ratio of the sticking by the sliding contact increases as the friction increases. For high friction, the sliding of the work material layer at the tool-chip interface is practically controlled by the shear flow stress of the work material (i.e., $\tau_f = \bar{\tau}$ in Equation (6)). According to Figure 15, the average contact length follows the same trend as the apparent friction coefficient. It increases until reaching a quasi-saturated level. So the BUE formation can be related mainly to the dominant sticking contact at the tool-chip interface.

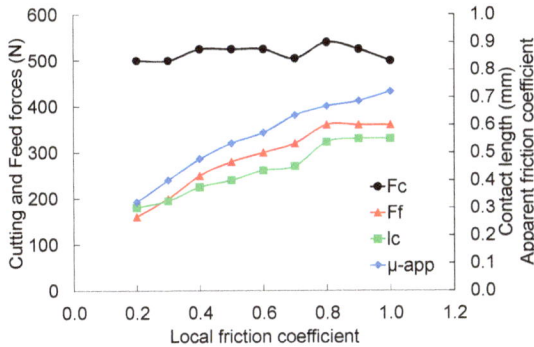

Figure 15. Evolution of average cutting and feed forces, apparent friction coefficient, and average contact length with COF.

4. Conclusions

Machining aeronautical aluminum alloy AA2024-T351 in dry conditions was investigated. Cutting forces, chip segmentation, and built-up edge formation were analyzed. The main concluding remarks are as follows:

(1) Measured cutting forces evolve highly at low cutting speeds for fixed feed and tend to stabilize rapidly at about 100 m/min. LAG-FE and ALE-FE models predict well cutting force, but LAG-FE model fails to predict feed force.
(2) Chip segmentation depends highly on the tool rake angle and the uncut chip thickness.

(3) Using LAG-FE model, it is shown that there is a close link between cutting forces evolution and chip segmentation intensity. So, chip segmentation phenomenon reduces the average cutting forces, but its fluctuation increases.

(4) The BUE can be explained by the contact/friction change at the tool–work material interface during cutting. The ductility of aluminum alloy AA2024-T351 also promotes BUE formation.

(5) The BUE can be modeled by making the friction coefficient time-dependent. This is done in the ALE-FE model. So, the sticking zone becomes larger with the increase of friction coefficient. This condition promotes the BUE formation, but there is no unique scenario of friction evolution as BUE occurs.

Author Contributions: M.N. performed the experiments; B.H., S.A., and A.M. analyzed the data; B.H. wrote the paper.

Conflicts of Interest: The authors declare no conflict of interest.

Abbreviations

The following abbreviations are used in this manuscript:

V_c	cutting speed (m/min)
f	feed (mm)
α, γ	tool-rake and clearance angles (°)
F_c, F_f	cutting and feed forces (N)
l_c	contact length (mm)
h_{min}	contact length (mm)
h_{max}	maximum chip thickness
l_s	chip segmentation length
f_s	chip segmentation frequency
SIR_g	global segmentation intensity ratio
SIR_l	local segmentation intensity ratio
σ	Cauchy stress tensor (MPa)
$\tilde{\sigma}$	effective stress tensor (not affected by damage)
ρ	material density (kg/m^3)
E, ν	Young modulus (GPa) and Poisson's ratio
A, B, C, m, n	Johnson-Cook flow stress parameters
$\bar{\varepsilon}^p$	von Mises equivalent plastic strain
$\dot{\bar{\varepsilon}}^p$	von Mises equivalent plastic strain-rate
$\dot{\bar{\varepsilon}}_0$	Reference equivalent plastic strain-rate
$\bar{\sigma}$	von Mises equivalent stress (MPa)
ω_d	damage initiation criterion
d	damage variable
G_f	fracture strain energy
σ_n	contact pressure (MPa)
τ_f	friction stress (MPa)
COF	local friction coefficient
μ-app	apparent friction coefficient
τ_{max}	shear stress limit (MPa)
V_s	sliding velocity at the tool-workpiece interface (m/s)
T	temperature (°C)
T_0	reference ambient temperature (°C)
T_m	melting temperature (°C)
T_{int-t}	tool temperature at the tool-workpiece interface (°C)
T_{int-w}	workpiece temperature of at the tool-workpiece interface (°C)
λ	thermal conductivity (W/m/°C)

c_p	specific heat capacity (J/kg/°C)
α	thermal expansion (μm/m/°C)
η_p	plastic work conversion factor (Taylor-Quinney factor)
η_f	frictional work conversion factor
f_f	heat partition coefficient
h	heat transfer coefficient (kW/m^2/°C)
$\dot{q}_p$	volumetric heat generation due to plastic work (W/m^3)
$\dot{q}_c$	heat conduction flux at the tool-workpiece interface (W/m^2)
$\dot{q}_{\rightarrow tool}$	heat flux going into the tool at the tool-workpiece interface (W/m^2)
$\dot{q}_{\rightarrow workpiece}$	heat flux going into the workpiece at the tool-workpiece interface (W/m^2)
Exp.	experiment
Num.	Numerical
Err.	error (%)
Max.	Maximum

References

1. Pauze, N. Fatigue Corrosion dans le Sens Travers Court de Tôles d'aluminium 2024-T351 Présentant des Défauts de Corrosion Localisée. Ph.D. Thesis, Ecole Nationale Supérieure des Mines de Saint-Etienne, Saint-Etienne, France, 2008. (In French)

2. Komanduri, R.; Brown, R.H. On the mechanics of chip segmentation in machining. *J. Eng. Ind.* **1981**, *103*, 33–51. [CrossRef]

3. Bayoumi, A.E.; Xie, J.Q. Some metallurgical aspects of chip formation in cutting Ti-6 wt. % Al-4 wt. % V alloy. *Mater. Sci. Eng. A* **1995**, *190*, 173–180. [CrossRef]

4. Barry, J.; Byrne, G. The mechanisms of chip formation in machining hardened steels. *J. Manuf. Sci. Eng.* **2002**, *124*, 528–535. [CrossRef]

5. Komanduri, R.; von Turkovich, B.F. New observations on the mechanism of chip formation when machining titanium alloys. *Wear* **1981**, *69*, 179–188. [CrossRef]

6. Atlati, S.; Haddag, B.; Nouari, M.; Zenasni, M. Analysis of a new segmentation intensity ratio "SIR" to characterize the chip segmentation process in machining ductile metals. *Int. J. Mach. Tools Manuf.* **2011**, *51*, 687–700. [CrossRef]

7. Kouadri, S.; Necib, K.; Atlati, S.; Haddag, B.; Nouari, M. Quantification of the chip segmentation in metal machining: Application to machining the aeronautical aluminium alloy AA2024-T351 with cemented carbide tools WC-Co. *Int. J. Mach. Tools Manuf.* **2013**, *64*, 102–113. [CrossRef]

8. Ernst, H.; Martellotti, M. The formation of the built-up edge. *ASME Mech. Eng.* **1938**, *57*, 487–498.

9. Shaw, M.C.; Usui, E.; Smith, P.A. Free Machining Steel: III—Cutting Forces; Surface Finish and Chip Formation. *J. Eng. Ind.* **1961**, *83*, 181–192. [CrossRef]

10. Philip, P.K. Built-up edge phenomenon in machining steel with carbide. *Int. J. Mach. Tool Des. Res.* **1971**, *11*, 121–132. [CrossRef]

11. List, G.; Nouari, M.; Géhin, D.; Gomez, S.; Manaud, J.-P.; LePetitcorps, Y.; Girot, F. Wear behaviour of cemented carbide tools in dry machining of aluminium alloy. *Wear* **2005**, *259*, 1177–1189. [CrossRef]

12. List, G. Etude des Mécanismes d'endommagement des Outils Carbure WC-Co par la Caractérisation de l'interface Outil Copeau: Application à l'usinage à sec de l'alliage d'aluminium Aéronautique AA2024-T351. Ph.D. Thesis, Ecole Nationale Supérieure des Arts et Métiers of Bordeaux, Saint-Etienne, France, 2004. (In French)

13. Abaqus/Explicit®. Dassault Systemes 2015. Available online: http://www.3ds.com/products-services/simulia/products/abaqus/abaqusexplicit (accessed on 17 August 2016).

metals

MDPI

Article

Substructural Alignment during ECAE Processing of an Al-0.1Mg Aluminium Alloy

Yan Huang

BCAST, Institute of Materials and Manufacturing, Brunel University London, Uxbridge UB8 3PH, UK;
yan.huang@brunel.ac.uk; Tel.: +44-1895-266-976

Academic Editor: Nong Gao
Received: 24 May 2016; Accepted: 5 July 2016; Published: 12 July 2016

Abstract: An investigation has been carried out into the microstructures developed during the early stages of equal channel angular extrusion (ECAE) in a polycrystalline single-phase Al-0.13Mg alloy, with emphasis on the substructural alignment with respect to the die geometry and the crystallographic slip systems, which is essentially related to the grain refinement and texture development during deformation. The material was processed by ECAE at room temperature to three passes, via a 90° die. Microstructures were examined and characterized by EBSD. It was found that dislocation cell bands and microshear bands were respectively the most characteristic deformation structures of the first and second pass ECAE. Both formed across the whole specimen and to align approximately with the die shear plane, regardless of the orientation of individual grains. This confirmed that substructural alignment was in response to the direction of the maximum resolved shear stress rather than to the crystallographic slip systems. However, a significant fraction of material developed preferred orientations during deformation that allowed the coincidence between the crystallographic slip systems and the simple shear geometry to occur, which governed texture development in the material. The third pass deformation was characterized with the formation of a fibre structure with a significant fraction of high angle boundaries, being aligned at an angle to the extrusion direction, which was determined by the total shear strain applied.

Keywords: ECAE; simple shear; deformation structure; substructural alignment; EBSD

1. Introduction

Equal channel angular extrusion (ECAE) has become a routine method for severe plastic deformation (SPD) for producing submicron-grained and nanocrystalline metals [1–4]. During ECAE material is extruded through a die comprising two connected, equally cross-sectioned channels, which are intersected at an angle 2φ (see Figure 1); plastic deformation thus occurs substantially by simple shear in a narrow region along the intersectional plane (die shear plane) of the two extrusion channels [1,5]. Similar to conventional processes, such as tension and rolling, dislocations are generated in ECAE deformation and they accumulate to form cell bands of aligned dense dislocation boundaries [6–8]. At larger scales, deformation banding due to orientation splitting may occur on top of the grain shape change in response to the strain applied [9]. A unique feature of ECAE is that intensive shear banding often takes place, particularly upon strain path change during repetitive processing [6]. Although numerous investigations of the microstructural development and the mechanisms of grain refinement during ECAE have been carried out in the past twenty years or so, it seems there are still some fundamental unclear. An essential issue is how characteristic deformation structures such as the dislocation boundaries and shear bands are aligned with respect to the die geometry and crystallographic slip planes, which is critical to the understanding of the deformation mechanisms and texture development in the process. Segal [6,7] suggests that grain refinement in ECAE is primarily a result of shear banding parallel to the die shear plane, which is macroscopically oriented in response to the maximum resolved shear stress, rather than the primary slip planes. We name this type of structural alignment as non-crystallographic. Such non-crystallographic shear banding in ECAE has been observed in several investigations [10,11]. On the other hand, Fukuda et al. [12] were able to show, from their ECAE experiments on high aluminium single crystals of predetermined orientations, that the alignment of dense dislocation boundaries formed is crystallographic, i.e., they are parallel to the primary slip planes. Certainly, shear bands and dislocation cell bands are different features of deformation structure, and shear banding, as a result of strain localisation, can only occur on top of the dislocation cell bands structure. Non-crystallographic shear banding is then not necessarily in conflict with the crystallographic alignment of the dislocation cell bands. However, controversial experimental results have been reported with the alignment of both the dislocation cell bands and shear bands. Therefore, it is necessary to carry out a comprehensive investigation in order to clarify the general nature of the substructrual alignment during deformation in ECAE.

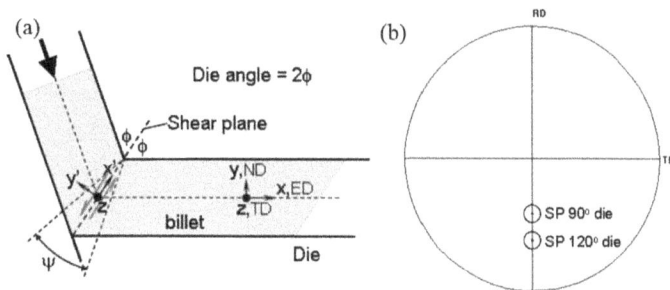

Figure 1. Schematic diagram illustrating (**a**) equal channel angular extrusion (ECAE) process and reference coordinates and (**b**) poles of (111) planes parallel to the dies shear plane in the 90° and 120° dies respectively in the (111) pole figure.

In fact, there has been an ongoing debate, in the last few decades, regarding whether aligned substructure boundaries are oriented along precise crystallographic slip planes in deformation of fcc (face centred cubic) and bcc (body centrerd cubic) metals with intermediate-to-high stacking fault energy [13,14]. This is because the nature of substructural alignment is of fundamental importance to the understanding of basic principles of deformation and mechanical properties of material. In the case of dislocation cell bands, evidence for crystallographic alignment (i.e., planar dislocation boundaries oriented along precise crystallographic slip planes) has been presented for cold deformed aluminium [15] and steels [16] while other work on the same materials [17] has pointed to alignment dictated primarily by the imposed flow field. The study of substructural alignment with the ECAE process has advantages over other processes, such as rolling and tension, because it has a constant and more strictly defined deformation zone and the plane and direction of the maximum resolved shear stress are predetermined by the extrusion angle 2φ, thus, effectively eliminating confusions about the actual deformation geometry as one may encounter in other processes.

In the present work, substructure alignment and crystallographic features of deformation during repetitive ECAE processing of a single-phase aluminium alloy were investigated using high resolution EBSD technique, with a focus on the deformation structure formation during the first three passes. A simple single-phase aluminium alloy was used for this study as dislocation slip in this alloy is limited to the primary slip systems {111}<110> at the testing temperature without twinning, which simplified the verification of fundamental principles.

2. Materials and Methods

A high purity Al-0.13 wt. % Mg alloy, supplied by Alcan International, was used for ECAE processing. The starting material was DC cast and homogenised, cold rolled to 50% in reduction and recrystallised at 400 °C for 1 h, giving a starting grain size of ~350 µm. Extrusion billets, 15 mm in diameter and 100 mm in length, were subsequently machined out in the rolling direction (RD) and processed by ECAE at room temperature with a ram speed of 50 mm/min through a 90° with a small blend radius between the die channels. Colloidal graphite was used as lubricant. The ECAE rig was designed such that the die split in half along their symmetrical plane, parallel to the plane of view in Figure 1. The billet orientation was maintained constant throughout repetitive deformation—commonly referred to as route-A. To facilitate microstructural examination, extrusion was performed half way through the extrusion dies, and samples with deformation structures of two consecutive passes were thus obtained in one sample.

Metallographic samples were cut through the centre of the die symmetrical plane defined by the normal direction (ND) and the extrusion direction (ED) (see Figure 1). Microstructural characterisation was carried out in a field emission gun scanning electron microscope (FEG-SEM) and characterised by backscatter imaging and high resolution EBSD techniques. For EBSD analysis the samples were mechanically polished, followed by electropolishing, to give a strain-free surface. EBSD orientation maps were acquired from samples at different strain levels, using a Philips/FEI Siron FEG-SEM (FEI, OR, USA), fitted with automated HKL-EBSD pattern collection systems. EBSD data was analysed using VMAP© EBSD analysis in-house software (The University of Manchester, Manchester, UK). In the data presented, high angle boundaries (HABs) are defined as having misorientations greater than, or equal to, 15° and low-angle boundaries (LABs) are defined as having misorientations of less than 15°. Due to misorientation noise, boundaries of less than 1.5° misorientation were cut off. Textures were determined from the EBSD maps using either VMAP©EBSD or HKL Channel 5 software (Oxford Instrument, High Wycombe, UK). In the EBSD maps presented, grains are coloured according to orientation, with red, green and blue levels proportional to the three Euler angles (Euler contrast). Low angle boundaries are depicted as white lines and high angle boundaries as black. In all the maps, the extrusion direction (ED) is horizontal.

3. Results

3.1. Cell Bands Alignment in the 1st Pass

Deformation was found to take place in a narrow region along the die shear plane. The deformation zone was roughly fan-shaped, but did not strictly converge to a single point at the die inner corner. A friction-affected layer in the surface was detected, with a depth of ~500 µm, where the microstructure was less defined than in the centre of the sample. All results presented in this paper were therefore obtained from the centre of the sample to avoid confusions.

Figure 2 shows the characteristic microstructures of the material at various scales after one pass ECAE with the 90° die. It is seen that initial grains are elongated in response to the strain applied (Figure 2a). Deformation banding took place due to orientation splitting and deformation bands with either regular slab-like shapes at a scale close to the average grain size (Figure 2b) or irregular geometry (Figure 2c) at a finer scale have formed. At the finest scale, dislocation cell bands are the most distinctive microstructural feature (Figure 2d). The substructural alignment was examined with respect to the deformation geometry and crystallographic orientations of individual grains. It was found that the overall elongation direction of the fine scale substructure is generally within ±5° of the die shear plane. The variation may occur within a grain but mostly across boundaries. The highest variation was found in the area where severe irregular deformation banding took place and the deviation from the die shear plane could be as high as 15°. Figure 4a shows the distribution of the inclination angle to the extrusion direction measured from 50 grains in the centre of the sample. The elongation direction of individual cells is in line with the substructural alignment direction they are in, although the direction of some cells is more diverse than the overall substructure. Figure 3 shows a statistical distribution of cell band alignment direction relative to the extrusion direction, based on image analysis, and it is clear from the figure that dislocation cell bands are strongly aligned with the die shear plane, i.e., at approximately 45° to the extrusion direction. The average misorientation across the cell bands, where they did not coincide with other features of the deformation structure that superimposed strong local orientation gradients, were ~4.1°. This suggests that cell bands did not generate new high angle boundaries. Aligned cell bands were found to have formed in all the grains throughout the billet and continue within deformation bands, with a clear correlation to the dies shear plane as described above. Figure 4 shows an example of die shear plane aligned cell band structure aligned across a few grains in an EBSD map (Figure 4a) and the orientations of the individual grains in a (111) pole figure (Figure 4b), demonstrating that the cell bands are closely aligned in the die shear plane regardless of the orientation of individual grains. It should be noted, however, that plastic flow tended to bring primary slip planes of {111} into the die shear plane. This is similar to the texture development during rolling in which {111} planes tend to rotate and coincide with the rolling plane.

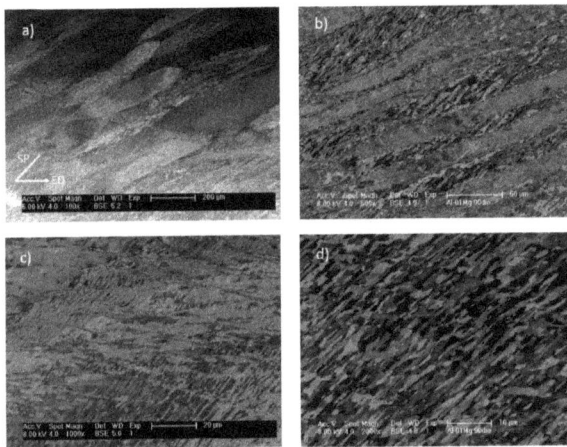

Figure 2. FEG-SEM back scattered images, showing features of characteristic deformation structures after the 1st pass with the 90° die: (**a**) overall elongated microstructure; (**b**) primary deformation bands; (**c**) secondary deformation bands; (**d**) aligned cell bands. ED—the extrusion direction; SP—the die shear plane.

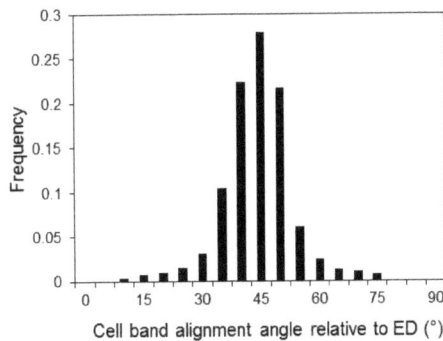

Figure 3. Distribution of cell band alignment angle relative to the extrusion direction (ED), showing that cell bands are substantially aligned with the die shear plane around 45° to ED.

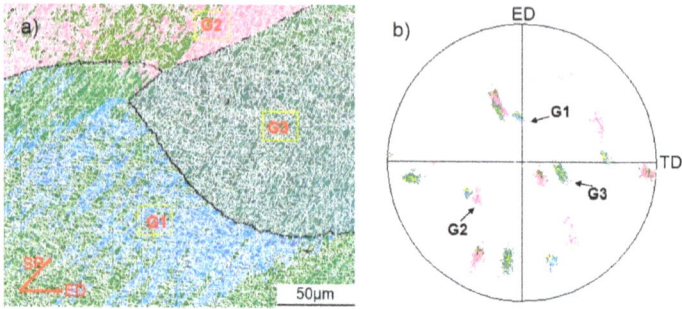

Figure 4. (a) EBSD map, showing the substructural alignment in grains of different orientations and **(b)** (111) pole figure, showing the orientations of the individual grains—after the 1st pass ECAE.

3.2. Structural Evolution in the 2nd Pass

In the 2nd pass of ECAE, although the original grain structure continues to distort in proportion to the shear strain with route A, there is always a strain path change depending on die angle and processing route. This involves a rotation of the idealized shear plane and the activation of latent slip systems, except for route C, which represents a Bauschinger type reversal. For route A used in this investigation, the theoretical shear plane alternates by π-2φ every cycle and should thus intercept the cell bands from the 1st pass by 90°. Such strain path changes are known to promote shear banding as the collapse of the lamellar cell bands, when subjected to an orthogonal shear on new slip systems, can lead to transient flow softening [18].

As expected, it was found that microshear banding parallel to the die shear plane dominated the deformation structure in the 2nd pass. Figure 5 shows shear bands formed in the 2nd pass cutting through the aligned cell band structure developed in the 1st pass. The shear bands were always aligned closely with the die shear plane across the whole sample, and are, therefore, not crystallographically orientated. Some shear bands were only one cell wide, but many were seen in packets of 2–5 bands and even extended across grain boundaries (Figure 5).

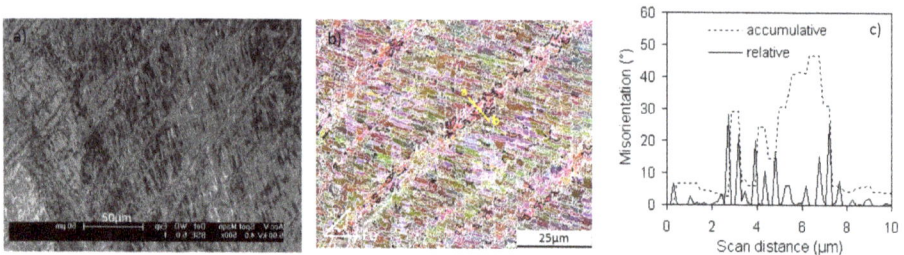

Figure 5. Characteristic features of shear bands formed in the 2nd pass; **(a)** FEGSEM backscatter image; **(b)** EBSD map and **(c)** a linescan of relative and accumulative boundary misorientations along ab in **(b)**, demonstrating the generation of medium and high angle boundaries by shear banding.

EBSD measurements showed that shear bands generated a significant orientation spread along the ND-ED great circle in the (111) pole figures, as a result of rotation around the transitional direction (TD) towards ED, relative to the matrix. The overall lattice rotation was generally in the range of 6 to 30° and in some cases reached 50°, which readily generated new high angle boundary (HAB) segments across the microshear band boundaries as can be seen from the misorientation linescan in Figure 5c (along line ab in Figure 5b). Given that the shear bands are relatively closely spaced, they are an important source of new HAB generation, which benefit from the strain path change during ECAE processing.

3.3. Lamellar Fibre Structure in the 3rd Pass

The fraction of HABs was found to have increased considerably after the 3rd pass. These HABs were predominantly aligned and formed a fibre structure, as shown in Figure 6a. EBSD results showed that the fraction of HABs reached 30% with an average spacing of 1.65 μm. The overall spacing of HABs and low angle boundary (LABs) across the fibre structure was 0.67 μm. The fibre structure was aligned at about 12° to ED. This is in close agreement with the theoretically expected values between the grain elongation direction and the ED (β), which is 9.5° for a 90° die, calculated according to the relationship

$$\beta_n = \arctan \Gamma^{-1} \tag{1}$$

where n is the number of ECAE passes and $\Gamma = 2n\cot\varphi$ is the total shear strain in the die shear plane after n passes. This suggests that the formation of the fibre structure was primarily a result of rigid structuralrotation, which is related to the non-symmetric material spin in simple shear deformation, and compression of the prior HABs in response to the total macroscopic shear strain. The overall fibre structure textures were determined from EBSD measurements over an area of ~500 × 500 μm, and were found to be dominated by {431}<527> component in the ECAE reference frame, as shown in Figure 6b or {111}<110> in the dis shear plane (x'y'z') frame.

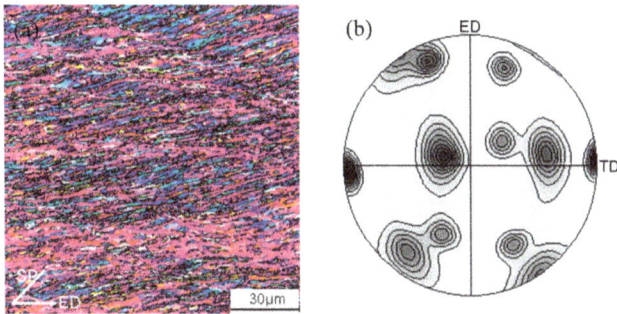

Figure 6. (**a**) EBSD map and (**b**) (111) pole figure, showing the formation of a fibre microstructure and textures developed after 3rd pass ECAE.

A detailed description of the deformation structure evolution to higher strains with route A is beyond the scope of this work, and has been previously been described in general terms [19,20]. Briefly, at larger strains, the HABs become dominant and the fibre boundary spacing approaches one subgrain in width, forming high aspect ratio "ribbon" grains with variable lengths. Further refinement then occurs through shortening of the ribbon grains by transverse LABs developing higher misorientations, and dynamic recovery [8,19,21]. Characteristically, the rate of increase in the fraction of HAB areas and the reduction in the average HAB spacing reduces greatly after the 3rd pass. In the meantime, the effect of crystallographic features on structural evolution became less important.

4. Discussion

The evolution of deformation structure during early stages of ECAE, for up to three passes, has been investigated, with emphasis on the alignment of important features of the deformation structure and their spatial correlation to preferred crystallographic orientations. The development of these characteristic features and their relationship to the texture formation during ECAE processing is discussed below. It should be noted that the deformation mode in ECAE, which approximates to simple shear and the strain path change that causes the shear plane to alternate each pass with route A, leads to some unique features of the deformation structure development. When compared to more conventional processes like rolling and torsion, however, there are also many similarities.

4.1. General Features of Deformation Structure

The experimental results have shown that the aligned cell bands, deformation bands and microshear bands are the main characteristic deformation structure features developed during the early stages of ECAE. In the 1st pass, the deformation structure was characterised by features forming on a range of length scales from fine ~1 m wide aligned cell bands to 40 m large scale primary deformation bands, and at an intermediate level by regions dominated by finer irregular deformation bands. In comparison, the 2nd pass was characterised by the introduction of intense microshear bands, with little further development of cell band structure. In the 3rd pass, a fibrous structure emerged with the HABs becoming aligned and rotated in the direction of grain distortion, with the cell band structure being replaced by subgrains. Orientation splitting was found to be largely established within the 1st pass associated with deformation banding. In rolling and other deformation processes, primary deformation banding usually takes place rapidly at low strains [22] and the present work has evidently confirmed this feature in ECAE.

The generation of new HAB area occurred, by orientation splitting, shear banding and the extension of boundaries due to the imposed shear. Of these mechanisms at low strains, orientation splitting involving fine scale irregular deformation banding is probably the most significant source of grain refinement, whereas at high strains with route A, the extension of boundaries with strain is much more important as previously shown [9]. Shear banding, being dominant in the 2nd pass, also provided an important mechanism for grain subdivision. However, the relative importance of this mechanism is sensitive to material and deformation conditions. In this study shear banding became less significant once the cell band structure broken down and the HAB spacing substantially reduced. The development of new HAB area required for grain refinement is clearly a highly heterogeneous process, which is dependent on localised shear and orientation instability, rather than the gradual uniform rotation of a cellular structure.

4.2. Substructural Alignment

In his review paper, Winther [13] pointed out that although approximate macroscopic alignment generally occurred, precise crystallographic alignment with the most highly stressed slip plane was frequent. In ECAE, investigations have highlighted some crystallographic features of the deformation structures formed but none has yet shown convincing results to clarify this fundamental substructural alignment issue. For example, Fukuda and co-workers [12], proposed that the alignment of cell bands and shear bands was crystallographically related, based on an investigation with aluminium single crystals. However, in their work the micrographs show that the deformation zone was extremely wide and the deformation mode is thus unlikely to be a pure simple shear, making it difficult to know the direction of the maximum shear stress.

In the present work, cell bands formed in the 1st pass have been clearly identified to be aligned substantially within a deviation of $\pm 10°$ with the die shear plane as shown in Figure 3, irrespective of individual grain orientations. In the 2nd pass, microshear bands also formed in alignment with the die shear plane (Figure 5a). Extensive microstructural examinations and EBSD measurements showed that cell bands and microshear bands were aligned with the die shear plane across the entire sample encompassing an unlimited amount of grains. Clearly they were macroscopically oriented in response to the maximum resolved shear stress in the die shear plane, rather than being crystallographically related. However, rotation of {111} planes towards the shear plane took place to facilitate dislocation slip, which increased the coincidence of planar dislocation boundaries with {111} slip planes in the deformation structure.

In rolling, the boundaries of aligned cell bands are transient features of the microstructure, which remain active and undergo continuous reorganisation during deformation, maintaining their alignment with respect to the deformation geometry [23]. These boundaries are the result of dynamic recovery of the dislocation debris produced during slip in which the dislocation walls are continually dissolving, reforming and re-orienting during deformation. In ECAE, however, the cell bands formed in the 1st pass were a permanent feature during the subsequent deformation pass. The change in strain path between passes suddenly realigns the cell bands relative to the shear plane, by φ-2π. For the 90° die this is equivalent to an orthogonal change in strain path and this dramatic change in the active slip systems results in the boundaries collapsing, which promotes shear banding; with no sign of reorganisation and reformation of the cell bands. The cell bands thus largely undergo a rigid body rotation in the 2nd pass, which removes their shear plane correspondence.

It has been recognized that the substructure formed in the 1st pass has a strong effect on the development of deformation structure in the following repetitive ECAE processing [8,24]. The study of the 3rd pass deformation structure showed that the generally reported fibre/lamellar structure formed at medium to high strains is the result of evolution and realignment of the 1st pass cell bands and the 2nd pass microshear bands, during which a significant amount of HABs developed. It should be noted that, although the boundaries generated by microshear bands have much higher average misorientations than cell band boundaries, the overall intensity of the latter is substantially higher. On average, the spacing of microshear bands is about 5 to 10 times of cell bands. Therefore, cell bands might have contributed more than microshear bands in structural refinement.

4.3. The Development of the Preferred Orientations and Their Effect on Deformation

The development of deformation textures in ECAE has previously been widely reported to be related to simple shear [3,9,25,26], although different authors have used either the shear plane or tool reference frame. Simulations have also been performed given reasonable agreement with experimental measurements [27]. Most authors have related the textures observed to the preferred orientations and $A_{\{111\}}$ and $B_{<110>}$ fibres seen in torsion tests. Simple shear occurs when direction of slip is in the plane of tilt. Coincidence between the crystallographic slip plane and the simple shear plane during ECAE not only allows the least number of slip systems to operate during deformation, but also requires less energy than non-crystallographically correlated shear. Therefore, once a crystallographic correlation is established, it tends to be maintained during subsequent deformation. For an idealised simple shear deformation path in ECAE, it is thus unsurprisingly to see the development of preferred orientations that provide the crystallographic coincidence with the die shear plane. However, deformation through most dies only approximates to a simple shear and frequently involves other strain components. Furthermore, simple shear textures tend to be relatively weak as, unlike in plane strain compression, there is a constant rigid body rotation which results in material element break away from preferred orientations.

The preferred orientations found in the present work were in agreement with the predictions of texture simulations using either full or relaxed constraints Taylor model [27–29] and bulk texture measurements findings [20,30]. The main texture components observed in the EBSD maps are summarised in Table 1. The orientation transformation from the ECAE die reference frame (ND-ED-TD), in which orientations are presented in the pole figures, to the idealised shear plane reference frame (x'y'z', see Figure 1a) was carried out by an anticlockwise-rotation about TD of 45° for the 90° die.

As given in Table 1, common orientations on $A_{\{111\}}$ fibres A1 and A2 are equivalent to {111}<1-10> and {111}<11-2> components in the shear plane reference frame. B-type orientations in the shear plane reference frame, i.e., B1 {100}<01-1> developed with both dies, an additional different B-type orientation B2 developed with the 90° die, which was observed in a high fraction in the 3rd pass fibre structure. Furthermore, the {111}<110> orientation in the ECAE die reference frame, found to be often associated with deformation bands in both the 90° and 120° dies, rotates, to a "random" near {521}<012> and {113}<174> with the 90° and 120° die is in the shear plane reference frame.

Table 1. A summary of the characteristic textures and associated microstructural features such as cell bands (CBs), primary deformation bands (1DBs) and secondary deformation bands (2DBs), and shear bands (SBs) and fibre structures (fibre) in both simple shear and ECAE reference systems x'y'z and xyz, including the related slip systems.

Notation	Orientations {hkl}<uvw>		Slip Systems in SSRS		Schmid Factor	Structural Features
	ECAERS	SSRS	Plane/θ_{SP} [*1]	Direction/θ_{SD} [*2]		
A1	(521)[01-2] (52-1)[012]	(111)[-110] (11-1)[-110]	(111)/0 (111)/0	[-110]/0 [-110]/0	1	CBs, SBs
A2	(81-1)[1-44]	(111)[-1-12]	(111)/0	[-101]/+35.26 [0-11]/−35.26	0.82	CBs, 1DBs, SBs
B1	(122)[-411]	(100)[01-1]	(11-1)/57.4 (1-11)/−57.4	[011]/0 [011]/0	~0.56	CBs, 2DBs
B2	(341)[-527] (3-14)[57-2]	(112)[1-10] (121)[10-1]	(111)/+19.42 (111)/−19.42	[1-10]/0 [10-1]/0	~0.94	SBs, fibre

[*1] θ_{SP}—rotation angle of (111) plane away from the die shear pane about −x'; [*2] θ_{SD}—rotation angle of <110> directions away from −x' about y' in the die shear plane.

The initial starting texture of the annealed Al-0.13Mg used in the present work was a recrystallised rolling texture, comprised primarily of weak P {110}<111> and ND-rotated cube {001}<012> components [31]. Grains of P {110}<111> texture require a rotation of 45° with the 90° die to be in the orientation parallel to the die shear plane, whereas grains of rotated cube {001}<012> orientation are only a few degrees away from B1, suggesting that the initial textures could contribute to the development of such texture in association with cell bands and irregular deformation bands. However, since the intensity of both P and rotated cube texture components was very low (less than 10% volume fraction in total), their contributions must be limited.

Shearing on A-type orientations occurs on a {111} plane, parallel to die shear plane. The ideal orientation is A1 with the 90° die with a <110> direction in the maximum resolved shear stress direction and Schmid factor of 1. However, A1 and A2 orientations were found to be equally dominant in the 1st pass. On the other hand, A2, despite being of a smaller Schmid factor of 0.82, was more frequently observed than A1 during shear banding in the 2nd pass. This could be due to that the A2 orientation contains two <110> directions in the slip plane symmetrical about ED at ±35.26° and this allows two slip systems to operate simultaneously, which may result in a more stable orientation than single system dominated slip. The B$_{<110>}$ fibres contain a <110> direction aligned with the shearing direction −x', although they have up to four (111) planes, which are at an angle to the die shear plane but parallel to the shear direction (−x'). Since the deformation zone is in reality a three dimensional volume with a near fan-shaped cross section on the ND-ED planes, the (111) planes of the B type orientations actually have a certain area of intersection with the deformation zone, which was found to have a dimension of ~1000–1500 μm, which is larger than the average grain size (~300 μm) in the present investigation. Thus slip on the B-type orientations was certainly possible and should be easier than on random orientations.

Upon strain path change, simple shear in the 2nd pass took place on new slip systems and the shear bands formed are seen to cut through the cell bands formed in the 1st pass, leading to a spread in orientation away from the matrix, whereas the matrix preferred orientations were still close to those expected from simple shear. The relative intensities of the A and B textures were different, with an increase in B type components relative to A. It is thus interesting to note that dominant starting A-type orientations in the shear plane reference frame developed in the 1st pass undergo an anticlockwise rotation of φ about Z' relative to the shear plane in the 2nd pass. Because by symmetry there are several variants possible for each preferred orientation this readily realigns a close to a favourable stable orientation in the 2nd pass.

The development of preferred orientations certainly reduces the energy consumption for deformation. It also has effect on the deformation structure and contributes directly to the textures. It has been shown that the intensity of the aligned cell bands is higher in the grains with preferred orientations (in particular A-type) than other grains. This is not difficult to understand because (1) the crystallographic coincided slip has the benefit of higher Schmid factor and (2) the cell band boundaries formed tend to have the same Burgers vector, which reduces the chance for the boundaries dislocations to be annihilated during dynamic recovery as may occur to dislocations of different signs. In the case of shear banding, the coincidence between crystallographic slip systems with the die geometry should effectively "soften" the material under shear and promote strain localisation.

4.4. Comparison with Torsion and Rolling

Simple shear in ECAE is basically one dimensional and symmetrical only to the central plane defined by ND-ED, whereas in torsion sample symmetry contains a rotation axis of infinite order and the simple shear is axial symmetrical, in which the orientation distribution is independent of rotation about the axis and textures are thus fibre textures [32]. Additionally, the simple shear direction changes constantly in torsion whereas in ECAE it remains constant and deformation occurs in a narrow region along the die shear plane. The coincidence of characteristic orientations between the two deformation modes was rather likely due to the fact that dislocations in fcc metals can only glide on (111) planes and in <110> directions. It is expected that coincidence between crystallographic slip systems and the deformation geometry should be less common in rolling than in ECAE because the dominant deformation mode in rolling is pure shear. In pure shear, no frames of reference remain unchanged, although directions of greatest compression and extension are constant. This means that with any increment of strain the crystallographic planes will rotate. In rolling, no stable crystallographic relationship with respect to the maximum resolved shear direction, or the principle shear direction, is maintained and more slip systems must operate to retain the continuity of material and cross-slip has to take an essential part during deformation in the fcc type of materials. This is in agreement with observations by Humphreys and co-workers [33].

5. Conclusions

Simple shear along the die shear plane was the dominant deformation mode. The aligned cell bands and shear bands, both being closely aligned in the die shear plane, are the most characteristic deformation structures for both the 1st and 2nd pass ECAE. The 3rd pass deformation was characterized with the formation of a fibre structure with a significant fraction of high angle boundaries. Deformation banding occurred in the 1st pass due to orientation splitting and two types of deformation bands—primary and irregular, were observed and both developed high angle boundaries. No further deformation banding was observed in the 2nd and 3rd pass.

Cell bands and microshear bands generated during deformatin were aligned to the maximum shear stress plane, i.e., die shear plane, regardless of the crystallographic slip planes of individual grains.

Despite that simple shear was non-crystallographically related in principle, a significant fraction of material developed preferred orientations during deformation that allowed the coincidence between the crystallographic slip systems and the simple shear geometry to occur. The aligned cell bands formed in the 1st pass exhibited a permanent nature and evolved into the fibre structures in the 3rd pass, with certain amount of boundaries developed high misorientations. The 2nd pass shear bands were primarily microshear bands and contained a certain fraction of high misorientation boundary segments upon formation and formed an important part of the 3rd pass fibre structure.

The developed preferred orientations showed a crystallographic relationship with the simple shear geometry—having a (111) plane parallel to the die shear plane, which resulted in the development of A type textures, although B type textures were also observed with only slip directions in the dis shear plane.

Acknowledgments: The author would like to acknowledge the financial support from EPSRC Light Alloys Portfolio Partnership (EP/D029201/1) for this project and the helpful discussion with Professor P.B. Prangnell.

Author Contributions: The work is done solely by the author.

Conflicts of Interest: The author declares no conflict of interest.

References

1. Segal, V.M.; Reznikov, V.I.; Drobyshevskiy, A.E.; Kopylov, V.I. Simple shear in equal channel angular extrusion. Russian Metallurgy. *Engl. Transl.* **1981**, *1*, 99–107.
2. Segal, V.M. Materials processing by simple shear. *Mater. Sci. Eng. A* **1995**, *197*, 157–164. [CrossRef]
3. Gholinia, A.; Prangnell, P.B.; Markushev, M.V. The effect of strain path on the development of deformation structures in severely deformed aluminium alloys processed by ECAE. *Acta Mater.* **2000**, *48*, 1115–1130. [CrossRef]
4. Casati, R.; Fabrizi, A.; Tuissi, A.; Xia, K.; Vedani, M. ECAP consolidation of Al matrix composites reinforced with in-situ γ-Al_2O_3 nanoparticles. *Mater. Sci. Eng. A* **2015**, *648*, 113–122. [CrossRef]
5. Valiev, R.Z.; Islamgaliev, R.K.; Alexandrov, I.V. Bulk nanostructured materials from severe plastic deformation. *Prog. Mater. Sci.* **2000**, *45*, 103–189. [CrossRef]
6. Segal, V.M. Equal Channel angular extrusion: From macromechanics to structure formation. *Mater. Sci. Eng. A* **1999**, *271*, 322–333. [CrossRef]
7. Segal, V.M. Engineering and commercialization of equal channel angular extrusion (ECAE). *Mater. Sci. Eng. A* **2004**, *386*, 269–276. [CrossRef]
8. Prangnell, P.B.; Bowen, J.R.; Apps, P.J. Ultrfine grain structures in aluminium alloys by severe deformation processing. *Mater. Sci. Eng. A* **2004**, *375–377*, 178–185. [CrossRef]
9. Huang, Y.; Prangnell, P.B. Orientation splitting and its contribution to grain refinement during equal channel angular extrusion. *J. Mater. Sci.* **2008**, *43*, 7273–7279. [CrossRef]
10. Zhang, D.; Li, S. Orientation dependencies of mechanical response, microstructure and texture evolution of AZ31 magnesium alloy processed by equal channel angular extrusion. *Mater. Sci. Eng. A* **2011**, *528*, 4982–4987. [CrossRef]
11. Sitdikov, O.; Avtokratova, E.; Sakai, T. Microstructural and texture changes during equal channel angular pressing of an Al-Mg-Sc alloy. *J. Alloy. Comp.* **2015**, *648*, 195–204. [CrossRef]
12. Fukuda, Y.; Oh-ishi, K.; Furukawa, M.; Horita, Z.; Langdon, T.G. Influence of crystal orientation on ECAP of aluminium single crystals. *Mater. Sci. Eng. A* **2006**, *420*, 79–86. [CrossRef]
13. Winther, G. Slip patterns and preferred dislocation boundary planes. *Acta Mater.* **2003**, *51*, 417–429. [CrossRef]
14. Hurley, P.; Humphreys, F.J.; Bate, P. An objective study of substructural boundary aliognment in aluminium. *Acta Mater.* **2003**, *51*, 4737–4750. [CrossRef]
15. Winther, G.; Juul Jensen, D.; Hansen, N. Dense dislocation walls and microbands aligned with slip planes—Theoretical considerations. *Acta Mater.* **1997**, *45*, 5059–5068. [CrossRef]
16. Haldar, A.; Huang, X.; Leffers, T.; Hansen, N.; Ray, R.K. Grai orientation dependence of microstructures in a warm rolled IF steel. *Acta Mater.* **2004**, *52*, 5405–5418. [CrossRef]
17. Humphreys, F.J.; Bate, P. Measuring the alignment of low angle boundaries formed during deformation. *Acta Mater.* **2006**, *54*, 817–829. [CrossRef]
18. Ferrasse, F.; Segal, V.M.; Hartwig, T.K.; Goforth, R.E. Microstructure and properties of copper and aluminium alloy 3003 heavily worked by equal angular extrusion. *Metall. Mater. Trans. A* **1997**, *28*, 1047–1057. [CrossRef]
19. Apps, P.J.; Bowen, J.R.; Prangnell, P.B. The effect of coarse second-phase particles on the rate of grain refinement during severe deformation processing. *Acta Mater.* **2003**, *51*, 2811–2822. [CrossRef]
20. Etter, A.L.; Baudin, T.; Rey, C.; Penelle, R. Microstructural and texture characterization of copper processed by ECAE. *Mater. Char.* **2006**, *56*, 19–25. [CrossRef]
21. Prangnell, P.B.; Huang, Y.; Berta, M.; Apps, P.J. Mechanisms of formation of submicron grain structures by severe deformation. *Mater. Sci. Forum* **2007**, *550*, 159–168. [CrossRef]

22. Kuhlmann-Wilsdorf, D. Q: Dislocations structures—How far from equilibrium? A: Very close indeed. *Mater. Sci. Eng. A* **2001**, *315*, 211–216. [CrossRef]

23. Humphreys, F.J.; Hatherly, M. *Recrystallization and Annealing Phenomena*, 2nd ed.; Pergamon Press: Oxford, UK, 2004; pp. 28–35.

24. Longdon, T.G. The principles of grain refinement in equal channel angular extrusion. *Mater. Sci. Eng. A* **2007**, *462*, 3–11. [CrossRef]

25. Zhilyaev, A.P.; Oh-ishi, K.; Raab, G.I.; McNelley, I.R. Influence of ECAP processing parameters on texture and microstructure of commercially pure aluminium. *Mater. Sci. Eng. A* **2006**, *441*, 245–252. [CrossRef]

26. Kliauga, A.M.; Bolmaro, R.E.; Ferrante, M. The evolution of texture in an equal channel pressed aluminium AA1050. *Mater. Sci. Eng. A* **2015**, *623*, 22–31. [CrossRef]

27. Li, S.; Gazder, A.; Beyerlein, I.J.; Davies, C.H.J.; Pereloma, E.V. Microstructure and texture evolution during equal channel angular extrusion of interstitial free steel—Effects of die angle and processing route. *Acta Mater.* **2007**, *55*, 1017–1032. [CrossRef]

28. Beyerlein, I.J.; Lebensohn, R.A.; Tomé, C.N. Modelling texture and microstructure in the equal channel angular extrusion process. *Mater. Sci. Eng. A* **2003**, *345*, 122–138. [CrossRef]

29. Signorelli, J.W.; Turner, P.A.; Sordi, V.; Ferrante, M.; Vieira, E.A.; Bolmaro, R.E. Computational simulation of texture and microstructure evolution in Al alloys deformed by ECAE. *Scr. Mater.* **2006**, *55*, 1099. [CrossRef]

30. Pithan, C.; Hashimoto, T.; Kawazoe, M.; Nagahora, J.; Higashi, K. Microstructure and texture evolution in ECAE processed A5056. *Mater. Sci. Eng. A* **2000**, *280*, 62–68. [CrossRef]

31. Gholinia, A.; Bate, P.; Prangnell, P.B. Modelling texture development during equal channel angular extrusion of aluminium. *Acta Mater.* **2002**, *50*, 2121–2136. [CrossRef]

32. Canova, G.R.; Kocks, U.F.; Jonas, J.J. Theory of torsion texture dev elopement. *Acta Metal.* **1984**, *32*, 211–226. [CrossRef]

33. Hurley, P.J.; Humphreys, F.J. Modelling the recrystallization in single phase aluminium. *Acta Mater.* **2003**, *51*, 3779–3793. [CrossRef]

metals

MDPI

Article

Synthesis and Characterization of Nanocrystalline Al-20 at. % Cu Powders Produced by Mechanical Alloying

Molka Ben Makhlouf [1], Tarek Bachaga [1,2,*], Joan Josep Sunol [2], Mohamed Dammak [1] and Mohamed Khitouni [1]

[1] Laboratory of Inorganic Chemistry, University of Sfax, Sfax 3000, Tunisia;
molkabenmakhlouf@yahoo.fr (M.B.M.); meddammak@yahoo.fr (M.D.); khitouni@yahoo.fr (M.K.)

[2] Department de Fisica, Universitat de Girona, Campus Montilivi, Girona 17071, Spain;
joanjosep.sunyol@udg.edu

* Correspondence: bachagatarak@yahoo.fr; Tel.: +216-950-607-15 (ext. 00216); Fax: +216-742-744-37

Academic Editor: Nong Gao
Received: 8 May 2016; Accepted: 15 June 2016; Published: 29 June 2016

Abstract: Mechanical alloying is a powder processing technique used to process materials farther from equilibrium state. This technique is mainly used to process difficult-to-alloy materials in which the solid solubility is limited and to process materials where nonequilibrium phases cannot be produced at room temperature through conventional processing techniques. This work deals with the microstructural properties of the Al-20 at. % Cu alloy prepared by high-energy ball milling of elemental aluminum and copper powders. The ball milling of powders was carried out in a planetary mill in order to obtain a nanostructured Al-20 at. % Cu alloy. The obtained powders were characterized using scanning electron microscopy (SEM), differential scanning calorimetry (DSC) and X-ray diffraction (XRD). The structural modifications at different stages of the ball milling are investigated with X-ray diffraction. Several microstructure parameters such as the crystallite sizes, microstrains and lattice parameters are determined.

Keywords: mechanical alloying; nanocrystalline; crystallite sizes; morphology

1. Introduction

Mechanical alloying (MA) is considered a powerful technique as it can facilitate true alloying materials. In general, both stable and metastable phases can be produced by ball milling [1–6]. Solid-state reactions induced by high-energy ball milling have recently attracted a large amount of research work [7,8]. This is because the high-energy ball milling approach has been recognized as a complex process which can be applied to the processing of advanced materials at low cost. Among these, mechanical alloying (MA) has often been reported to be a powerful and relatively simple technique that allows for the preparation of nanostructured alloys [8]. It is commonly known that during MA, powders undergo a severe plastic deformation, which introduces a number of defects into the material, and it is worth noting that this causes a gradual change in the state of the powder mixtures and hence their properties [8,9]. Further, Eckert et al. [10] found that the final grain size is determined by the competition between the deformation produced by a milling process, and the dynamic recovery in the milled material. On the other hand, it has been suggested that the stacking fault energy (SFE) has a strong influence on the evolution of the dislocation structure, which precedes and results in the nanocrystalline structure formation [11].

The Al-Cu system is an example of a binary system with a low solid miscibility at room temperature (a miscibility of approximately 0.1 at. % [12]). According to the phase diagram, the solubility of Cu in Al is about 1 at. % near 350 °C and reaches a maximal value (about 2.5 at. %) at 548 °C [13]. However, by means of MA, Al-Cu solid solutions can be obtained. At these temperatures, the solid solution is in equilibrium with the chemical compound Al_2Cu. In general, the mechanical strengthening of Al metal was usually achieved by the impurity doping of 0.5%–4% Cu [14,15]. Further studies on Al-Cu have reported that θ-Al_2Cu is the first nucleus intermetallic compound [16]. Premkumar et al. [15] added that this intermetallic becomes γ-Al_4Cu_9 or η-AlCu as the purity of the Cu wire increases. Generally, depending on the thermodynamics, the diffusion couples of Al-Cu can be produced in several intermediate phases and in many ways. Earlier studies reported by Li et al. [17] and by Chattopadhyay [18] have indicated that mechanical alloying yields a metastable bcc solid solution in the composition range of Al-35 to 65 at. % Cu. Moreover, Murray indicated that the maximum solid solubility of copper in Al by mechanical alloying is estimated to be 2.7 at. % Cu, which is larger than the solubility of 0.1 at. % Cu in the equilibrium state at room temperature. In the present work, we are interested the compound Al-20 at. % Cu prepared by the mechanical milling technique. The microstructure changes as a function of milling time were investigated by means of X-ray diffraction (XRD) and scanning electron microscopy (SEM). Furthermore, special attention will be paid to thermal stability by using differential scanning calorimetry (DSC).

2. Materials and Methods

Al (99.5% purity, mean particle size <50 μm, 325 mesh) and Cu (99.95% purity, mean particle size <40 μm, 200 mesh) elemental powders were used as starting materials. The initial powders with the nominal compositions of Al-20 at. %Cu were milled up to 20 h using a planetary ball mill (Pulverisette P7, Fritsch, Industriestraße 8, Idar-Oberstein, Germany) under argon atmosphere. The ball-to-powder weight ratio was maintained as 1:5. The milling was repeated for different milling times (2 h, 4 h, 6 h, 10 h, 16 h, and 20 h) at 600 rpm. To avoid the local temperature rise inside the vials during milling, each 10 min of milling was followed by a pause of 5 min. The structural changes of the milled samples were investigated by X-ray diffraction (XRD) by means of a Bruker D8 Advance diffractometer (Bruker D8; Manning Park Billerica, MA, USA) in a (2θ) geometry using Cu-Kα radiation ($\lambda_\alpha Cu = 0.15406$ nm). The XRD data was collected at a slow scan rate of $0.016°/4$ s. The microstructural parameters were taken out from the refinement of the XRD patterns by using the MAUD program [19] which is based on the Rietveld method. The evolution of the particle morphology during MA was carried out by means of a scanning electron microscope (SEM) (DSM960A ZEISS, Norman, OK, USA) with energy dispersive X-ray microanalysis (EDX, Norman, OK, USA). Thermal analyses were performed by means of differential scanning calorimetry (DSC, DSC822 apparatus of Mettler Toledo; Columbus, OH, USA) instrument with a heating rate of 20 °C/min up to 700 °C under constant Ar flow.

3. Results and Discussion

3.1. X-ray Diffraction

Evidence of the continuous refinement of the microstructure and the introduction of several structural defects (grain boundaries, dislocations, vacancies, stacking faults, etc.), with increasing milling time, was provided by the decrease of the diffraction peak intensities and their broadening. The disappearance and/or the appearance of some peaks can be assigned to the mixing of the elemental powders and, therefore, to the formation of new phases [8].

Figure 1 presents the XRD patterns of the powders milled for various milling times. The unmilled sample exhibits a pattern consistent with the structure of fcc-Al (space group Fm3m; $a_0 = 0.4046(4)$ nm) and fcc-Cu (space group Fm3m; $a_0 = 0.3611(4)$ nm) precursors. As shown in Figure 1, after 2 h milling, the peaks specific to the Al and Cu diffraction peak profiles became asymmetric and started to broaden and no significant mechano-reaction occurred during this initial period of milling. However, after 4 h milling, one can see the decrease of the main Cu diffraction peak and the appearance of new ones for 2θ ~43.98; 64.14; 80.96 and 97.40°. These peaks can be indexed as nonequilibrium body-centered cubic (bcc) phase with space group Immm and lattice parameters $a_0 = 0.2897(4)$ nm. The same results have been found by Chattopadhyay et al. [5] after MA of the Al-Cu system in the composition range $Al_{65}Cu_{35}$ to $Al_{35}Cu_{65}$. After 6 h milling, tetragonal-Al_2Cu with space group I4/mcm and lattice parameters $a = 0.9107(4)$ nm and $c = 0.4460(4)$ nm started to form.

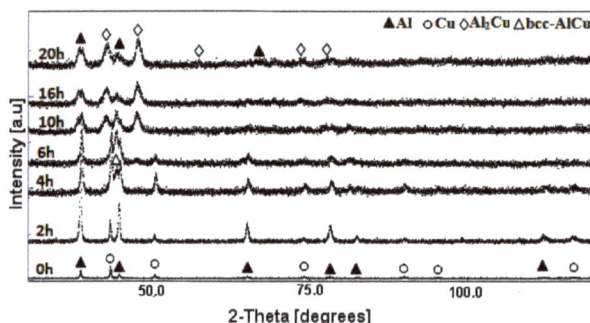

Figure 1. XRD patterns of Al-20 at. % Cu powders collected at different milling times.

Figure 2 gives the Rietveld refinement for patterns obtained before (corresponds to 0 h) and after mechanical milling for 4 h, 6 h, and 20 h. The best Rietveld refinement (GOF = 1.12) for the pattern of the unmilled powder is obtained with two crystalline phases as well as the fcc-Al phase with the lattice parameter $a = 0.4051(1)$ nm and the fcc-Cu phase with the lattice parameter $a = 0.3617(5)$ nm (Figure 2a). The best Rietveld refinement (GOF = 1.3) for the pattern of the mixture milled for 4 h is obtained with both crystalline phases and the apparition of a bcc-AlCu solid solution with the lattice parameter $a = 0.2897(4)$ nm (Figure 2b). Increasing the milling time up to 6 h, the diffusion of the Cu atoms into the Al matrix leads to the formation of two supersaturated solid solutions, tetragonal-Al_2Cu and bcc-AlCu. This result was confirmed by the best Rietveld refinement (GOF = 1.27) of the pattern corresponding to MA powder for 6 h (Figure 2c). In a previous work, Onuki et al. [20] reported that the formation of the tetragonal-Al_2Cu phase during mechanical milling is due to the negative enthalpy of the mixing of the Al-Cu system. Furthermore, the solubility of the solutes is enhanced with a grain refinement on the nanometer scale. After 20 h milling, the Rietveld refinement of the powder pattern was successfully obtained with the fcc-Al and tetragonal-Al_2Cu phases (Figure 2d). These phases correspond to the Al-20 at. % Cu composition in the equilibrium phase diagram of the Al-Cu system at room temperature.

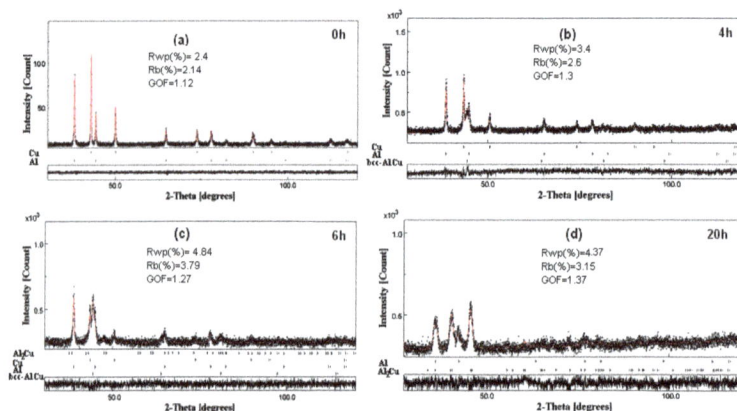

Figure 2. Rietveld refinement of the XRD patterns of the Al-20 at. % Cu powders at different milling times: (**a**) 0 h; (**b**) 4 h; (**c**) 6 h; (**d**) 20 h.

The lattice parameter of Cu increases from 0.3611(4) to 0.3617(5) nm after 2 h of milling. The relative deviation of the lattice parameter from that of the perfect crystal, which is defined by $\Delta a/a_0 = (a - a_0)/a_0$, reaches as much as 0.16%. The lattice parameter of the Al phase is enhanced by 0.20% after 4 h of milling. Sui et al. [21] attributed this lattice distortion (lattice expansion or contraction) to the supersaturation of point defects or vacancies inside the nanometer crystallites due to their higher energetic solution.

The observed broadening of diffraction peaks suggests the accumulation of lattice strain and a reduction in crystallite size. Figure 3 presents the evolutions of the average crystallite size and microstrains deduced from the Rietveld refinement as a function of milling time. As shown, one can observe an important decrease of the crystallite size and an increase of the microstrains during the first stage of milling (0 to 4 h milling). For a prolonged milling time, both the crystallite size and microstrains become less dependent on the milling time. After 10 h of milling, the bcc-AlCu is characterized by a smaller crystallite size and higher microstrains as compared to the Al_2Cu. The final values of the average crystallite size of the Al and Al_2Cu phases calculated after 20 h of milling were 7 nm and 13 nm, respectively. The high degree of microstrains in the Al_2Cu (1.08%) may be due to a high concentration of stacking faults and a high dislocation density. In general, microstrains may arise from a mismatch in the size of the constituents, an increase in the grain boundary fraction, or a mechanical deformation [22]. The microstrains caused by MA have also been previously reported in the literature and have commonly been attributed to the generation and movement of dislocations [23,24]. In order to investigate the stage's mechanical stabilities during milling, we have the calculated phase's proportions of the identified phase as a function of the milling times. Note that there are some phases that progressively decrease (Al and Cu) and others which arise in the form of solid solutions (bcc-AlCu and Al_2Cu). In addition, we observe that the percentage of the bcc-AlCu stage reaches a maximum (70%) after 12 h of milling, and then it progressively decreases to a value of 8.80% after 16 h of milling, while the percentage of the Al_2Cu phase increases continuously with the milling time to reach its maximum value at a milling time of 20 h (82%). So the final system obtained after 20 h of milling is biphasic with two phases: fcc-Al and tetragonal-Al_2Cu phases.

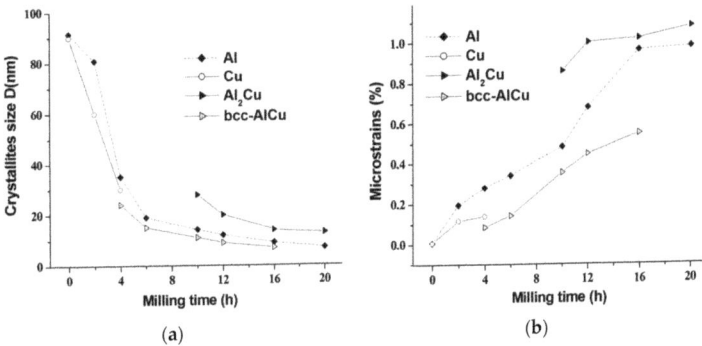

Figure 3. Dependences of refined microstructural parameters of Al-20 at. % Cu powder mixtures on milling time: (a) Crystallite size and (b) Microstrains.

3.2. Scanning Electron Microscopy

The morphologies of as-received Al and Cu powders are shown in Figure 4. Before milling, the Al particles have a spherical-like morphology while the Cu particles have elongated forms (Figure 4a). The changes in morphology during the milling process are due to the competition between the fracturing, cold welding, agglomeration and de-agglomeration of the powder particles. After 4 h of milling, as is normal during the milling of ductile-ductile systems, the mixture is only composed of big particles (Figure 4b). Since the powder particles are soft during the early stage of milling, they tend to weld together and form big particles. With the increase of the milling time (6 h), fine particles aggregate to shape flake-like powders. A broad range of a particle size can then be seen (Figure 4c). Due to the hardening of the powder under the effect of the repeated shocks of the balls during continued milling (20 h), the particles become fractured, and are hence fine and fairly homogeneous in size and shape (Figure 4d). The induced heavy plastic deformation in the powder particles during the milling process gives rise to the creation of a great amount of crystal defects such as dislocations, vacancies, interstitials and grain boundaries which promote a solid-state reaction at ambient temperature. Depending on the initial mixture, changes in structures of mechanically alloyed powders can occur as follows: grain refinement, solid solution diffusion and/or formation of new phases.

Figure 4. SEM morphologies of the Al-20 at. % Cu powders for different milling times: (**a**) 0 h; (**b**) 4 h; (**c**) 6 h; (**d**) 20 h.

3.3. Thermal Stability

Nanostructured and disordered structures obtained by MA are metastable and, therefore, they will experience an ordering transition during heating. Hence, the thermal stability of the alloy is dependent on the structural state after each milling time. Several thermal effects are revealed in the DSC curves of the Al-20 at. % Cu powders milled several times (Figure 5). Before milling, the analysis of the mixture of the powders in the temperature range 25 °C–700 °C shows an endothermic peak at 660 °C, attributed to the melting of aluminum particles [25]. As shown in Figure 5b, the DSC trace of 4 h milled powder exhibited an endothermic peak at 560 °C followed by an exothermic peak at 580 °C. This later was followed by a small endothermic peak at 610 °C. The endothermic peaks might be caused by the melting of the bcc-AlCu phase identified by XRD (see Figure 1) and another fine Al-rich phase formed through the diffusion of Cu into Al during heating. After the powder was milled for 6 h, the DSC trace presented an exothermic peak at 580 °C followed by an endothermic peak at 650 °C, which might be related to the formation and dissolution of the Al-rich phase. The same results have been reported by Ying et al. [26] in the case of Cu-Al alloy with an Al composition of 35 at. % for different times. They attributed these endothermic peaks to the melting of Al-rich phases formed during heating. The endothermic peaks identified for the powder milled for 12 h were in the temperature range of 560 °C and 580 °C, and they were likely caused by the melting of Al-rich phases initially formed through the diffusion of Cu into Al during MA as well as bcc-AlCu and Al_2Cu. The DSC trace of the powders milled for 20 h exhibited an endothermic peak at 550 °C followed by an exothermic one at 600 °C, attributed to the melting of the eutectic Al-Cu binary phase and the formation of the Al-Cu intermetallic, respectively.

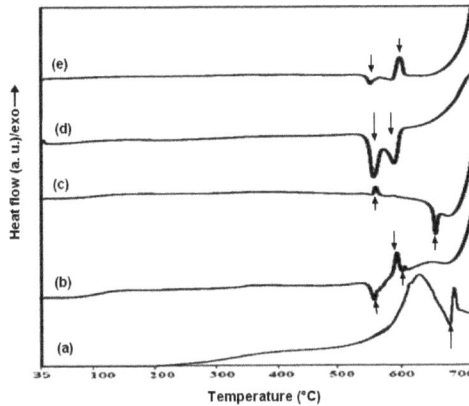

Figure 5. DSC pattern of the Al-20 at. % Cu powders milled for different milling times: (**a**) 0 h; (**b**) 4 h; (**c**) 6 h; (**d**) 16 h; (**e**) 20 h.

4. Conclusions

Structural, morphological and thermal properties of mechanically alloyed Al-20 at. % Cu powders have been carefully studied as a function of milling time. The interdiffusion of Cu and Al leads to the formation of bcc-AlCu and tetragonal Al_2Cu phases. The formation of the bcc-AlCu phase was observed in the early stage of milling (4 h of milling), while the tetragonal Al_2Cu was revealed from 6 h of milling. It was also found that the crystallite size of the relatively milled powder was decreased with the increasing milling duration. The crystallite size and microstrain of the milled powder were estimated to be in the range of 10–15 nm and 1%–1.1%, respectively. It was shown that during the milling process, Al and Cu particles underwent severe plastic deformation, which can lead to grain refinement, solid solution diffusion and/or the formation of new phases. The thermal stability of the mechanically alloyed Al-20 at. % Cu powders was found to be dependent on the structural state after each milling time. Endothermic and exothermic reactions are revealed in the DSC curves; they are attributed to the melting and the formation of Al-rich phases.

Acknowledgments: Authors would like to thank Xavier Fontrodona Gubau for her XRD support and Leila Mahfoudhi from the English Language Unit at the Faculty of Sciences of Sfax (Tunisia) for accepting to proofread and polish the language of this paper.

Author Contributions: Joan Joseph Sunol, Mohamed Dammak and Mohamed Khitouni conceived, designed the experiments and wrote the SEM, DSC and XRD results; Molka Ben Makhlouf and Tarek Bachaga performed the experiments and analyzed the data. All authors discussed the results and worked on preparing the manuscript.

Conflicts of Interest: The authors declare no conflict of interest.

References

1. Bachaga, T.; Daly, R.; Escoda, L.; Suñol, J.J.; Khitouni, M. Amorphization of $Al_{50}(Fe_2B)_{30}Nb_{20}$ mixture by mechanical alloying. *J. Metall. Mater. Trans. A* **2013**, *44*, 4718–4724. [CrossRef]
2. Esparza, R.; Rosas, G.; Ascencio, J.A.; Pérez, R. Effects of minor element additions to the nanocrystalline FeAl intermetallic alloy obtained by mechanical alloying. *Mater. Manuf. Process.* **2005**, *20*, 823–832. [CrossRef]
3. Sharma, P.; Sharma, S.; Khanduja, D. On the use of ball milling for the production of ceramic powders. *Mater. Manuf. Process.* **2015**, *30*, 1370–1376. [CrossRef]

4. Chittineni, K.; Bhat, D.G. X-ray Diffraction investigation of the formation of nanostructured metastable phases during short-duration mechanical alloying of Cu-Al powder mixtures. *Mater. Manuf. Process.* **2006**, *21*, 527–533. [CrossRef]
5. Chattopadhyay, P.P.; Mann, I. Effect of Partial Substitution of Cu in $Al_{65}Cu_{35}$ by Transition Metal in Mechanical Alloying of $Al_{65}Cu_{20}TM_{15}$. *Mater. Manuf. Process.* **2002**, *17*, 583–594. [CrossRef]
6. Qiu, Y.; Gu, M.L.; Zhang, F.G.; Wei, Z. Influence of Tool Inclination on Micro-Ball-End Milling of Quartz Glass. *Mater. Manuf. Process.* **2014**, *29*, 1436–1440. [CrossRef]
7. Archana, M.S.; Ramakrishna, M.; Ravi, C.; Gundakaram, V.V.; Srikanth, S.S.; Joshi, S.V. Nanocrystalline Phases during Mechanically Activated Processing of an Iron (Fe)-Aluminium(40 at.% Al) Alloy. *Mater. Manuf. Process.* **2014**. [CrossRef]
8. Suryanarayana, C. Mechanical alloying and milling. *Prog. Mater. Sci.* **2001**, *46*, 1–184. [CrossRef]
9. Qingquan, K.; Lian, L.; Liu, Y.; Zhang, J. Fabrication and Characterization of Nanocrystalline Al-Cu Alloy by Spark Plasma Sintering. *Mater. Manuf. Process.* **2014**, *29*, 1232–1236.
10. Eckert, J.; Holzer, J.C.; Krill, C.E.; Johnson, W.L. Mechanically driven alloying and grain size changes in Non crystalline Fe-Cu powders. *J. Appl. Phys.* **1993**, *73*, 2794–2802. [CrossRef]
11. Slimi, M.; Azabou, M.; Escoda, L.; Suñol, J.J.; Khitouni, M. Stacking faults and structural characterization of mechanically alloyed $Ni_{50}Cu_{10}(Fe_2B)_{10}P_{30}$ powders. *Eur. Phys. J. Plus* **2015**, *130*, 1–8. [CrossRef]
12. Murray, J.L. The aluminium-copper system. *Int. Met. Rev.* **1985**, *30*, 211–233. [CrossRef]
13. Massalski, T.B.; Okamoto, H.; Subramanian, P.R.; Kacprzak, L. *Binary Alloy Phase Diagrams*, 2nd ed.; ASM International: Novelty, OH, USA, 1990.
14. Nguyen, L.T.; McDonald, D.; Danker, A.R.; Ng, P. Optimization of copper wire bonding on Al-Cu metallization. *IEEE Trans. Compon. Packag. Manuf. Technol. A* **1995**, *18*, 423–429. [CrossRef]
15. Premkumar, J.; Kumar, B.S.; Madhu, M.; Sivakumar, M.; Song, K.Y.J.; Wong, Y.M. Key factors in Cu wire bonding reliability: Remnant aluminum and Cu/Al IMC thickness. In Proceedings of the 10th Electronics Packaging Technology Conference, Singapore, 9–12 December 2008.
16. Xu, H.; Liu, C.; Silberschmidt, V.V.; Chen, Z. A re-examination of the mechanism of thermosonic copper ball bonding on aluminium metallization pads. *Scr. Mater.* **2009**, *61*, 165–168. [CrossRef]
17. Li, F.; Ishihara, K.N.; Singu, P.H. The Formation of Metastable Phases by Mechanical Alloying in the Aluminum and Copper System. *Metall. Trans. A* **1991**, *22*, 2849–2854. [CrossRef]
18. Chattopadhyay, P.P. Ph.D. Thesis, Indian Institute of Technology, Kharagpur, India, 2000.
19. Lutterotti, L. MAUD, Int. Union of Crystallography CPD Newsletter (IUCr), No. 24. 2000.
20. Onuki, J.; Koizumi, M. Investigation of the Reliability of Copper Ball Bonds to Aluminum Electrodes. *IEEE Trans. Compon. Hybrids Manuf. Technol.* **1987**, *12*, 550–555. [CrossRef]
21. Sui, M.L.; Lu, K. Variation in lattice parameters with grain size of a nanophase Ni-3P compound. *Mater. Sci. Eng. A* **1994**, *541*, 179–180.
22. Eckert, J.; Holzer, J.C.; Johnson, W.L. Thermal stability and grain growth behavior of mechanically alloyed nanocrystalline Fe-Cu alloys. *J. Appl. Phys.* **1993**, *73*, 131–141. [CrossRef]
23. Mohamed, F.A. A dislocation model for the minimum grain size obtainable by milling. *Acta Mater.* **2003**, *51*, 4107–4119. [CrossRef]
24. Kalita, M.P.C.; Perumal, A.; Srinivasan, A. Structure and magnetic properties of nanocrystalline $Fe_{75}Si_{25}$ powders prepared by mechanical alloying. *J. Magn. Magn. Mater.* **2000**, *320*, 2780–2783. [CrossRef]
25. Das, S.; Peperezko, J.; Wu, R.; Wilde, G. Under cooling and glass formation in Al-based alloys. *Mater. Sci. Eng. A* **2001**, *159*, 304–306.
26. Ying, D.Y.; Zhang, D.L. Solid-state reactions between Cu and Al during mechanical alloying and heat treatment. *J. Alloy. Compd.* **2000**, *311*, 275–282. [CrossRef]

metals

MDPI

Article

Detecting Milling Deformation in 7075 Aluminum Alloy Aeronautical Monolithic Components Using the Quasi-Symmetric Machining Method

Qiong Wu *, Da-Peng Li and Yi-Du Zhang

State Key Laboratory of Virtual Reality Technology and Systems, School of Mechanical Engineering and Automation, Beijing University of Aeronautics and Astronautics, Beijing 100191, China; lidapeng@buaa.edu.cn (D.-P.L.); yidzhang@buaa.edu.cn (Y.-D.Z.)
* Correspondence: wuqiong@buaa.edu.cn; Tel.: +86-10-8231-7756

Academic Editor: Nong Gao
Received: 25 February 2016; Accepted: 29 March 2016; Published: 7 April 2016

Abstract: The deformation of aeronautical monolithic components due to CNC machining is a bottle-neck issue in the aviation industry. The residual stress releases and redistributes in the process of material removal, and the distortion of the monolithic component is generated. The traditional one-side machining method will produce oversize deformation. Based on the three-stage CNC machining method, the quasi-symmetric machining method is developed in this study to reduce deformation by symmetry material removal using the M-symmetry distribution law of residual stress. The mechanism of milling deformation due to residual stress is investigated. A deformation experiment was conducted using traditional one-side machining method and quasi-symmetric machining method to compare with finite element method (FEM). The deformation parameters are validated by comparative results. Most of the errors are within 10%. The reason for these errors is determined to improve the reliability of the method. Moreover, the maximum deformation value of using quasi-symmetric machining method is within 20% of that of using the traditional one-side machining method. This result shows the quasi-symmetric machining method is effective in reducing deformation caused by residual stress. Thus, this research introduces an effective method for reducing the deformation of monolithic thin-walled components in the CNC milling process.

Keywords: thin-walled component; residual stress; deformation; CNC machining

---

1. Introduction

Whole aeronautic thin-walled structures produced by various methods are widely used in aerospace technology, especially produced by machining methods. Schubert *et al.* [1] posed that the light-weight components are of crucial interest for all branches that produce moving masses. The aim to reduce weight has to be accompanied by high production efficiency and component performance. Mangalgiri [2] proposed that the complicated materials and composite materials would been used extensively.

However, when oversize whole aeronautic structure is machined, most of material is removed causing its deformation, which is thought to be a hard work universally in the aerospace manufacturing field. Dong and Ke [3] summarized that the main cause of deformation is that when a large amount of material is removed, the residual stress equilibrium in the blank is broken. To re-equilibrate it, the residual stress is redistributed, and the distortion of the monolithic component is generated at the same time. At the same time, Wang *et al.* [4] made a similar point that when machining thin-walled aeroplane parts, more than 90% of the material would be removed, resulting in severe distortion of the parts due to the weakened rigidity and the release of residual stress. This might also lead to stress concentration and damage of the parts. Thus, the residual stress is one of the most important reasons for machining deformation of aeronautic thin-walled structure. Meanwhile, a great number of workpieces are manufactured by high-speed milling, where problems can arise related to chatter vibration, an instable process. Chatter phenomena appear in the high removal rate roughing, as well as in the finishing of low-rigidity airframe components [5–7].

A number of studies have focused on the influence of different machining process of thin-walled plates to aid in the optimization of mechanical manufacturing processes. Guo *et al.* [8] established a three-dimensional (3-D) finite element model with consideration of an initial residual stress field to compute the machining deformation of thin-walled frame shape workpiece. Jomaa *et al.* [9] presented an experimental study of the surface finish and residual stress induced by the orthogonal dry machining of AA7075-T651 alloys and investigated the surface damage mechanisms in detail. Liu [10] posed the 3-D finite element models of a helical tool and a thin-walled part with a cantilever to predict the cutting deformation of a Ti6Al4V titanium alloy thin-walled part in milling process. Using a Lagrangian formulation with an explicit solution scheme and a penalty contact algorithm, Maurel-Pantel *et al.* [11] investigated the simulations of shoulder milling operations on AISI 304L stainless steel using the commercial software LS-Dyna. Eslampanah *et al.* [12] employed the thermal elastic-plastic finite element method to predict residual stress and deformation in a T-Fillet welded joint, and developed an uncoupled thermal-mechanical 3-D model. Ocana *et al.* [13] presented a model to provide a predictive estimation of the residual stress and surface deformation induced by laser action relevant for analysis the influence of different parameters in the milling process. In order to increase productivity and tool life in machining of titanium alloys, Nouari and Makich [14] studied the poor machinability of titanium alloys, especially Ti-55531, which exhibits extreme tool wear and unstable cutting forces. Denkena *et al.* [15] presented an approach for the identification and modeling of these process damping effects in transient milling simulations, and a simulation- and experiment-based procedure for the identification of required simulation parameters depending on the tool chamfer geometry is introduced and evaluated. Abe and Sasahara [16] explored the relationship between residual stress and temperature distribution in the shell structure after lamination and measured the deformation caused by residual stress release. This article focuses on reducing deformation caused by residual stress during the milling process.

Other studies concentrated on the different research method for analysis the relationship between deformation and residual stress, especially the computer simulation and experimental study method. Wei and Wang [17] established a finite element model of original residual stress to analyze the corresponding deflection by machining aerospace thin-walled parts, especially in machining large aerospace parts, and simulation results were validated approximately consistent with experimental results. Yaghi *et al.* [18] discussed the residual stress in thin- and thick-walled stainless steel pipe-welded components and presented a brief review of weld simulation, and analyzed more FE models with an inside radius to wall thickness ratio to investigate the effect of pipe diameter on residual stress. Wu and Li [19] proposed a numerical approach to predict the surface residual stress and strain gradients resulting from a 3D milling process. The finite element simulation of residual stresses is compared with experiments based on X-ray measurements on 7075 samples machined under different cutting conditions. The effects of cutting conditions on surface residual stress distribution are investigated. Husson *et al.* [20] presented an approach to estimate the influence of some factors on the distortion, based on the idea of a distortion potential taking into account not only geometry but also the residual stress. To study the effect of blank initial residual stress on component deformation, Huang *et al.* [21] used chemical milling to remove the machining-induced residual stress on the machined surface of the components and proved that the initial residual stress in the blank was the main factor of deformation for a three-frame monolithic beam. Taking into account the deformation of thin-walled parts being significantly affected by the residual stress generated after the material is cut away, Li *et al.* [22] analyzed the effects of cutting depth on the redistribution of residual stress and demonstrated that the magnitude of distortion and residual stress can be decreased and optimized efficiently by controlling and optimizing the depth of cut in the roughing and finishing. The current paper presents the quasi-symmetric machining method and FEM to reduce deformation caused by residual stress.

Some researchers concentrated on finding the approaches for analyzing and controlling residual stress. Muñoz-Sánchez et al. [23] developed and validated a numerical model to analyze the tool wear effect in machining induced residual stresses. The model was applied to predict machining induced residual stresses in AISI 316 L. Ballestra et al. [24] studied the dynamic characterization of gold micro beams by electrostatic excitation in the presence of residual stress gradient, experimentally. Additionally, a comparison with different numerical FEM models and experimental results has been carried out. Chen et al. [25] investigated the effect of electrical contact on the thermal contact stress of a microrelay switch. The results showed that the residual stress increased as the number of switching cycles increased. Based on the uncut chip thickness (UCT) model, Jiang et al. [26,27] found that residual tangential stress is influenced by the UCT and it is possible to optimize the residual stress distribution by controlling UCT (feed rate and tool diameter) with high-speed milling, and in order to control the material removal rate, they proposed a method by optimizing the milling tool diameters based on the relation between residual stress and UCT. Zhang et al. [28] provided some insight into the uniformity of compressive residual stress generated by massive overlapping laser shock peening impacts, which has a lot of practical use in engineering application. Mohammadpour et al. [29] conducted a finite element analysis based on the nonlinear finite element code MSC. Superform for investigating the effect of cutting speed and feed rate on surface and subsurface residual stress induced after orthogonal cutting. Toribio et al. [30] dealt with the effect of several residual stress profiles on the fatigue crack propagation in pre-stressing steel wires subjected to tension loading or bending moments. Considering the cutter radius of the milling process, Li et al. [31] created a model to analyze cutting force and thermal properties in the high-speed milling process and, combined with experimental validation, indicate that mechanical forces play an essential part on the formation of residual stress. Palkowski et al. [32] presented a 3-D model to calculate the residual stress state of drawn tubes in cold drawing processing. Krottenthaler [33] presented a simple method to determine residual stress of thin films, locally, by using stress relaxation tests by means of focused ion beam (FIB) milling and digital image correlation (DIC) and the proposed method offered a simple way for analyzing residual stresses in thin amorphous coatings. Additionally, nanohardness of the thin films was measured by nanoindentation, and residual stress was determined using grazing incidence X-ray diffraction [34].

However, the traditional one-side machining method will produce oversized deformation caused by residual stress. The current paper presents the quasi-symmetric machining method to reduce deformation by symmetrical material removal using the M-symmetry distribution law of residual stress. The obtained results are compared with those of FEM. The quasi-symmetric machining method is validated as a reliable and effective method in reducing deformation caused by residual stress.

2. Analyses of Deformation Caused by Residual Stress Releasing

The initial residual stress in pre-stretched 7075aluminum alloy plate, produced in the process of rolling, heat-treatment, and stretching, need to meet the universal hypothesis. The initial residual stress is a self-balanced force, and the resultant force and resultant moment in the cross-section that is perpendicular to the stress are zero; that is:

$$\int_{-\frac{e}{2}}^{\frac{e}{2}} \sigma dx = 0 \tag{1}$$

$$\int_{-\frac{e}{2}}^{\frac{e}{2}} \sigma x dx = 0 \tag{2}$$

where e is the thickness of the thin-walled plate, and σ is the initial residual stress of the thin-walled plate.

Residual stress of each layer that is segmented, on average, along the thickness direction and is showed in the Figure 1. The thickness of each layer is t, and the average residual stress of each layer is. $\sigma_1, \sigma_2, \ldots \ldots \ldots, \sigma_n$.

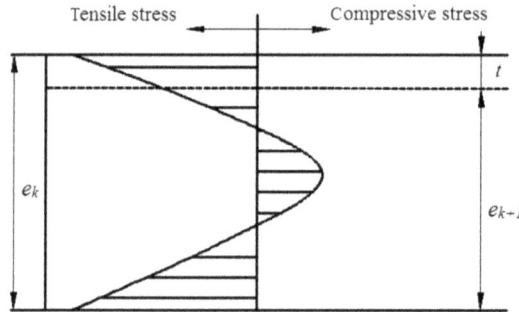

Figure 1. Residual stress distribution in pre-stretched 7075 aluminum alloy plate.

When materials are removed in milling, the initial residual stress equilibration is broken. To re-equilibrate it, the residual stress is redistributed, and the distortion of the plate is generated at the same time. The curvature relation of the workpiece before and after milling a layer is as follows:

$$\frac{1}{R_k} - \frac{1}{R_{k+1}} = \frac{6te_k\sigma_j}{E(e_{k+1})^3} \tag{3}$$

where k is the number of the layer, R_k and R_{k+1} are the curvature radius before and after the kth layer is stripped, e_k and e_{k+1} are the thickness of workpiece before and after the kth layer is stripped, σ_j is the stress before the kth layer is stripped, and E is the elastic modulus.

In this equation, $j = k - 1$ (when $k = 1$, $\sigma_j = \sigma_1$). When $k = 1$, σ_j represents the initial stress.

The remaining stress σ_{kr} in each layer after the kth layer being stripped is obtained by vector operations between the stress σ_{im} before the kth layer being stripped and the generated stress S_{ki}. The computational process is:

$$a_k = \frac{e_{k+1}(3e_k + e_{k+1})}{6e_k} \tag{4}$$

$$S_k = -a_k E\left(\frac{1}{R_k} - \frac{1}{R_{k+1}}\right) \tag{5}$$

$$S_{ki} = \frac{S_k}{a_k}[a_k - (i - k + 0.5)t] \tag{6}$$

$$\sigma_{kr} = \sigma_{lm} - S_{ki} \tag{7}$$

where $l = k - 1$, $m = r + 1$, $i = r + k - 1$, $r = 1,2, \ldots n - k$, $k \geqslant 2$.

When $k = 1$, then:

$$\sigma_{lr} = \sigma_{l+r} - S_{lr} \tag{8}$$

After the kth layer is stripped, a_k and b_k are the distance from upper surface and lower surface to the neutral plane, respectively. S_k and S_k' are the stress generated on the upper surface and lower surface when the rest of workpiece recovers to the shape before the kth layer being stripped, respectively. This is shown in Figure 2.

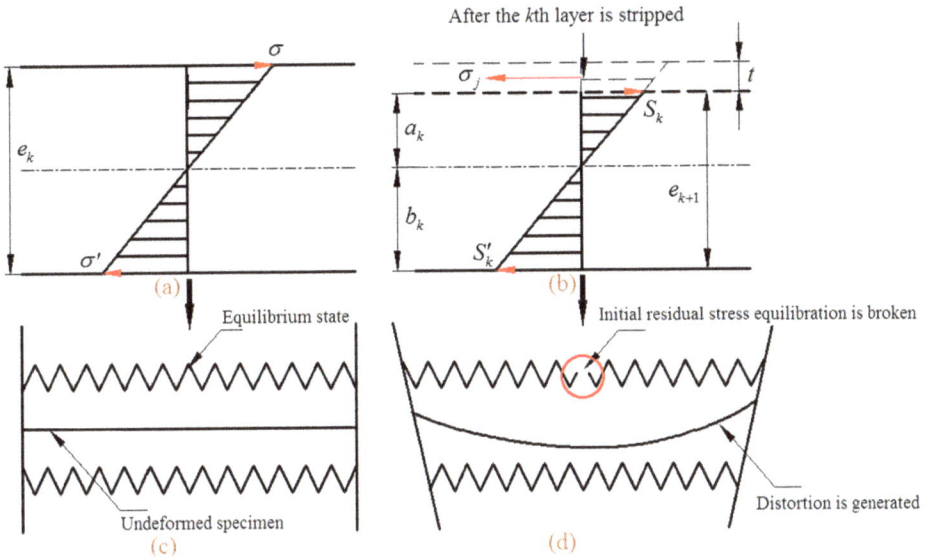

Figure 2. Stress state and deformed condition of the specimen before and after the *k*th layer is stripped. (**a**) and (**c**) show the equilibrium stress state and undeformed condition; (**b**) and (**d**) the re-equilibrium stress state and deformed condition.

For a more intuitive description of deformation principle, the stress state and deformation condition have been simplified, as shown in Figure 2. As depicted in Figure 2a,c, they are the equilibrium stress state and undeformed condition of the specimen before the *k*th layer is stripped. Moreover, the re-equilibrium stress state and deformed condition, after the *k*th layer is stripped, are exhibited in Figure 2b,d. The residual stress release and redistribute in the process of *k*th layer material removed, and the distortion of monolithic component is generated.

3. Analyses of Quasi-Symmetric Processing Technique

The sectioning technique is a destructive measurement method for material that relies on the measurement of deformation due to the release of residual stress upon removal of material from the specimen. The strains released during the cutting process are generally measured using electrical or mechanical strain gauges [35]. As is shown in Figure 3, using the sectioning technique, the excellent residual stress measurement results are obtained, and the distribution of residual stress in pre-stretched 7075 aluminum alloy plate present M-symmetry distribution along the thickness (Figure 4). The residual stress is tensile stress at the surface and in the middle, and compressive stress in the upper and lower part, respectively. This uneven tensile or compressive stress distribution along the thickness is the major cause of subsequent deformation. After some adjustment, the initial residual stress of 7075 aluminum alloy plate only partially reduced. Nonetheless, the trend of the M-symmetric distribution law will not change any more.

(a)　　　　　　　　　　　　　　　　　(b)

Figure 3. Quasi-nondestructive test for measuring distribution of residual stress. (**a**) Strain gauges and wire connecting; and (**b**) test equipment.

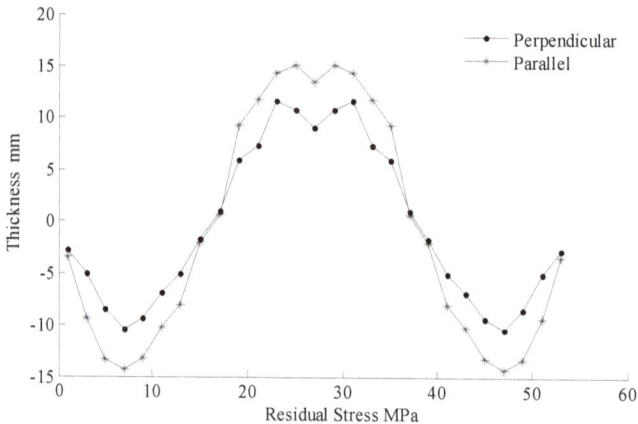

Figure 4. M-symmetric distribution law of initial residual stress.

Since the distribution of residual stress in pre-stretched 7075 aluminum alloy plate present M-symmetry distribution along the thickness, the symmetric machining method can reduce the deformation caused by the releasing and redistribution of the initial residual stress. For double-sided machining, the three-stage CNC machining method processes the upper surface to small allowances, then the lower surface to the precise size, then the allowance of the upper surface to an accurate size. Based on the three-stage CNC machining method, the quasi-symmetric machining method is developed to reduce deformation by symmetry material removal using the M-symmetry distribution law of residual stress. The plate is processed several times on both sides, repeatedly, to the accurate size.

4. Solution for the Plates Using FEM

Due to the high cutting speed and small cutting depth, the cutting force and cutting heat of pre-stretched 7075 aluminum alloy plate produced in the process of CNC machining on high-speed machine tools can be negligible. In addition, the influence of the clamping scheme for machining distortion is negligible in CNC machining processes. Therefore, this paper mainly focuses on workpiece deformation caused by releasing and redistribution of residual stress for aeronautical monolithic components in the CNC milling process.

The FEM that provides an approximate solution to continuum problems is a powerful tool to assess potential distortion caused by residual stress in the machining process. The FEM software ANSYS is applied to measure the deformation of aeronautical monolithic components in the traditional one-side machining process and quasi-symmetric machining process for further validation in this paper.

The pre-stretched 7075 aluminum alloy plate is assumed to be an elastic-plastic material. Its physical specifications are listed in Table 1, and the dimensions are 300 mm × 200 mm × 26 mm. The worktable is considered as a rigid body in the ANSYS model. In this paper, a two-sided constraint has been used for the solution in order to simulate the real fixture more authentically. Additionally, the stiffness is assumed to be infinite and displacement is assumed to be zero, respectively. Meanwhile, the influence of temperature, milling force, and clamping force are ignored.

Table 1. Physical parameters of 7075 aluminum alloy.

Properties	Young's Modulus/GPa	Poisson's Ratio	Density/kg·m³	Thermal Conductivity/W·m⁻¹·°C⁻¹	Specific Heat/J·kg⁻¹·°C⁻¹
Value	71	0.33	2800	155	960

On the basis of the M-symmetry distribution law of residual stress that was mentioned in Section 2, we assign the corresponding residual stress values into the workpiece layer by layer. Using the birth and death element method, we simulate the real situation of materials removed in the milling process. To achieve the "element death" effect, the ANSYS does not actually remove "killed" elements. Instead, it deactivates them by multiplying their stiffness by a severe reduction factor (10^{-6} by default) [36,37]. Then, the deformation process caused by residual stress release were simulated.

When simulate the traditional one-sided machining process, the materials of the upper surface are removed in a one-time processing. By contrast, for quasi-symmetric machining process, the materials of two surfaces are removed successively.

To facilitate description, the aeronautical monolithic component machined by the traditional one-side machining method and the quasi-symmetric machining method are labelled as Specimen 1 and 2, respectively.

Using the simulation procedure described above, the deformation distribution after the milling process are calculated by ANSYS. The model generated for this analysis is depicted in Figure 5. It can be seen from Figure 5a that the deformation trend of Specimen 1 is convex upward at both ends and concave downward at the middle, while the maximum deformation is located at the end of the specimen. On the contrary, from the Figure 5b, the deformation trend of Specimen 2 is concave downward at both ends and convex upward at the middle, while the maximum deformation is located at the middle of the specimen. As depicted in the deformation nephogram, the maximum deformation decreases by approximately 0.28 mm by the application of quasi-symmetric machining method compared to that of the traditional one-sided machining method. This result is confirmed by experimental investigation in Section 4 of this paper regarding deformation measurements.

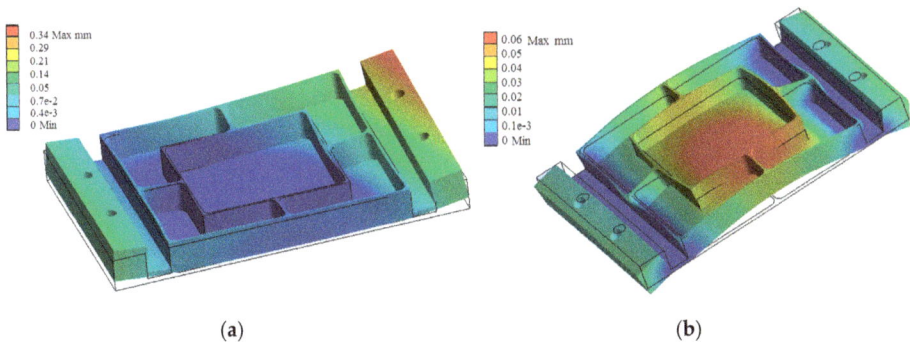

<table>
<tr><td>(a)</td><td>(b)</td></tr>
</table>

Figure 5. Measured deformation results of (**a**) Specimen 1 and (**b**) Specimen 2 obtained using ANSYS.

5. Analyses of the Experimental and Simulation Results

Consider the quantity of stiffener plate and the characteristics of the chamber profile, a reasonable simplification of an aeronautical monolithic component in view of the practical processing is made. The structure can not only reflect the real deformation, but is also in favor of the data acquisition.

In the machining process, the axial feed is 0.8 mm each time. After three passes, we removed specimens from fixture and measured the deformation. The clamping scheme of Specimen 2 in the reverse-side machining process is similar to Specimen 1, while that in the front-side machining process is shown in Figure 6.

Figure 6. Measured deformation by an infrared automatic sensor.

The staggered inner-milling form of the machining sequence and feed route were designed for the experiment. On a WF74CH-mikron CNC machine tool (Mikron Group, Boudry, Switzerland), a φ10 mm, four-tooth carbide milling cutter was used in the front side machining process, and a φ20 mm, two-tooth cemented carbide end milling cutter was used in the reverse side machining process. The processing parameters are shown in Table 2. Figure 6 illustrates the process of using MP8 infrared automatic sensor head (Wafer, Shanghai, China) to measure the specimen axial deformation.

Table 2. Machining parameters in the milling process.

Milling Cutter Diameter/mm	Rotate Speed of Milling Cutter/r· min^{-1}	Feeding Speed/mm· m^{-1}	Axial Feed/mm	Method of Cooling and Lubrication
10	3650	700	0.8	Oil spray cooling

The overall deformation of Specimen 2 is smaller than that of Specimen 1, intuitively. The maximum deformation of Specimen 1 occurs at the end of the workpiece. As is shown in Figure 7a, a 0.32 mm thick stalloy insert into maximum deformation place of Specimen 1. By contrast, the maximum deformation of Specimen 2 occurs at the middle of the specimen. As is shown in Figure 7b, a 0.045 mm thick stalloy insert into maximum deformation place of Specimen 2.

The CNC machining process is a dynamic deformation process. To facilitate the description, Specimen 1 and 2 are divided into seven cross-sections along the x-coordinate, namely cross-sections A–G, and all of the seven cross-sections are perpendicular to the X-Y plane, as shown in Figure 8. The abscissa X of cross-sections A–G are 100 mm, 70 mm, 35 mm, 0 mm, −35 mm, −70 mm, −100 mm, respectively. The size and structure of the specimen are shown in the Figure 8 at the same time.

In the milling process, the materials are removed layer by layer. Figure 9 illustrates the dynamic deformation curves of cross-sections A–G for Specimen 1 and 2 at different process stage. From Figure 9, the deformation trend and values in the process of the last six layers stripped are easily observed.

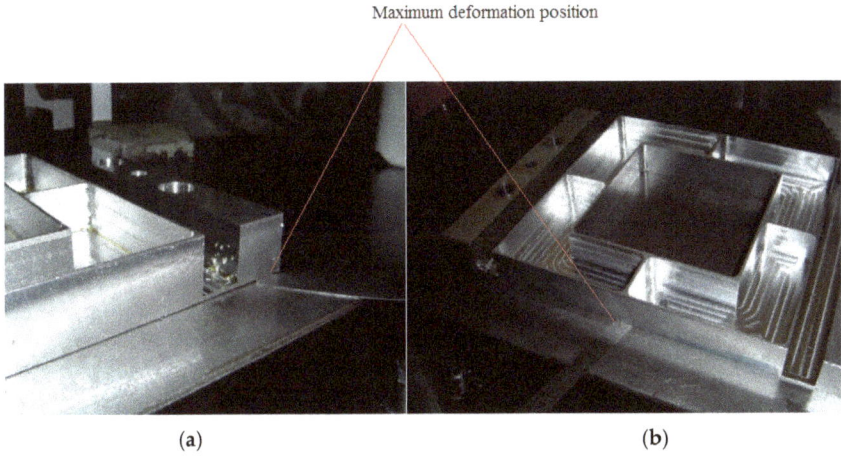

Figure 7. Stalloy insert into the maximum deformation place of (**a**) Specimen 1and (**b**) Specimen 2.

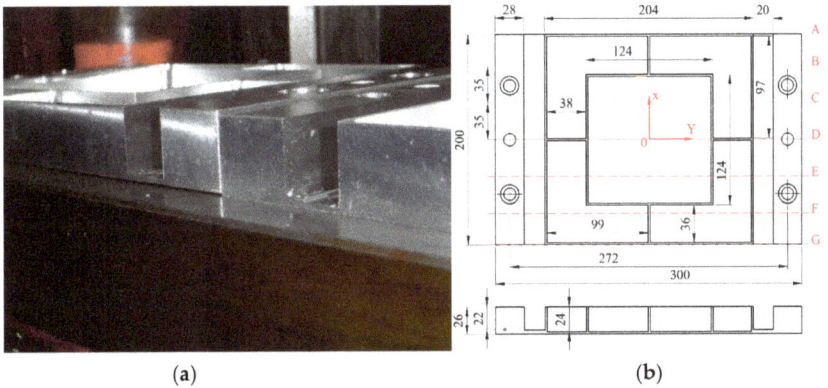

Figure 8. The (**a**) comparison of the maximum deformation of the Specimen 1 and (**b**) size and structure of the specimen.

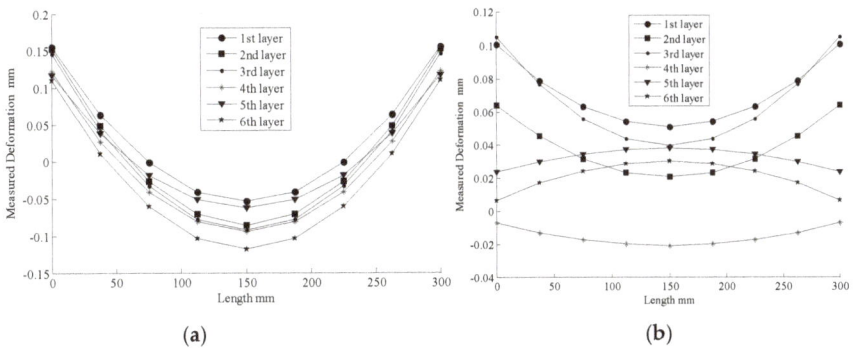

Figure 9. *Cont.*

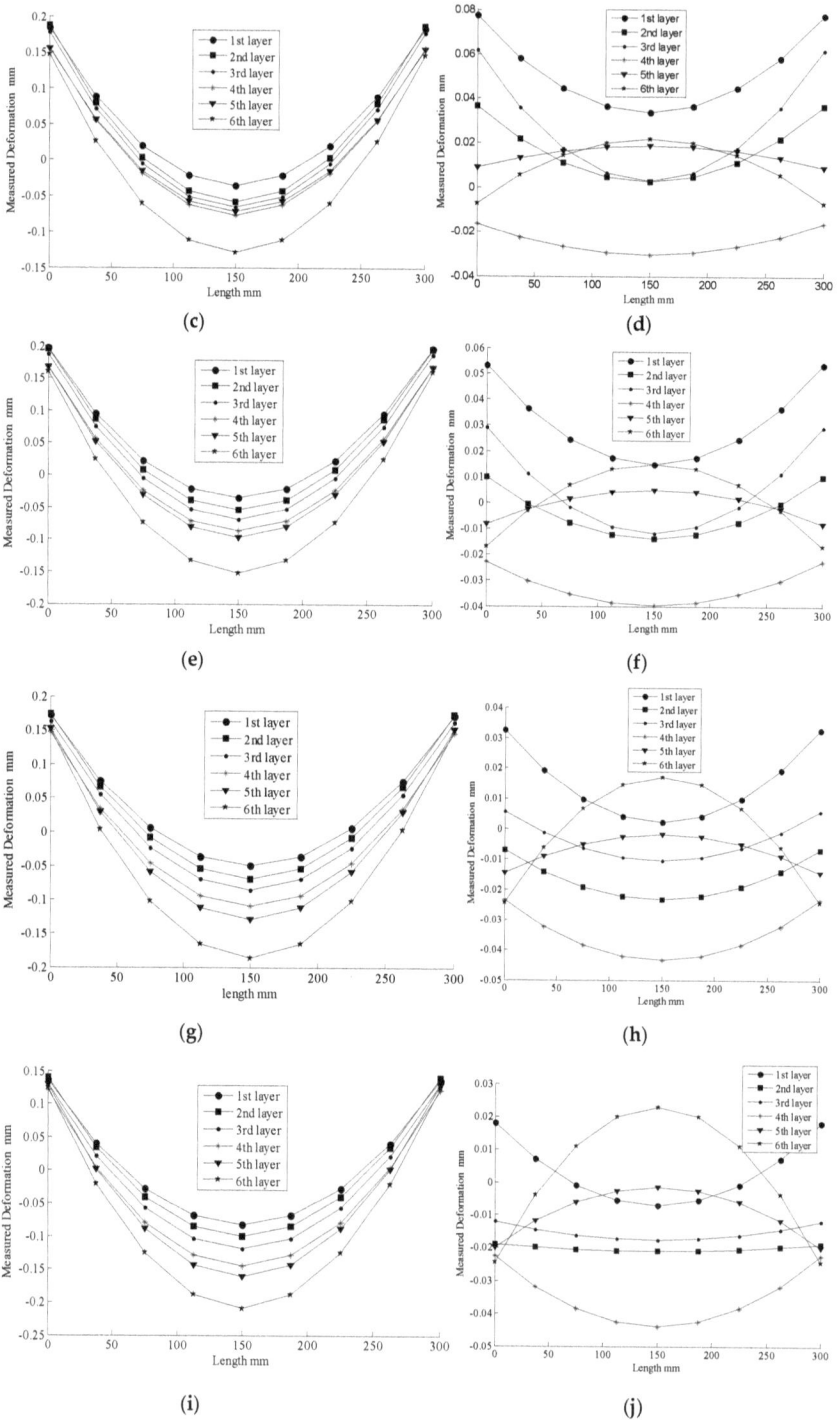

Figure 9. *Cont.*

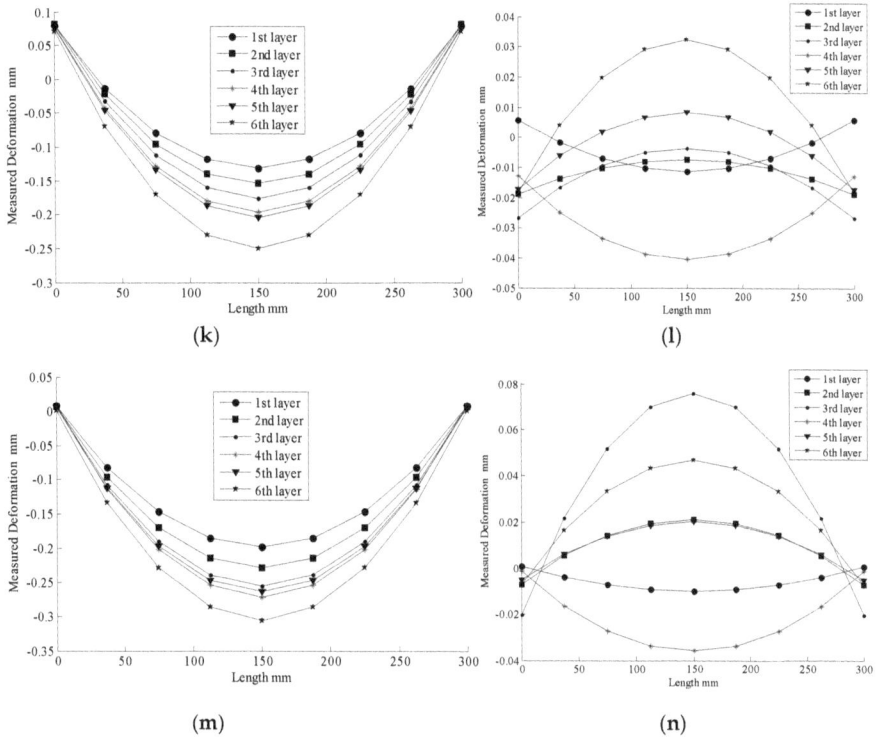

Figure 9. Deformation curves of cross-sections A-G for Specimen 1 and 2 at different process stage.
(**a**) Cross-section A of Specimen 1; (**b**) cross-section A of Specimen 2; (**c**) cross-section B of Specimen 1;
(**d**) cross-section B of Specimen 2; (**e**) cross-section C of Specimen 1; (**f**) cross-section C of Specimen 2;
(**g**) cross-section D of Specimen 1; (**h**) cross-section D of Specimen 2; (**i**) cross-section E of Specimen 1;
(**j**) cross-section E of Specimen 2; (**k**) cross-section F of Specimen 1; (**l**) cross-section F of Specimen 2;
(**m**) cross-section G of Specimen 1; (**n**) cross-section G of Specimen 2.

The curves of Figure 9 indicate that the deformation trend of cross-sections A–G for Specimen 1 are all convex upward at both ends and concave downward at the middle for the entirety of the machining time, while the maximum deformation occurs at the end of the specimen. Meanwhile, the deformation value increased during the materials be removed layer by layer. By contrast, the curves of Figure 9 also indicate that the deformation trend of cross-sections A–G for Specimen 2 are either concave downward at both ends and convex upward at the middle or convex upward at both ends and concave downward at the middle, which is a symmetric, repeated process. Especially, the deformation value is increased and decreased during the process, alternatively. Additionally, the maximum deformation occurs at the end of the specimen.

For the traditional one-side machining method, the residual stress releasing and redistribution happened at one side of the specimen all of the time, and the deformation caused by residual stresses were accumulated. By contrast, for the quasi-symmetric machining method, the residual stress releasing and redistribution happened on two sides of the specimen, alternatively. The deformation caused by residual stress were offset.

Meanwhile, the values of maximum deformation of cross-sections A–G for the Specimen 1 and 2 are displayed in Figure 9. The comparison of those is presented in Table 3.

Table 3. Comparison of maximum deformation of cross-sections A–G for Specimens 1 and 2.

	Cross-Section A/mm	Cross-Section B/mm	Cross-Section C/mm	Cross-Section D/mm	Cross-Section E/mm	Cross-Section F/mm	Cross-Section G/mm
Specimen 1	0.2125	0.2655	0.3060	0.3212	0.3246	0.3232	0.3069
Specimen 2	0.0201	0.0270	0.0314	0.0451	0.0502	0.0537	0.0589

Furthermore, the maximum deformation of cross-sections A–G is compared on the basis of the results of Specimen 1 divided by those of Specimen 2, respectively. Percentages of the maximum deformation are 9.459%, 10.17%, 10.26%, 14.04%, 15.47%, 16.62%, and 19.19%, successively. Obviously, the maximum deformation of Specimen 1 is much larger than that of Specimen 2. The maximum percentage of maximum deformation is only 19.19%. Thus, the quasi-symmetric machining method is effective in reducing deformation of monolithic thin-walled components caused by residual stress.

Figure 10 illustrates the measured deformation results and deformation trend of experiment and simulation for cross-section D of an aeronautical monolithic component in the traditional one-sided machining process and the quasi-symmetric machining process. The change curves indicate the relationship between the measured deformation results of FEM and the experiment. The original data curves and quadratic fit curves presented in Figure 10 are suited, thus suggesting that the FEM results are consistent with the experimental results. These findings confirm that FEM and the experiment exhibit high accuracy in the milling process.

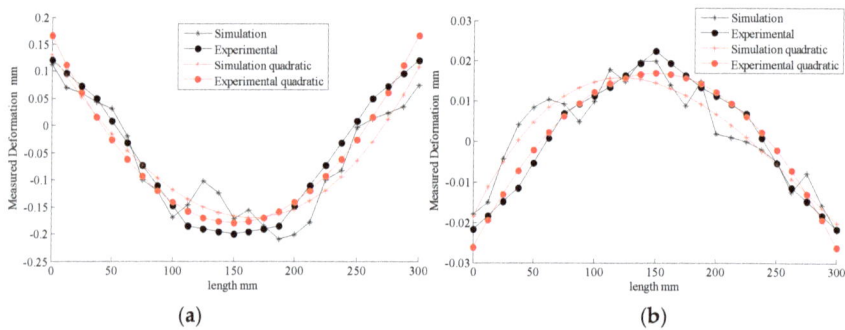

Figure 10. Measured deformation results of experiment and simulation for cross-section D. (**a**) Specimen 1; and (**b**) Specimen 2.

For Specimen 1, the experiment maximum deformation is 0.32 mm, while that of the simulation is 0.34 mm. By contrast, the experimental maximum deformation is 0.045 mm and the simulation maximum deformation is 0.06 mm for Specimen 2. Despite a slight deviation results comparing with the simulation and experiment, the results are very similar. Nonetheless, results of the simulation are inferior to those of the experiment, all of which remain within the allowable error. The main causes of this deviation are as follows:

1. FDM is based on the approximations that facilitate the replacement of differential equations with finite difference equations. As such, the calculation method inevitably displays systematic errors.
2. In practice, the boundary conditions are not identical in the simulation and in the experiment. The ideal condition has yet to be determined.
3. The residual stress state between simulation and experiment are not consistent. FEM simulating the true condition of residual stress releasing and redistribution, totally, is impossible. Although there are some objective reasons that lead to some certain errors, most of them are within the permitted error range.

6. Conclusions

1. The maximum deformation value of using a quasi-symmetric machining method is within 20% of that of using a traditional one-side machining method. This result shows the quasi-symmetric machining method is feasible and effective in reducing deformation of monolithic thin-walled components caused by residual stress.

2. Errors are low, and most are within 10% for modifying the comparative results of FEM and experimentation. These results confirm that the quasi-symmetric machining method is a reliable and suitable method for releasing deformation.

3. In the quasi-symmetric machining process, the deformation trend of Specimen 2 is concave downward at both ends, and convex upward at the middle, while the maximum deformation occurs at the middle of the specimen.

Acknowledgments: This study was financially supported by National Natural Science Foundation of China (Grant No. 51105025), China National Key Technology R&D Program (Grant No. 2014BAF08B01), National Science and Technology Major Project (Grant No. 2014ZX04001011).

Author Contributions: Q.W., D.-P.L. and Y.-D.Z. have conceived and designed the experiments; Q.W. and D.-P.L. performed the experiments; Q.W. and D.-P.L. analyzed the data; Q.W. and D.-P.L. wrote the paper.

Conflicts of Interest: The authors declare no conflict of interest.

References

1. Schubert, E.; Klassen, M.; Zerner, I.; Walz, C.; Sepold, G. Light-weight structures produced by laser beam joining for future applications in automobile and aerospace industry. *J. Mater. Process. Technol.* **2001**, *115*, 2–8. [CrossRef]
2. Mangalgiri, P.D. Composite materials for aerospace applications. *Bull. Mater. Sci.* **1999**, *22*, 657–664. [CrossRef]
3. Dong, H.Y.; Ke, Y.L. Study on machining deformation of aircraft monolithic component by FEM and experiment. *Chin. J. Aeronaut.* **2006**, *19*, 247–254. [CrossRef]
4. Wang, Z.J.; Chen, W.Y.; Zhang, Y.D. Study on the machining distortion of thin-walled part caused by redistribution of residual stress. *Chin. J. Aeronaut.* **2005**, *18*, 175–179. [CrossRef]
5. Herranz, S.; Campa, F.J.; Lacalle, L.N.L.D.; Rivero, A.; Lamikiz, A.; Ukar, E.; Sánchez, J.A.; Bravo, U. The milling of airframe components with low rigidity: A general approach to avoid static and dynamic problems. *Proc. Inst. Mech. Eng. B J. Eng. Manuf.* **2005**, *219*, 789–802. [CrossRef]
6. Bravo, U.; Altuzarra, O.; Lacalle, L.N.L.D.; Sánchez, J.A.; Campa, F.J. Stability limits of milling considering the flexibility of the workpiece and the machine. *Int. J. Mach. Tools. Manuf.* **2005**, *45*, 1669–1680. [CrossRef]
7. Campa, F.J.; Lacalle, L.N.L.D.; Lamikiz, A.; Sánchez, J.A. Selection of cutting conditions for a stable milling of flexible parts with bull-nose end mills. *J. Mater. Process. Technol.* **2007**, *191*, 279–282. [CrossRef]
8. Wang, Z.J.; Chen, W.Y.; Zhang, Y.D.; Chen, Z.T.; Liu, Q. The application of FEM technology on the deformation analysis of the aero thin-walled frame shape workpiece. *Key Eng. Mater.* **2006**, *315*, 174–179. [CrossRef]
9. Jomaa, W.; Songmene, V.; Bocher, P. Surface Finish and Residual Stresses Induced by Orthogonal Dry Machining of AA7075-T651. *Materials* **2014**, *7*, 1603–1624. [CrossRef]
10. Liu, G. Study on deformation of titanium thin-walled part in milling process. *J. Mater. Process. Technol.* **2009**, *209*, 2788–2793.
11. Maurel-Pantel, A.; Fontaine, M.; Thibaud, S.; Gelin, J.C. 3D FEM simulations of shoulder milling operations on a 304L stainless steel. *Simul. Model. Pract. Theory* **2012**, *22*, 13–27. [CrossRef]
12. Eslampanah, A.H.; Aalami-aleagha, M.E.; Feli, S.; Ghaderi, M.R. 3-D numerical evaluation of residual stress and deformation due welding process using simplified heat source models. *J. Mech. Sci. Technol.* **2015**, *29*, 341–348. [CrossRef]
13. Ocana, J.L.; Morales, M.; Molpeceres, C.; Torres, J. Numerical simulation of surface deformation and residual stresses fields in laser shock processing experiments. *Appl. Surf. Sci.* **2004**, *238*, 242–248. [CrossRef]
14. Nouari, M.; Makich, H. On the Physics of Machining Titanium Alloys: Interactions between Cutting Parameters, Microstructure and Tool Wear. *Metals* **2014**, *4*, 335–358. [CrossRef]

15. Denkena, B.; Bickel, W.; Grabowski, R. Modeling and simulation of milling processes including process damping effects. *Prod. Eng.* **2014**, *8*, 453–459. [CrossRef]

16. Abe, T.; Sasahara, H. Residual Stress and Deformation after Finishing of a Shell Structure Fabricated by Direct Metal Lamination using Arc Discharge. *Int. J. Automat. Technol.* **2012**, *6*, 611–617.

17. Wei, Y.; Wang, X.W. Computer simulation and experimental study of machining deflection due to original residual stress of aerospace thin-walled parts. *Int. J. Adv. Manuf. Technol.* **2007**, *33*, 260–265. [CrossRef]

18. Yaghi, A.; Hyde, T.H.; Becker, A.A.; Sun, W.; Williams, J.A. Residual stress simulation in thin and thick-walled stainless steel pipe welds including pipe diameter effects. *Int. J. Pres. Vessels Pip.* **2006**, *83*, 864–874. [CrossRef]

19. Wu, Q.; Li, D.P. Analysis and X-ray measurements of cutting residual stresses in 7075 aluminum alloy in high speed machining. *Int. J. Precis. Eng. Manuf.* **2014**, *15*, 1499–1506. [CrossRef]

20. Husson, R.; Baudouin, C.; Bigot, R.; Sura, E. Consideration of residual stress and geometry during heat treatment to decrease shaft bending. *Int. J. Adv. Manuf. Technol.* **2014**, *72*, 1455–1463. [CrossRef]

21. Huang, X.; Sun, J.; Li, J. Finite element simulation and experimental investigation on the residual stress-related monolithic component deformation. *Int. J. Adv. Manuf. Technol.* **2014**, *77*, 1035–1041. [CrossRef]

22. Li, B.; Jiang, X.; Yang, J.; Liang, S.Y. Effects of depth of cut on the redistribution of residual stress and distortion during the milling of thin-walled part. *J. Mater. Process. Technol.* **2015**, *216*, 223–233. [CrossRef]

23. Muñoz-sánchez, A.; Canteli, J.; Cantero, J.; Miguélez, M. Numerical Analysis of the Tool Wear Effect in the Machining Induced Residual Stresses. *Simul. Model. Pract. Theory* **2011**, *19*, 872–886. [CrossRef]

24. Ballestra, A.; Somà, A.; Pavanello, R. Experimental-Numerical Comparison of the Cantilever MEMS Frequency Shift in presence of a Residual Stress Gradient. *Sensors* **2008**, *8*, 767–783. [CrossRef]

25. Chen, Y.; Tsai, H.; Lu, W.; Chen, L. Effect of Electrical Contact on the Contact Residual Stress of a Microrelay Switch. *Sensors* **2007**, *7*, 2997–3011. [CrossRef]

26. Jiang, X.; Li, B.; Yang, J.; Zuo, X.Y.; Li, K. An approach for analyzing and controlling residual stress generation during high-speed circular milling. *Int. J. Adv. Manuf. Technol.* **2013**, *66*, 1439–1448. [CrossRef]

27. Jiang, X.; Li, B.; Yang, J.; Zuo, X.Y. Effects of tool diameters on the residual stress and distortion induced by milling of thin-walled part. *Int. J. Adv. Manuf. Technol.* **2013**, *68*, 175–186. [CrossRef]

28. Zhang, W.; Lu, J.; Luo, K. Residual Stress Distribution and Microstructure at a Laser Spot of AISI 304 Stainless Steel Subjected to Different Laser Shock Peening Impacts. *Metals* **2016**, *6*, 6. [CrossRef]

29. Mohammadpour, M.; Razfar, M.R.; Saffar, R.J. Numerical investigating the effect of machining parameters on residual stresses in orthogonal cutting. *Simul. Model. Pract. Theory* **2010**, *18*, 378–389. [CrossRef]

30. Toribio, J.; Matos, J.; González, B.; Escuadra, J. Influence of Residual Stress Field on the Fatigue Crack Propagation in Prestressing Steel Wires. *Materials* **2015**, *8*, 7589–7597. [CrossRef]

31. Li, B.Z.; Jiang, X.H.; Jing, H.J.; Zuo, X.Y. High-speed milling characteristics and the residual stresses control methods analysis of thin-walled parts. *Adv. Mater. Res.* **2011**, *223*, 456–463. [CrossRef]

32. Palkowski, H.; Brück, S.; Pirling, T.; Carradò, A. Investigation on the Residual Stress State of Drawn Tubes by Numerical Simulation and Neutron Diffraction Analysis. *Materials* **2013**, *6*, 5118–5130. [CrossRef]

33. Krottenthaler, M.; Schmid, C.; Schaufler, J.; Durst, K.; Göken, M.A. Simple method for residual stress measurements in thin films by means of focused ion beam milling and digital image correlation. *Surf. Coat. Technol.* **2013**, *215*, 247–252. [CrossRef]

34. Hernández, L.; Ponce, L.; Fundora, A.; López, E.; Pérez, E. Nanohardness and Residual Stress in TiN Coatings. *Materials* **2011**, *4*, 929–940. [CrossRef]

35. Rossini, N.S.; Dassisti, M.; Benyounis, K.Y.; Olabi, A.G. Methods of measuring residual stresses in components. *Mater. Des.* **2011**, *35*, 572–588. [CrossRef]

36. Wang, L.; Wang, Y.; Sun, X.G.; He, J.Q.; Pan, Z.Y.; Wang, C.H. Finite element simulation of residual stress of double-ceramic-layer $La_2Zr_2O_7/8YSZ$ thermal barrier coatings using birth and death element technique. *Comput. Mater. Sci.* **2012**, *53*, 117–127. [CrossRef]

37. Chen, J.Q.; Shen, W.L.; Yin, Z.X.; Xiao, S.H. Simulation of Welding Temperature Distribution Based on Element Birth and Death. *Hot Work. Technol.* **2005**, *7*, 64–65.

metals

MDPI

Article

Creep Aging Behavior Characterization of 2219 Aluminum Alloy

Lingfeng Liu [1], Lihua Zhan [1,2,]* and Wenke Li [1]

[1] Light Metal Research Institute, Central South University, Changsha 410083, China;
llfeng2014@csu.edu.cn (L.L.); lwk1992@csu.edu.cn (W.L.)

[2] National Key Laboratory of High Performance Complex Manufacturing, Central South University,
Changsha 410083, China

* Correspondence: yjs-cast@csu.edu.cn; Tel.: +86-731-8883-0254

Academic Editor: Nong Gao
Received: 11 May 2016; Accepted: 11 June 2016; Published: 29 June 2016

Abstract: In order to characterize the creep behaviors of 2219 aluminum alloy at different temperatures and stress levels, a RWS-50 Electronic Creep Testing Machine (Zhuhai SUST Electrical Equipment Company, Zhuhai, China) was used for creep experiment at temperatures of 353~458 k and experimental stresses of 130~170 MPa. It was discovered that this alloy displayed classical creep curve characteristics in its creep behaviors within the experimental parameters, and its creep value increased with temperature and stress. Based on the creep equation of hyperbolic sine function, regression analysis was conducted of experimental data to calculate stress exponent, creep activation energy, and other related variables, and a 2219 aluminum alloy creep constitutive equation was established. Results of further analysis of the creep mechanism of the alloy at different temperatures indicated that the creep mechanism of 2219 aluminum alloy differed at different temperatures; and creek characteristics were presented in three stages at different temperatures, i.e., the grain boundary sliding creep mechanism at a low temperature stage ($T < 373$ K), the dislocation glide creep mechanism at a medium temperature stage (373 K $\leqslant T < 418$ K), and the dislocation climb creep mechanism at a high temperature stage ($T \geqslant 418$ K). By comparative analysis of the fitting results and experiment data, they were found to be in agreement with the experimental data, revealing that the established creep constitutive equation is suitable for different temperatures and stresses.

Keywords: 2219 aluminum alloy; creep; creep mechanism; constitutive modeling

1. Introduction

Creep aging forming is a forming technique combining creep and aging heat treatment, which utilizes the creep and stress relaxation characteristics of materials to partially transform the elastic pre-strain of the component to be formed into plastic strain after a certain length of time and to provide aging strengthening in the meantime to obtain the required shape and properties of the component, so as to realize synchronization of part forming and formation of the properties [1,2]. The technique of creep aging forming can be dated back to the beginning of the 1950s and is currently considered one of the most important forming techniques in modern large aircraft manufacturing. In comparison with other forming methods such as shot peen forming and roll bending forming, creep aging forming is characterized by better mechanical properties, higher forming precision, and lower residual stress.

Establishing a creep constitutive equation is to accurately predict the properties and shape of the formed component. Abroad, Kowalewski et al. [3] established a metallic material creep unified constitutive model, which described the creep deformation behaviors of the material from the initial stage to the third stage of creep induced by dislocation hardening, nucleation at grain boundary holes, etc. K.C. Ho and Jianguo Lin [4,5] established a macro-micro coupling unified creep aging constitutive model based on aging dynamics and creep unity theory. However, neither of these models introduced the influence of temperature changes on creep behaviors of the material. In recent years, Jing Zhang [6] introduced the influence of temperature changes at different constant temperature aging stages on creep behaviors of the material from the perspective of multi-level (second-level) aging, Guan Chun-long [7] studied the creep behavior of 2024 aluminum alloy at cryogenic temperature. Further, An et al. [8] studied the influence of pre-deformation amount upon its mechanic performance and organization of 2219 aluminum alloy panels during two instances of thermo-mechanical treatment. However, as to the creep aging forming process of large aerospace components, the actual heating rate under the action of autoclave-tooling system is far lower than the heating rate of the specimens on the creep testing machine when establishing a material scale constitutive model. While conducting experimental research on the creep aging at an earlier stage in which a lower heating rate (0.75 K/min) was applied to reach to the aging temperature and then stayed for a period of time, the author discovered that, within the aging time of 13 h and under the conditions of experimental stresses at 150 and 210 MPa, respectively, the creep value at the heating stage reached 29.28% and 21.56% of the total creep value, respectively. In consideration of the influence of heating stages on material creep behaviors, this paper employed a RWS-50 Electronic Creep Relaxation Testing Machine (Zhuhai SUST Electrical Equipment Company, Zhuhai, China) for systematic research on the creep behaviors of 2219 aluminum alloy at different temperatures and stress states. Stress exponent, creep activation energy, and other parameters were analyzed and calculated to judge the alloy's creep mechanism under different experimental conditions. A creep unified constitutive model was established as well that can apply to different temperatures and stresses.

2. Materials and Methods

The 2219 aluminum alloy used in this experiment was hot rolling stripe steel provided by an institution and was cut into 2-mm standard specimens along the rolling direction in accordance with GB/T2039-1997. The exact chemical composition is given in Table 1. See Figure 1 for specimen dimension. The solution temperature of 2219 aluminum alloy is 808 K, and the solution time 36 min. The solid solution furnace temperature was controlled to maintain the tolerance within ±3 K as far as possible; the alloy was then treated by water quenching at room temperature before conducting the creep experiment. The quenching time was more than 35 s, and the specimens were kept in a refrigerated condition to reduce the influence of natural aging. Later, the RWS-50 Electronic Creep Machine was adopted for the creep experiment, which was produced at Zhuhai SUST Electrical Equipment Company in Zhuhai, China.

Table 1. Main chemical constituents of 2219 aluminum alloy.

Chemical Composition	Cu	Mg	Mn	Si	Fe	Ni	Zr	Ti	Al
Mass fraction	5.24	0.028	0.27	0.042	0.13	0.03	0.14	0.065	Bal

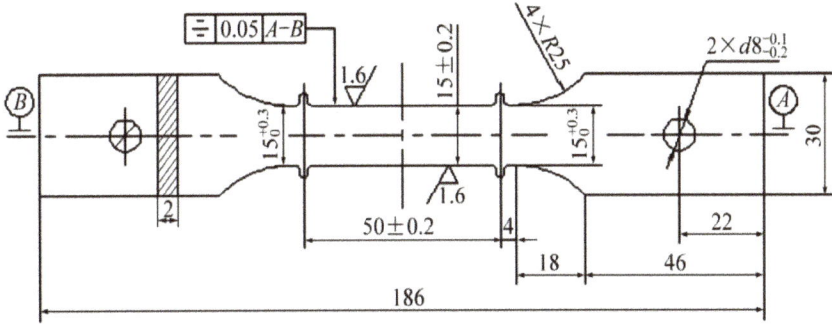

Figure 1. Creep specimen dimension (unit: mm). A and B: datum plane.

3. Results and Discussion

3.1. Alloy Creep Behaviors

After solid solution and quenching, the experimental materials were put on the creep testing machine for creep tensile test. The stress conditions were set at three states of 130, 150, and 170 MPa, respectively, with an aging time of 15 h, and the experimental temperatures were set in order at 353, 373, 393, 418, 438, and 458 K. Figures 2 and 3 show the creep curves under different experimental conditions.

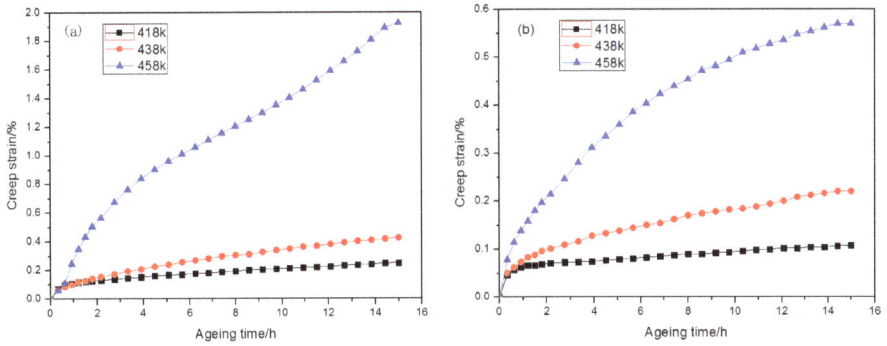

Figure 2. Creep curves of 2219 aluminum alloy at different experimental temperatures under the same stress. (**a**) 150 MPa; (**b**) 130 MPa.

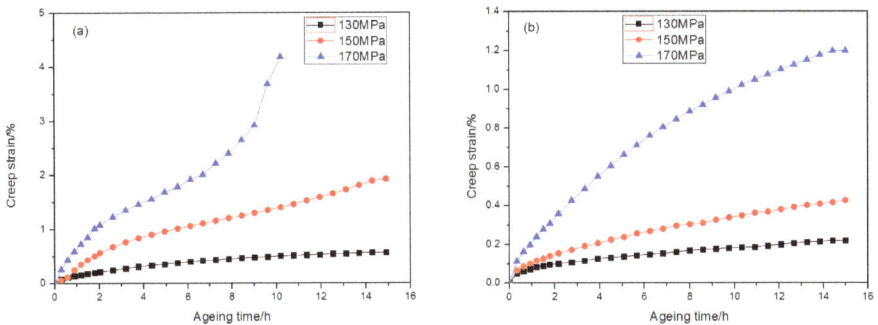

Figure 3. *Cont.*

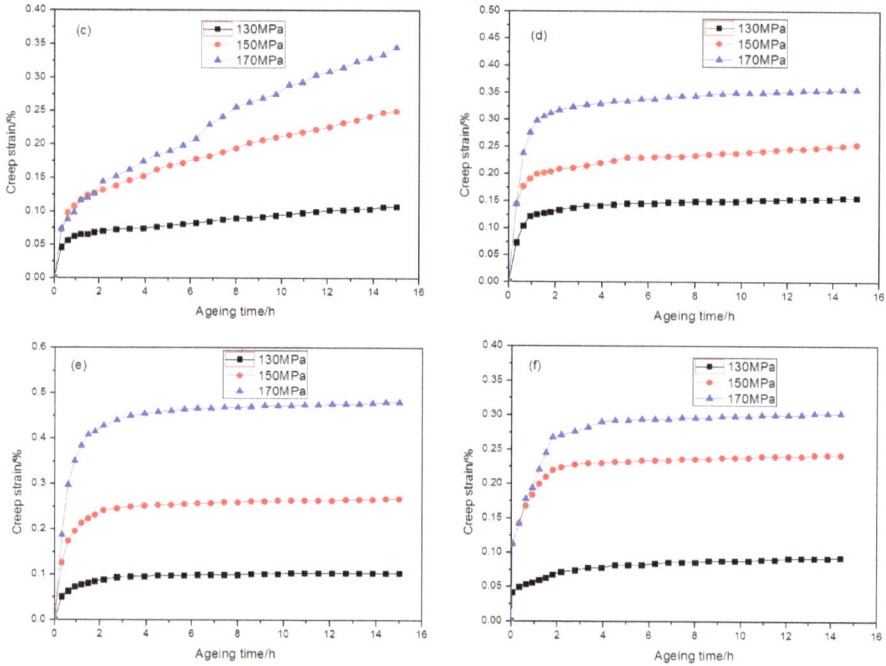

Figure 3. Creep curves of 2219 aluminum alloy at different temperatures and stress states. (**a**) 458 K; (**b**) 438 K; (**c**) 418 K; (**d**) 393 K; (**e**) 373 K; (**f**) 353 K.

From Figures 2 and 3, it can be observed that experimental temperature and stress state are two principal factors that influence creep behaviors: the higher the temperature, the greater the experimental stress and hence the larger the creep deformation value. It can be discovered from Figure 2a,b that, as temperature increases, creep deformation value increases. For example, when the experimental stress was 150 MPa with an aging time of 15 h, the creep deformation values at experimental temperatures of 458, 438 and 418 k were 1.911%, 0.426%, and 0.251%, respectively. This is because an increase in temperature provides the atoms and vacancies with a possibility of thermal activation so that dislocation can continue with activity by overcoming certain short-range obstructions, giving rise to a continual increase in plastic deformation and rapid progression of creep [9]. As shown in Figure 3b, when the experimental temperature was 438 K, the creep deformation value under an experimental stress of 150 MPa was 0.426%, while that under an experimental stress of 170 MPa was 1.198%, which might have been due to a great deal of dislocation generated inside the material upon loading. The major obstruction of dislocation was the long range stress field caused by the dislocation; the overcoming of which a shearing stress must be relied on [10]. Therefore, the greater the applied stress, the easier it is for dislocation to go through its obstruction. When the stress is constant and consistent with the aging time, creep deformation value increases as temperature increases.

In general, the creep process can be divided into three stages: the first creep stage (decelerated creep stage); the second stage (steady creep stage); and the third stage (accelerated creep stage). Within the selected temperatures and stresses in the experiment, the first and second creep stages can be clearly observed on the creep curves, most of which failed to enter the third stage. However, when the experimental conditions reached a certain degree, such as in Figure 3a, at an aging temperature of 458 k when the experimental stress reached 170 MPa, the creep curve presented an S-shape. When it came to the aging time of 9 h, creep deformation value accelerated; around the aging time of 11 h, a fracture to the creep specimen was observed.

It can be seen from Figure 3 that, when the aging temperature was below a certain degree, it ceased to be the principal factor affecting creep deformation value. For instance, in Figure 3d–f, when the experimental stress was 150 MPa, the creep deformation values at temperatures of 393, 373 and 353 K, with an aging time of 15 h, were 0.254%, 0.268%, and 0.242%, respectively. At that point, the condition of stress amounted to be the principal factor affecting creep aging. Within the scope of conditions set in this experiment, the creep deformation value at the highest experimental temperature was more than 20 times that at the lowest aging time, from which it can be determined that, within the temperature range under 353 K, there is basically no creep aging behavior in 2219 aluminum alloy.

3.2. Computational Analysis of Creep Mechanism and Constitutive Equation Setup

Based on the creep deformation characteristics of the material, the creep process generally consists of a dislocation glide, dislocation climb, grain boundary sliding and diffusion, and other creep mechanisms. By the difference in stress exponent, the corresponding creep mechanism can be roughly determined [11–14]. Creep aging can be regarded as a process of thermal activated deformation, in which the constitutive equation models that describe the flow stress include [15]:

$$\text{Low stress state}: \ \dot{\varepsilon} = A_1 \sigma^{n_1} \exp[-Q/(RT)], \tag{1}$$

$$\text{High stress state}: \ \dot{\varepsilon} = A_2 \exp(\beta\sigma) \exp[-Q/(RT)], \text{ and} \tag{2}$$

$$\text{All stress states}: \ \dot{\varepsilon} = A \sinh(\alpha\sigma)^n \exp[-Q/(RT)], \tag{3}$$

where, α, n, and β are generally believed to have the following correlation: $\alpha = \beta/n_1$. In the above equations, A_1, A_2, A, n_1, n, α, and β are all material parameters, Q denotes the apparent activation energy for creep, R denotes molar gas constant, which is 8.314 J/mol, σ is the experimental stress, and T is the thermodynamic temperature.

Logarithms were taken on both sides in Equations (1) and (2):

$$ln\dot{\varepsilon} = lnA_1 - Q/RT + n_1 ln\sigma; \tag{4}$$

$$ln\dot{\varepsilon} = lnA_2 - Q/RT + \beta\sigma. \tag{5}$$

Treated by linear regression, the relationship graphs between $ln\dot{\varepsilon} - ln\sigma$ and $ln\dot{\varepsilon} - \sigma$ at different aging temperatures were obtained; the slope of line of the former is n_1 and that of the latter is β. Thereby, it is deduced that $\alpha = \beta/n_1$. Specific parameters at different temperatures are presented in Table 2.

Table 2. Experimental parameters of 2219 aluminum alloy at different aging temperatures.

Temperatures	n_1	α	β
458 K	7.76	0.00699	0.0543
438 K	6.84	0.00676	0.0543
418 K	6.22	0.00671	0.0416
393 K	4.93	0.00704	0.0350
373 K	4.44	0.00673	0.0296
353 K	1.32	0.00695	0.00834

Put the obtained α into Equation (3) and logarithm was taken on both sides:

$$ln\dot{\varepsilon} = lnA - Q/RT + nln[\sinh(\alpha\sigma)]. \tag{6}$$

By using the data obtained from the previous experiment and calculated parameters, the relationship graph between $ln\dot{\varepsilon}$ and $ln[\sinh(\alpha\sigma)]$ was plotted, as shown in Figure 4. In the graph, the slope of line is stress exponent n, the specific value of which is given in Table 3.

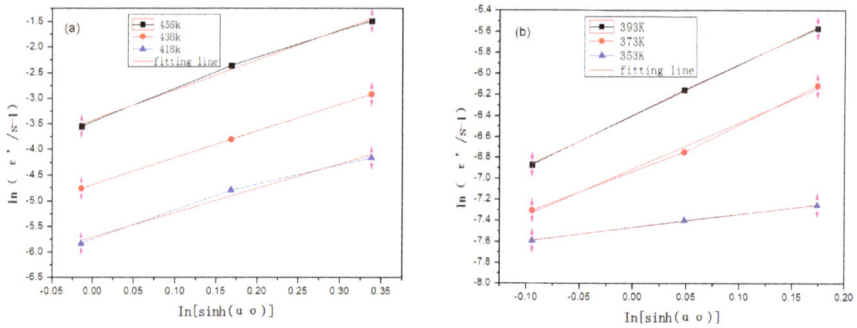

Figure 4. Relationship between steady creep rate $\dot{\varepsilon}$ and experimental stress σ of 2219 aluminum alloy at different aging temperatures. (**a**) 458 K, 438 K and 418 K; (**b**) 393 K, 373 K and 353 K.

Table 3. Stress exponents of 2219 aluminum alloy at different aging temperatures.

Aging Temperatures	Stress Exponent n
458 K	5.91
438 K	5.17
418 K	4.74
393 K	3.81
373 K	3.35
353 K	1.23

Stress exponents at different aging temperatures can be obtained from Table 3. Based on the creep characteristics parameters of the alloy [16], when $T \geqslant 418$ K, stress exponents n = 4~6 and fell in the category of dislocation climb mechanism; when 373 K $\leqslant T <$ 418 K, n = 4~6 and fell in the category of dislocation glide mechanism; $n \approx 1$ around 353 K and fell in the category of grain boundary sliding mechanism. It can be seen from the obtained data that, during the process of creep aging, there were usually multiple creep mechanisms, and, under certain conditions, the process of creep aging displayed single dominating creep mechanism. When calculating creep activation energy, corresponding to different creep mechanisms, two temperature ranges—$T \geqslant 418$ K and $T <$ 418 K—were used to solve creep activation energy Q_1 and Q_2 (it is recognized that the alloy showed no creep behavior when $T <$ 353 K; therefore, this temperature range was not considered).

Under a certain stress condition, temperature is considered as a variable and natural logarithm difference was taken for Equation (1) as follows:

$$Q = -R[(dln\dot{\varepsilon})/d(1/T)]. \tag{7}$$

From this, the slope of line K was obtained and multiplied by $-R$ to obtain creep activation energy of the alloy as Q_1 = 96.2 kJ/mol and Q_2 = 36.2 kJ/mol.

The aluminum alloy constitutive model proposed by Kowalewski [3,17,18] was adopted in this paper to describe creep aging behaviors:

$$\dot{\varepsilon} = A\sinh[B(\sigma - \sigma_0)(1 - H)^{m_0}], \text{ and} \tag{8}$$

$$\dot{H} = \frac{h}{\sigma^{m_1}}(1 - \frac{H}{H^*})\dot{\varepsilon}, \tag{9}$$

where $\dot{\varepsilon}$ is creep strain rate, A, B, h, H^*, m_0, σ_0, and m_1 are all material constants that are independent of the experiment process of creep aging, among which h, m_1, and H are parameters to describe the first stage of creep, H^* is the maximum of H, the value range of H is 0~H^*, H^* indicates the influence of strain strengthening at the first stage of creep, and A and B are parameters that describe the whole creep stage. Due to relatively short aging time for the experiment, except for individual conditions, most of the materials failed to enter the third stage of creep. Equations (8) and (9) can effectively reflect the aging behaviors at the first and second stages.

The relationship between creep strain rate and temperature [12]:

$$\dot{\varepsilon} = a\exp(-Q/RT), \tag{10}$$

where $\dot{\varepsilon}$ is the creep strain rate, and Q is the creep activation energy. R refers to a molar gas constant of 8.314 J/mol, and T is the thermodynamic temperature.

Considering the influence of stress and temperature upon creep aging formation, a creep constitutive model able to uniformly reflect the influence of aging temperature and stress conditions is established:

$$\dot{\varepsilon} = A\sinh[B(\sigma - \sigma_0)(1 - H)^{m_0}]\exp(-Q/RT), \text{ and} \tag{11}$$

$$\dot{H} = \frac{h}{\sigma^{m_1}}(1 - \frac{H}{H^*})\dot{\varepsilon}. \tag{12}$$

Since the parameters involved in this constitutive model are more than one, numerical optimization algorithm was used for the determination of parameters. Moreover, the particle swarm optimization algorithm boasts advantages of high precision, easy realization, and quick convergence, having displayed its superiority in solving practical problems. Therefore, this constitutive equation adopted the particle swarm optimization algorithm [12,19]. According to previously obtained experiment data, 2219 aluminum alloy displays different creep mechanisms at different aging temperatures, as well as different creep activation energies calculated at high and low temperatures. Based on the above, the creep activation energy and calculations in this experiment were respectively fitted at different temperature ranges in order. Related parameters are shown in Tables 4 and 5, with fitting results in Figures 5 and 6.

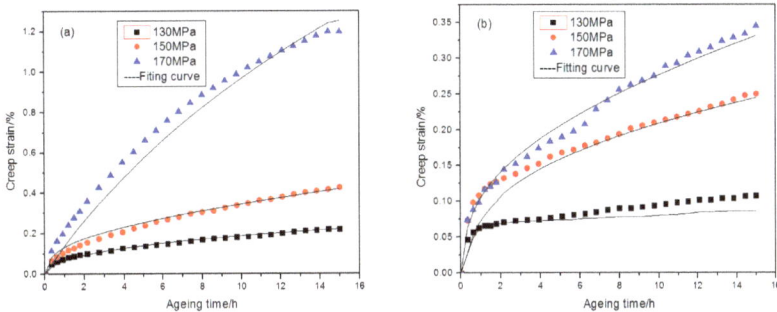

Figure 5. Comparison of fitted values and measured values of 2219 aluminum alloy creep curves at (a) 438 K and (b) 418 K.

Table 4. Creep model parameters of 2219 aluminum alloy (Q_1 = 96.2 kJ/mol).

Material Constants	A	B	σ_0	m_0	h	H^*	m_1	R^2
Number	4×10^3	0.14998	128.1	8.214	239.76	0.3971	0.004223	0.9727

Table 5. Creep model parameters of 2219 aluminum alloy (Q_2 = 36.2 kJ/mol).

Material Constants	A	B	σ_0	m_0	h	H^*	m_1	R^2
Number	8×10^3	0.01968	129.4	9.96	491.82	0.8499	0.164473	0.9028

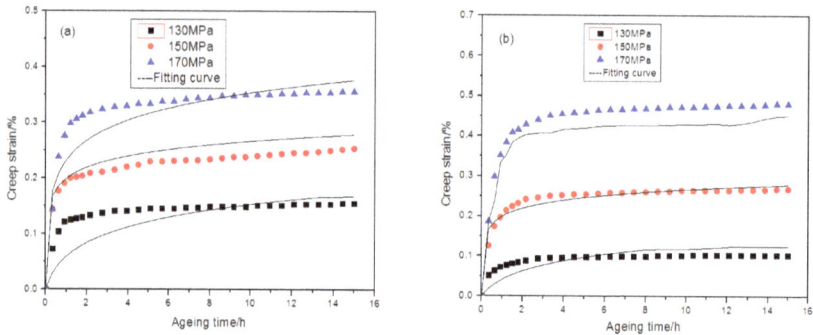

Figure 6. Comparison of fitted values and measured values of 2219 aluminum alloy creep curves at (**a**) 393 K and (**b**) 373 K.

It can be seen from the fitting results that this creep model enjoys a good fitting effect in general. When the experimental temperature exceeded or equaled 418 K, the equation showed a very high fitting precision with a coefficient of determination R^2 of 0.9727; when the experimental temperature was lower than or equaled 393 K, the coefficient of determination of this equation R^2 was 0.9028. Only individual curves presented slight deviation at the creep aging initial stage, because under a relatively low temperature, the creep deformation of aluminum alloy became less and mostly concentrated on the first stage, hence the certain deviation in fitting results. In general, there is a good consistency between the predicted value and the experiment value for this constitutive equation, thus demonstrating that this constitutive equation can apply to different temperatures and stresses and express the material's creep aging processes.

4. Conclusions

(1) This paper carried out creep experiments for 2219 aluminum alloy at different temperature and stress conditions separately. It was found that stress state at a lower temperature is the principal factor affecting the creep aging behaviors of 2219 aluminum alloy.

(2) At different aging temperatures, 2219 aluminum alloy displayed different creep mechanisms. When $T \geqslant 418$ K, stress exponents $n = 4 \sim 6$ and fell in the category of the dislocation climb mechanism; when 373 K $\leqslant T < 418$ K, $n = 4 \sim 6$ and fell in the category of the dislocation glide mechanism; when the temperature was around 353 K, $n \approx 1$ and fell in the category of the grain boundary sliding mechanism. When the temperature was lower than 353 K, the alloy can be basically considered with no creep.

(3) By processing and analysis of the experimental data, the stress exponent and activation energy under different conditions were calculated. Coupled with revision of the classical creep equation, a creep constitutive model applicable to different temperatures and stress conditions was established, which proved itself with great consistency with the experimental data.

Acknowledgments: This work was supported by Major State Basic Research Development Program of China (2014CB046602), National Natural Science Foundation Key Program of China (51235010).

Author Contributions: Lingfeng Liu and Lihua Zhan conceived and designed the experiments; Lingfeng Liu performed the experiments; Lingfeng Liu and Lihua Zhan analyzed the data; Wenke Li contributed reagents, materials and analysis tools; Lingfeng Liu wrote the paper.

Conflicts of Interest: The authors declare no conflict of interest.

References

1. Sallah, M.; Peddieson, J., Jr.; Foroudastan, S. A mathematical model of autoclave age forming. *J. Mater. Process Technol.* **1991**, *28*, 211–219. [CrossRef]

2. Zeng, Y.-S.; Huang, X.; Huang, S. The research situation and the developing tendency of creep age forming technology. *J. Plast. Eng.* **2008**, *15*, 1–8.

3. Kowalewski, Z.L.; Hayhurst, D.R.; Dyson, B.F. Meachanisms-based creep constitutive equations for aluminum alloy. *J. Strain Anal.* **1994**, *29*, 309–316. [CrossRef]

4. Ho, K.C.; Lin, J.; Dean, T.A. Constitutive modelling of primary creep for age forming an aluminium alloy. *J. Mater. Process Technol.* **2004**, *153*, 122–127. [CrossRef]

5. Ho, K.C.; Lin, J.; Dean, T.A. Modelling of springback in creep forming thick aluminum sheets. *Int. J. Plast.* **2004**, *20*, 733–751. [CrossRef]

6. Zhang, J.; Deng, Y.; Yang, J.; Zhang, X. Experimental studies and constitutive modeling for creep aging of 2124 aluminum alloy. *Acta Metall. Sin.* **2013**, *49*, 379–384. [CrossRef]

7. Guan, C.; He, W.; Zhao, Z.; Wang, G. Creep characteristics of 2024 alloy at cryogenic temperatures. *J. Aeronaut. Mater.* **2014**, *34*, 93–96. [CrossRef]

8. An, L.; Cai, Y.; Liu, W.; Yuan, S.; Zhu, S.; Meng, F. Effect of pre-deformation on microstructure and mechanical properties of 2219 aluminum alloy sheet by thermomechanical treatment. *Trans. Nonferr. Met. Soc. China* **2012**, *22*, s370–s375. [CrossRef]

9. Zhan, L.; Li, J.; Huang, M. Creep aging behavior and constitutive equation of 2524 aluminum alloy. *Mater. Mech. Eng.* **2013**, *37*, 92–96.

10. Wang, J. *A Study on Creep Behavior of Titanium Alloy*; Harbin Institute Technology: Harbin, China, 2008; pp. 39–41.

11. Somekawa, H.; Hirai, K.; Watanabe, H.; Takigawa, Y.; Higashi, K. Dislocation creep behavior in Mg-Al-Zn alloys. *Mater. Sci. Eng.* **2005**, *A407*, 53–61. [CrossRef]

12. Li, Y.G. Experimental Study on Creep Aging Behavior and Constitutive Modeling of 2124 Aluminum Alloy. Master's Thesis, Central South University, Changsha, China, 2012.

13. Mckamey, C.G.; Maziasz, P.J.; Jones, J.W. Effect of addition of molybdenum or niobium on creep rupture properties of Fe_3Al. *J. Mater. Res.* **1997**, *7*, 2089–2106.

14. Mary, A.W. *Properties of Materials*; Oxford University Press: New York, NY, USA, 1999.

15. Yang, S.; Li, J.; Wei, S.; Xu, L.; Zhang, G.; Zhang, E. Pyroplastic deformation behavior of pure molybdenum plate slab and constitutive equation. *Trans. Nonferr. Met. Soc. China* **2011**, *21*, 2126–2131.

16. Frost, H.J.; Ashby, M.F. *Deformation-Mechanism Maps: The Plasticity and Creep of Metals and Ceramics*; Pergamon Press: Oxford, UK, 1982; pp. 102–112.

17. Li, B.; Lin, J.; Yao, X. Characteristics of optimization of creep constitutive equations. *Chin. J. Mech. Eng.* **2003**, *39*, 37–39. [CrossRef]

18. Zhan, L.; Li, Y.; Huang, M.; Zhang, M. Constitutive equation describing creep ageing of 2124 aluminum alloy. *J. South China Univ. Technol.* **2012**, *40*, 107–111.

19. Shi, X.H.; Liang, Y.C.; Lee, H.P.; Lu, C.; Wang, L.M. An improved GA and a novel PSO-GA-based hybrid algorithm. *Inform. Process. Lett.* **2005**, *9*, 255–261. [CrossRef]

Article

Influence of Laser Welding Speed on the Morphology and Phases Occurring in Spray-Compacted Hypereutectic Al-Si-Alloys

Thomas Gietzelt [1],*, Torsten Wunsch [1], Florian Messerschmidt [1], Holger Geßwein [2] and Uta Gerhards [1]

[1]   Karlsruhe Institute of Technology, Institute for Micro Process Engineering, P.O. Box 3640, 76021 Karlsruhe, Germany; torsten.wunsch@kit.edu (T.W.); florian.messerschmidt@kit.edu (F.M.); uta.gerhards@kit.edu (U.G.)

[2]   Karlsruhe Institute of Technology, Institute for Applied Materials, P.O. Box 3640, 76021 Karlsruhe, Germany; holger.gesswein@kit.edu

*   Correspondence: thomas.gietzelt@kit.edu; Tel.: +49-721-608-23314

Academic Editor: Nong Gao
Received: 27 June 2016; Accepted: 17 November 2016; Published: 24 November 2016

Abstract: Normally, the weldability of aluminum alloys is ruled by the temperature range of solidification of an alloy according to its composition by the formation of hot cracks due to thermal shrinkage. However, for materials at nonequilibrium conditions, advantage can be taken by multiple phase formation, leading to an annihilation of temperature stress at the microscopic scale, preventing hot cracks even for alloys with extreme melting range. In this paper, several spray-compacted hypereutectic aluminum alloys were laser welded. Besides different silicon contents, additional alloying elements like copper, iron and nickel were present in some alloys, affecting the microstructure. The microstructure was investigated at the delivery state of spray-compacted material as well as for a wide range of welding speeds ranging from 0.5 to 10 m/min, respectively. The impact of speed on phase composition and morphology was studied at different disequilibrium solidification conditions. At high welding velocity, a close-meshed network of eutectic Al-Si-composition was observed, whereas the matrix is filled with nearly pure aluminum, helping to diminish the thermal stress during accelerated solidification. Primary solidified silicon was found, however, containing considerable amounts of aluminum, which was not expected from phase diagrams obtained at the thermodynamic equilibrium.

Keywords: hypereutectic aluminum alloy; DISPAL; laser welding; nonequilibrium solidification

---

1. Introduction

Aluminum is a lightweight construction material possessing sufficient strength for many applications in automotive industries. Multiple alloying elements can influence and tailor the properties in a wide range. Silicon, for example, improves the wear resistance without increasing the density considerably. However, usually the weldability of aluminum alloys is influenced by the temperature range of solidification, and hot cracks are formed due to thermal shrinkage [1].

To take advantage of the improved properties of aluminum materials with very high silicon contents, spray-compacting is a process for the manufacturing of nonequilibrium materials with improved mechanical and tribological properties [2,3]. Alloys not producible by conventional casting processes can be made by spray-compacting at high temperature gradients of more than 1000 K/s during solidification. Insoluble alloying components can be distributed homogeneously in a matrix material since only small amounts hit a cold substrate and solidify rapidly. Indirect rod extrusion for compaction and occurring phases is described in [4]. These materials are known, e.g., under the trade name DISPAL™ and are supplied by Erbslöh Aluminium GmbH in different compositions and conditions.

However, the question arises of how these materials can be joined since the dimension of semi-finished products is limited. In the literature, special welding techniques like friction stir welding are described, avoiding the formation of a liquid and segregation of nonequilibrium phases [5,6]. Therefore, laser welding could be an additional welding technique preserving the nonequilibrium state of the microstructure due to its high welding velocity. Especially since the development of multi-kilowatt solid state lasers and decreasing investment costs, laser welding has become widespread in, for example, the automotive industries [7–9].

Additionally, for many applications, a huge increase in productivity could be obtained by laser welding due to its high welding velocities, e.g., it replaces spot welding and fixes many issues combined with reduced heat input per unit of length.

In this paper, different hypereutectic Al-Si alloys were investigated at welding velocities between 0.5 to 10 m/min using a 3 kW disc laser to supply an additional joining technique to stir friction welding with high velocities. Changes in morphology and composition of occurring phases are discussed in respect to nonequilibrium conditions. The mechanical and especially the dynamic durability of welded parts made of hypereutectic Al-Si-alloys will depend on the size and number of precipitations in the microstructure of the weld seam. Welding parameters should be optimized to obtain small precipitations and to prevent the formation of pores. This is supported by high welding speeds and accelerated solidification of the weld pool.

2. Materials and Methods

2.1. Materials Used and Sample Preparation

Hypereutectic spray-compacted Al-Si-alloys were chosen by different silicon contents and additional alloying elements (see Table 1). The material was obtained by Erbslöh Aluminium GmbH as round stock of different diameters. Discs, approximately 10 mm in thickness, were cut using a BRILLANT 250 by ATM cutting machine (Mammelzen, Germany) with a 300 mm disk, type C with a thickness of 2 mm. Then, both sides of the samples were grinded using grained abrasive paper with 600 meshes per square inch in the SAPHIR 550 grinding machine by ATM (Mammelzen, Germany).

Table 1. Composition of DISPAL-alloys used for laser welding [10].

Alloy	Composition	Si (wt. %)	Fe (wt. %)	Ni (wt. %)	Cu (wt. %)	Mg (wt. %)	Zr (wt. %)	Co (wt. %)	Ti (wt. %)	Al
S220	AlSi35	35								balance
S225	AlSi35Fe2Ni	35	2	1						balance
S232 [11]	AlSi17Fe4Cu2.5MgZr	17	4		3	1	<1			balance
S263	AlSi25Fe2.5Cu2.5Ni2.5MgCo	25	3	3	3	1		1	1	balance

2.2. Experimental Procedure

For laser welding, a TruCell 3010 machine and a TruDisk 3001 solid state disc laser by TRUMPF (Ditzingen, Germany) were combined. A light conducting cable (LCC) 100 μm in diameter was used to transfer the laser radiation from the disk laser to the machine tool. The optics had a focal length of 150 mm.

For the cross-section of a laser weld seam, the focal position of the focal spot may have a huge impact: To avoid the formation of pores, the energy density at the surface level must be sufficient to prevent solidification here first. Especially for small LCC-diameters the variation of the spot area at the surface level for a given focal position, in regards to the focal length, is more pronounced than for larger LCC-diameters.

Tests showed that different operators set the focal position using the image of a CCD-camera within a z-range of about 1.6 mm. Hence, a distance ruler was used to adjust the focal position related to the surface of the work piece (Figure 1). A focal position of $F = 0$ mm was used for all experiments.

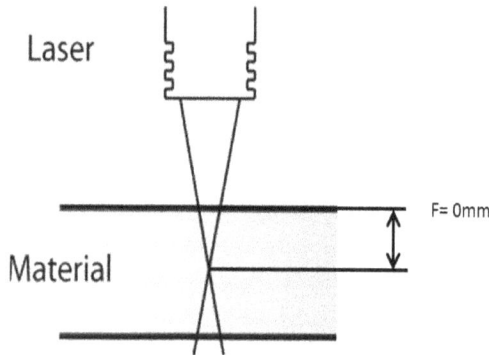

Figure 1. Focal position and its impact on the energy density at the surface, depending on the focal length.

All welding experiments were performed using 15 L/min of Ar 5.0. The inert gas supply was adjusted for backhand welding to shield from the melting bath. A length of 30 mm was welded for each welding velocity at a focal position of $F = 0$ mm using the maximum power of 3 kW.

Afterwards, the samples were cut perpendicular to the weld seams into two halves using the cutting machine BRILLANT 250. Both parts were hand grinded starting at 600 meshes per square inches down to 2400 abrasive paper using the SAPHIR 550 grinding machine, and polished using monocrystalline diamond suspension of 9, 6 and 1 µm, respectively.

From Figure 2, the impact of the welding speed on the cross-section can be seen: For low welding speed, the deepest weld seams were obtained; however, also the width at the surface is high due to slow solidification and heat conduction. For high welding speeds, narrower weld seams at reasonable welding depths are possible.

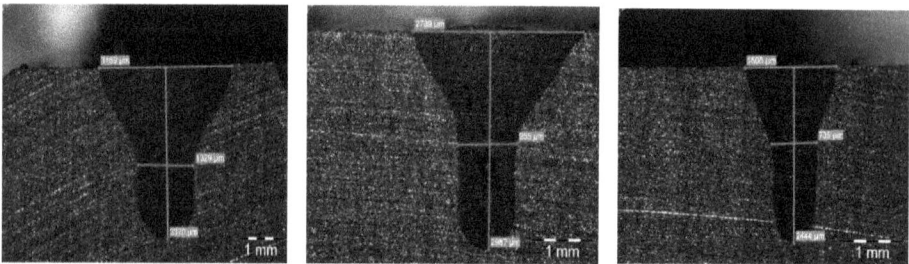

Figure 2. From **left** to **right**: Cross-section views for DISPAL S220 for welding velocities of 3, 5 and 10 m/min for $P = 3$ kW and a focal position of $F = 0$ mm. For 3 m/min the positions of EDX measurements are marked for base material, transition zone and weld seam.

Photographs were taken using a stereomicroscope SZX12 by OLYMPUS, Hamburg, Germany. SEM pictures were taken by a JSM 6300 by JEOL (Freising, Germany) at different spots of the weld seams. A microprobe JXA 8530F by JEOL (Freising, Germany) with field emitting cathode was used for EDX and WDX investigations. For EDX a voltage of 15 kV, a working distance of 11 mm and a current of 1.5 nA to ensure a count rate of at least 2500 count/s were used.

3. Results and Discussion

3.1. Microstructure of Spray-Compacted and Welded DISPAL-Alloys

As expected, primary silicon precipitations in the delivery condition of all materials are globular and uniformly distributed all over the matrix for all kinds of materials. In laser-welded areas, however, spiky precipitations are generally observed according to the local solidification rate. Additionally, depending on the welding velocity, a network of near-eutectic composition is formed, embedding inclusions of nearly pure aluminum.

SEM pictures were taken in the transition section to the base material as well as in the center of the weld seam. EDX measurements were performed in the middle of the weld seams at a depth where the funnel-shaped area changes to a straight one, representing the lowest solidification velocity.

Whereas the compositions in all tables are given in percent by weight, the evaluation of the results refers to percent by atom to recognize phase compositions. In case of aluminum and silicon, the differences between both are small; however, for metals with higher atomic number, e.g., iron, nickel and copper, distinction is obvious.

From Figure 3 it can be seen that the solubility of silicon in aluminum is very low and iron is not soluble in silicon. For other DISPLA-alloys containing up to five additional elements, the conditions concerning solubility and formation of intermetallic compounds are conspicuous.

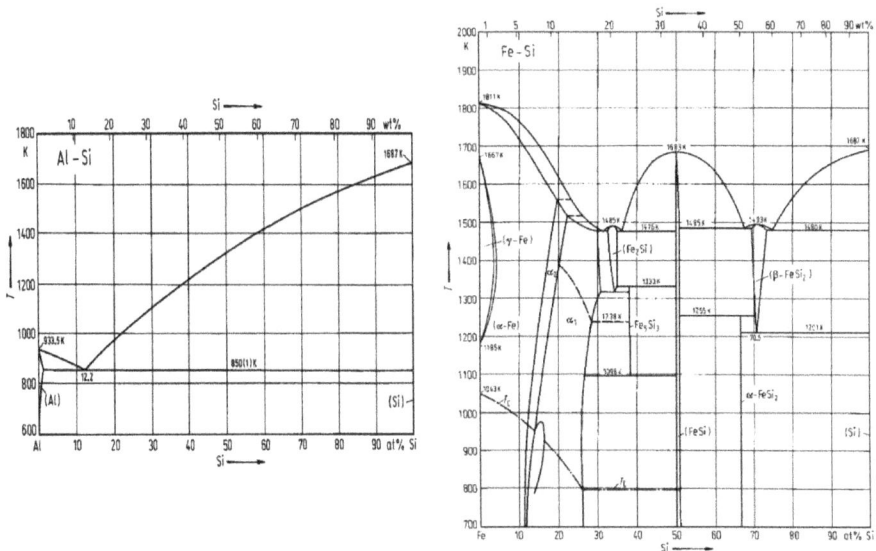

Figure 3. Binary phase diagrams of Si-Al (**left**) and Si-Fe (**right**) [12].

Figures 4, 6, 8 and 10 show four different SEM pictures of all materials at different spots according to the increasing silicon content of the alloys: top left of the original base material; top right the transition zone of the weld seam to the base material; and below the microstructure in the center of the weld seams for welding velocities of 0.5 and 10 m/min, respectively. All pictures for the transition zone were taken from the 10 m/min weld seam due to a more pronounced and narrower crossover. For 0.5 m/min, however, the transition was more blurred.

3.2. DISPAL S220: Comparision of Comosition Measured by WDX and EDX

In Figure 4, SEM pictures of S220 with the highest silicon content of 35 wt. % without additional alloying elements are shown. Some small bright precipitations are found, probably due to impurities by other elements originating from impurities of the recycled material.

Compared to other alloys, coarser and predominantly silicon-containing precipitations are observed.

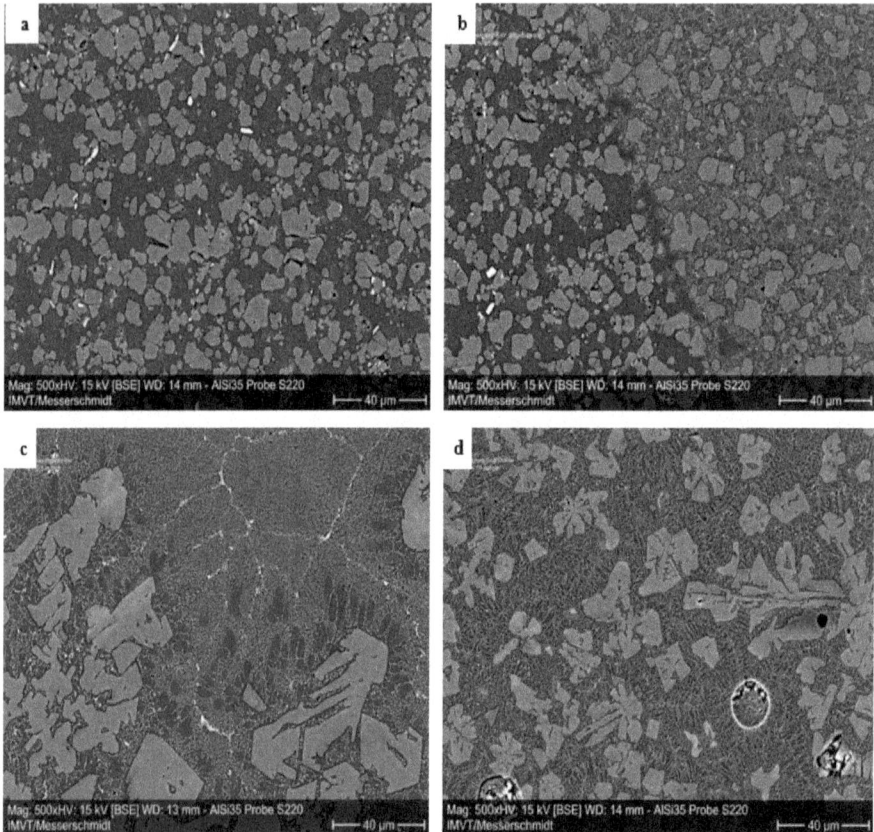

Figure 4. S220: (**a**) Base material; (**b**) Transition of weld seam to base material; (**c**) Center of the weld seam for $v = 0.5$ m/min; (**d**) Center of the weld seam for $v = 10$ m/min.

Light elements like carbon or oxygen may be overestimated by EDX since its specific energy is superimposed to the "bremsstrahlung". Hence, citing S220 as an example, WDX was employed for a welding speed of 0.5 m/min to test if the content of oxygen measured by EDX (Table 2) is in a reliable range. The results in wt. % obtained from measurements employing elemental standards showed a deviation within ±1% to 100% for different sample points. For Table 3, the values were scaled to 100% and given in at. %. From this it can be concluded that the values of oxygen content obtained by EDX for other samples are reasonable. All alloys contain noticeable amounts of oxygen, in the base material as well as in the weld seams, likely from passivation layers and impurities during spray-compacting. A significant increase of oxygen content from as-delivered condition and welded material is not detected.

Table 2. Composition of S220 measured using EDX for a welding speed of 0.5 m/min (at. %).

Spot	No.	Al	Si	Fe	Cu	O	Suggested Phase
	1	96.8	1.57	0.18	0.01	1.45	
	2	53.89	44.63	0.2	/	1.28	
	3	1.55	96.66	0.03	0.08	1.69	
Base mat.	4	1.55	96.66	0.03	0.08	1.69	
	5	86.38	12.02	0.01	0.03	1.56	eutectic
	6	93.37	4.39	0.05	0.14	2.51	
	1	90.0	7.93	0.44	0.04	1.59	
	2	86.26	11.09	0.94	0.1	1.61	eutectic
Welding speed	3	3.31	95.58	0.12	/	0.99	
0.5 m/min	4	41.65	57.29	/	/	1.06	
	5	68.21	29.31	0.04	0.14	2.29	
	6	92.6	5.92	0.02	0.07	1.4	
	1	77.47	20.07	0.53	0.14	1.8	
	2	74.79	20.66	0.93	0.15	3.46	
Welding speed	3	1.77	97.11	0.05	0.03	1.04	
10 m/min	4	2.99	95.82	/	0.19	1.0	
	5	85.33	12.45	0.35	/	1.87	eutectic
	6	78.95	18.92	0.05	0	2.08	

Table 3. Composition of S220 measured by WDX for a welding speed of 0.5 m/min (at. %).

Spot/Phases		No.	Al	Si	Fe	Cu	O
		1	2.80	96.53	0.02	0.01	0.64
		2	2.22	97.53	0.00	0.01	0.24
	"bright" phase	3	3.20	96.45	0.01	0.01	0.34
		4	1.27	98.20	0.01	0.01	0.51
		5	1.56	97.98	0.00	0.02	0.45
Base mat.		1	98.57	0.22	0.01	0.02	1.18
		2	98.20	0.24	0.01	0.01	1.53
	"dark" phase	3	97.44	0.85	0.23	0.00	1.48
		4	98.30	0.32	0.01	0.01	1.36
		5	98.41	0.26	0.00	0.01	1.32
		1	2.76	96.91	0.01	0.00	0.32
		2	3.25	96.24	0.01	0.01	0.50
	"bright" phase	3	2.68	96.82	0.00	0.00	0.51
		4	3.71	95.78	0.00	0.00	0.51
Welding speed		5	2.88	96.60	0.00	0.00	0.52
0.5 m/min		1	81.74	16.03	0.02	0.04	2.17
		2	87.73	10.67	0.14	0.06	1.40
	"dark" phase	3	86.67	11.90	0.01	0.01	1.41
		4	80.29	17.90	0.00	0.00	1.81
		5	79.16	17.62	0.03	0.02	3.16

EDX measurements showed for the base material as well as the weld seams in Figure 5 and Table 2 that the precipitations consist mainly of silicon for more than 95 at. %. The silicon content in the aluminum matrix of the base material is 12 and 4.4 at. % for No. 5 and 6, respectively, making it difficult to derive a statement about the influence of the solidification speed on the silicon content in the supersaturated aluminum matrix.

It is noteworthy that the silicon content in the eutectic network for 10 m/min (No. 1, 2 and 6) is around 19–20 at. %. Globular grey islands (No. 5) are embedded, having exactly the eutectic composition of 12.5 at. % without the typically lamellar segregation. For 0.5 m/min, however, the silicon content in the matrix (No. 5) is 6 at. %, but the influence of welding velocity on the segregation of silicon from a homogeneous melt cannot be surely stated due to contradictory results from the base material.

Additionally, pearl-shaped, small precipitations forming a loose network are formed for 0.5 m/min. For 10 m/min, however, their size is clearly much smaller due to faster solidification and insular areas. The composition is difficult to determine due to the small size for both welding speeds. It seems to contain impurities of iron and copper.

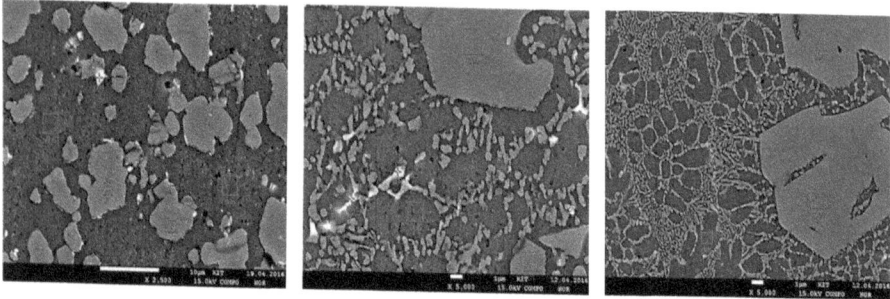

Figure 5. EDX investigation of phases in S220: **Left**: base material; **Middle**: welding speed 0.5 m/min; **Right**: welding speed 10 m/min.

3.3. Microstructures of DISPAL-Alloys Containing Additional Alloying Elements

3.3.1. DISPAL S225

In contrast to S220, for S225, 2 wt. % of iron and about 1 wt. % of nickel are added. This alloy was chosen to evaluate differences in phase compostion and morphology in relation to S220.

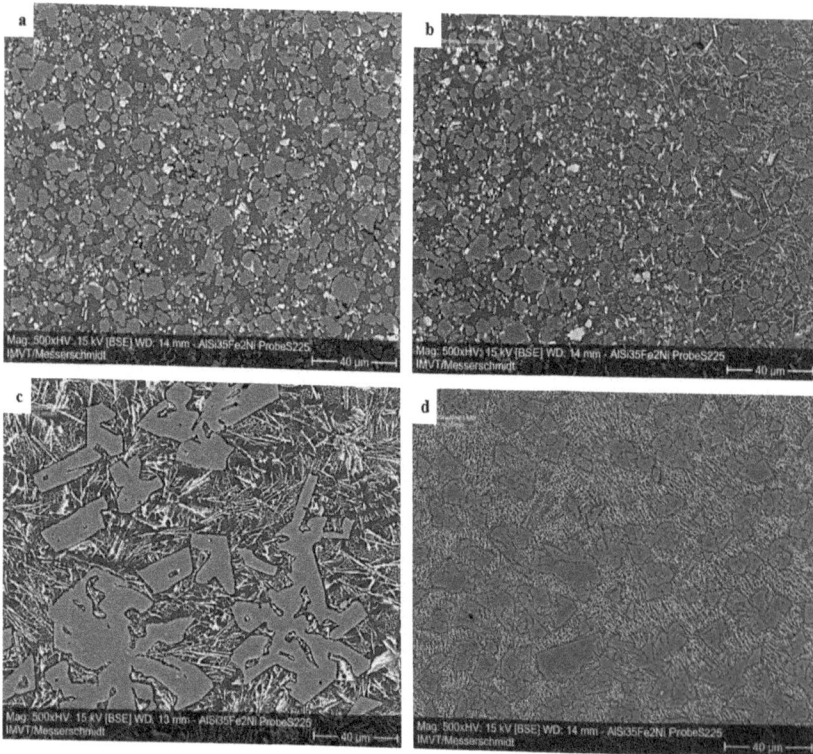

Figure 6. S225. (**a**) Base material; (**b**) Transition of weld seam to base material; (**c**) Center of the weld seam for v = 0.5 m/min; (**d**) Center of the weld seam for v = 10 m/min.

Actually, as shown in Figure 7, already in the base material more bright precipitations containing metals of higher atomic number are found. Shape and size of silicon-rich precipitations are similar to S220. However, in the weld seam, the morphology of the precipitation network is changed drastically: iron and nickel change the shape of bright precipitations between the large, primary silicon-rich precipitations from pearl-like to spicular, even for the low welding speed of 0.5 m/min. From the EDX-investigations in Figure 7 and Table 4, it can be seen that iron is concentrated in these spicular precipitations. The distance between these spicular precipitations decrease considerably for 10 m/min welding speed, whereas the size of primary silicon-rich precipitations is only slightly affected. Especially the microstructure for 10 m/min should have favorable mechanical properties due to its homogeneous distribution of the precipitations.

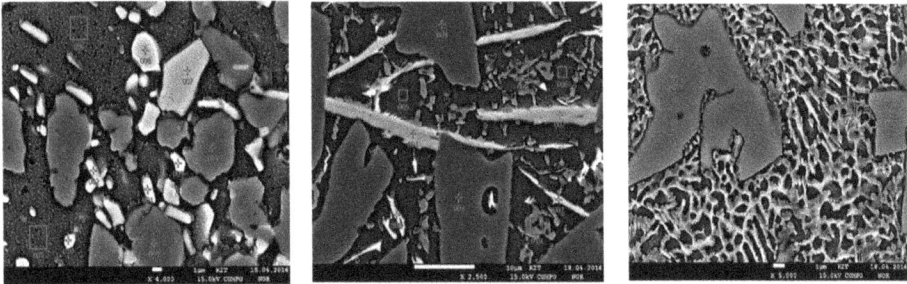

Figure 7. EDX investigation of phases in S225: **Left**: base material; **Middle**: welding speed 0.5 m/min; **Right**: welding speed 10 m/min.

Table 4. Composition of S225 measured by EDX (at. %).

Spot	No.	Al	Si	Fe	Ni	Mg	O
	1	54.44	30.79	9.35	0.73	0.18	4.51
	2	51.67	34.29	8.43	0.83	0.08	4.69
	3	12.53	85.83	0.28	0.18	0.12	1.06
	4	4.47	93.32	0.17	0.21	/	1.83
Base mat.	5	96.64	0.73	0.18	/	0.22	2.23
	6	96.73	0.56	0.14	0.12	0.17	2.28
	7	66.64	3.03	6.61	8.59	0.1	15.04
	8	65.42	3.29	7.4	10.88	0.05	12.97
	9	72.09	14.56	4.77	5.91	0.04	2.63
	10	93.6	1.45	1.0	1.24	0.18	2.54
	1	64.92	20.58	10.89	1.41	0.05	2.15
	2	81.27	10.67	5.15	0.67	0.21	2.03
	3	0.94	97.92	0.07	0.21	0.06	0.8
	4	1.49	97.29	/	/	0.02	1.2
Welding speed	5	95.02	2.83	0.22	0.05	0.11	1.76
0.5 m/min	6	90.83	6.54	0.2	0.32	0.19	1.91
	7	93.33	3.99	0.96	0.14	0.03	1.55
	8	77.56	11.7	6.59	2.37	0.3	1.47
	9	69.11	28.33	0.15	0.05	0.13	2.22
	10	64.49	33.5	0.1	/	0.08	1.83
	1	4.67	93.47	0.14	0.18	0.02	1.53
	2	5.95	92.96	0.02	0.01	0.02	1.04
Welding speed	3	80.61	11.66	2.5	1.04	0.23	3.97
10 m/min	4	80.36	12.6	1.78	0.86	0.07	4.33
	5	69.58	22.39	2.78	0.83	0.18	4.29
	6	67.03	26.36	1.46	0.57	0.12	4.46

3.3.2. DISPAL S232

For S232 in Figure 8, it can be seen that there are different globular precipitations in the base material rich in silicon and iron. Compared to the other alloys, the content of precipitations corresponds to the silicon content.

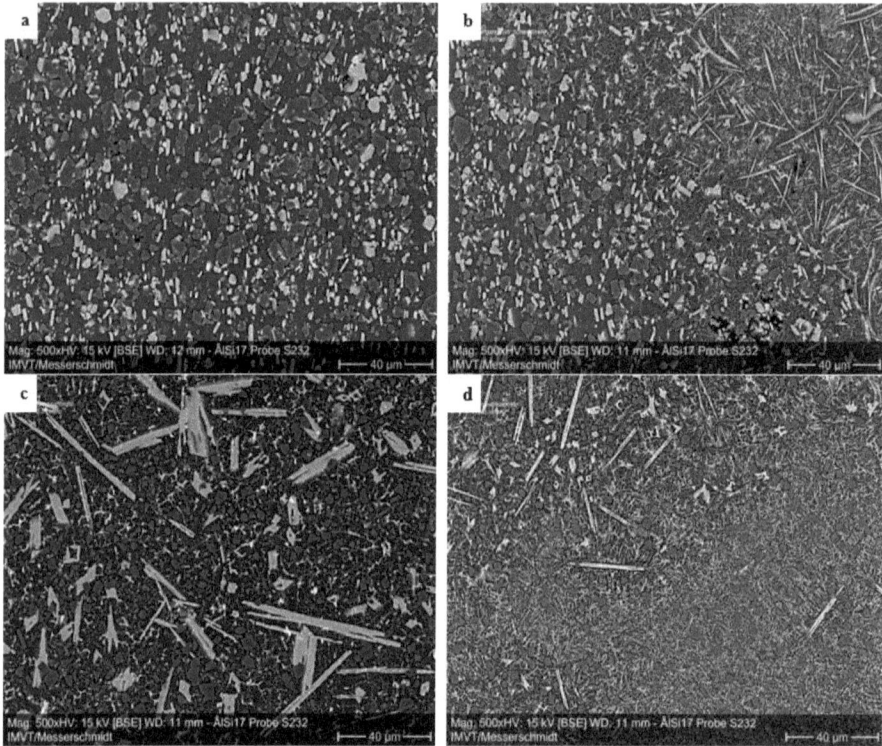

Figure 8. S232: (**a**) Base material; (**b**) Transition of weld seam to base material; (**c**) Center of the weld seam for $v = 0.5$ m/min; (**d**) Center of the weld seam for $v = 10$ m/min.

In the transition area, the morphology of the matrix is already changed to network-like meshes, pointing to a near-eutectic phase, whereas the globular silicon precipitations still exist.

The alloying elements are solved in the melting bath and new precipitations, at least differently shaped ones, are formed. Inside the weld seams for welding velocities of 0.5 and 10 m/min, respectively, grey angular precipitations are visible, similar to the matrix, although smaller in size. Bright spicular precipitations, containing most of the iron and copper, are formed. Length and thickness of these precipitations seem to grow for lower welding speeds since the solidification is retarded in relation to high velocities.

In between with increasing welding speed, as matrix an eutectic microstructure is developed [4]. However, zirconium forming Al3Zr-dispersoids could not be found in this alloy (Table 5).

In any case, these different phases solidifying at different temperatures and maintaining short distances in between help to reduce thermal contraction strains and to avoid hot cracks as known for alloys with a wide solidification range like for some aluminum alloys.

Table 5. Composition of S232 measured by EDX for a welding speed of 0.5 m/min (at. %).

Spot	No.	Al	Si	Fe	Cu	Mg	O	Suggested Phase
	1	68.31	14.25	11.49	0.77	0.2	4.97	Al5FeSi
	2	87.33	3.94	1.42	2.11	0.46	4.74	
	3	12.82	84.24	0.69	0.3	0.1	1.85	
Base mat.	4	42.64	53.8	0.14	0.76	0.2	2.46	
	5	87.8	4.15	2.0	1.6	0.55	3.90	
	6	91.29	1.64	0.25	1.97	0.64	4.21	
	7	65.01	16.91	14.32	0.35	0.11	3.3	Al5FeSi
	8	81.52	12.53	0.25	2.01	0.59	3.08	
	1	67.23	14.17	14.85	0.07	0.03	3.66	
	2	71.47	13.04	11.2	0.77	0.14	3.38	
	3	5.72	92.06	0.14	0.26	0.1	1.71	
	4	28.93	68.92	0.1	0.16	0.24	1.64	
Welding speed	5	74.76	22.1	0.13	0.92	0.47	1.62	
0.5 m/min	6	85.76	5.2	0.33	4.99	0.56	3.17	
	7	63.85	6.23	1.74	12.76	1.57	13.84	
	8	86.29	3.62	0.59	3.25	0.89	5.36	
	9	65.05	30.93	0.24	0.62	0.29	2.89	
	10	77.48	18.72	0.25	0.91	0.43	2.21	
	1	58.84	23.81	12.03	1.65	0.39	3.28	
	2	67.61	13.81	14.0	0.57	0.4	3.61	
	3	27.24	71.02	0.23	0.19	0.16	1.16	
Welding speed	4	31.96	65.39	0.38	0.29	0.19	1.8	
10 m/min	5	71.58	24.63	0.31	0.68	0.42	2.38	
	6	71.43	23.66	1.12	0.81	0.6	2.38	
	7	83.34	6.94	2.06	4.31	1.17	2.17	
	8	76.43	14.48	1.84	2.06	1.32	3.87	

EDX-investigations were done at different spots for the different phases in the base material as well as inside the weld seams. Figure 9 gives an overview of the morphologies and the spots where the composition was determined. Noteworthy is the higher magnification for the SEM-picture for 10 m/min welding speed, indicating a finer microstructure than for 0.5 m/min. In the base material, three compositions of precipitations different in phase contrast were found: bright particles, e.g., No. 1, 2, 7 and 8 of the left picture of Figure 9, respectively, contain nearly the whole amount of iron and copper. No. 1 and 7 corresponds to the intermetallic phase Al5FeSi, mentioned in [4]. For ternary or quaternary alloys, even more complex phases are reported, e.g., in [13]. Darker precipitations, e.g., No. 3 and 4, contain silicon in excess of the eutectic composition. The matrix, however, for No. 5 and 6, consists mainly of aluminum with some iron and copper, in which copper is soluble in aluminum and iron is not.

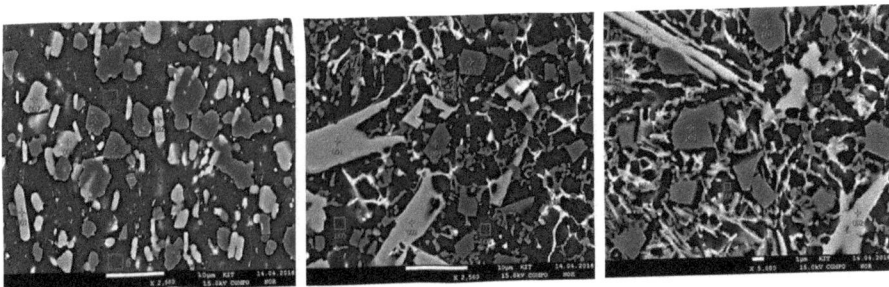

Figure 9. EDX investigation of phases in S232: **Left**: base material; **Middle**: welding speed 0.5 m/min; **Right**: welding speed 10 m/min.

For the weld seams, the composition of the precipitations is maintained despite different species in terms of aluminum and silicon can be found. The silicon content in the matrix, however, rises from less than 5 at. % for more than 20 at. %. The bright network contains about 6 at. % silicon, 2 at. % iron and copper between 4.3 and 12.8 at. % copper (No. 7 for 0.5 and 10 m/min, respectively). No. 8 for 0.5 m/min, however, contains much more aluminum. Probably it is due to the small dimension of the network, so the measurement is affected by the material below.

3.3.3. DISPAL S263

In Figure 10, for the base material, there are more precipitation—albeit less bright—corresponding to the higher silicon content of 25 wt. %. In the weld seams, the network in the matrix is more developed than for S232 due to the higher silicon content, forming an eutectic phase. Obviously, the dark precipitations for 0.5 m/min welding speed are larger than for 10 m/min, in contradiction to S232, and bright precipitations containing metals with higher atomic numbers are thicker than for 10 m/min.

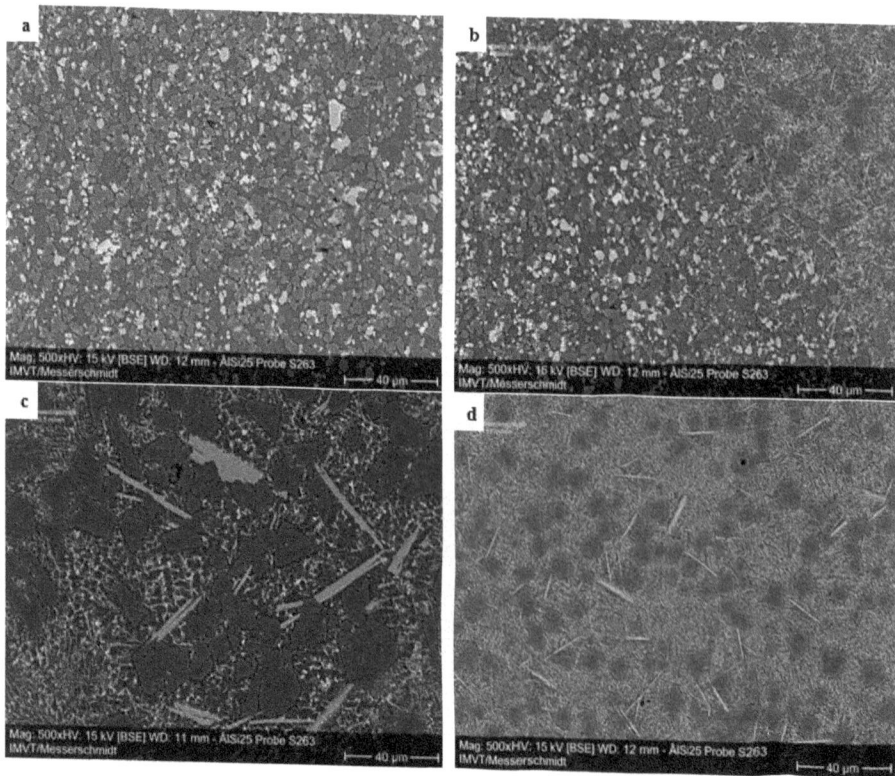

Figure 10. S263: (**a**) Base material; (**b**) Transition of weld seam to base material; (**c**) Center of the weld seam for v = 0.5 m/min; (**d**) Center of the weld seam for v = 10 m/min.

Different phases were investigated as shown in Figure 11 and Table 6, respectively. Whereas the silicon content inside the dark precipitations in the base material are between 84.5 and 76 at. %, for No. 3 and 4, respectively, the silicon content in these precipitations in the weld seams is higher than 90 at. % (same numbers as for the base material). Again, the precipitations are a little larger for the lower speed. Whereas the bright precipitation (No. 1 and 2 for both welding speeds) for 0.5 m/min contains about 24 at. % silicon, 11 to 14 at. % iron, about 0.5 at. % copper and 2.1 to 2.6 at. % nickel, the silicon content for 10 m/min is near-eutectic between 8.5 and 11.9 at. % and iron, nickel and copper are around 1 at. %.

Table 6. Composition of S263 measured by EDX for a welding speed of 0.5 m/min (at. %).

Spot	No.	Al	Si	Fe	Co	Ni	Cu	Mg	O	Suggested Phase
Base mat.	1	73.05	5.76	9.11	1.15	7.16	0.54	0.11	3.13	
	2	76.86	4.83	7.43	0.64	5.8	0.89	0.11	3.44	
	3	12.88	84.65	0.32	0.01	0.32	0.47	0.11	1.23	
	4	22.7	75.92	0.18	0	0.12	0.16	0.02	0.89	
	5	94.61	1.61	0.27	0.02	0.05	0.87	0.19	2.37	
	6	90.19	6.49	0.12	0.02	0.27	0.8	0.27	1.84	
Welding speed 0.5 m/min	1	57.49	23.78	10.94	0.91	2.61	0.17	0.05	4.03	Fe2Al3Si3?
	2	55.3	25.63	13.91	0.89	2.1	0.34	0.11	1.76	Fe2Al3Si3?
	3	1.12	96.86	0.03	0.05	0.13	0.29	/	1.5	
	4	2.67	95.33	0.07	/	0.18	0.19	/	1.56	
	5	77.65	13.88	1.3	0.1	1.53	1.42	0.8	3.32	
	6	85.9	10.76	0.22	0.01	0.08	0.7	0.25	2.08	
	7	80.93	11.99	2.13	0.21	1.63	0.9	0.7	1.5	
	8	83.07	8.98	0.21	0.02	1.78	2.96	0.35	2.63	
Welding speed 10 m/min	1	85.95	8.47	1.15	0.07	0.9	0.86	0.48	2.12	
	2	82.15	11.92	1.09	0.04	1.07	0.98	0.72	2.02	
	3	6.94	91.07	0.18	0.06	0.03	0.23	0.04	1.45	
	4	5.0	93.44	0.12	0.12	0.15	0.11	0.03	1.02	
	5	75.05	16.27	1.54	0.06	1.5	1.34	0.48	3.76	
	6	79.21	12.59	0.76	0.05	1.12	1.22	0.41	4.65	

The bright network (No. 7 for 0.5 m/min) is nearly eutectic with some content of iron, nickel and copper. For 10 m/min, the dimension is too small to determine the composition of the network. The matrix (No. 6 for both welding speeds) possesses near-eutectic composition without showing an additional lamellar microstructure at a smaller scale. It is likely that the silicon content is frozen in a supersaturated solid solution.

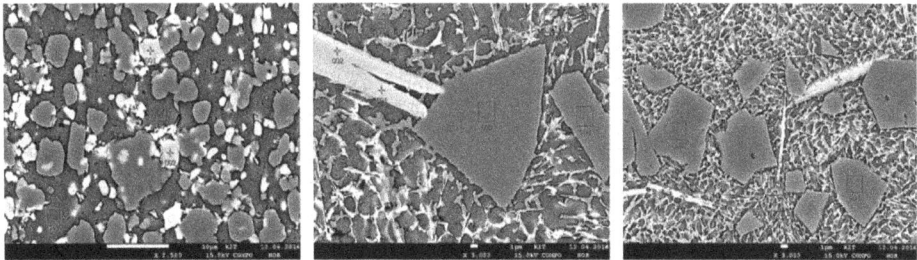

Figure 11. EDX investigation of phases in S263: **Left**: base material; **Middle**: welding speed 0.5 m/min; **Right**: welding speed 10 m/min.

3.4. Micro Hardness Measurements

For S 225, S232 and S263, containing additional alloying elements except silicon, microhardness measurements were performed at weld seams at a velocity of 1 m/min. A Matsuzawa Model MMT-X7B and HV0.1, corresponding to a load of 0.98 N, was employed. The distance between single indentations was 0.5 mm. Different grey levels for background in Table 7 were used to distinguish between as-delivered conditions of spray-compacted materials, heat-affected zones and weld seams. Figure 12a displays the schema of single indentations. Taking into account the scale of the inhomogeneity of the microstructure and the indentations, a certain variance of hardness values is plausible (Figure 12b).

Table 7. Microhardness measurements for S 225, S232 and S263. White background: base material, middle grey: heat-affected zone, dark: weld seam.

Material	$HV_{0.1}$									
S225	121	118	127	185	208	146	165	120	138	119
	120	120	150	220	233	225	231	134	126	121
S232	176	174	181	216	215	198	176	181	174	170
	170	177	170	195	192	180	171	175	165	171
S263	134	143	199	228	226	228	235	213	174	147

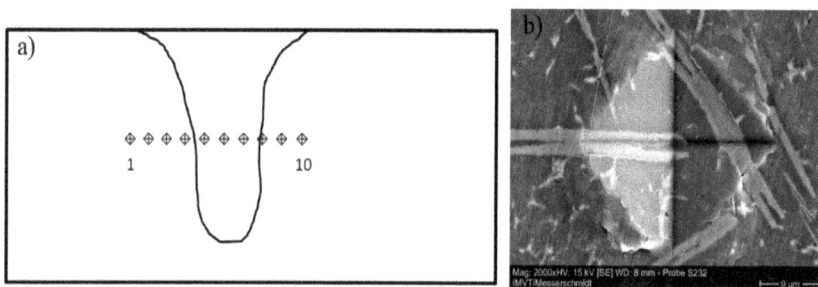

Figure 12. (**a**) Schema of microhardness measurements; (**b**) Detail of micro indentation.

For S225 and S232, two microhardness measurements were made at different depths, both in the area of constant width of the weld seam.

Despite the highest silicon content for S225 resulting in the highest density of silicon precipitations, the hardness of the spray-compacted matrix material is lowest. It is probable that the contribution of solid solution hardening is low since it contains only 2% iron, and because nickel is insoluble in aluminum. Inside the weld seam, higher microhardness values were obtained due to spicular and more homogeneous distributed precipitations.

For S232, containing higher contents of additional alloying elements contributing to solid solution hardening, a higher hardness of the as-delivered state is found. Since the silicon content is only half in comparison to S225, the precipitations are smaller. The microstructure appears finer than for S225, and hence microhardness values scatter less.

S263 has a silicon content of 25% and contains the highest amount and most alloying elements. Only for nickel is no solubility in aluminum found, and solubility for cobalt and titanium is very low. All metals form intermetallic compounds with aluminum. However, the content is possibly too low and solidification time during laser welding too short to reach these compositions and equilibrium conditions. Surprisingly, the microhardness of the as-delivered material is lower than for S232 but, in the weld seam, the hardness level is slightly increased compared to S232.

3.5. X-ray Diffraction (XRD) Measurements of S263

XRD measurements were performed using a Bruker D8 Discover with GADDS and a Hi-STAR area detector in reflection geometry. The sample to detector distance was 13.7 cm. A 500 μm pinhole collimator was used and the sample was oscillated with 1 mm steps to radiate a larger sample area. Cu-Kα-radiation was used and the exposure time was 30 min. Figure 13 shows spectra of as-delivered material and weld seam for a velocity of 1 m/min. Data were evaluated using an online database [14]. S263 contains seven elements and is a very complex system. Despite the obvious additional peaks, most could not be fitted to known intermetallic compounds. Additionally, the database PDF-2,

Release 2015, by the International Centre for Diffraction Data, containing more than 700 phases for the Al-Si-Fe-Ni-Cu-system, was used. Here, phases like Fe3Al2Si4 (PDF 1-087-1921), Al3FeSi2 (PDF 52-0917), AlCu3 (PDF 1-74-6895), Al5.4Fe2 (PDF 1-73-8846), Al5.6Fe2 (PDF 1-71-9849), NiSi (PDF 1-70-9169) and Ni31Si12 (PDF 17-222) were suggested; however, positions of specific reflexes are in poor accordance.

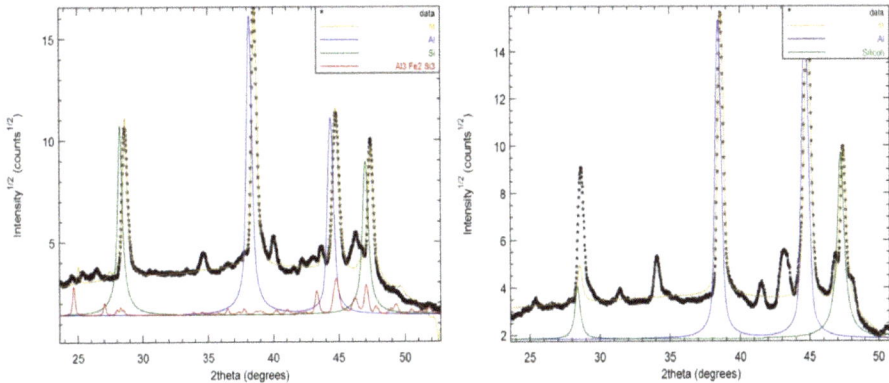

Figure 13. Left: XRD pattern of S263 as-delivered state; **Right**: As welded at 1 m/min.

Although the solidification rate for the weld seam should be lower and closer to thermodynamically equilibrium than for spray-compacted material, no additional intermetallic compound was found. However, one intermetallic compound was suggested for the as-delivered state. A phase similar to Al3Fe2Si3 can be supposed according to [15,16].

In general, XRD investigations delivered no clarity about additional intermetallic compounds due to complexity of the alloy S263.

4. Conclusions

By spray-compacting, material combinations not producible by the conventional melting route can be manufactured. Therefore, advantageous properties including increased wear resistance and mechanical properties can be realized. Due to the production route, a consolidation step to densify remaining pores is necessary to improve crack-propagation behavior. Extruding is usually used, thereby resulting in round stock and ensuring a constant natural strain of the spray-compacted cone. It also restricts available semi-finished products of spray-compacted alloys.

By laser welding, reasonable microstructures for all investigated hypereutectic Al-Si-alloys could be obtained for a wide range of welding speeds, implying reasonable production costs. Despite the large temperature range of solidification, no hot cracks were observed. The formation of equally distributed small precipitations annihilates thermal contraction strains on a microscopic scale. This is in opposition to conventionally casted materials.

With laser welding, an alternative joining technology to stir-friction welding, has become available. Flat semi-finished products of spray-compacted Al-Si-materials could thus be produced at high speed and reasonable cost.

Acknowledgments: We acknowledge the support of Erbslöh Aluminium GmbH for supplying all materials used for this work. The contribution of H. Leiste from the Institute for Applied Materials by evaluating XRD measurements regarding specific phases employing up-to-date database of the Al-Si-Fe-Ni-Cu-system is appreciated. The support by Bolich from the Institute for Applied Materials regarding microhardness measurements is acknowledged.

Author Contributions: Thomas Gietzelt conceived this work, interpreted EDX measurements and morphologies, depending on different alloys, and wrote the paper (corresponding author). Torsten Wunsch conducted the sample preparation and practical laser welding experiments. Florian Messerschmidt was responsible for preparation, metallography and SEM pictures. Uta Gerhards performed microprobe analysis and interpretation of EDX measurements. Holger Geßwein performed XRD measurements and interpretation.

Conflicts of Interest: The authors declare no conflict of interest.

References

1. Wolf, M. Zur Phänomenologie der Heißrissbildung beim Schweißen und Entwicklung aussagekräftiger Prüfverfahren. Ph.D. Thesis, University of Hamburg, Hamburg, Germany, September 2006.
2. Uhlenwinkel, V.; Achelis, L.; Bauckhage, K. *Symposium Sprühkompaktieren*; SFB 372: Bremen, Deutschland, 2004; Volume 7, pp. 123–136.
3. Kainer, K.U. *Metallische Verbundwerkstoffe*; Wiley-VCH: Weinheim, Germany, 2003; pp. 232–240.
4. Rose, A. Verzugsreduzierung beim Wärmebehandeln von Bauteilen aus sprühkompaktierten und gießtechnisch hergestellten Aluminiumlegierungen durch Gasabschrecken. Ph.D. Thesis, University of Bremen, Bremen, Germany, July 2014; pp. 29–34.
5. Roos, A. Grundlegende Untersuchung über ein neues Schweißverfahren namens HFDB. Ph.D. Thesis, GKSS-Forschungszentrum Geesthacht, Geesthacht, Germany, August 2010.
6. Su, J.Q.; Nelson, T.W.; Sterling, C.J. Grain refinement of aluminum alloys by friction stir processing. *Philos. Mag.* **2006**, *86*, 1–24. [CrossRef]
7. Kaufmann, S.; Fleckenstein, M. *Lasermaterialbearbeitung im Automobilbau*; Springer-VDI-Verlag GmbH: Duesseldorf, Germany, 2001; Volume 143, pp. 31–35.
8. Sefler, P.; Wallmeroth, K.; Mann, K. *Stab, Faser und Scheibe—Die Geschichte des Festkörperlasers, Laser Magazin*; Magazin Verlag Hightech Publications KG: Bad Nenndorf, Germany, 2010; pp. 6–9.
9. Prange, W. Maßgeschneiderte Werkstoffe für den Automobilbau. *Automob. Z.* **2001**, *2*, 140–141. [CrossRef]
10. Material Data for DISPAL-Alloys. Available online: http://www.lookpolymers.com/ (accessed on 24 June 2016).
11. Tschegg, S.E.S.; Mayer, H.; Schuller, R.; Przeorski, T.; Krug, P. Fatigue properties of spray formed hypereutectic aluminium silicon alloy DISPAL® S232 at high and very high numbers of cycles. *Mat. Sci. Eng. A Struct.* **2012**, *538*, 327–334. [CrossRef]
12. Binary Phase Diagrams, Landolt-Börnstein. Available online: http://materials.springer.com/ (accessed on 24 June 2016).
13. Bestimmung von α- und β-AlFeSi-Phasen in Aluminium-Knetlegierungen; Trimet Aluminium Ag. Available online: http://www.trimet.eu/de/fe-veroeffentlichungen/trimet_phasen_in_aluminium-knetlegierungen.pdf (accessed on 19 October 2016).
14. XRD-Online Data Base. Available online: http://cod.iutcaen.unicaen.fr/ (accessed on 26 October 2016).
15. Gueneau, C.; Servant, C.; D'Yvoire, F.; Rodier, N. $Fe_2Al_3Si_3$. *Acta Cryst.* **1995**, *51*, 2461–2464. [CrossRef]
16. Gupta, S.P. Intermetallic Compound Formation in Fe-Al-Si Ternary System: Part I. *Mater. Charact.* **2002**, *49*, 269–291. [CrossRef]

metals

MDPI

Article

Welding Distortion Prediction in 5A06 Aluminum Alloy Complex Structure via Inherent Strain Method

Zhi Zeng [1,2,*], **Xiaoyong Wu** [1], **Mao Yang** [1] and **Bei Peng** [1,2]

[1] School of Mechatronics Engineering, University of Electronic Science and Technology of China, Chengdu 611731, China; xywuestc@163.com (X.W.); yangmao2016@yahoo.com (M.Y.); beipeng@uestc.edu.cn (B.P.)

[2] Center for Robotics, University of Electronic Science and Technology of China, Chengdu 611731, China

* Correspondence: zhizeng@uestc.edu.cn; Tel.: +86-28-6183-0229

Academic Editor: Nong Gao

Received: 9 June 2016; Accepted: 11 August 2016; Published: 6 September 2016

Abstract: Finite element (FE) simulation with inherent deformation is an ideal and practical computational approach for predicting welding stress and distortion in the production of complex aluminum alloy structures. In this study, based on the thermal elasto-plastic analysis, FE models of multi-pass butt welds and T-type fillet welds were investigated to obtain the inherent strain distribution in a 5A06 aluminum alloy cylindrical structure. The angular distortion of the T-type joint was used to investigate the corresponding inherent strain mechanism. Moreover, a custom-designed experimental system was applied to clarify the magnitude of inherent deformation. With the mechanism investigation of welding-induced buckling by FE analysis using inherent deformation, an application for predicting and mitigating the welding buckling in fabrication of complex aluminum alloy structure was developed.

Keywords: welding distortion; residual stress; inherent strain method

1. Introduction

Residual stresses and distortions are two of the major concerns in welded structures, especially for aluminum alloy thin-walled structures [1,2]. Welding stresses and distortion cause dimensional deviation due to the highly localized, non-uniform, transient heating and subsequent cooling of the welded material, and the non-linearity of aluminum material properties [3]. These stresses lead to the crucial cracking after welding. Particularly tensile residual stresses near the weld area cause stress rising, fatigue failure and brittle fracture [4,5].

Validating methods for predicting welding stresses and distortion are desirable because of the complexity of the welding process. Accordingly, finite element (FE) simulation has become a popular tool for the prediction of welding residual stresses and distortion [6–8]. Many investigators have developed the analytical and experimental methods to predict the welding residual stresses. Da Nóbrega et al. evaluated the temperature field and residual stresses in a multi-pass weld of API 5L X80 steel using the finite element method [9]. Zeng et al. predicted the thermal elasto-plastic analysis using finite element techniques to analyze the thermo-mechanical behavior and evaluate the residual stresses and distortion of 5A06 aluminum alloy structure in discontinuous welding [10]. Syahroni and Hidayat focused on numerical simulation of welding sequence effect on temperature distribution, residual stresses and distortions of T-joint fillet welds [11]. Normally, the commercial welding software SYSWELD Weld Planner provides access to welding-induced distortion simulation even for people unfamiliar with finite element simulation at the early stage of preliminary design and planning, and ESI Distortion Engineering can comprise the former and offer services to solve welding problems [12]. However, new heat source models or material properties for novel materials need to be built up based on basic interdisciplinary studies rather than direct application of commercial software. For large and complex welded structures applied in the marine, ship and aerospace fields, the models developed by the thermal elastic-plastic finite element method need to be divided into a large number of grids and time steps, which are not applicable in the actual work. Therefore, the inherent strain method is an alternative for estimating the overall welding distortions [13–15]. It is noted that the inherent strain method induces the transient effect of the welding process to inherent strain key parameters, avoiding the transient analysis of numerical simulation and the computational difficulty in high temperature, reducing the calculation time and obtaining the residual stress and deformation value with a certain degree of accuracy [16]. Few studies have explored the welding multi-physics mechanism in detail despite its high efficiency in calculating the size and distribution of inherent strain with a certain accuracy in the numerical simulation of complex welded structures, especially for fillet joint angular distortion. The prerequisite of this method is that the inherent deformations (i.e., longitudinal shrinkage, transverse shrinkage, angular distortion and longitudinal bending) in each joint should be known beforehand [17].

In this paper, the purpose is to investigate the quantitative relationship between the inherent strain and the structural factor during the welding process and analyze the welding residual stresses and distortion of the 5A06 aluminum alloy structure. The method combined initial thermo-elastic-plastic analysis with the inherent strain method developed in this paper, illustrating much more feasibility and accuracy for a complex welding structure in terms of the effects of welding sequence and weldment size, compared to the empirical formula or data of the commercial welding simulation software. Two models, a multi-pass V-type butt weld and T-type fillet-welded joint, were used to analyze the size and distribution of the inherent strain. The inherent strain mechanism was discussed by analyzing the angular distortion of the T-type joint. Moreover, the effect of buckling deformation was also investigated.

2. Model Analysis

A 5A06 aluminum alloy structure with an outer diameter of 538 mm, thickness of 9 mm, and length of 250 mm was examined in this paper. There are five identical cylinder substructures combined to form the structure. In order to control the distortion, stiffeners were welded onto the structure. Welding condition is shown in Table 1, and the composition of the 5A06 aluminum alloy used in this investigation is shown in Table 2. Figure 1a,b showed the welding structure. The central axis of the welding fixture was parallel to the workplace radial direction. Meanwhile, there were several claws supporting rigidly on the workplace wall in the circumferential direction during the welding process. At the same time, the force on the claw acted on the inclined plane below through a lower supporting structure and a rolling wheel. The inclined plane was connected to the central axis. The welding stress could transfer to the claws, then to the tooling, which could prevent large deformation during the welding process.

Table 1. Welding condition parameters.

Welding Parameters	U (V)	I (A)	Welding Speed (cm·min^{-1})	Wire Feed Rate (cm·min^{-1})
Value	26.2	286	50–60	15.7–20

Table 2. Chemical composition of 5A06 aluminum alloy.

Composition	ω (Si)	ω (Cu)	ω (Mg)	ω (Zn)	ω (Mn)	ω (Ti)	ω (Fe)	ω (Al)
Mass fraction	0.004	0.001	0.058–0.068	0.002	0.005–0.008	0.0002–0.001	0.004	balance

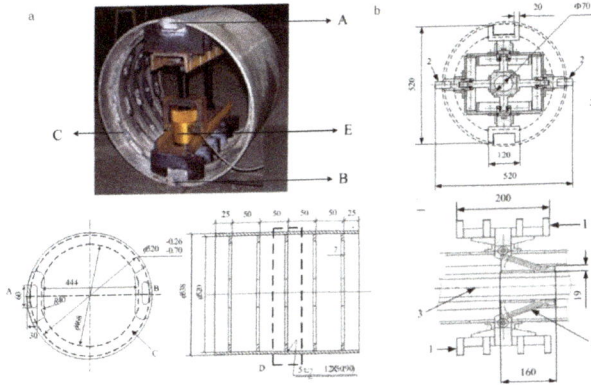

Figure 1. The welding structure and finite element (FE) model. (**a**) Welding structure: first, the stiffeners (C) were welded onto the cylinder substructures (D) by 24 separate fillet welds; the welding fixtures (E) were installed on the whole structure to make sure it would not collapse; then, parts A and B were removed by machining; finally, the five cylinder substructures were welded together by four butt welds. (**b**) Welding fixture, 1-Vertical supporting claws; 2-Horizontal supporting claws; 3-Central spindle; 4-Linkage and rolling wheel.

As no metallurgical phase transformation occurs in the aluminum alloy used in this work, the volumetric effects due to the phase transformation on residual stress evolution are not considered here; the procedure seems acceptable to achieve accurate stress and distortion distribution in references [18]. Since welding processes undergo a high temperature cycle and exhibit material properties that are temperature dependent, the thermal and mechanical properties of the 5A06 aluminum alloy in Table 3 were determined by the Probability Design System (PDS) in the finite element software ANSYS [19]. The FE model and the butt welds for inherent strain estimation are shown in Figure 2. In the mechanical analysis, the fillet welds' thermal stresses and distortion were calculated from the temperature distribution determined by thermal elasto-plastic FE model in references [10,20]. The material was assumed to follow the Von Mises yield criterion and flow rule here. The 20-node hexahedral element SOLID185 was applied in this investigation. These thermal strains of SOLID185 element change linearly for stress analysis. Moreover, the element thermal strain can be confirmed if each node's mutative temperature and anisotropic thermal expansion coefficient matrix are defined, i.e., inherent strain components could be mapped to the elasto-plastic model in forms of t equivalent thermal strain.

Table 3. Material properties of 5A06 aluminum alloy.

Materials Properties	Temperature, °C					
Name	20	100	200	500	587	630
Young's modulus (GPa)	70	70	61	41.	10	1
Linear expansion coefficient ($10^{-6} \cdot K^{-1}$)	0.93×10^{-4}	1.91	4.50	13.3	15.9	17.6
Poisson's ratio	0.35	0.35	0.35	0.35	0.35	0.35
Density (kg·m^{-3})	2750	2730	2710	2640	2630	2450
Specific heat (J·kg^{-1}·K^{-1})	898	951	1003	1150	1195	1165
Yield stress (MPa)	130	100	54	10	5	5

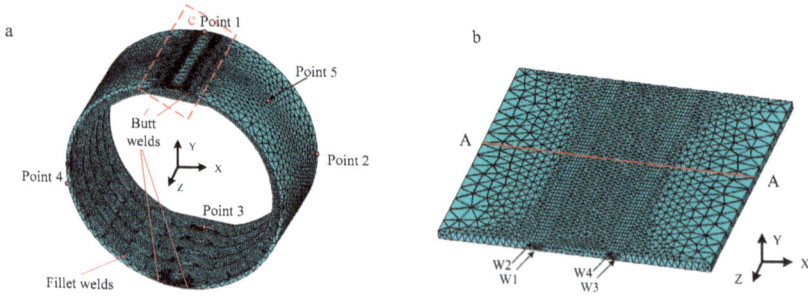

Figure 2. Finite model of 5A06 aluminum alloy structure (a) FE model. (b) Butt weld FE model, two layers of welding were applied to each butt and fillet weld.

3. Welding Distortion Prediction in Thin Plate Fabrication by Means of Inherent Strain FE Method

In the FE elements, the Mindlin plate theory was employed and the geometrical nonlinear effect was also considered. Considering transverse shear strain components, the total strains could be expressed as follows [16].

$$\varepsilon_x = \varepsilon_x^i + \varepsilon_x^b = \left[\frac{\partial u}{\partial x} + \frac{1}{2}\left(\frac{\partial w}{\partial x}\right)^2\right] + \left[-z\frac{\partial^2 w}{\partial x^2} + z\frac{\partial \theta_y}{\partial x}\right] \tag{1}$$

$$\varepsilon_y = \varepsilon_y^i + \varepsilon_y^b = \left[\frac{\partial v}{\partial y} + \frac{1}{2}\left(\frac{\partial w}{\partial y}\right)^2\right] + \left[-z\frac{\partial^2 w}{\partial y^2} - z\frac{\partial \theta_x}{\partial y}\right] \tag{2}$$

$$\gamma_{xy} = \gamma_{xy}^i + \gamma_{xy}^b = \left[\frac{\partial u}{\partial y} + \frac{\partial v}{\partial x} + \left(\frac{\partial w}{\partial x}\right)\left(\frac{\partial w}{\partial y}\right)\right] + \left[-2z\frac{\partial^2 w}{\partial x \partial y} + z\frac{\partial \theta_y}{\partial y} - z\frac{\partial \theta_x}{\partial x}\right] \tag{3}$$

$$\gamma_{xz} = \theta_y + \frac{\partial w}{\partial x} \tag{4}$$

$$\gamma_{yz} = -\theta_x + \frac{\partial w}{\partial y} \tag{5}$$

where u and v are in-plane displacements at mid-plane; w is out-of-plane displacements; θ_x and θ_y are rotations; ε_x, ε_y, γ_{xy} are total strains, and ε_x^i, ε_y^i and γ_{xy}^i are in-plane strains; ε_x^b and ε_y^b and γ_{xy}^b are bending strains, γ_{xz} and γ_{yz}, are transverse shear strains.

The curvature K_x in a plane parallel to the x–z plane and the curvature K_y in a plane parallel to the y–z plane and the twisting curvature K_{xy}, which represents the warping of the x–y plane, can be defined as follows.

$$K_x = -\frac{\partial^2 w}{\partial x^2} \tag{6}$$

$$K_y = -\frac{\partial^2 w}{\partial y^2} \tag{7}$$

$$K_{xy} = -\frac{\partial^2 w}{\partial x \partial y} \tag{8}$$

In the present FEM, three types of inherent deformations, namely longitudinal shrinkage, transverse shrinkage and angular distortion are introduced into the elastic FEM. When a welding line is arranged parallel to the x-axis, longitudinal shrinkage can be transformed into in-plane strain component ε_x in longitudinal direction. As mentioned above, another equivalent method is to use tendon force to represent longitudinal shrinkage. Transverse shrinkage can be changed into in-plane strain component ε_y in transverse direction. In a similar way, angular distortion can be converted into curvature K_x along the x-axis. These inherent strain components are introduced into the elastic FEM as initial strains.

4. Results and Discussion

In this paper, the thermo-elastic-plastic FE method was used to estimate the welding stress and distortion in the welds including the butt welds and fillet welds with inherent stain.

4.1. Inherent Strain Analysis of Butt Weld and Fillet Welds

As mentioned above, the welding stress and distortion in complex welding structures were mostly caused by longitudinal and transverse inherent strains. In order to obtain the size and distribution of inherent strain after welding and determine the quantitative relationship between the inherent strain and the primary influence factors, a sequential coupling of thermal-elasto-plastic analytical method was employed to acquire the value of inherent strain of two typical welded joints in Figure 2b. Figure 3 showed the distribution of longitudinal and transverse residual stresses of butt-joint with multi-passes. It was found that the tensile stress was widely distributed in the center of welded seams and localized in the edge region. Most of the welding area was under in the tensile stress. With the increase of heat input, the longitudinal and transverse tensile stresses of the second layer's welded seam rose up gradually. When reaching the yield limit, the value of residual tensile stresses was stable, but their distribution area increased. This was due to residual thermal contraction deformation caused by the cooling down of filler metal [21]. Herein, residual compressive plastic deformation and residual thermal contraction deformation are both attributed to inherent strain.

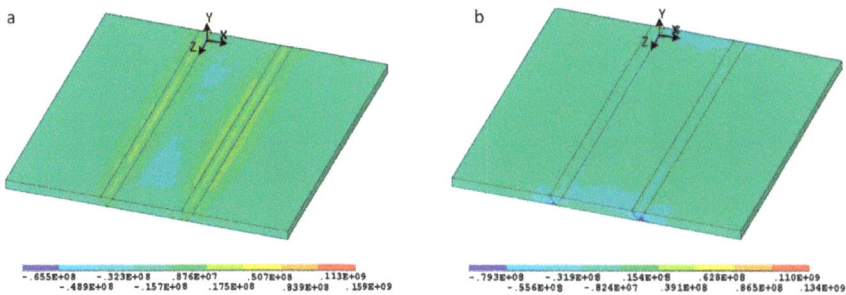

Figure 3. Residual stress distribution of 5A06 aluminum alloy butt welds (**a**) Longitudinal residual stress (**b**) Transverse residual stress.

Figure 4 explains the distribution of longitudinal and transverse residual plastic strains of the butt joint. Obviously, most of the weld zone was plastically deformed in a compressive state except for both ends. The residual strain in the center of the welded seams was stable but their distribution area increased similar to the stress distributions. Figure 5 shows the distribution of the welding residual stress and plastic strain of path A-A (as shown in Figure 2b). The residual strain was concentrated in the vicinity of the welded seams. Moreover, both longitudinal and transverse directions were under compressive deformation. Meanwhile, the residual plastic strain in the transverse direction was larger than that in the longitudinal direction, whereas the strain away from the seam was almost zero. In the welding zone, the welding residual stress was tensile stress, and the longitudinal residual stress was much greater than the transverse direction. Correspondingly, the adjacent area of welded seams displayed compressive stress. When moving backwards to the welded seams, the stress decayed. When the weldment exceeded a certain length, the longitudinal and transverse inherent strain coefficients were insignificantly changed except for the arc initiation and closing positions [22].

The influence of the weld length on the W3 and W4-welded seams was similar to that of the W1 and W2. In addition, the inherent strain coefficients of the W3 and W4-welded seams were slightly decreased because of the reduced thermal gradients.

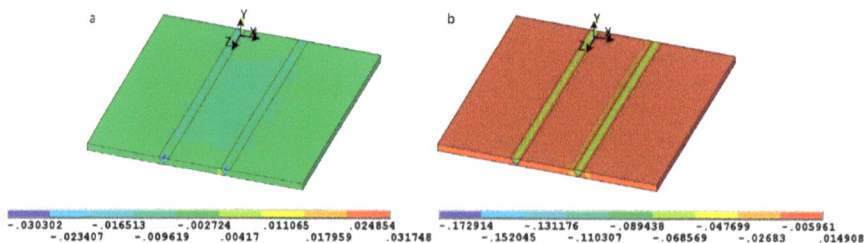

Figure 4. Residual strain distribution of 5A06 aluminum alloy butt welds. (a) The longitudinal residual strain. (b) The transverse residual strain.

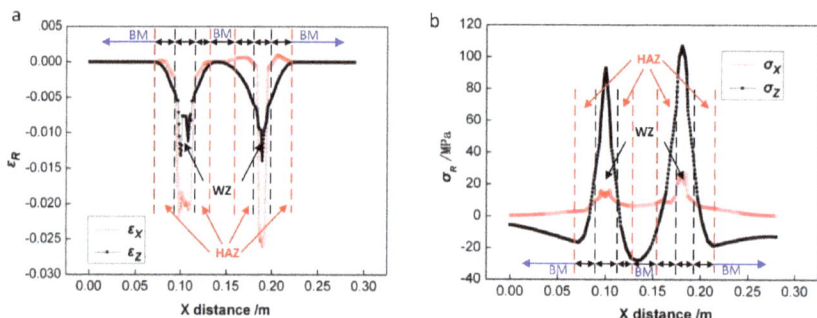

Figure 5. Residual stress and strain distribution of A-A path on butt welds. (a) Residual strain ε_R. (b) Residual stress σ_R. (BM-base material, HAZ-Heat Affected Zone, WZ-welded zone).

The three-dimensional uncoupled thermo-mechanical FE analysis was carried out to produce the temperature distribution, residual stresses and distortion on fillet welds of the aluminum alloy structure in the discontinuous welding, illustrating factors of angular distortion and the effects of weld sequence on residual stress distributions. The simulated results demonstrated that the temperature gradient through thickness was a main factor that strongly governs the generation of angular distortion in a fillet-welded joint. The angular distortion decreased because of the preheating and reheating during the discontinuous welding. However, the temperature lagging between the external surface and internal surface aggravated the angular distortion. Welding sequences and weld length are greatly important to residual stresses and distortion [10,20].

In the prediction methods of welding residual stress and distortion, such as the inherent strain method and the volumetric shrinkage method, the boundary condition imposed by the welds and the effect of actual inherent strain distribution on the structure directly determine the prediction accuracy. Like all strain tensors, the inherent strain had six strain components in different directions. The residual stress and distortion were from the comprehensive interaction of all these six components. The simplification in current prediction methods was based on the correspondence of residual stresses and distortion with the chosen components of inherent strain. For example, we used the inherent strain method to predict longitudinal residual stress, longitudinal shrinkage deformation and bending deformation by exerting boundary conditions corresponding to longitudinal inherent strain in the welds through ignoring inherent strains in other directions. Similarly, the boundary values related to transversing the inherent strain on welded seams could be set to predict the transverse residual stress.

Through this, the prediction of transverse residual stress, the transverse shrinkage deformation and the angular distortion were possible and could be used to predict welding shrinkage distortion and butt-welding angular distortion of the plate. However, this was invalid for angular distortion of the fillet weld [23].

The deformation deviation caused by insufficient data of inherent strain and uncertain volumetric shrinkage in the welded seam could be reduced by improving the database or empirical formula [24]. However, the simplification method is attributed to the determination of the relationship between the inherent strain component and welding deformation. The mechanism of welding deformation is one of the key aspects of this simplification method. In this work, the mapping of the inherent strain component was used to analyze the relationship between angular distortion and the inherent strain component of the fillet weld.

Here, the data of the inherent strain component existing in the welded structure was analyzed through the thermal elasto-plastic analysis using finite element techniques. The deformation of different inherent strain components acting alone could then be obtained by linear elasticity calculation. Compared to such angular distortion with standard welding angular distortion (angular distortion from thermal elasto-plastic analysis), the effect of inherent strain on welding angular distortion could be analyzed. For example, the conversion formula corresponding to the temperature change of inherent strain components $\Delta T_{xx} = (x, y, z)$ can be written as,

$$\Delta T_{xx}(x, y, z) = \frac{\varepsilon_{xx}(x, y, z)}{\alpha_{xx}} \tag{9}$$

where $\alpha_{xx} = C$, C is a constant. The corresponding material with anisotropic thermal expansion coefficient is defined as: $\alpha_{xx} = C$, $\alpha_{yy} = \alpha_{zz} = \alpha_{xy} = \alpha_{xz} = \alpha_{yz} = 0$. Other mechanical parameters and physical parameters are assumed as constant values at room temperature. The same method works on other five inherent strain components.

The standard deformation measured using elasto-plastic analysis is $\delta^{EPA}(x, y, z)$. The inherent strain component is expressed as $\varepsilon_{ij}(x, y, z)$, and then the deformation, taking welding structure into account, only is $f(\varepsilon_{ij}(x, y, z))$. Theoretically, $\delta^{EPA}(x, y, z) = \delta^{sum}(x, y, z)$. The relative error of numerical results is as follows.

$$\text{error} = \frac{\delta^{sum} - \delta^{EPA}}{\delta^{EPA}} \times 100\% \tag{10}$$

As shown in Figure 4, the end of the joint was with an upward displacement because of a positive component of transverse inherent strain ε_{xx}. However, the actual displacement was smaller than the standard value. From the angular distortion results, including the mapping of the inherent positive strain component ε_{yy}, ε_{zz}, T-joint panels produced a very small amount of negative angular distortion under the influence of positive inherent strain in two directions, i.e., the larger angle of fillet weld after deformation. The calculation results of angular distortion using the mapping of inherent shear strain component ε_{xy} indicated that the end of the T-joint panel warped upwards and the magnitude of displacement was close to the standard value. In addition, other strains ε_{yz} ε_{xz} were very small suggesting that the inherent shear strain components from these two directions almost had no effect on the angular distortion of the fillet weld.

In the study of the mapping of strain components in Figure 6, the sum of average value of angular distortion Φ_0 of the panel's free end caused by six separate inherent strain components was 0.448 mm. The standard value, however, was 0.496 mm, meaning the error was only 1.6% according to the simulation.

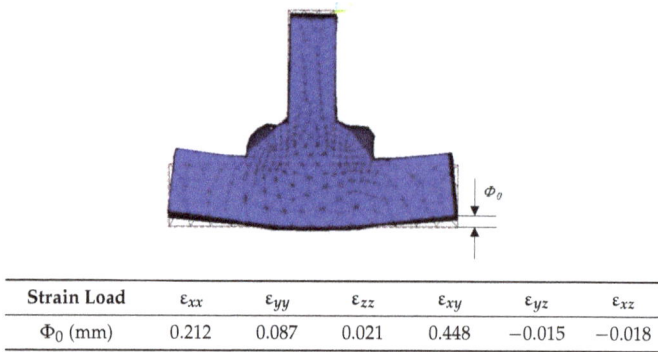

Strain Load	ε_{xx}	ε_{yy}	ε_{zz}	ε_{xy}	ε_{yz}	ε_{xz}
Φ_0 (mm)	0.212	0.087	0.021	0.448	−0.015	−0.018

Figure 6. Angular distortion under different stain loads.

The relationship between the angular distortion and the inherent strain component was inconsistent. It depended on the welding structure and joint configuration including workplace thickness, height of weld leg and welding conditions, etc. From the thermo-elastic-plastic analysis, the component of inherent shear strain ε_{xy} had an obvious effect on the angular distortion of fillet weld close to standard value. The effect of a positive component of transverse inherent strain on angular distortion ε_{xx} was smaller than ε_{xy}. The other two components ε_{yy}, ε_{zz} would cause a small negative angular distortion. Also, the influence of the components of inherent shear strain ε_{yz} and ε_{xz} on angular distortion could be ignored. This was also in agreement with the report in references [25]. Therefore, ε_{xy} was the greatest influence factor on angular distortion of fillet weld among the six components of inherent strain, which determined the final angular distortion.

4.2. Analysis of the Aluminum Alloy Structure

The SOLID 185 was used to mesh the model into 172,680 elements and 66,513 nodes. As shown in Figure 1, aluminum ring 1 to 5 were specified from the z-axis-positive direction to negative direction and calculated through the inherent strain method according to analysis of the multi-pass longitudinal welded seam and discontinuous fillet weld.

In ANSYS, the values of inherent strain cannot be directly loaded. However, the anisotropic thermal expansion coefficient can reflect different contractions in both longitudinal and transverse directions. Therefore, it is associated with the temperature load of units that can be used to apply loads. The thermal expansion coefficient is the value of strain.

$$\varepsilon = W/F = \alpha \cdot \Delta T \tag{11}$$

where, W stands for the total amount of welding shrinkage in per unit length. F is the cross-sectional area of unit locating in inherent strain. α represents the thermal expansion coefficient equal to the numerical value of inherent strain. T is the temperature load of each unit. The strain load was applied to welded seam and adjacent units. Before computing, a few aspects should be addressed: Thermal expansion coefficients of longitudinal s and transverse strain corresponded with the direction of welded seam; longitudinal and transverse strains were negative with an opposing symbol of temperature load; the directions without inherent strains set the thermal expansion coefficient as zero while the other elements were the same.

Since the fillet weld was not linear, converting the data according to the spatial direction of the weld and coordinate system of angular orientation was needed. In Figure 7, we assumed that welded seams extended along the X_0 direction with an angle of θ in the local coordinate system XOY. The formulas to calculate the welded seams were,

$$\varepsilon_x = \varepsilon_{x0} \cos\theta - \varepsilon_{y0} \sin\theta \tag{12}$$

$$\varepsilon_y = \varepsilon_{y0} \cos\theta - \varepsilon_{x0} \sin\theta \tag{13}$$

The calculated inherent strains of different welding joints were thus obtained as shown in Table 4.

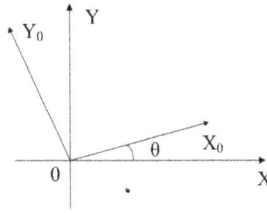

Figure 7. Heat expansion coefficient transfer of the fillet welding.

Table 4. Inherent strain on each weld joints.

Weld No.	Location	Joint Type	Cross-Section (mm^2)	W_X (mm^2)	W_Y (mm^2)	ε_x	ε_y	ε_{xy}
1	Stiffer front	T-joint	32	0.651	2.16	0.0204	0.067	0.0319
2	Stiffer front	T-joint	32	0.668	2.26	0.0208	0.071	0.0361
3	Stiffer front	T-joint	32	0.673	2.27	0.0210	0.072	0.0475
4	Stiffer front	T-joint	32	0.680	2.33	0.0212	0.073	0.0526
5	Stiffer front	T-joint	32	0.688	2.38	0.0215	0.074	0.0655
6	Stiffer front	T-joint	32	0.695	2.41	0.0217	0.076	0.0734
7	Stiffer front	T-joint	32	0.702	2.42	0.0219	0.076	0.0752
8	Stiffer front	T-joint	32	0.710	2.43	0.0222	0.076	0.0808
9	Stiffer front	T-joint	32	0.719	2.44	0.0225	0.077	0.0851
10	Stiffer front	T-joint	32	0.733	2.45	0.0229	0.077	0.0862
11	Stiffer front	T-joint	32	0.746	2.48	0.0233	0.078	0.0877
12	Stiffer front	T-joint	32	0.751	2.50	0.0235	0.079	0.0890
13	Stiffer back	T-joint	32	0.290	0.96	0.0090	0.030	0.0165
14	Stiffer back	T-joint	32	0.297	1.01	0.0092	0.031	0.0187
15	Stiffer back	T-joint	32	0.299	1.01	0.0093	0.031	0.0247
16	Stiffer back	T-joint	32	0.303	1.03	0.0094	0.032	0.0273
17	Stiffer back	T-joint	32	0.306	1.06	0.0095	0.033	0.0340
18	Stiffer back	T-joint	32	0.309	1.07	0.0096	0.034	0.0388
19	Stiffer back	T-joint	32	0.312	1.08	0.0097	0.034	0.0391
20	Stiffer back	T-joint	32	0.316	1.08	0.0099	0.034	0.0420
21	Stiffer back	T-joint	32	0.320	1.09	0.0100	0.034	0.0442
22	Stiffer back	T-joint	32	0.326	1.09	0.0102	0.034	0.0448
23	Stiffer back	T-joint	32	0.332	1.10	0.0104	0.035	0.0456
24	Stiffer back	T-joint	32	0.334	1.10	0.0105	0.035	0.0463
25	W1 (in Figure 1)	butt weld	18	0.729	2.67	0.0405	0.148	-
26	W2 (in Figure 1)	butt weld	40	0.756	2.75	0.0419	0.153	-
27	W3 (in Figure 1)	butt weld	18	0.960	3.11	0.0240	0.078	-
28	W4 (in Figure 1)	butt weld	40	0.934	3.01	0.0233	0.075	-

To verify the reliability of the simulation results, it is necessary to compare the results with experimental ones. A customized experimental facility (in Figure 8) was used to verify the FEM results in this paper. The residual stress measurement was mainly carried out by the hole-drilling method, cutting method, and X-ray diffraction method [26–28]. However, nearly all the stress experiments are not economically or conveniently suitable for such large-scale, complex structural components in this investigation. Thus, we developed a new method to verify the numerical simulation results with the welding compression load and strain tests. Normally, for thin-wall cylinder structure, significant contraction and wave distortion occur after welding. The stiffeners were welded separately for minimizing the contraction distortion. It is noted that the welding contraction load corresponds to the strain, which could be measured by the static resistance tester installed on the welding fixture. It means that the welding contraction load could be attained by measuring the strain on the welding fixture, which could also be predicted as residual stress by FEM using the inherent strain method. The relationship between strain and welding shrinkage strain is calibrated first which follows as Equation (14):

$$F = 0.427 + 0.10169\mu\varepsilon \tag{14}$$

In the equation, F stands for depressive load corresponding to the related strain. Thus, the welding shrinkage load or residual stress could be confirmed if the stains are measured before and after welding.

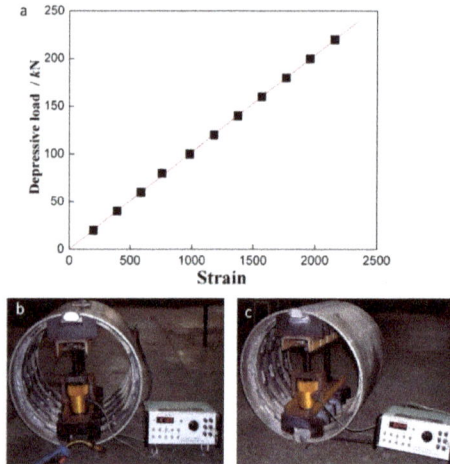

Figure 8. Depression load and strain measurement of strain before and after welding. (**a**) Relation of welding depression load and strain (**b**) Depression strain before reinforcing plate welding (**c**) Depression strain after reinforcing plate welding.

Before welding the stiffener, the strain was 105 µε. According to the calibrated curve in Figure 8, the corresponding contraction load was acquired as 11 kN/240 mm, i.e., 3.833 MPa. Similarly, the strain value changed to 249 µε after welding and the contraction load was 26 kN/240 mm, i.e., 11.479 MPa. Hereby, the welding strain and contraction load could be measured under different conditions.

According to the actual welding process, the inherent strains in Table 4 were imposed on each weld joints sequentially, and the depression load and welding stresses of the workplace were estimated through the thermal elasto-plastic finite element analysis. In Figure 9, After Fillet welds welding on the stiffer, the simulated residual stress in the Y direction (radial direction) of the workplace corresponding to the depression load on vertical supporting claws was calculated before and after stiffened plate welding. Obviously, the results of numerical simulation using the inherent strain method coincided with the experiment study.

Figure 10a,b shows the welding deformation of the structure at room temperature. With the welding process, the radial shrinkage of the structure decreased. The radial shrinkage at the lower edge of the last aluminum ring decreased about 30% more than the upper edge of aluminum ring. As above, the entire structure deformed within 2 mm in the condition of appropriate welding process and welding sequence. However, the position of stiffened plate was not web-plated; radial deformation at the position of the stiffened plate was significant, reaching up to 8 mm.

High residual stresses of the welding structure were shown in Figure 10c,d. The longitudinal residual stress of the welded seam reached about 120 MPa, while the residual stress of the 5A06 aluminum alloy cylinder reached about −40 MPa. The other positions such as the stiffened plate approached the yield strength of the 5A06 aluminum alloy.

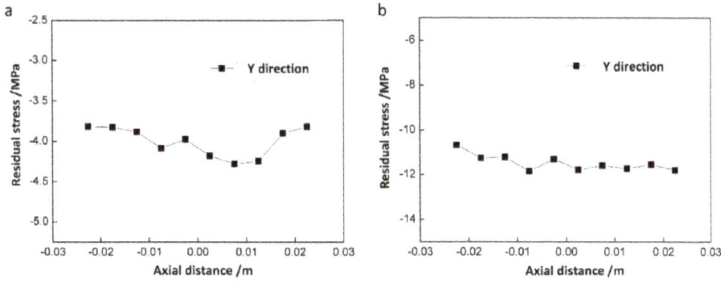

Figure 9. Radial residual stress before and after the reinforcing plate. Machining part A in Figure 1a welding. (**a**) Residual stress before butt welding. (**b**) Residual stress after butt welding.

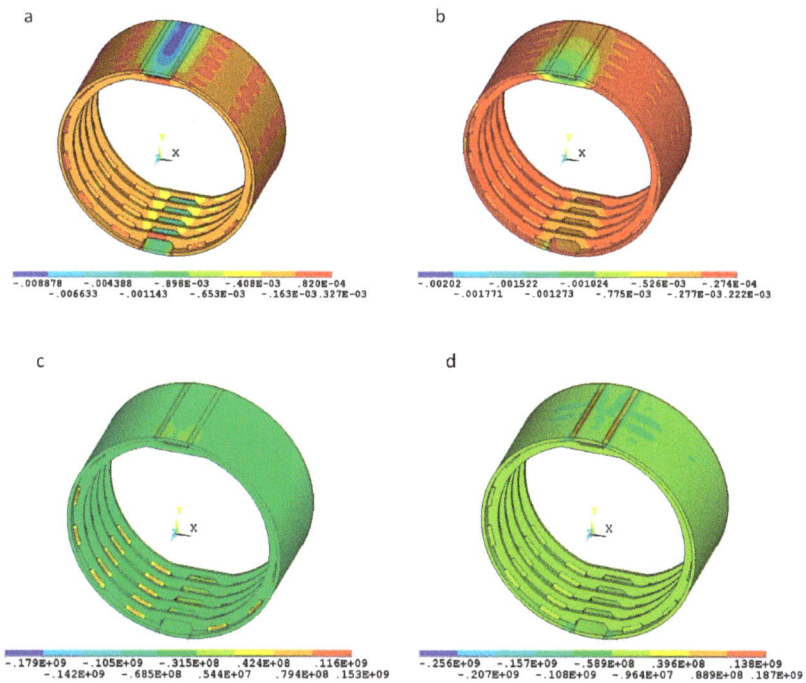

Figure 10. Residual distortion and stress of the 5A06 aluminum alloy cylinder. (**a**) Radial distortion (m). (**b**) Axial distortion (m). (**c**) Radial residual stress (MPa). (**d**) Axial residual stress (MPa).

4.3. Buckling Deformation Analysis

From the above analysis, the residual stress field in the thin plate developed gradually and formed tensile stress approaching the welded seams and compressive stress away from the welded seams. It was caused by the effect of the plastic deformation. This residual stress field was due to the uncoordinated weld zone (unevenness) instead of the external load, and it was a balanced stress field. If the value of residual compressive stress reached a critical load of the structure, the plate had a greater warpage out of the plane, namely buckling deformation.

In the foregoing analysis, the maximum deformation of the workplace was 8.8 mm (in Figure 10a), attributed to the category of small deflection theory analysis. Therefore, the balance method was adopted to analyze the critical yield load.

During the numerical simulation process, five testing points were selected to analyze whether the buckling deformation generated. These chosen points away from the welded seam were to avoid the result being influenced by the welding residual stress. The stress and strain were measured as shown in Figure 11. From points 2, 4, 5 of the cylinder, a conclusion can be drawn that the stress and strain on the cylinder were not uniform during the welding process including heating and cooling stages. Compared to the stress-strain curves of measured points 1 and 3, the mutation was produced by the deformation in which the stress changed smoothly. That is, the buckling deformation generated at the cooling time of 100 s after the stiffened plate was welded. Using the eigenvalue buckling analysis, the minimum eigenvalues of the structure could be calculated and its value was 0.676 (i.e., the critical loading of instability was equal to 1739 N).

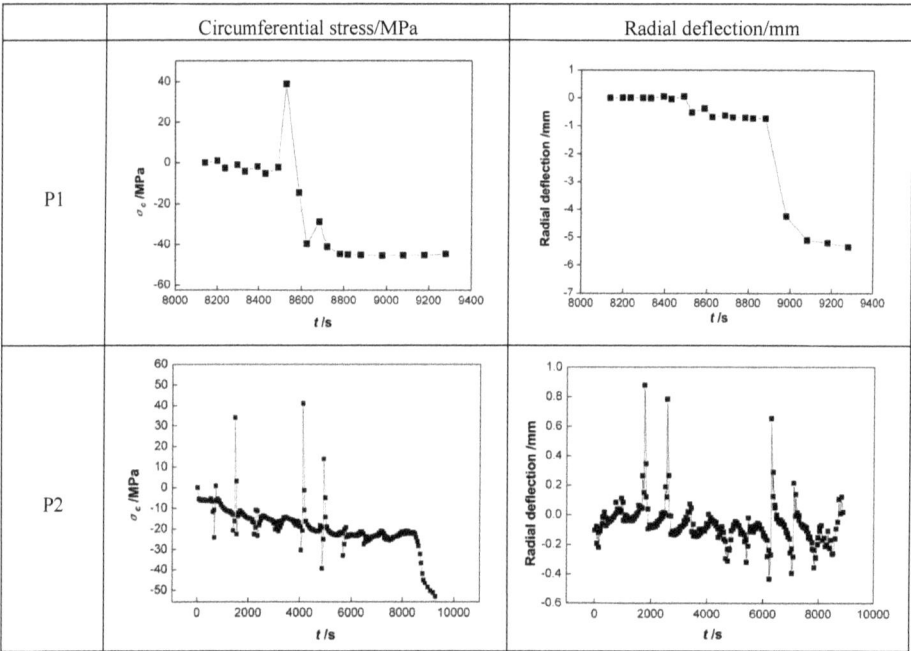

Figure 11. *Cont.*

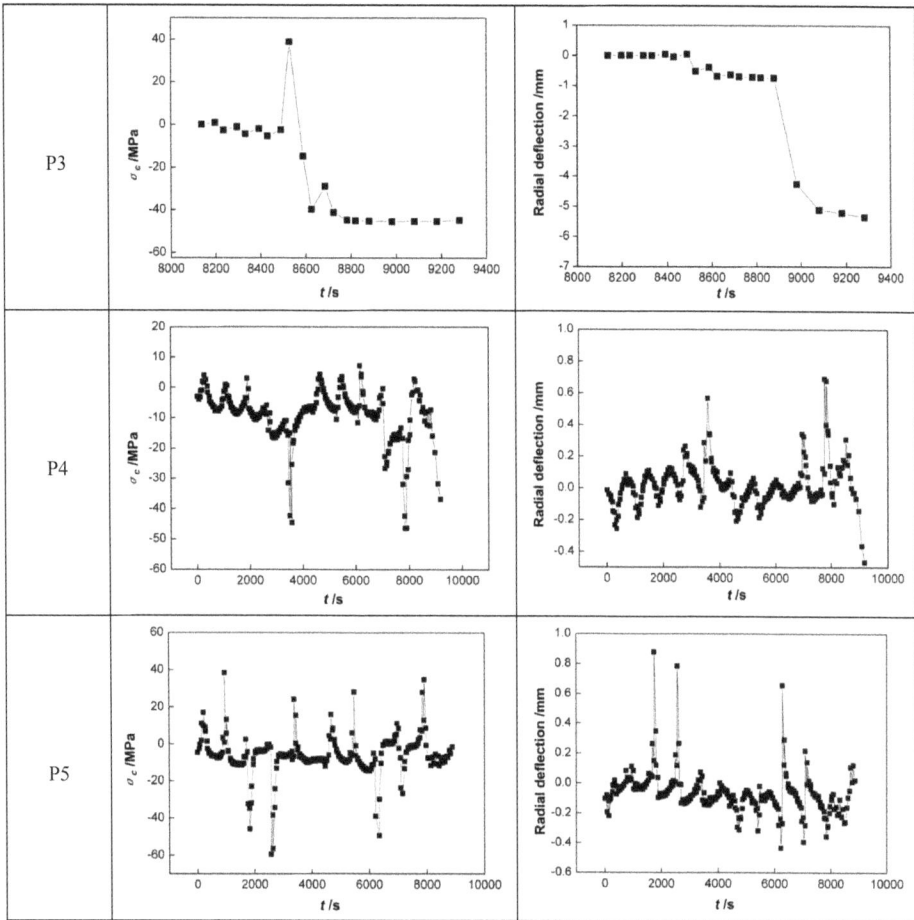

Figure 11. Buckling deformation of different points (in Figure 2a) on the workplace.

Buckling deformation of a thin-walled structure was caused by two reasons. The first was the free deformation of aluminum metal located at the welded seams, whereby the near area was blocked at a high temperature, resulting in a compressive plastic deformation. The second was the angular distortion caused by the angle deformation of the fillet weld because of the uneven transverse shrinkage deformation in the thickness direction. Moreover, the deformation of the front side of the welded seam was large while its backside was small. It caused the plane component's warpage. For the thin-plate welding with reinforced rib, the weld metal area was affected by compressive stress. During the cooling stage, the metals on both sides of the welded seam were influenced by compressive stress generating instability deformation. The final residual deformation depended on the final residual stress of the thin plate. Then the temperature during the cooling stage in the workplace tended to be uniform with balanced internal stresses. Therefore, the residual deformation also tended to be stable [29].

5. Conclusions

Based upon the thermal-elastic-plastic analysis of the discontinuous fillet welding, the inherent strain method was applied to the welding residual stress and deformation of a 5A06 aluminum alloy-reinforced cylinder structure, also taking into consideration the effects of welding technology and weldment size on the multi-pass V-type butt weld. It put great importance on the inherent strain distribution of joint type. For fillet welding, the relationship between the angular distortion and inherent strain component was inconsistent, depending on the welding structure and joint configuration. The shear inherent strain was the greatest influence on angular distortion of fillet weld among the six components of inherent strain, which determined the final angular distortion. It is noted that the wave propagation appeared in the stiffened plate during the welding, which caused the buckling deformation. The buckling deformation of this thin-walled structure was 8.8 mm, which mainly results from the non-uniform welding temperature distribution and angle deformation of the fillet weld. In order to ensure the dimensional accuracy, a post heat treatment might be needed to relieve stress of the thin-walled aluminum alloy structure. Compared to the experimental results of the welding depression load and strain, the inherent strain method has highly efficient and can ensure definite precision in the numerical simulation of the complex welding structure.

Acknowledgments: Financial support by the National Natural Science Foundation of China (51205047) and Science and Technology Planning Project of Guangdong Province (2016A010102002).

Author Contributions: Zeng Z. and Peng B. conceived and designed the simulation and experiments; Wu X. performed the numerical simulation; Yang M. analyzed the data; Zeng Z. wrote the paper.

Conflicts of Interest: The authors declare no conflict of interest.

References

1. Masubuchi, K.; Bryan, J.J.; Muraki, T. Analysis of thermal stresses and metal movement during welding. *ASME J. Eng. Mater. Technol.* **1975**, *97*, 81–91.
2. Zhang, H.; Wang, M.; Zhang, X.; Zhu, Z.; Yu, T.; Yang, G. Effect of welding speed on defect features and mechanical performance of friction stir lap welded 7B04 aluminum alloy. *Metals* **2016**, *6*, 87. [CrossRef]
3. Subramanian, J.; Seetharaman, S.; Gupta, M. Processing and properties of aluminum and magnesium based composites containing amorphous reinforcement: A review. *Metals* **2015**, *5*, 743–762. [CrossRef]
4. Zeng, Z.; Li, X.B.; Miao, Y.G.; Wu, G.; Zhao, Z.J. Numerical and experiment analysis of residual stress on magnesium alloy and steel butt joint by hybrid laser-TIG welding. *Comput. Mater. Sci.* **2011**, *50*, 1763–1769. [CrossRef]
5. Zhao, Y.B.; Lei, Z.L.; Chen, Y.B.; Tao, W. A comparative study of laser-arc double-sided welding and double-sided arc welding of 6 mm 5A06 aluminium alloy. *Mater. Des.* **2011**, *32*, 2165–2171. [CrossRef]
6. Carlone, P.; Citarella, R.; Lepore, M.; Palazzo, G.S. A FEM-DBEM investigation of the influence of process parameters on crack growth in aluminum friction stir welded butt joints. *Int. J. Mater. Form.* **2015**, *8*, 591–599. [CrossRef]
7. Jiang, W.; Fan, Q.; Gong, J. Optimization of welding joint between tower and bottom flange based on residual stress considerations in a wind turbine. *Energy* **2010**, *35*, 461–467. [CrossRef]
8. Kuo, H.C.; Wu, L.J. Prediction of deformation to thin ship panels for different heat sources. *J. Ship Prod.* **2001**, *17*, 52–61.
9. Da Nóbrega, J.A.; Diniz, D.S.; Silva, A.A.; Maciel, T.M.; Albuquerque, V.H.C.; Tavares, J.M.R.S. Numerical evaluation of temperature field and residual stresses in an API 5L X80 steel welded joint using the finite element method. *Metals* **2016**, *6*, 28. [CrossRef]
10. Zeng, Z.; Wang, L.J.; Du, P.A.; Li, X.B. Determination of welding stress and distortion in discontinuous welding by means of numerical simulation and comparison with experimental measurements. *Comput. Mater. Sci.* **2010**, *49*, 535–543. [CrossRef]
11. Syahroni, N.; Hidayat, M.I.P. 3D Finite Element Simulation of T-Joint Fillet Weld: Effect of Various Welding Sequences on the Residual Stresses and Distortions. In *Numerical Simulation—From Theory to Industry*; Andriychuk, M., Ed.; InTech: Rijeka, Croatia, 2012.
12. ESI Group. Available online: http://www.esi-group.com (accessed on 11 August 2016).
13. Park, J.U.; An, G.B.; Woo, W.C.; Choi, J.; Ma, N. Residual stress measurement in an extra thick multi-pass weld using initial stress integrated inherent strain method. *Mar. Struct.* **2014**, *39*, 424–437. [CrossRef]
14. Wang, J.; Rashed, S.; Murakawa, H. Mechanism investigation of welding induced buckling using inherent deformation method. *Thin Walled Struct.* **2014**, *80*, 103–119. [CrossRef]

15. Takeda, Y. Prediction of but welding deformation of curved shell plates by inherent strain method. *J. Ship Prod.* **2002**, *18*, 99–104.
16. Murakawa, H.; Deng, D.; Ma, N.; Wang, J. Applications of inherent strain and interface element to simulation of welding deformation in thin plate structures. *Comput. Mater. Sci.* **2012**, *51*, 43–52. [CrossRef]
17. Kim, T.J.; Jang, B.S.; Kang, S.W. Welding deformation analysis based on improved equivalent strain method to cover external constraint during cooling stage. *Int. J. Naval Archit. Ocean Eng.* **2015**, *7*, 805–816. [CrossRef]
18. Fu, G.; Lourenço, M.; Duan, M.; Estefen, S.F. Influence of the welding sequence on residual stress and distortion of fillet welded structures. *Mar. Struct.* **2016**, *46*, 30–55. [CrossRef]
19. Zeng, Z.; Wang, L.J.; Zhang, H. Efficient estimation of thermo physical parameters for LF6 aluminum alloy. *Mater. Sci. Technol.* **2008**, *24*, 309–314. [CrossRef]
20. Zeng, Z.; Wang, L.J.; Wang, Y.; Zhang, H. Numerical and experimental investigation on temperature distribution of the discontinuous welding. *Comput. Mater. Sci.* **2009**, *44*, 1153–1162. [CrossRef]
21. Wang, R.; Zhang, J.; Serizawa, H.; Murakawa, H. Three-dimensional modelling of coupled flow dynamics, heat transfer and residual stress generation in arc welding processes using the mesh-free SPH method. *Mater. Des.* **2009**, *30*, 3474–3481. [CrossRef]
22. Kim, T.J.; Jang, B.S.; Kang, S.W. Welding deformation analysis based on improved equivalent strain method considering the effect of temperature gradients. *Int. J. Naval Archit. Ocean Eng.* **2015**, *7*, 157–173. [CrossRef]
23. Wang, J.; Yuan, H.; Ma, N.; Murakawa, H. Recent research on welding distortion prediction in thin plate fabrication by means of elastic FE computation. *Mar. Struct.* **2016**, *47*, 42–59. [CrossRef]
24. Jung, G.H.; Tsai, C.L. Plasticity-based distortion analysis for fillet welded thin-plate T-joints. *Weld. Res. Abroad* **2004**, *81*, 177–187.
25. Bachorski, A.; Painter, M.J.; Smailes, A.J.; Wahab, M.A. Finite element prediction of distortion during gas metal arc welding using the shrinkage volume approach. *J. Mater. Process. Technol.* **1999**, *92–93*, 405–409. [CrossRef]
26. Von Mirbach, D.; Schluter, A. Influence of measurement point preparation by rough grinding on the residual stress determination using hole drilling method. *Mater. Test.* **2016**, *58*, 585–587. [CrossRef]
27. Yang, Z.R.; Kang, H.; Lee, Y. Experimental study on variations in charpy impact energies of low carbon steel, depending on welding and specimen cutting method. *J. Mech. Sci. Technol.* **2016**, *30*, 2019–2028. [CrossRef]
28. Matesa, B.; Kozuh, Z.; Dunder, M.; Samardzic, I. Determination of clad plates residual stresses by X-ray diffraction method. *Teh. Vjesn. Tech. Gaz.* **2015**, *22*, 1533–1538.
29. AsleZaeem, M.; Nami, M.R.; Kadivar, M.H. Prediction of welding buckling distortion in a thin wall aluminum T joint. *Comput. Mater. Sci.* **2007**, *38*, 588–594. [CrossRef]

metals

MDPI

Article

Effect of Welding Speed on Defect Features and Mechanical Performance of Friction Stir Lap Welded 7B04 Aluminum Alloy

Huijie Zhang *, Min Wang, Xiao Zhang, Zhi Zhu, Tao Yu and Guangxin Yang

State Key Laboratory of Robotics, Shenyang Institute of Automation, Chinese Academy of Sciences, Shenyang 110000, China; mwangsia@sina.com (M.W.); xiaozhang_sia@sina.com (X.Z.); zhizhu_winner@sina.com (Z.Z.); btx6622@sina.com (T.Y.); yangguangxin1@sia.cn (G.Y.)
* Correspondence: zhanghuijie@sia.cn; Tel.: +86-24-2397-0722

Academic Editor: Nong Gao
Received: 9 March 2016; Accepted: 11 April 2016; Published: 15 April 2016

Abstract: Friction stir lap welding of 7B04 aluminum alloy was conducted in the present paper, and the effect of welding speed on the defect features and mechanical performance of lap joints was investigated. The results indicate that the hook defect at the advancing side (AS) can reduce the effective thickness of the top sheet, and the sheet thinning level is gradually lowered by increasing the welding speed. The cold lap defect at the retreating side (RS) can result in effective thickness reduction in both top and bottom sheets, and the total height of the cold lap defect varies slightly with the welding speed. The tensile properties of the lap joints are largely related to the sheet thinning levels caused by the defects. The fracture strength of AS-loaded lap joints is progressively increased with increasing welding speed, while that of RS-loaded lap joints evolves slightly with welding speed. It is found that the affecting characteristic of loading configuration on the joint performance is also dependent on the welding speed. At lower welding speeds, the AS-loaded lap joints show lower fracture strength than the RS-loaded lap joints. When the welding speed is high, the AS-loaded lap joints present superior tensile properties to RS-loaded lap joints.

Keywords: friction stir lap welding; aluminum alloy; welding speed; defect feature; mechanical performance

1. Introduction

Friction stir welding (FSW) has been extensively utilized to weld various aluminum alloys since its invention in 1991 [1–3]. The focuses of previous investigations on FSW were mainly placed on the butt joint configuration. In fact, the lap joint configuration is also widely used in joining aluminum alloy structures, particularly in automotive and aerospace industries. Thus, a number of studies have also been conducted on friction stir lap welding (FSLW) of aluminum alloys [3].

In FSLW, two welding samples are overlapped by a certain width. A rotating tool is plunged into the top sheet and traversed along the centerline of the overlap. After the tool removal, a lap linear weld is then produced. Compared with friction stir seam welding, FSLW defects, resulting from the movement of the initial faying surface during the tool stirring, are inevitably present on both advancing and retreating sides of the lap joints [4]. On the advancing side (AS), the faying surface of the lap joint generally remains outside the weld nugget and folds upwards along the nugget boundary, which is known as the hook defect. On the retreating side (RS), the faying surface first lifts up and then penetrates into the nugget, which is known as the cold lap defect. The hook and cold lap defects are actually reflections of material flow patterns at the faying surface of the lap joint, and thus their profiles are significantly influenced by the tool's geometrical features and process parameters. For FSLW of aluminum alloys, developing specific shapes of tools and optimizing the process parameters would contribute to the minimization of defect levels in lap joints [5–8].

The hook and cold lap defects act to reduce the effective thickness of the lap joints, which exerts significant effect on their mechanical properties. The correlations between the hook defect at AS and joint performance have been extensively investigated. In Yadava *et al.*'s research [6], the load-carrying ability of lap joints was found to be linearly decreased with the increase of the height of the hook defect, indicating that the height of the hook defect had remarkable influence on the joint properties. A similar phenomenon was also observed in our previous study on FSLW of 7B04 aluminum alloy where the large height of the hook defect would lead to the poor tensile properties of the joints [9]. Nevertheless, Yazdanian *et al.* [7] suggested that for the FSLW of 3-mm-thick 6060-T5 aluminum alloy, the strength reduction due to the hook defect only became significant when the height of the hook defect reached a critical value of ~0.9 mm. Furthermore, the author pointed out that the hook defect was not the only factor affecting the joint properties. The local softening occurring in the heat affected zone could be another important factor determining the fracture location and the final fracture strength. Until now, the effect of the hook defect on properties of FSLW joints has been investigated in a large amount of papers, whereas only a few investigations are related to the correlation between the cold lap defect and the performance of lap joints.

Because of the different characteristic profiles of the hook and cold lap defects, the FSLW joints can be produced in two different loading configurations, which are AS loading and RS loading, respectively [10–13]. In AS loading, the AS of a lap joint is loaded on the upper plate, while in RS loading, the RS of a lap joint is loaded on the upper plate (see Figure 1). Several researchers have pointed out that different loading configurations could lead to different mechanical properties of lap joints for a given set of process parameters. In Cederqvist *et al.*'s research [4], the 2024-T3 and 7075-T6 aluminum alloys were friction stir lap welded, and the effect of loading configuration on joint properties was investigated. The results indicated that the initial faying surface on AS did not exhibit detectable uplift, and the AS-loaded lap joints showed higher strength than the RS-loaded lap joints. Buffa *et al.* [11] carried out the FSLW of AA2198-T4 aluminum alloy at three different heat input levels. It was found that the joints obtained in the AS loading configuration always showed definitively larger tensile properties than the ones welded in the RS loading configuration. The author argued that this was because the hook defects on AS did not negatively affect the transmitting load capability of the welded joints, but a deeper explanation was lacking. On the other hand, the phenomenon of higher fracture strength in RS loading compared with AS loading has been reported by other researchers [12,13]. The smaller extent of sheet thinning and lower stress concentration on the RS were considered to be the main reasons for the higher fracture strength of RS loading. Understanding the effect of loading configuration on joint performance is of great significance because it can assist in choosing the appropriate lap configuration for FSLW from the perspective of structure security design. However, this problem is not quite clear and is currently still controversial.

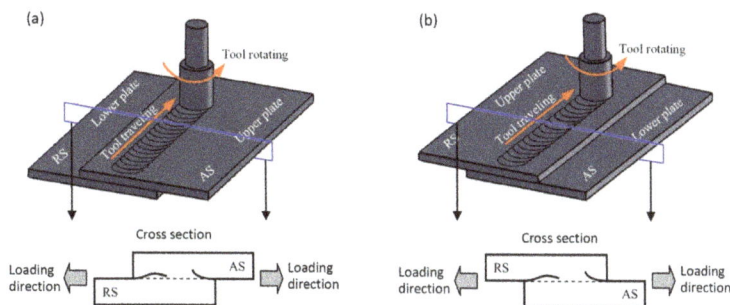

Figure 1. Different loading configurations in friction stir lap welding (FSLW): (**a**) advancing side (AS) loading; (**b**) retreating side (RS) loading [9].

The above statements have shown that the hook and cold lap defects have negative effects on mechanical properties of lap joints; therefore, the defect size should be minimized by process optimization in order to improve the joint properties. Since the correlations between process parameters and joint performances are relatively complex, illuminating the affecting characteristics of process parameters on the defect features and mechanical performance of lap joints is of importance and significance, and this would provide guidance for process optimization.

The 7B04 aluminum alloy has high specific strength and good corrosion resistance, and has been widely utilized for lightweight structures in the aviation industry. In this study, a 7B04-T74 aluminum alloy is friction stir lap welded, and the effect of welding speed on the defect features and mechanical performance of the lap joints is investigated. The focuses are mainly placed on the evolutions of the defect features, the load-carrying ability of the lap joints and the affecting characteristics of the loading configuration on the joint properties with welding speed. The present study is expected to provide guidance for the optimization of the FSLW process and the security design of the lap structure.

2. Materials and Methods

A 7B04-T74 aluminum alloy with a thickness of 2 mm was used for both the top and bottom sheets of the lap joints, whose chemical compositions and mechanical properties are listed in tab:metals-06-00087-t001. Both top and bottom sheets were 150 mm long and 100 mm wide. The longitudinal direction of the plates was perpendicular to the plate rolling direction.

Table 1. Chemical compositions and mechanical properties of 7B04-T74 aluminum alloy.

Chemical Compositions (wt. %)							Mechanical Properties	
Al	Zn	Mg	Cu	Mn	Fe	Cr	Tensile Strength	Elongation
Bal.	5.75	2.51	1.68	0.26	0.20	0.15	486 MPa	11%

After being cleaned by acetone, two welding samples were overlapped by a width of 50 mm along the longitudinal direction of the plates and tightly clamped on the work table by the welding fixture. FSLW was performed along the centerline of the overlap using a FSW machine. In the present study, FSLW experiments were conducted in both AS loading and RS loading configurations. The lap joints produced from both loading configurations were named as the AS-loaded lap joint and RS-loaded lap joint, respectively. The welding tool used for the experiments had a 12-mm-diameter shoulder and a conical right-hand screwed pin. The tool pin had a diameter of 4.5 mm (at the shoulder) and a pin length of 2.8 mm. Note that the pin length of the welding tool is the optimal result of our previous work [9]. During the welding, an axial load of 4.6 kN and a tilting angle of 2.5° were applied to the welding tool. The rotation speed was fixed at a constant value of 600 rpm and the welding speed varied from 50 to 200 mm/min.

After welding, the joints were cross-sectioned perpendicular to the welding direction for metallographic analysis and tensile testing. The cross-sections of the metallographic specimens were polished using a diamond paste, etched with Keller's reagent and observed by an optical microscopy (KEYENCE VHX-100). The Vickers microhardness profiles were measured on the middle of the upper sheets throughout the cross-sections with a load of 500 g and a dwell time of 10 s. Weld strengths of the joints were evaluated by lap shear tests. Rectangular test specimens with the width of 15 mm were cut from friction stir lap welds perpendicular to the welding direction. During the test, samples were aligned by using 2-mm-thick 7B04-T74 aluminum spacers in the clamping grip in order to ensure that an initial pure shear load was applied to the interfacial plane. Lap shear testing was conducted at a rate of 1.0 mm/min. Three samples were tested for each combination of experimental parameters. The fracture strength of a sample, defined by dividing the fracture load by the sample width (N/mm), was utilized to determine the load-carrying ability of FSLW joints. This has been commonly used to evaluate the mechanical properties of a friction stir lap welded joint in previous studies [5–8].

After tensile test, the optical microscopy mentioned above was utilized to analyze the fracture features of the joints.

3. Results

3.1. Defect Features of Lap Joints

Figure 2 shows the cross-sections of FSLW joints. Note that the RS is on the left while the AS is on the right for each cross-section in the figure and throughout the whole paper, and zones A–E shown in Figure 2a represent the locations where microstructure analyses were conducted at 50 mm/min. Two types of welding defects are present in all the lap joints, *i.e.*, the hook defect and cold lap defect, as marked in Figure 2a. In essence, the profiles of defects result from the material flow patterns at the faying surface during FSLW (see Figure 3). In FSLW, the material around the tool at the RS first exhibits an upward flow trend under the shearing effect of the threaded tool pin (*i.e.*, Flow I in Figure 3), and then flows downward to fill the pin cavity at the rear of the tool (*i.e.*, Flow II in Figure 3). Driven by the rotation of the tool pin, the initial faying surface adjacent to the nugget zone at the AS is directed upwards at an inclination angle towards the weld surface and then arrested at the nugget extremity (see Flow III in Figure 3), inducing the formation of the hook defect.

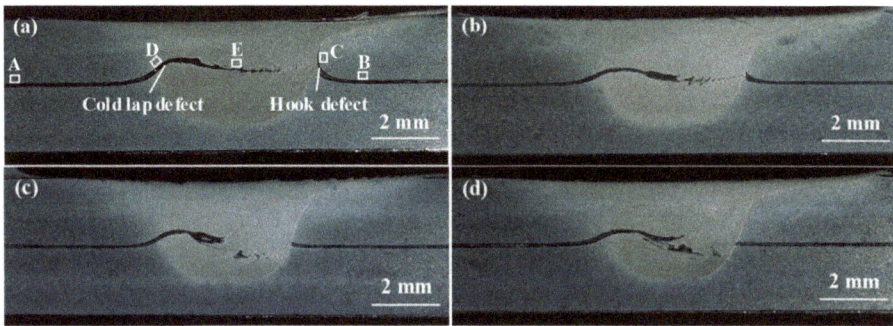

Figure 2. Cross-sections of the lap joints obtained at different welding speeds: (**a**) 50 mm/min; (**b**) 100 mm/min; (**c**) 150 mm/min; (**d**) 200 mm/min; A–E marked in (**a**) reflect the locations where microstructure analyses were conducted.

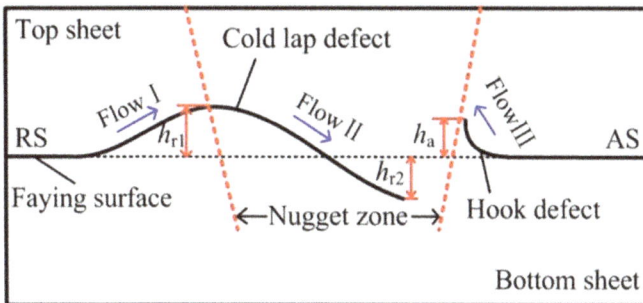

Figure 3. Schematic view of material flow during FSLW; note that the border of nugget zone is represented by the red dashed lines.

The formation mechanism of the defects in lap joints can be further illustrated from the microstructure characteristics adjacent to the defects. Figure 4 gives the grain structures extracted from zones A–E in Figure 2a, which are corresponding to the base metal (BM), heat affected zone (HAZ), thermo-mechanically affected zone (TMAZ) on the AS, TMAZ on the RS and the weld nugget zone (WNZ) of the lap joint welded at 50 mm/min, respectively. The initial base material is characterized by elongated grain structures, as seen in Figure 4a. The HAZ only experiences welding thermal cycles during FSW and no deformation occurs in the faying surface at this zone, thus the HAZ exhibits similar grain structures as the BM (see Figure 4a,b). The material around the tool is plastically deformed during tool rotation and forms the TMAZ on both sides of the weld. The similar flow trends of the extruded grains and the lap defects in TMAZ at each side are both reflections of the material flow patterns by passage of welding tool (see Figure 4c,d). The WNZ undergoes intense plastic deformation during welding and is featured by fine equiaxed grain structures due to dynamic recrystallization (see Figure 4e).

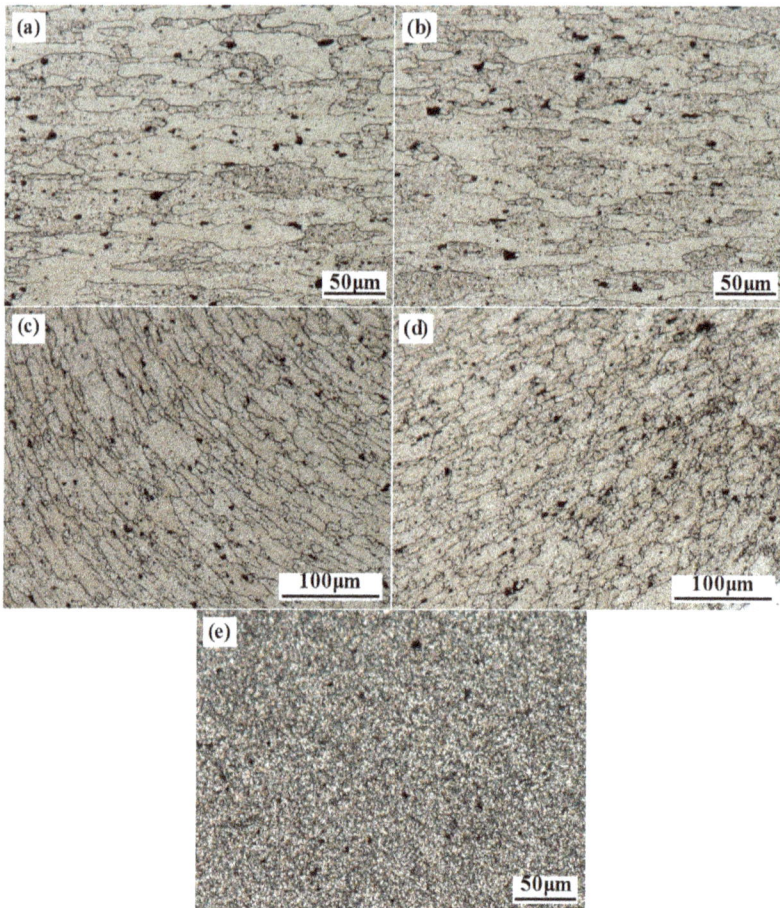

Figure 4. Microstructures extracted from different zones of the FSLW joint welded at 50 mm/min: (**a**) base metal (BM); (**b**) heat affected zone (HAZ); (**c**) thermo-mechanically affected zone (TMAZ) at AS; (**d**) TMAZ at RS; (**e**) weld nugget zone (WNZ).

The formation of the defects mentioned above results in the sheet thinning of lap joints. The hook defect only reduces the effective thickness of the top sheet; however, the cold lap defect can not only lead to the effective thickness reduction in the top sheet, but it can also result in the thinning of the bottom sheet when the welding speed is above 150 mm/min (see Figure 2c,d). This is different from that commonly observed in previous investigations, where only the top sheet thinning occurs due to the cold lap defect [6,13–15]. In order to quantify the levels of sheet thinning caused by the both defects, Figure 5 plots the height of the hook and cold lap defects, which was measured in the optical microscope. The defect height in both the top and bottom sheets is considered in this paper, and the total height of the hook defect or cold lap defect is defined as the sum of the maximum vertical distance from the defect to the initial faying surface in the top and bottom sheets (*i.e.*, h_a for the hook defect and $h_{r1} + h_{r2}$ for the cold lap defect, as marked in Figure 3).

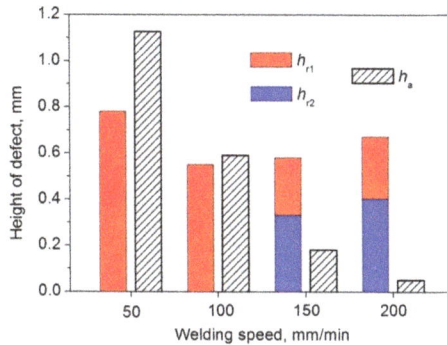

Figure 5. Height values of the hook and cold lap defects formed at different welding speeds; h_a refers to the height of the hook defect at AS, while h_{r1} and h_{r2} represent the heights of the cold lap defects in the top and bottom sheets at RS, respectively.

When the welding speed is increased, the upward movements of the initial faying surface caused by Flow I and Flow III are both limited due to the weakening of the pin shearing effect, leading to the notable decrease of h_a and h_{r1}. On the other hand, an increase in welding speed can not only increase the pin cavity volume at per tool rotation, but it can also weaken the tool pin shearing effect at per unit length of weld. These two factors are both favorable for the downward flow of the pin sheared material, and thus the downward motion of the cold lap defect profile caused by Flow II is progressively improved with the welding speed. Above 150 mm/min, the cold lap defect even extends into the bottom sheet of the lap joints (see Figure 2c,d). Since the height of the cold lap defect is decreased in the top sheet and simultaneously increased in the bottom sheet when the welding speed is increased, the total height of the cold lap defect ($h_{r1} + h_{r2}$) does not vary so evidently with the welding speed as that of the hook defect does (see Figure 5).

3.2. Mechancial Performance of Lap Joints

The lap shear test was conducted to evaluate the mechanical performance of FSLW joints. During the test, two fracture modes are observed. For Mode I, fracture occurs in the hook defect on the AS, and the final fracture path is nearly normal to the weld top surface (see Figures 6 and 7a). In Mode II, however, the fracture just occurs in the cold lap defect (see Figure 7b). Apparently, the pre-crack is more favorable for occurring at the hook and cold lap defects under tensile loading, and thus the two fracture modes are actually introduced on the basis of fracture mechanics definitions.

In the AS loading configuration, when the welding speed is increased from lower values (50 and 100 mm/min) to higher values (150 and 200 mm/min), the fracture feature of the lap joints is changed from Mode I to Mode II, and the fracture strength of AS-loaded lap joints is progressively increased with the welding speed, as revealed in Figures 6 and 8. For the RS loading configuration, however, all the joints are fractured in Mode II during the tensile test, and the fracture strength of the RS-loaded lap joints varies slightly with the welding speed.

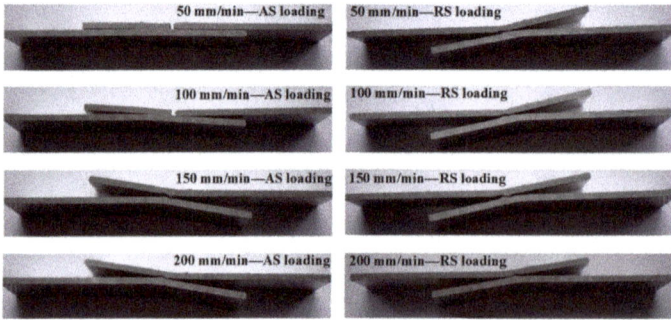

Figure 6. Fracture features of the lap joints.

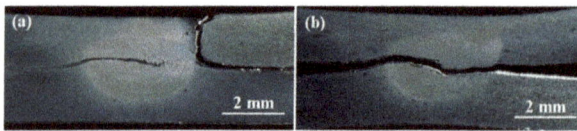

Figure 7. Enlarged views of the failed joint (100 mm/min): (**a**) AS loading; (**b**) RS loading [9].

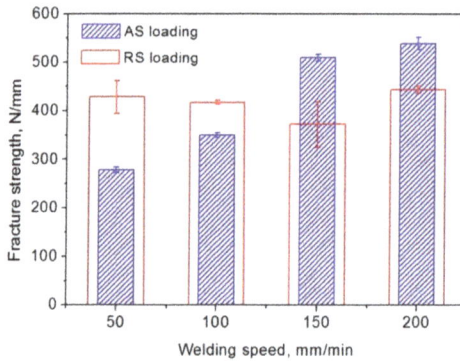

Figure 8. Fracture strength of the lap joints obtained at different welding speeds.

During the tensile test, the tensile force whose direction is fixed is applied on the original faying surface of the lap joints. In contrast, the hook and cold lap defects, where the crack is first generated and the relative movement of both sheets takes place, exhibit curved features (see Figure 2). Apparently, the tensile direction and the movement direction of the lap sheets do not take place at the same surface, the additional bending moment then leads to the occurrence of local bending and the rotation of both sheets, as shown in Figure 9. This phenomenon was also emphasized in previous studies, and has been simulated in the research of Yazdanian [7], Babu [8] and Fersini [16] *et al.* The present result reveals that the joints with larger rotating angles require more time to reach failure during lap shear tests, and thus the rotating level of the lap sheets actually indicates the deformation ability of the lap joints, *i.e.*, the difficult degree for failure to occur during the tensile test. Owing to this, the rotating angle and the fracture strength exhibit similar evolving trends with the welding speed for both AS- and RS-loaded lap joints (see Figures 8 and 9). As the welding speed is increased, the trend of a rapid increase in the rotating angle is clear for the AS-loaded joints; in comparison, the rotating angle of the RS-loaded lap joints shows a slight variation.

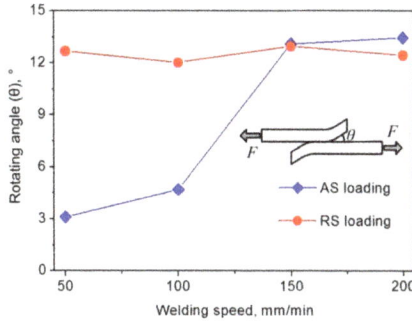

Figure 9. Rotating angle of the lap sheets during the lap shear test.

The effect of loading configuration on joint performance also varies with the welding speed. At lower welding speeds of 50 and 100 mm/min, the RS-loaded lap joints show superior fracture strength to the AS-loaded lap joints. When the welding speed is high (150 and 200 mm/min), the fracture strength of AS-loaded lap joints is larger than that of RS-loaded lap joints.

4. Discussion

4.1. Evolution of Tensile Properties with Welding Speed

The mechanical properties of FSLW joints can be influenced by several factors. The sheet thinning level caused by the hook and cold lap defects has been demonstrated to play a prominent role in determining the performance of lap joints [6–13]. Yuan *et al.* [13] reported that the crystallographic texture was another factor that influenced the fracture load and fracture path of FSLW joints of AZ31 magnesium alloy. Meanwhile, as far as the heat-treatable aluminum alloy is concerned, the thermal effect during FSW can cause local softening to occur in the joint, leading to the reduction in mechanical properties of the weld relative to the BM [17,18]. The local softening phenomenon is also observed in the present study, as seen in Figure 10. The 7B04 aluminum alloy possesses the hardness of 152–155 Hv, while the weakest region of the weld only presents the hardness value of 130–135 Hv, much lower than that of the BM. Yazdanian *et al.* [7] pointed out that the softening of the HAZ could be an important factor in determining the tensile properties of lap joints of heat-treatable aluminum alloys, because some of the FSLW joints were found to fail in the HAZ rather than along the hook or cold lap defect during the tensile test. However, for the present study, it should be noted that all the lap joints are fractured in the hook and cold lap defects under tensile loading. The fracture strength and the level of sheet thinning induced by the defects exhibit similar evolving trends as the welding speed. Therefore, the tensile properties of FSLW joints are believed to be largely related to the sheet thinning level, *i.e.*, the height of the defects.

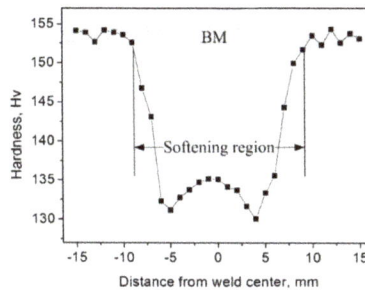

Figure 10. Hardness profile measured across the middle of the top sheet of a FSLW joint (100 mm/min).

When the welding speeds are low (50 and 100 mm/min), the hook defects at the AS are characterized by a large height and show a sharp curved feature (see Figure 2a,b). In this case, the AS-loaded lap joints fail in the hook defects and present lower fracture strength with a lesser extent of local bending and rotation (see Figures 6, 8 and 9). As the welding speed is increased, the height of the hook defect is dramatically decreased (see Figure 5). The cold lap defect that causes the larger extent of sheet thinning is then more favorable for a fracture to occur during AS loading, and an increase of fracture strength with welding speed is observed (see Figure 8). For the case of RS loading, the failure of all the joints progresses along the cold lap defect (see Figure 6). The level of sheet thinning resulting from the cold lap defect varies slightly with the welding speed, and thus the RS-loaded lap joints do not vary significantly in fracture strength and exhibit a nearly similar rotating angle during lap shear tests as the welding speed is increased (see Figures 8 and 9).

4.2. Effect of Loading Configuration on Joint Properties

Previous investigations have reported that the FSLW joints showed different mechanical performances between AS and RS loading configurations due to the different characteristics of hook and cold lap defects. In the research of Cederqvist [4] and Buffa [11] *et al.*, the AS-loaded lap joints showed higher fracture strength than the RS-loaded lap joints. In contrast, Yang [12] and Yuan [13] claimed that the joint properties obtained in RS loading were superior to those obtained in AS loading. In the present study, both of the cases are observed, and the results indicate that the affecting characteristic of loading configuration on joint performance is actually variable with the welding speed.

To illustrate the intrinsic reason for the affecting characteristics, Figure 11 plots the schematic views of the tensile behavior of lap joints under different loading configurations. The hook and cold lap defects are described by red and blue lines, respectively. The big gray arrow indicates the path of load applied on the initial faying surface of the lap joints. The level of tensile stress is marked by the dash line, which is progressively decreased from a maximum (σ_{max}) at the loaded end to a minimum (σ_{min}) at the unloaded end in the top and bottom sheets.

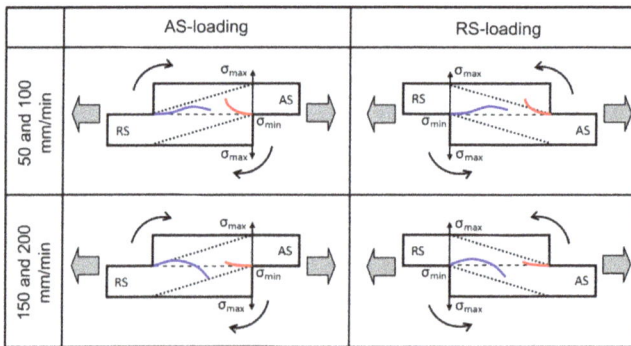

Figure 11. Schematic views of tensile behavior of lap joints under different loading configurations.

When the welding speeds are low (50 and 100 mm/min), the hook defects with great height are formed at the AS (see Figures 2 and 5). During AS loading, the sharp curved character of the hook feature introduces a significant stress concentration at the tip of the hook. The lap joints are fractured in the hook defects during the tensile test and exhibit poor tensile properties. In contrast, for the case of RS loading, the maximum stress is achieved at the RS of the top sheet (see Figure 11). Therefore, the cold lap defect is the favorable location for fracture during the lap shear test. The slightly upward then downward movement of the cold lap defect can remarkably reduce the stress concentration and thus retard the crack generation and propagation. Consequently, greater fracture strength in RS loading is obtained in contrast to AS loading.

When the welding speed is increased to large values (150 and 200 mm/min), the height of the hook defect is dramatically decreased. Under such a condition, the cold lap defect is more favorable for crack generation and propagation than the hook defect, and thus the AS- and RS-loaded lap joints both fail along the cold lap defect. The stress applied to the cold lap defect in AS loading is smaller than that in RS loading during the tensile test (see Figure 11), which makes the crack propagation along the cold lap defect more difficult during AS loading. As a consequence, the AS loading configuration leads to stronger joint properties than the RS loading configuration.

Above all, FSLW joints tend to be fractured at the location that experiences the largest extent of sheet thinning or where the maximum tensile stress is located. From this point of view, two aspects should be considered in order to obtain the high performances of lap joints. Firstly, a relatively high welding speed should be applied in FSLW on the premise of avoiding welding defects such as grooves and voids, since the high welding speed tends to lower the sheet thinning level on the AS of the lap joint. Secondly, the loading side of the joint, AS or RS, should possess a larger effective thickness, because in such a case the stress acting on the weakest location of joint is relatively low. By taking these two measures, the FSLW joint needs to experience a large force of local bending and rotation for crack generation, which improves the load-carrying ability of the lap joint and finally leads to superior joint properties.

5. Conclusions

Based on the present investigation, the results of significance are drawn as follows:

(1) The increase of welding speed can remarkably limit the upward motion of the initial faying surface, which lowers the level of top sheet thinning induced by the hook defect. The cold lap defect can result in the reduction of effective thickness in both the top and bottom sheets. The height of the cold lap defect is decreased in the top sheet but gradually increased in the bottom sheet when the welding speed is increased, leading to a slight variation in the total height of the cold lap defect with the welding speed.

(2) The tensile properties of FSLW joints are largely related to the level of sheet thinning caused by the hook and cold lap defects. In the AS loading configuration, the fracture strength of AS-loaded lap joints is significantly increased with increasing welding speed, while for the RS loading configuration, the fracture strength of RS-loaded lap joints varies slightly with the welding speed. Local bending and rotating occurred in the lap joints during the tensile test. The rotating angle of the lap sheets and the fracture strength of the lap joints exhibit similar evolving trends with the welding speed in both AS and RS loading configurations.

(3) The affecting characteristic of loading configuration on joint performance is dependent on the welding speed. At lower welding speeds, the AS-loaded lap joints show lower fracture strength than the RS-loaded lap joints. When the welding speed is high, the AS-loaded lap joints then present larger tensile properties than RS-loaded lap joints. In order to obtain the high performance of lap joints, a relatively high welding speed should be applied during FSLW. Meanwhile, the loading side of the joint, AS or RS, should possess a larger effective thickness in the joint.

Acknowledgments: The authors are grateful to be supported by National Natural Science Foundation of China (Grant No. 51505471) and Special Fund for Youth Innovation Promotion Association, Chinese Academy of Sciences (Grant No. 2015162).

Author Contributions: Huijie Zhang and Min Wang conceived and designed the experiments; Huijie Zhang, Xiao Zhang and Guangxin Yang performed the experiments; Zhi Zhu and Tao Yu analyzed the experimental data; Huijie Zhang wrote the paper.

Conflicts of Interest: The authors declare no conflict of interest.

References

1. Mishra, R.S.; Ma, Z.Y. Friction stir welding and processing. *Mater. Sci. Eng. Rep.* **2005**, *50*, 1–78. [CrossRef]

2. Buffa, G.; Fratini, L.; Ruisi, V. Friction stir welding of tailored joints for industrial applications. *Int. J. Mater. Form.* **2009**, *2*, 311–314. [CrossRef]

3. Cam, G.; Mistikoglu, S. Recent developments in friction stir welding of Al-alloys. *J. Mater. Eng. Perform.* **2014**, *23*, 1936–1953. [CrossRef]

4. Cederqvist, L.; Reynolds, A.P. Factors affecting the properties of friction stir welded aluminum lap joints. *Weld. J.* **2001**, *80*, 281–287.

5. Lee, C.Y.; Lee, W.B.; Kim, J.W.; Choi, D.H.; Yeon, Y.M.; Jung, S.B. Lap joint properties of FSWed dissimilar formed 5052 Al and 6061 Al alloys with different thickness. *J. Mater. Sci.* **2008**, *43*, 3296–3304. [CrossRef]

6. Yadava, M.K.; Mishra, R.S.; Chen, Y.L.; Carlson, B.; Grant, G.J. Study of friction stir joining of thin aluminium sheets in lap joint configuration. *Sci. Technol. Weld. Join.* **2010**, *15*, 70–75. [CrossRef]

7. Yazdanian, S.; Chen, Z.W.; Littlefair, G. Effects of friction stir lap welding parameters on weld features on advancing side and fracture strength of AA6060-T5 welds. *J. Mater. Sci.* **2012**, *47*, 1251–1261. [CrossRef]

8. Babu, S.; Ram, G.D.J.; Venkitakrishnan, P.V.; Reddy, G.M.; Rao, K.P. Microstructure and mechanical properties of friction stir lap welded aluminum alloy AA2014. *J. Mater. Sci. Technol.* **2012**, *28*, 414–426. [CrossRef]

9. Wang, M.; Zhang, H.J.; Zhang, J.B.; Zhang, X.; Yang, L. Effect of pin length on hook size and joint properties in friction stir lap welding of 7B04 aluminum alloy. *J. Mater. Eng. Perform.* **2014**, *23*, 1881–1886. [CrossRef]

10. Urso, G.D.; Giardini, C. The influence of process parameters and tool geometry on mechanical properties of friction stir welded aluminum lap joints. *Int. J. Mater. Form.* **2010**, *3*, 1011–1014.

11. Buffa, G.; Campanile, G.; Fratini, L.; Prisco, A. Friction stir welding of lap joints: Influence of process parameters on the metallurgical and mechanical properties. *Mater. Sci. Eng. A* **2009**, *519*, 19–26. [CrossRef]

12. Yang, Q.; Li, X.; Chen, K.; Shi, Y.J. Effect of tool geometry and process condition on static strength of a magnesium friction stir lap linear weld. *Mater. Sci. Eng. A* **2011**, *528*, 2463–2478. [CrossRef]

13. Yuan, W.; Carlson, B.; Verma, R.; Szymanski, R. Study of top sheet thinning during friction stir lap welding of AZ31 magnesium alloy. *Sci. Technol. Weld. Join.* **2012**, *17*, 375–380. [CrossRef]

14. Xu, X.D.; Yang, X.Q.; Zhou, G.; Tong, J.H. Microstructures and fatigue properties of friction stir lap welds in aluminum alloy AA6061-T6. *Mater. Des.* **2012**, *35*, 175–183. [CrossRef]

15. Dubourg, L.; Merati, A.; Jahazi, M. Process optimization and mechanical properties of friction stir lap welds of 7075-T6 stringers on 2024-T3 skin. *Mater. Des.* **2010**, *31*, 3324–3330. [CrossRef]

16. Fersini, D.; Pirondi, A. Fatigue behavior of Al 2024-T3 friction stir welded lap joints. *Eng. Fract. Mech.* **2007**, *74*, 468–480. [CrossRef]

17. Starink, M.J.; Seschamps, A.; Wang, S.C. The strength of friction stir welded and friction stir processed aluminum alloys. *Scr. Mater.* **2008**, *58*, 377–382. [CrossRef]

18. Fratini, L.; Buffa, G.; Shivpuri, R. Mechanical and metallurgical effects of in process cooling during friction stir welding of AA7075-T6 butt joints. *Acta Mater.* **2010**, *58*, 2056–2067. [CrossRef]

metals

MDPI

Article

Influence of Post Weld Heat Treatment on Strength of Three Aluminum Alloys Used in Light Poles

Craig C. Menzemer [1], Eric Hilty [2], Shane Morrison [3], Ray Minor [4] and Tirumalai S. Srivatsan [2,*]

[1] Civil Engineering and Associate Dean, College of Engineering, The University of Akron, Akron, OH 44325, USA; ccmenze@uakron.edu
[2] Department of Civil Engineering, the University of Akron, Akron, OH 44325, USA; hilty.eric@gmail.com
[3] Research and Development, HAPCO, Abingdon, VA 24210, USA; Shane.Morrison@hapco.com
[4] HAPCO, Abingdon, VA 24210, USA; ray.minor@hapco.com
* Correspondence: tsrivatsan@uakron.edu; Tel.: +1-330-608-8355; Fax: +1-330-972-6027

Academic Editor: Nong Gao
Received: 8 December 2015; Accepted: 22 February 2016; Published: 3 March 2016

Abstract: The conjoint influence of welding and artificial aging on mechanical properties were investigated for extrusions of aluminum alloy 6063, 6061, and 6005A. Uniaxial tensile tests were conducted on the aluminum alloys 6063-T4, 6061-T4, and 6005A-T1 in both the as-received (AR) and as-welded (AW) conditions. Tensile tests were also conducted on the AR and AW alloys, subsequent to artificial aging. The welding process used was gas metal arc (GMAW) with spray transfer using 120–220 A of current at 22 V. The artificial aging used was a precipitation heat treatment for 6 h at 182 °C (360 °F). Tensile tests revealed the welded aluminum alloys to have lower strength, both for yield and ultimate tensile strength, when compared to the as-received un-welded counterpart. The beneficial influence of post weld heat treatment (PWHT) on strength and ductility is presented and discussed in terms of current design provisions for welded aluminum light pole structures.

Keywords: light poles; aluminum alloy; welding; gas metal arc welding; artificial aging; post weld heat treatment; microstructure; tensile properties

1. Introduction and Background

Over the last four decades, *i.e.*, since the early 1970s, structural aluminum alloys have been used in a myriad of applications, primarily because they can offer an attractive combination of strength, are light in weight, have a high strength-to-weight (σ/ρ) ratio, and, most importantly, are cost efficient [1]. Many products are being increasingly fabricated from 6XXX-series aluminum alloys due to their innate ability to be extruded into complex shapes, coupled with their receptiveness to welding and their notable resistance to environment-induced degradation or corrosion [2]. Understanding the weldability and resultant mechanical properties is important in an attempt to put these alloys to efficient use. It is uncommon for an aluminum alloy to be welded with no influence on microstructure and resultant mechanical properties, such as strength. However, precipitation heat treatment does offer the promise of minimizing the negative effects of welding on the mechanical properties of the family of 6XXX alloys. A product of considerable practical interest and significance is welded aluminum light poles [3,4].

A widely-used method for joining the alloys of aluminum is welding. A few noteworthy examples related to the commercial industry include the following: (i) fabrication of rail vehicles; (ii) marine structures; (iii) pressure vessels; (iv) automotive components; and (v) structures in the civil construction industry. A few noteworthy advantages of the welding process include the following: (a) high joint efficiencies; (b) flexibility; (c) speed; and (d) a low fabrication cost [5]. Welding involves "localized" melting of the base material; as a consequence of which, both the microstructure and resultant mechanical properties will be different from those of the base material [6]. To obtain improved properties for the welded material, component, or structure, a heat treatment is both necessary and essential [7].

Section 2.5 of the 2010 Aluminum Design Manual (ADM) provides mechanical property information for welded and, subsequently, heat treated alloys that are chosen for use in aluminum light poles [8]. Aluminum alloys 6005 and 6063 have been widely used for welded light poles. For poles manufactured from aluminum alloy 6005, welded in the T1 temper, and having a thickness less than or equal to 6.4 mm, the specifications allow the engineer to make effective use of 85% of the strength of the base metal (6005-T5) in the un-welded condition, provided that the assembly is artificially aged for 6 h at 182 °C (360 °F). Light poles fabricated from AA6063, welded in the T4 temper, and having a thickness either equal to or less than 9.5 mm, the specifications allow for use of 85% of strength of the base metal, *i.e.*, AA6063-T6, provided the welded assembly is artificially aged for 6 h at 182 °C (360 °F). It is important to note that the 85% percent "rule" is permissible and allowed when welding aluminum alloy 6005 and aluminum alloy 6063 using aluminum alloy 4043 as the filler material. The basis for these provisions within the Aluminum Design Manual (ADM) was the result of round robin tests carried out in the early 1960s and up to the late 1970s [9,10]. Most importantly, results of these studies were never published in the open literature, and some test records have been either misplaced or lost over the years. A careful review of the available test data from an earlier round robin test program, using statistical techniques inscribed within the 2010 Aluminum Design Manual (ADM), was considered to be both incomplete and inconclusive [9,10].

The focus of the current study was a determination of the mechanical properties of welded and artificially aged aluminum alloys 6061, 6063 and 6005A having thicknesses commensurate with what is currently being used in aluminum alloy light poles. Further, the study provided an opportunity to carefully examine the extrinsic influence of welding on intrinsic microstructural effects in an attempt to characterize the microstructure-mechanical property relationships. Much of the work examining the relationships can be found elsewhere [11,12]. This paper focuses on the mechanical properties of Post Weld Heat Treated (PWHT) aluminum alloys that are preferentially chosen for use in light poles. In recent years, the influence of post-weld heat treatment subsequent to hybrid welding of aluminum alloy 5754 was carefully studied and the test results and observations documented in the open literature [13].

2. Specimen Preparation and Mechanical Testing

The parent materials chosen for use in this research study were the three aluminum alloys: (i) AA6063-T4, (ii) AA6061-T4, and (iii) AA 6005A-T1. All the three aluminum alloys were obtained in the as-extruded form. Blanks were then saw cut from the extruded sections. The tensile test specimens were precision machined from the as-extruded sections, the test specimen measured 6.4 mm (1/4 in) and 9.5 mm (3/8 in) in thickness, for both aluminum alloy 6063 and aluminum alloy 6061. Samples prepared for aluminum alloy 6005A were taken from extruded sections that measured 3.2 mm (1/8 in) in thickness. Different material thicknesses were selected based on their historical use in light pole applications, as well as provisions documented in the Aluminum Design Manual (ADM) for the purpose of enabling a scientific comparison of the tensile response.

The nominal compositions of the three alloys are given in Table 1 [14].

Table 1. Nominal composition (weight pct.) of the chosen 6XXX aluminum alloys [8].

Element	Al	Cr	Cu	Fe	Mg	Mn	Si	Ti	Zn
6063	Balance	0.1	0.1	0.35	0.9	0.1	0.6	0.1	Max 0.1
6005A	Balance	0.3	0.3	0.35	0.4	0.5	0.9	0.1	0.2
6061	Balance	0.04	0.15	0.7	1.2	0.15	0.8	0.15	Max 0.05

The filler metal used for this study was AA4043, as is common in the fabrication of light poles and for purpose of welding the 6XXX series alloys [15]. Fillet welds were used to form heat-affected zones across samples in an attempt to examine the influence of "localized" heating on the base material. The test samples, with short cover plates and a fillet weld, are shown in Figure 1. Gas metal arc welding (GMAW) was successfully used to deposit the weld metal using the technique of spray transfer, with the current varying from 120 to 220 A at a voltage of 22 V. The filler wire was 2.4 mm (3/32 in) in diameter with the shielding gas was 100 percent argon. Gas metal arc welding was chosen and used since it continues to be acceptable and preferentially chosen for use in a wide spectrum of industrial-related applications.

Figure 1. Pictorial view of test sample with fillet welded lap joint.

After the welds were placed, a Bridgeport vertical axis milling machine was used to remove the fillet and cover plate, leaving only the parent metal strip containing the heat affected zone (HAZ). Upon removal of the fillet welds and cover plates, each metal strip was transformed into a tensile specimen using a computer numerical control (CNC) machine (Model: HAAS) (Figure 2). A number of un-welded strips were also machined to provide tensile samples. A sizeable number of both the welded and parent metal tensile samples were chosen for the purpose of subsequent heat treatment.

Figure 2. A finished tensile test specimen containing a fillet weld subsequent to machining on a Computer Numerical Control (CNC) machine.

The heat treatment process followed procedures outlined in ASTM B918-01 [16]. Essentially, a precipitation type heat treatment was used, and this is referred to as post-weld heat treatment (PWHT). For the three chosen aluminum alloys (6005A, 6061 and 6063), the specifics of the treatment involved an initial soak at a temperature of 182 °C (360 °F) for six hours, followed by cooling in ambient air (27 °C). In this research study, both the welded (PWHT) and unwelded (PHT) tensile samples were subject to artificial aging at 182 °C (360 °F) for six hours. The resultant tempers were expected to be near the T6 condition for aluminum alloy 6063 and aluminum alloy 6061. Test specimens fabricated from aluminum alloy 6005A were expected to be close to a T5 temper following artificial aging. The guaranteed minimum strengths for extrusions of AA6005A-T5, AA6061-T6 and AA 6063-T6 were 262 MPa, 262 MPa and 207 MPa, respectively. The three aluminum alloys chosen for this research study had strengths that exceeded the guaranteed minimum yield strength and ultimate tensile strength that is recommended for use in conventional structural design.

A few tensile samples consisting of welded and parent metal AA6005A were re-solution heat treated and subsequently aged. The re-solution heat treatment followed guidelines given in Volume 4 of the ASM Handbook "Heat Treating" [17]. The re-solution treatment process consisted of an initial soaking of the test specimens at 529 °C (985 °F) for a full 60 min. This was followed by a rapid quench in a solution mixture of 60%/40% water/glycol. This treatment was expected to place both the coarse and intermediate-size constituent particles back in solution. If left unattended, the 6XXX-series alloys used in this research study would be expected to naturally age with time at an ambient temperature (27 °C). Following re-solution heat treatment, a selected number of specimens were subjected to artificial aging. The primary purpose of testing the re-solution heat treated and aged AA6005A samples was to compare the tensile properties with the 6005A counterpart.

Tensile samples of aluminum alloys 6061, 6063 and 6005A were tested in uniaxial tension in each of the following conditions:

1. As-Received (AR-6063, AR-6061 and AR-6005A).
2. As-Received and artificially aged (AR+PHT-6063, AR+PHT-6061, AR+PHT-6005A).
3. As-Welded (AW-6063, AW-6061, AW-6005A).
4. Welded and subsequently heat treated (PWHT-6061, PWHT-6063, PWHT-6005A).
5. As-Received and As-Welded subjected to re-solution heat treatment and aging (SHT+PHT-6005A).

Individual test specimens were placed in a universal test machine (Model: Warner-Swasey) and deformed in uniaxial tension up until failure by separation. An extensometer was fixed along the gage section of each sample to obtain a record of the axial strain during loading. Data from each test were recorded on a PC-based data acquisition system and subsequently used to develop the stress *versus* strain response. The engineering stress *versus* engineering strain curves were compared to provide an understanding of the response of the chosen test specimens when subjected to uniaxial deformation. Both yield strength and ultimate tensile strength values were obtained for each test sample, and the lower bounds statistically determined. The lower bound strengths were compared with the minimum guaranteed design values for the chosen aluminum alloy.

3. Results and Discussion

3.1. Microstructure

Light optical micrographs were taken over a range of low magnifications, revealed the initial microstructure of AA6061 in the (a) as-received (AR); (b) as-welded (AW); (c) as-received plus precipitation heat treated (AR+PHT); and (d) post weld heat treated (PWHT) conditions. The as-received alloy revealed a random distribution of both coarse and intermediate size intermetallic particles (Figure 3a,b). These intermetallic particles result from the presence and availability of residual elements, such as iron and silicon [18,19]. As documented elsewhere, these particles were identified to be $Al_{12}Fe_3Si$, $Al_{15}(FeMn)_3Si$ and Al_5FeSi [18–20]. The iron-rich intermetallic particles range in size from 1 to 10 microns and are potential sites for the early initiation of microscopic damage during plastic deformation. The manganese-rich particles in the chosen aluminum alloys help in controlling both grain size and grain growth during solidification. Micrographs of the alloy in the as-welded (AW) condition revealed fine recrystallized grains in the region of the weld bead (Figure 4a). By comparison, the weld bead in the post weld heat treated condition is shown in Figure 4b. A noticeable difference in microstructure between the weld bead and base metal is evident along the interface between the two regions (Figure 5a,b). The microstructure of aluminum alloy 6061, in the as-received plus precipitation heat treated condition revealed a significant volume fraction of both coarse and intermediate size second-phase particles in the base metal (Figure 6a). These particles were randomly dispersed throughout the microstructure. The as-welded and post weld heat treated samples revealed very well defined grains that could be classified as being: (i) small in size and of varying shape (Figure 6b), and (ii) distributed randomly through the microstructure of the base metal. The microstructure at the interface of the base metal and weld bead is shown in Figure 12. Fine microscopic cracks are evident and can be attributed to melting of the low melting point constituents both at and along the grain boundaries.

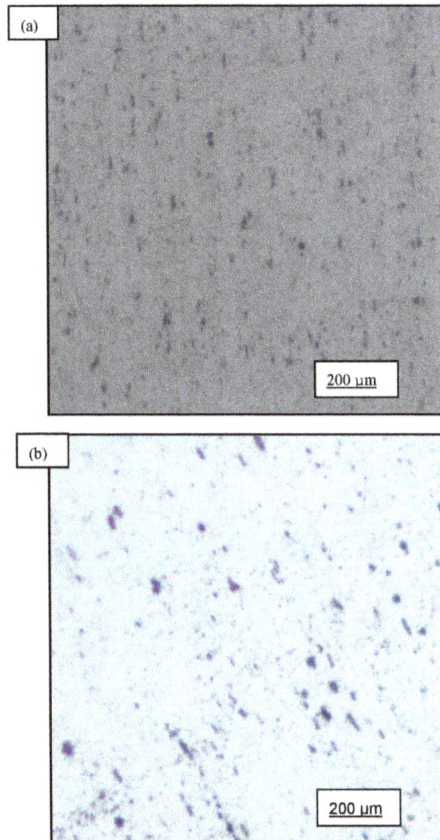

Figure 3. Light optical micrographs of aluminum alloy 6061-T4 showing microstructure the following: (**a**) Coarse and intermediate second phase particles in the base metal of the as-received or as-provided metal; (**b**) distribution of intermetallic particles in the heat-treated sample.

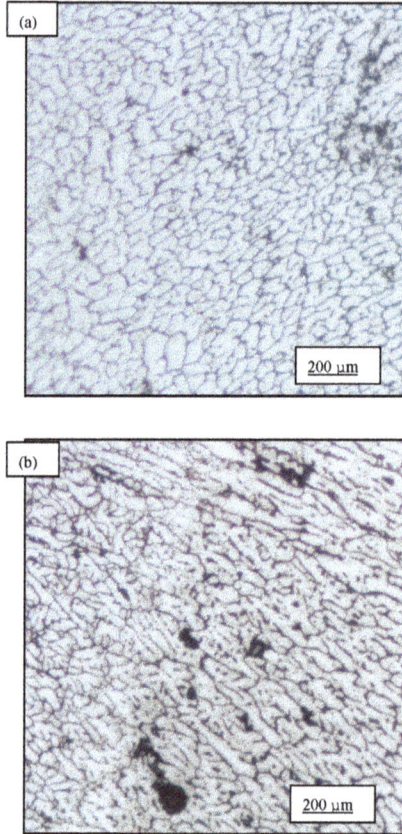

Figure 4. Optical micrograph of the weld pool showing fine grains of varying size and shape. (**a**) Grain size and morphology in the weld pool in the as-welded condition; (**b**) weld pool in the post weld heat-treated condition.

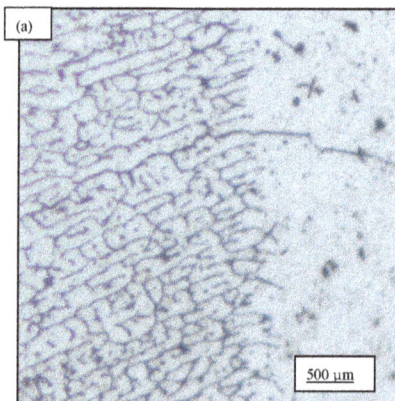

Figure 5. *Cont.*

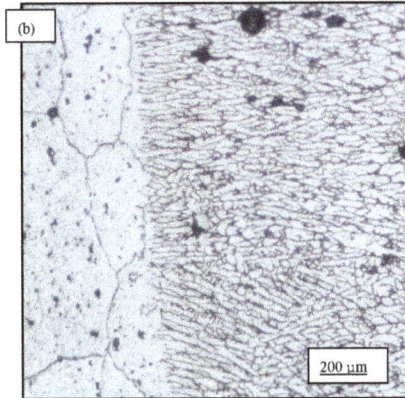

Figure 5. Optical micrographs showing the following: (**a**) Microstructure at the weld-base metal interface of the as-welded Aluminum alloy 6061-T4; (**b**) microstructure of the weld-base metal interface in the post weld heat treated aluminum alloy 6061.

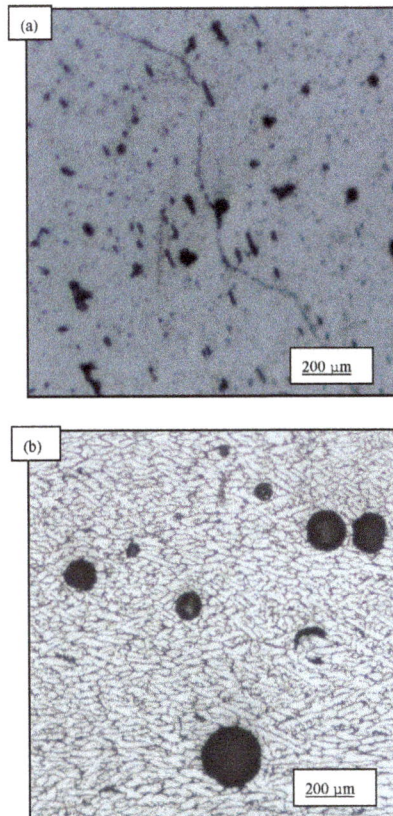

Figure 6. Optical micrographs of AA6061 showing the following: (**a**) Distribution of intermetallic particles in the base metal adjacent to the weld bead; and (**b**) microstructure of the weld pool of the heat-treated alloy.

3.2. Typical Stress-Strain Response

In Figures 7–11 the typical stress *versus* strain behaviors for the aluminum alloys and test conditions employed used this study are shown. Each plot shows the stress *versus* strain variation for either the as-received (AR) and as-received and aged (AR+PHT) or the as-welded (AW) and the Post Weld Heat Treated (PWHT) counterpart. Figure 11, shows the stress *versus* strain response for the re-solution heat treated and aged 6005A (SHT+PHT or SHT+PWHT).

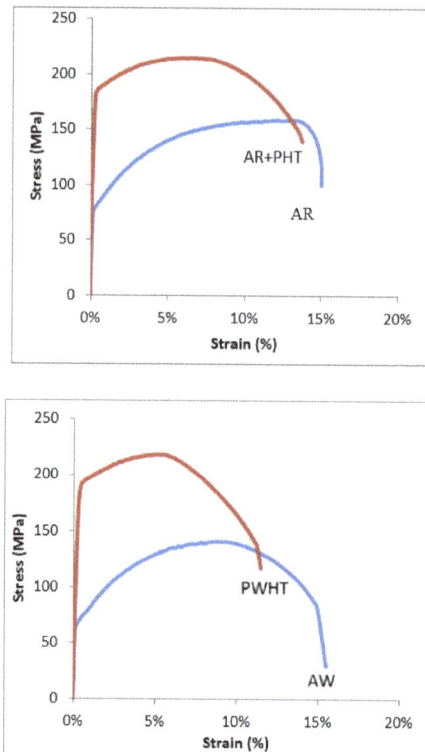

Figure 7. Stress *versus* strain response of 6.4 mm (1/4 in) thick AA6063 in the as-received (AR), as-received plus precipitation heat treated (AR+PHT), as-welded (AW) and post weld heat treated (PWHT) conditions.

In general, artificial aging increased the strength of aluminum alloy 6063 for both the as-received (AR) and as-welded (AW) conditions. The observed increase in strength was far more pronounced for the aluminum alloy material that was 6.4 mm (1/4 in) thick. This is not unexpected, primarily because a thicker material necessitates the need for additional heat input during welding.

As in the case of AA6063, all tensile specimens of AA6061 responded positively to heat treatment, showing an observable gain in strength. The increase in strength was evident even for the 9.5 mm (3/32 in) thick specimens that were initially welded and subsequently aged. However, the ultimate tensile strength obtained for the thicker specimen (t = 9.5 mm) was found to be lower than the test specimens that measured 6.4 mm (1/4 in) in thickness.

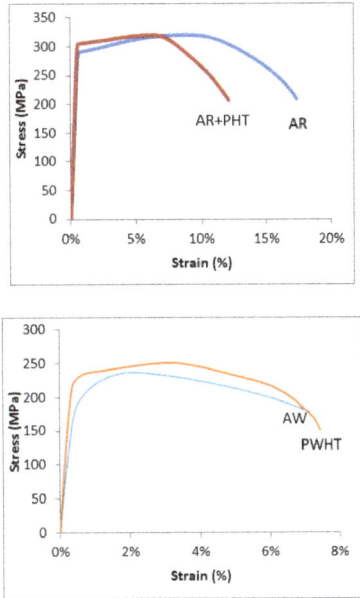

Figure 8. Stress *versus* Strain response of 9.5 mm (3/8 in) thick AA6063 in the as received (AR), as-received plus precipitation heat treated (AR+PHT), as-welded (AW) and post weld heat treated (PWHT) conditions.

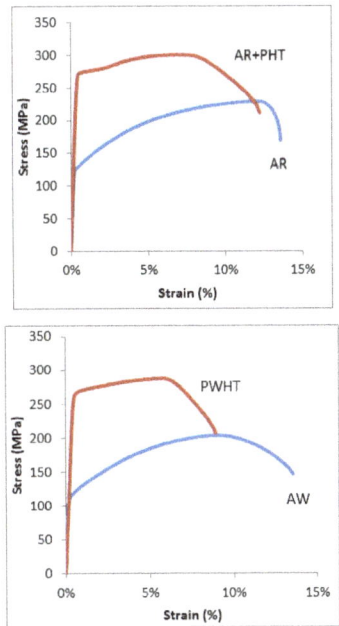

Figure 9. Stress *versus* strain response of 6.4 mm (1/4 in) thick extrusion of AA6061 in the as-received (AR), as-received plus precipitation heat treated (AR+PHT), as-welded (AW), and post weld heat treated conditions.

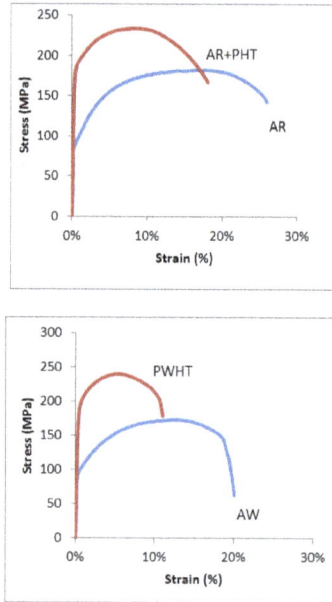

Figure 10. Stress-strain response of 9.5 mm (3/8 in) thick extrusion of AA6061 in the as-received (AR), as-received plus precipitation heat treated (AR+PHT), as-welded (AW) and post weld heat treated (PWHT) conditions.

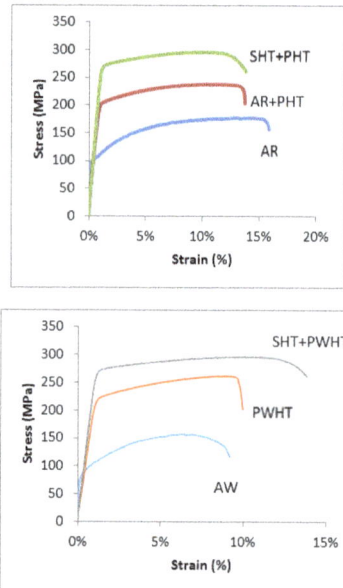

Figure 11. Stress *versus* strain response of 3.2 mm (1/8 in) AA6005A in the as-received (AR), as-received plus precipitation heat treated (AR+PHT), solution heat trearted plus precipitation heat treated (SHT+PHT), as-welded (AW), post weld heat treared (PWHT) and solution heat treated plus post weld heat treated (SHT+PWHT) conditions.

Both as-received (AR) and as-welded (AW) specimens of AA6005A, having a thickness of 3.2 mm (1/8 in), responded favorably to heat treatment, with a significant increase in both yield strength and tensile strength. Test specimens of aluminum alloy 6005A, in both the as-received (AR) and as-welded (AW) conditions, that were solution heat treated and subsequently aged showed the largest gain in strength when deformed in uniaxial tension.

3.3. Analysis of the Results

Welding did have an influence on both microstructure and mechanical properties of the chosen 6XXX series aluminum alloys. A similar influence of welding, *i.e.*, hybrid welding, was observed to have a noticeable influence on weld quality, microstructure and mechanical properties of aluminum alloy 5754 [13] and an experimental Al–Mg alloy [21]. The influence of welding differs depending on the following: (a) the alloy chosen; (b) the welding process used; (c) the parameters employed; and (d) overall quality of the weld. The type of joint and thickness of the starting material does have an influence on heat input, microstructure and resultant strength. Not surprisingly, this study of 6063, 6061 and 6005A aluminum alloys revealed that selective artificial aging or heat treatment increased the mechanical strength of the alloys. The observed increase in strength for the PWHT samples can be attributed to the existence of diffusion-assisted mechanisms that favor an initial increase in Guinier Preston (GP) zones coupled with a hinderance caused to the movement of dislocations as a consequence of the formation and presence of matrix strengthening precipitates. The precipitates in the PWHT alloy are finer and more uniformly distributed in the aluminum alloy metal matrix. This favors an increase in dislocation density, which contributes to the observed improvement in both yield strength and tensile strength.

When subjected to "localized" heat input as a direct consequence of welding, the chosen aluminum alloys experienced a decrease in strength when compared to strength of the as-received condition. The decrease in strength can be essentially attributed to changes in intrinsic microstructural features of the starting material as a consequence of the heat input during welding. A majority of the as-welded (AW) samples broke in an area adjacent to the weld; normally on the side of the weld to which more heat was applied. The heat-affected zone (HAZ) was observed to have lower strength when compared to the base metal.

A statistical analysis, using the guidelines established in the 2010 edition of the Aluminum Design Manual, coupled with the published guaranteed minimum strengths for AA6061, AA6063 and AA6005A, was used to determine reasonable design minimum strength for the samples that were subject to post weld heat treatment. The minimum tensile strength and yield strength for the PWHT tensile samples was established using the following equation:

$$\sigma_{min} = \sigma_{avg} - kS_\sigma \tag{1}$$

where

σ_{min} = calculated minimum stress for the PWHT specimens

σ_{avg} = average tensile or yield strength for a given alloy in the PWHT condition

k = statistical coefficient based on the number of tests, n

S_σ = standard deviation of the test results for the particular alloy

Results of the analysis of the strength of the PWHT specimens are summarized in Table 2. Detailed in Table 2 are: (a) the alloys studied; (b) thicknesses; (c) average yield strength and average ultimate tensile strength; (d) the number of tests in the data set; (e) the standard deviation; (f) the calculated minimum yield strength and ultimate tensile strength; (g) the ratio of the calculated minimum yield strength to the guaranteed minimum yield strength for the base alloy; as well as (h) ratio of the calculated minimum ultimate strength to the guaranteed minimum strength for the base alloy. Ratio of the calculated minimum yield strength to guaranteed minimum yield strength varied from 0.5 to a high of 0.99. Ratio of the calculated minimum ultimate strength to guaranteed strength varied from 0.66 to a high of 0.98. The lower values (0.5 and 0.68) correspond to 6061 having a thickness of 9.5 mm (3/8 in) that was post weld heat treated and tested. In order to obtain test specimens of aluminum alloy 6061 that were 9.5 mm (3/8 in) thick, blanks were removed from an extrusion that had a initial thickness of 12.7 mm (1/2 in), and subsequently milled to the required thickness of 9.5 mm (3/8 in). While this may have had some influence on the test results, Table 2 shows that for both AA6063 and AA6061, the data for the 9.5 mm (3/8 in) thick specimens had noticerably larger standard deviations for both yield strength and ultimate tensile strength when compared to the specimens that measured 6.4 mm (1/4 in) in thickness. This is attributed to the increased heat input used during the welding of the thicker materials. Presented and discussed in detail elsewhere [18], an examination of the microstructure of the area immediately around the fillet revealed a few instances of visible secondary melting (Figure 12).

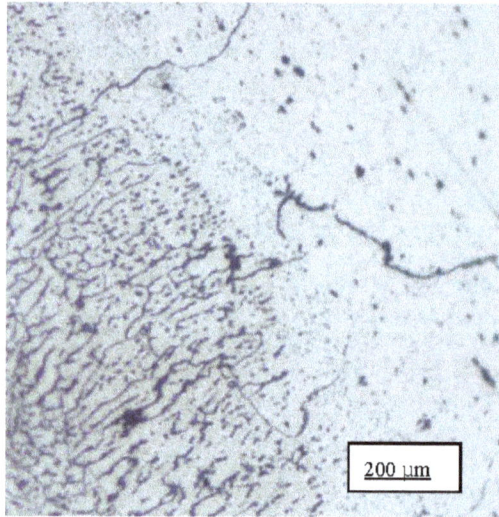

Figure 12. Optical micrograph showing region experiencing secondary melting in the weld-base metal fusion line in aluminum alloy 6061.

For the remaining alloys and thicknesses chosen, the provisions within the 2010 Aluminum Design Manual allowing the use of 85%. of the parent metal strength for PWHT light poles are confirmed. In addition, using 85%. of the base metal strength for extruded aluminum alloy AA6061, for thicknesses up to 6.4 mm (1/4 in) and welded in the T4 temper using AA4043 as filler followed by post weld heat treatment (PWHT), also conforms with the 85% rule. In summary the alloys and thicknesses include the following:

1.	6063-T4, PWHT up to 9.5 mm (3/8 in) thick and welded using AA4043.
2.	6005A-T1, PWHT up to 6.4 mm (1/4 in)thick and welded using AA4043.
3.	6061-T4, PWHT up to 6.4 mm (1/4 in) thick, and welded using AA4043.

Table 2. Data analysis summary—post weld heat treated (PWHT) yield and ultimate strengths.

Alloy	Thick (mm)	Average Yield (MPa)	Average Ultimate (MPa)	Number of Tests	Standard Deviation Yield/Ultimate	Min. Yield (MPa)	Min. Ultimate (MPa)	Min. Yield/ADM	Min. Ult./ADM
6063	6.4	190	219	10	0.92/0.95	165	193	0.96	0.93
6063	9.5	204	234	21	2.44/1.57	150	199	0.87	0.96
6005A	3.2	227	272	12	1.0/1.05	201	245	0.83	0.93
6005A	6.4	217	250	12	0.44/0.82	205	229	0.85	0.87
6061	6.4	261	281	15	0.93/0.98	239	258	0.99	0.98
6061	9.5	241	241	24	4.1/3.1	122	173	0.5	0.66

Note: Minimum Yield/ADM = calculated minimum yield strength from the testing program divided by the guaranteed minimum yield strength for the alloy. Minimum Ultimate/ADM = calculated minimum ultimate strength from the testing program divided by the guaranteed minimum ultimate strength for the alloy.

4. Conclusions

This study examined the influence of post weld heat treatment (PWHT) on strength of three aluminum alloys that are commonly chosen for use in welded light poles. Findings of the study are as follows:

1. Heat treating (aging) the as-received (AR) material increased both the yield strength and ultimate tensile strength of all the alloys.
2. Post weld heat treating increased both the yield strength and ultimate tensile strength of the three alloys studied.
3. Re-solution heat treating the as-received material increased the yield strength and tensile strength of aluminum alloy 6005A.
4. Re-solution heat treating subsequent to welding, followed by post weld heat treatment was observed to increase the tensile strength and yield strength of aluminum alloy 6005A.
5. With the exception of aluminum alloy 6061 having a thickness of 9.5 mm (3/8 in), design provisions permitting use of 85% of the parent metal strengths (in T6 temper) for post weld heat treated (PWHT) light poles are confirmed. The alloys and thicknesses include: (i) 6063-T4 PWHT up to 9.5 mm thick; (ii) 6005A-T1 PWHT up to 6.4 mm thick; and (iii) 6061-T4 PWHT up to 6.4 mm thick.

Acknowledgments: This research was made possible through funds provided by the University of Akron for the graduate student and with material and technical support from Hapco Inc (Abingdon, VA, USA).

Author Contributions: Eric Hilty performed the specimen fabrication and conducted the tests. Menzemer conceived and developed the program as well as advised Hilty. He also performed the final data analysis and assisted with manuscript preparation. Srivatsan conducted examination of the material microstructures, examined the fracture surfaces and assisted with the manuscript preparation. Morrison and Minor provided the material and guidance to the study and assisted with manuscript preparation.

Conflicts of Interest: The authors declare no conflict of interest.

References

1. Sharp, M.L. *Behavior and Design of Aluminum Structures*; McGraw-Hill: New York, NY, USA, 1992.
2. Kissell, R.; Ferry, R. *Aluminum Structures—A Guide to Their Specifications and Design*; John Wiley and Sons: London, UK; Washington, DC, USA, 1995.
3. Elangovan, K.; Balasubramanian, V. Influence of post weld heat treatment on tensile properties of friction stir welded AA6061 aluminum alloy joints. *Mater. Charact.* **2008**, *59*, 1168–1177. [CrossRef]
4. Demir, H.; Gunduiz, S. The effects of aging on the machinability of 6061 aluminum alloy. *J. Mater. Des.* **2009**, *30*, 1480–1483. [CrossRef]
5. Xu, W.; Gittos, M.F. *Materials and Structural Behavior of MIG Butt Welded in 6005-T6 Aluminum Alloy Extrusions under Quasi Static and Impact Loading*; TWI Report Number 14054/1/04; The Welding Institute: Cambridge, UK, 2004.
6. Groover, M.P. *Fundamentals of Modern Manufacturing, Materials, Processes and Systems*, 3rd ed.; John Wiley and Sons: Hoboken, NJ, USA, 2007.
7. Hatch, J.E. *Aluminum: Properties and Physical Metallurgy*, 10th ed.; American Society for Metals: Metals Park, OH, USA, 1983.
8. The Aluminum Association. *Aluminum Design Manual*; Aluminum Association of America: Washington, DC, USA, 2010.
9. Maedler, J.R. *NEMA to Gordon Allison*; The Aluminum Association: Washington, DC, USA, 12 February 1971.
10. Wagoner, N. *Reynolds Metals Company to R*; Hartmann, Hapco: Abingdon, VA, USA, 5 August 1963.
11. Hilty, E.; Menzemer, C.; Srivatsan, T. Influence of Welding and Heat Treatment on Microstructural Development and Properties of Aluminum Alloy 6005. In Proceedings of the 22nd PFAM Conference, Singapore, 18–20 December 2013.

12. Hilty, E.; Menzemer, C.; Manigandan, K.; Srivatsan, T. Influence of welding and heat treatment on microstructure, properties and fracture behavior of a wrought aluminum alloy. *Emerg. Mater. Res.* **2014**, *3*, 230–242. [CrossRef]

13. Leo, P.; D'Ostuni, S.; Casalino, G. Hybrid welding of AA5754 annealed alloy: Role of post weld heat treatment on microstructure and mechanical properties. *Mater. Des.* **2016**, *90*, 777–786. [CrossRef]

14. The Aluminum Association. *2009 Aluminum Standards and Data*; Aluminum Association of America: Washington, DC, USA, 2009.

15. Hobart Filler Metals. *Guide for Aluminum Welding*; Illinois Tool Works: Chicago, IL, USA, 2013.

16. ASTM B918-01. *Standard Practice for Heat Treatment of Wrought Aluminum Alloys*; ASTM International: West Conshohocken, PA, USA, 2001. [CrossRef]

17. *ASM Materials Handbook*; ASM International: Materials Park, OH, USA, 1991; Volumes 4 and 6.

18. Warmuzek, M.; Mrowkaand, G.; Sieniawski, J. Influence of Heat Treatment on Precipitation of Intermetallic Phases in Commercial AlMnFeSi Alloy. *J. Mater. Process. Technol.* **2004**, *157–158*, 624–632. [CrossRef]

19. Gallais, C.; Simar, A.; Fabreque, D.; Denquin, A.; Lapasset, G.; de Meester, B.; Brechet, Y.; Pardoen, T. Multiscale Analysis of the Strength and Ductility of AA 6056 Aluminum Friction Stir Welds. *Metall. Mater. Trans.* **2007**, *38*, 964–981. [CrossRef]

20. Starke, E.A., Jr. *Fatigue and Microstructure*; Meshii, M., Ed.; ASM International: Materials Park, OH, USA, 1979.

21. Leo, P.; Renna, G.; Casalino, G.; Olabi, A.G. Effect of power distribution on weld quality during hybrid laser welding of an Al–Mg alloy. *Opt. Laser Technol.* **2015**, *73*, 118–126. [CrossRef]

Article

The Effect of Creep Aging on the Fatigue Fracture Behavior of 2524 Aluminum Alloy

Wenke Li [1], Lihua Zhan [1,2,*], Lingfeng Liu [1] and Yongqian Xu [1]

[1] Light Metal Research Institute, Central South University, Changsha 410083, China;
     lwk1992@csu.edu.cn (W.L.); goodlucky321@yeah.net (L.L.); 143812027@csu.edu.cn (Y.X.)
[2] National Key Laboratory of High Performance Complex Manufacturing, Central South University,
     Changsha 410083, China
*   Correspondence: yjs-cast@csu.edu.cn; Tel.: +86-731-88830254

Academic Editor: Nong Gao
Received: 26 July 2016; Accepted: 31 August 2016; Published: 7 September 2016

Abstract: Normal temperature tensile and fatigue tests were adopted to test the mechanical performance and fatigue life of 2524 aluminum alloy under the three states of T3, artificial aging, and creep aging, and scanning electron microscope and transmission electron microscope were also used to observe the fatigue fracture morphology and aging precipitation features of the alloy under the above three states. Results showed that the alloy treated by creep aging can obtain higher fatigue life, but that treated by artificial aging is lower than T3; T3 alloy is mainly dominated by GPB region. Meanwhile, the crystal boundary displays continuously distributed fine precipitated phases; after artificial aging and creep aging treatment, a large amount of needle-shaped S′ phases precipitate inside the alloy, while there are wide precipitated phases at the crystal boundary. Wide precipitation free zones appear at the crystal boundary of artificial-aging samples, but precipitation free zones at the alloy crystal boundary of creep aging become narrower and even disappear. It can be seen that creep aging can change the precipitation features of the alloy and improve its fatigue life.

Keywords: 2524 aluminum alloy; creep aging; fatigue fracture behavior; aging precipitation

1. Introduction

Creep aging forming, as a processing method that utilizes the creep deformation of metal and the aging features enhancement of aluminum alloy, has been mainly applied to manufacturing plane wallboard and other whole-piece wallboard members [1,2]. Compared with conventional plastic forming, creep aging enjoys high forming accuracy and repeatability, hence reducing the risk of materials cracking in processing and residual stress of components [3,4]. Therefore, scholars have conducted in-depth research on the resilience prediction and performance of creep aging forming. Zhan et al. [5] built a constitutive model that could simulate the change rules of the strain changes of creep aging, precipitated phase changes, dislocation strengthening, solution strengthening, aging strengthening, and material property in the process of forming by combining uni-directional tensile tests. Hargarter et al. [6] studied the influence of a precipitated phase position under stress upon the yield strength of Al-Cu-Mg-Ag and Al-Cu alloys, discovering that the yield strength of materials of stress aging was lower than that of non-stress under the same heat treatment condition. Li et al. [7] studied the influence of different aging forming parameters (aging time, temperature) upon the organizational property of 2124 aluminum alloy and pointed out that the presence of stress accelerated the precipitation and transition of the strengthening phases of 2124 aluminum alloy. Chen et al. [8] explored the creep aging behaviors of 7050 aluminum alloy under a solution hardening state and pre-treatment state and discovered that, after pre-deformation, creep aging deformation of the material became larger. Higher temperature and stress led to greater creep aging deformation.

However, the service environment of modern aviation equipment has been expanding rapidly. Thus, apart from the requirements for material strength and toughness, sound service performance is also demanded, especially fatigue life. Therefore, there has been much research on the fatigue performance of aviation materials. Zabett et al. [9] studied the influence of micro organization upon production and expansion behaviors of the cracks of 2024-T351 in three directions and found out that cracks mainly appeared at a secondary phase fracture of Al_7Cu_2Fe and crack production would reduce fatigue life of the alloy. Carte et al. [10] explored the fatigue crack expansion performance of 7475 aluminum alloy and concluded that, with the increase of crystal particle size, the fatigue crack expansion speed dropped. Bray et al. [11] studied the influence of aging treatment upon the fatigue crack expansion speed of 2024 aluminum alloy and discovered that high-density solution clusters at the beginning of aging improved alloy fatigue performance.

To date, there has been much research on creep aging forming and aviation materials performance test, but few have involved materials performance after creep aging, especially service performance. 2524-T3 aluminum alloy, as a new type of highly strong Al-Cu-Mg aluminum alloy developed after 2024 and 2124 aluminum alloys [12], has been applied to Boeing 777 and Airbus A380 [13]. Therefore, this paper regards 2524 aluminum alloy as a research object and compares its micro structure, mechanical property, and fatigue performance after artificial aging and creep aging, so as to provide an experimental foundation for a creep aging forming technique of 2524 aluminum alloy.

2. Materials and Methods

The material used in this experiment was provided by Southwest Aluminum (Group) Co., Ltd., (Chongqing, China), namely 2.5-mm 2524-T3 aluminum alloy, whose chemical components (wt. %) include 4.26Cu-1.36Mg-0.57Mn-0.037Fe-0.024Zn-0.01Ti-0.002Cr-0.089Si-(bal.)Al. The heat treatment state of T3 refers to cold processing after solution treatment followed by natural aging to a basically stable state. First, hardness samples were taken for artificial aging under 180 °C and a hardness test under a Huayin microscopic hardness meter (Huayin Testing Instrument Co., Ltd., Yantai, China), so as to obtain an aging hardening curve; by analyzing the curve, it was obtained that the time for reaching peak aging was 12 h. Then, conventional artificial aging treatment and creep aging treatment were conducted on a plate, for which a RWS50 electron creep slackness tester (Changchun Research Institute for Machanical Science Co., Ltd., Changchun, China) was adopted for creep aging (CA: 180 °C × 180 MPa × 12 h), while airing dryer machine was used for artificial aging (AA: 180 °C × 12 h).

A room temperature tensile performance test and fatigue life test were conducted on the alloys at the three states, namely, the 2524-T3 state, after artificial aging, and after creep aging. Both of the tests were conducted on the MTS810 tester machine (MTS Systems Corporation, Eden Prairie, MN, USA). There were three horizontal tests of mechanical performance, with a tensile speed of 2 mm/min. Sine wave loading was adopted for the fatigue life test samples, with a frequency of 10 Hz and stress ratio of 0.1 (30/300 MPa). For each state, five horizontal samples were taken.

Fatigue sample fracture was captured to be observed under a TESCAN scanning electron microscope (Tescan company, Brno, Czech), so as to study the production and expansion of cracks. Micro organizations were observed under TecnaiG220 transmission electron microscope (United States FEI limited liability company, Hillsboro, OR, USA). Samples were thinned to 0.08 mm first and then thinned again on a TenuPol-5 electrolyzation double spraying thinner machine (Struers, Copenhagen, Denmark). The electrolyzation solution was a mixed liquid of 30% nitric acid and 70% carbinol, with the double spraying temperature being −35 °C to −25 °C, and the voltage being 15 V.

3. Results and Discussion

3.1. Mechanical Properties

Table 1 lists the normal temperature tensile performance data of 2524 aluminum alloy under the three treatment states of T3, AA, and CA, and Figure 1 presents the fatigue life data of alloy under the three states.

Table 1. Static tensile properties of examined alloy.

Sample	Tensile Strength/MPa	Yield Strength/MPa	Elongation/%
T3	477.65	339.72	18.53
AA	490.22	439.39	8.25
CA	503.17	462.17	8.56

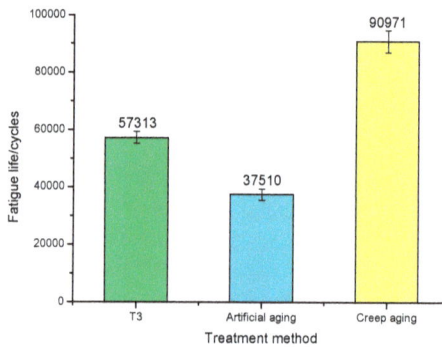

Figure 1. Fatigue lives of samples.

From Table 1, it can be seen that the strength of the alloys treated by artificial aging and creep aging enjoys a significant improvement over that of the alloy of the T3 state, with the tensile strength increasing by 5.5% and the yield strength increasing by 36%. Moreover, the strength of the alloy treated by creep aging is higher than that of the alloy treated by artificial aging, while the elongation of the alloys treated by artificial aging and creep aging is lower than that of the alloy of the T3 state.

From Figure 1, it can be seen that there are obvious differences among the three sample groups in terms of fatigue life. Fatigue life of the samples after creep aging treatment improves to the cycle number of 90971, almost 2 times more than the T3 alloy; and the fatigue life of the samples after artificial aging treatment decreased by the cycle number of about 20,000 when compared with T3 alloy.

3.2. Fatigue Fracture Analysis

Figure 2 presents the overall morphology of the fatigue fractures of the aluminum alloy at different states. It can be seen that all their macro fractures are similar, which can be divided into three areas, namely, fatigue source area, fatigue crack expansion area, and instant fracture area. Fatigue crack source area is located at the corner in a semi-circle; the following is a fatigue crack expansion area in the pattern of beach waves, where different interval arches and radial lines with the source area as the center can be observed; the morphology of the instant fracture area is similar to the regular tensile crack. Through comparing Figure 2a–c, it can be seen that the fatigue source area and the fatigue crack expansion area of the CA samples are the largest, followed by those of the T3 state. Those of the AA samples are the smallest. Given that the process of fatigue is a process of crack production and expansion, it can be concluded that the fatigue life of the CA samples is the best, followed by the T3 state and the AA samples successively, which is consistent with the test results of fatigue life.

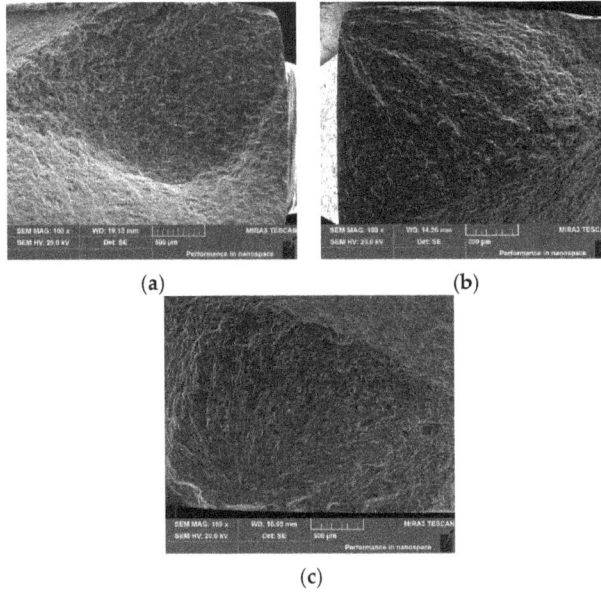

(a)   (b)

(c)

Figure 2. The overall morphology of the fatigue fractures: (**a**) T3; (**b**) AA; (**c**) CA.

Figure 3 presents the SEM images after enlarging fatigue source area of three kinds of samples. It can be seen that the three kinds samples are very similar, revealing cleavage-like feather-shaped morphology. Meanwhile, in some places, there are pits left by falling of rough second-phase particles where cracks concentrate.

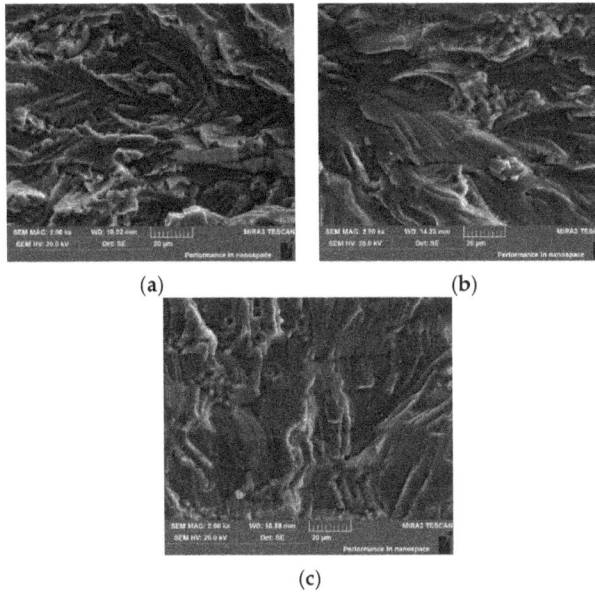

(a)   (b)

(c)

Figure 3. Fatigue source area: (**a**) T3; (**b**) AA; (**c**) CA.

Figure 4 presents the SEM images after enlarging the fatigue crack expansion area of three kinds of samples. It can be seen that they all display obvious fatigue striation, which are micro traces left by crack expansion. Each belt can be regarded as an expansion trace of one instance of stress circle [14]. By comparing the distance among three sets of fatigue striation, it can be seen that, within the distance of 2 μm, there are about 11 fatigue striation on the T3 samples, 8 on the AA samples, and 15 on the CA samples. Besides, micro cracks can also be found on the fatigue striation of the AA samples, which shows that CA samples enjoy higher fatigue life followed by T3 samples, and AA samples have the lowest fatigue life.

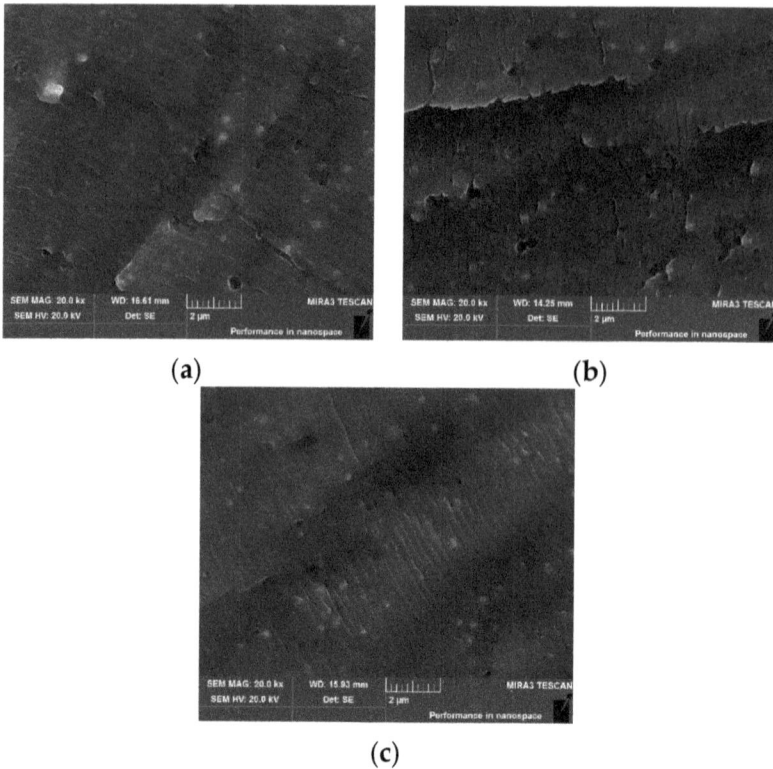

(a)

(b)

(c)

Figure 4. Fatigue crack expansion area: (**a**) T3; (**b**) AA; (**c**) CA.

Figure 5 presents the SEM images after enlarging the instant fracture area of the three kinds of samples. The three kinds of samples are similar, all showing morphology similar to the fracture at the normal-temperature static tensile test, and most of the parts on the fracture surface are dimples.

(a)   (b)

(c)

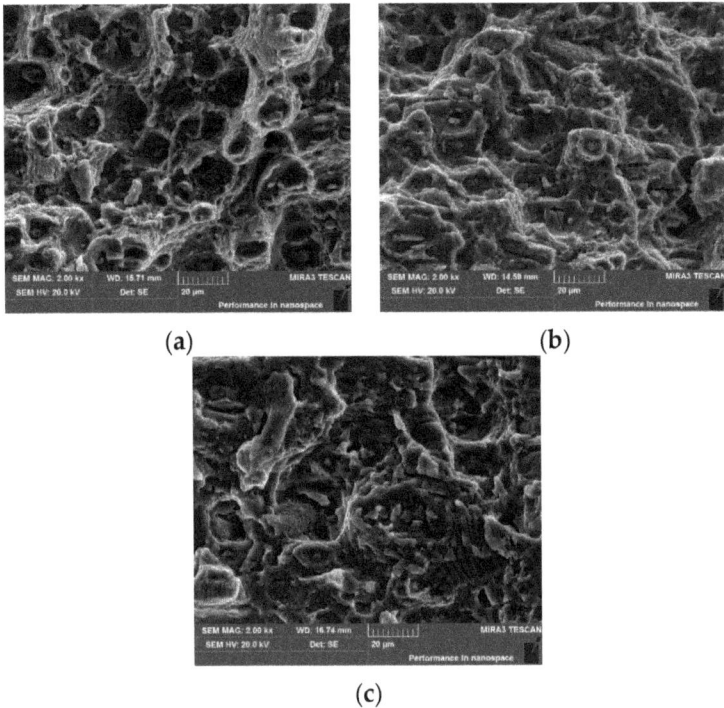

Figure 5. Instant fracture area: (**a**) T3; (**b**) AA; (**c**) CA.

3.3. Microstructure

Figure 6 presents the TEM micro structure of the 2524 aluminum alloy under the three states of T3, artificial aging, and creep aging. It can be discovered that there is no precipitated phase in the 2524-T3 aluminum alloy, showing that the enhanced phase mainly concentrates on the GPB area. Meanwhile, there are many baton-shaped T phases in the alloy crystal, results of the incomplete dissolution of solid solution [15]. There are needle-shaped S phases and S precipitated phases in the CA and AA samples, and these precipitations are not large, which can pin the dislocations. Therefore, the strength of the AA and CA samples is higher than that of the T3 samples. By comparing CA and AA samples, it can be seen that the precipitated phases in the CA samples slightly outnumber those in the AA samples, mainly because the CA samples are treated by creep aging, in which the presence of stress triggers partial plastic deformation in the substrate of the alloy, hence increasing the dislocation density, offering more space for second-phase core and facilitating the precipitation of a transgranular second phase. Figure 6 presents the TEM images of the crystal boundary of the samples in the three states. Phases precipitated at the crystal boundary of the T3 samples are continuously distributed, where even some T phases occur. However, there are already changes to precipitated phases at the boundary of the AA and CA samples, because the precipitated phase precipitation, combination, and aggregation at the crystal boundary of samples lead to the discontinuous distribution of the precipitated phases at the crystal boundary. Meanwhile, there is also a precipitation-free zone at a width of about 60 nm at the crystal boundary of the CA samples, but this cannot be found in the AA samples. The reason lies in the fact that, without stress, there is a huge difference in energy between the crystal boundary and inside the crystal boundary. After the introduction of stress, the energy difference drops, which leads to even precipitation momentum inside the boundary and at the crystal boundary; hence, no precipitation-free zone at the crystal boundary of the CA samples is narrow or even absent.

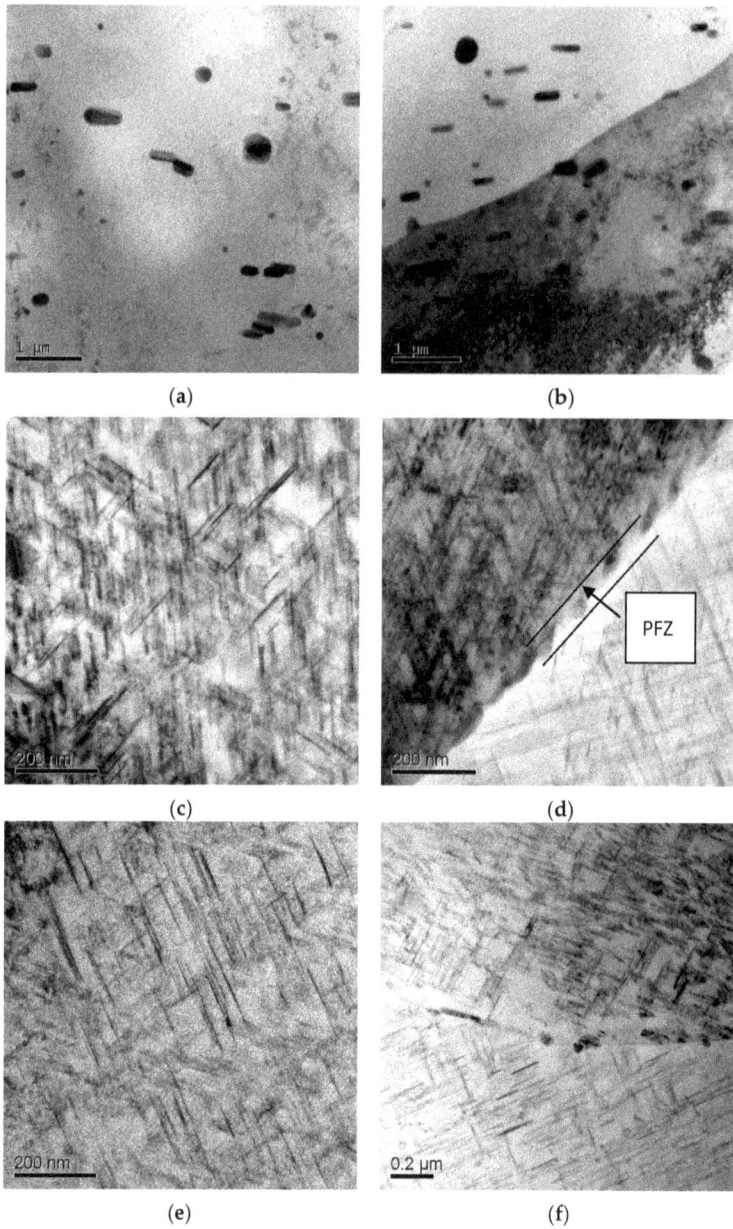

Figure 6. Aging precipitation characteristics of samples: (**a,b**) T3; (**c,d**) AA; (**e,f**) CA.

3.4. Discussion

The fatigue process of an alloy is the process of crack production and expansion, so fatigue life is the sum of the fatigue cycle number of crack production, crack expansion, and instant fracture. In this research, all the initial states of the alloys are T3, so the major difference among them lies in the aging stage after solid solution. Due to a low aging temperature, no great impact will be exerted upon crystal particle structure or the rough second phase of the alloy, so the major influence of alloy fatigue performance lies in the aging precipitation features under different states.

For the Al-Cu-Mg alloy, its desolvation phase in the aging process is mainly θ phase and S phase, which turn out to be a competitive precipitation process. From the Al-Cu-Mg ternary phase diagram and relevant literature [16], when the mass fraction ratio between Cu and Mg reaches 1.5–4, S phase will be the main enhanced phase of the alloy. The Cu/Mg ratio of the 2524 aluminum alloy is about 3.13, so the precipitation sequence at 180 °C is GPB–S″–S′–S [17].

The influence of aging precipitation features upon alloy fatigue performance is mainly displayed in three aspects. Firstly, the presence of inside crystal precipitated phase can increase the strength of the alloy and enhance its anti-plastic deformation ability, so as to reduce its deformation damage. Secondly, the precipitated phase is semi-coherent or non-coherent with the base body, which leads to the dislocation in the fatigue process when going through the precipitated phase, leaving behind dislocation rings and thus increasing the resistance of reverse slipping of dislocation when loading decreases and disabling dislocation to slip repeatedly within the crystal. Meanwhile, the presence of precipitation-free zone (PFZ) decreases the strength of the crystal boundary, enabling the dislocation to slip at the boundary and increasing the stress concentration upon it. Therefore, the alloy is more likely to crack along the crystal boundary.

The GPB in the crystal of the T3 alloy is coherent with aluminum substrate, and it only has a common effect upon the enhancement of the base body, so its ability of anti-plastic deformation is not remarkable. However, in fatigue deformation, slipping dislocation can cut these GPB areas, causing minor fatigue damages. Meanwhile, there are no rough precipitated phases or PFZ at the crystal boundary of the T3 alloy, so the strength between the crystal boundary and inside the crystal is the same, without causing dislocation to slip at the boundary. There are many S′ phases inside crystal of the AA samples that fixed the dislocation, increasing its strength and the anti-plastic deformation ability. However, these precipitated phases are semi-coherent or non-coherent with the base body, the dislocation in the fatigue process going through the precipitated phase and leaving behind dislocation rings, so the deformation damage is big. There is about 60-nm PFZ at the crystal boundary of the AA samples and the precipitated phases are rough, decreasing the combination degree between the crystal particles and ultimately inviting fatigue crack. The transgranular precipitated phases of CA samples outnumber those of the AA samples, so the mechanical performance of CA samples is higher than that of the AA samples; and the anti-plastic deformation ability of the CA samples is increased as well. Meanwhile, due to the narrowness or absence of PFZ at the crystal boundary for the existence of stress, its morphology of the CA samples is similar to T3 samples, and the strength between the crystal boundary and inside the crystal is the same, so dislocation will not slip at the boundary.

Based on the above, the CA samples enjoy the highest fatigue life followed by the T3 state, and the AA samples are the lowest.

4. Conclusions

(1) Mechanical performance of samples after artificial aging and creep aging has enjoyed significant improvement over that of the T3 alloy. Fatigue life of the alloy through creep aging has also enjoyed improvement over that of the T3 alloy, and the fatigue life of samples after artificial aging treatment drops.

(2) The fatigue fracture morphology of samples in three states is divided into the fatigue source area, the fatigue crack expansion area, and the instant fracture area. Among them, the morphological difference of the fatigue crack expansion area is the largest, fatigue striation of the creep aging alloy is relatively narrow, and there are micro cracks appearing at the fatigue striation of the artificial aging alloy.

(3) All samples treated by artificial aging and creep aging precipitate S′ phases, about 60-nm PFZ appears at the crystal boundary of the samples treated by artificial aging, and the PFZ of the samples treated by creep aging is narrow or even absent.

(4) Transgranular phase precipitation can improve the anti-plastic deformation ability of materials, hence raising the fatigue life of alloy. The presence of rough second phases and PFZ at the crystal boundary is likely to make the alloy crack at the boundary, lowering its fatigue life.

Acknowledgments: This work was supported by the Fundamental Research Funds for the Central Universities of Central South University (No. 2016zzts318), the Key Program of the National Science Foundation of China (No. 51235010) and the National Key Basic Research Development Pan Funded Project of China (No. 2014cb046602).

Author Contributions: Wenke Li and Lihua Zhan conceived and designed the experiment; Lingfeng Liu and Wenke Li performed the experiments; Wenke Li and Yongqian Xu analyzed the data; Lingfeng Liu contributed reagents, materials and analysis tools; Wenke Li wrote the paper.

Conflicts of Interest: The authors declare no conflicts of interest.

References

1. Zeng, Y.; Huang, X. Forming Technologies of Large Integral Panel. *Acta Aeronaut. ET Astronaut. Sin.* **2008**, *3*, 721–727.
2. Zhan, L.; Lin, J.; Dean, T.A. A review of the development of creep age forming: Experimentation, modelling and applications. *Int. J. Mach. Tools Manuf.* **2011**, *51*, 1–17. [CrossRef]
3. Holman, M.C. Autoclave age forming large aluminum aircraft panels. *J. Mech. Work. Technol.* **1989**, *20*, 477–488. [CrossRef]
4. Levers, A.; Prior, A. Finite element analysis of shot peening. *J. Mater. Process. Technol.* **1998**, *80–81*, 304–308. [CrossRef]
5. Zhan, L.; Lin, J.; Dean, T.A.; Huang, M. Experimental studies and constitutive modelling of the hardening of aluminium alloy 7055 under creep age forming conditions. *Int. J. Mech. Sci.* **2011**, *53*, 595–605. [CrossRef]
6. Hargarter, H.; Lyttle, M.T.; Starke, E.A. Effects of preferentially aligned precipitates on plastic anisotropy in Al-Cu-Mg-Ag and Al-Cu alloys. *Mater. Sci. Eng. A* **1998**, *257*, 87–99. [CrossRef]
7. Zhan, L.; Li, Y.; Huang, M. Microstructures and properties of 2124 alloy creep ageing under stress. *J. Cent. South Univ. Sci. Technol.* **2012**, *3*, 926–931.
8. Chen, Y.; Deng, Y.; Wan, L.; Jin, K.; Xiao, Z. Microstructure and Properties of 7050 Aluminum Alloy Sheet During creep aging. *J. Mater. Eng.* **2012**, *1*, 71–76.
9. Zabett, A.; Plumtree, A. Microstructural effects on the small fatigue crack behavior of an aluminum alloy plate. *Fatigue Fract. Eng. Mater. Struct.* **1995**, *18*, 801–809. [CrossRef]
10. Carter, R.D.; Lee, E.W.; Starke, E.A.; Beevers, C.J. The effect of microstructure and environment on fatigue crack closure of 7475 aluminum alloy. *Metall. Mater. Trans. A* **1984**, *15*, 555–563. [CrossRef]
11. Bray, G.H.; Glazov, M.; Rioj, R.J.; Lib, D.; Gangloffb, R.P. Effect of artificial aging on the fatigue crack propagation resistance of 2000 series aluminum alloys. *Int. J. Fatigue* **2001**, *23*, 265–276. [CrossRef]
12. Williams, J.C.; Starke, E.A. Jr. Progress in structural materials for aerospace systems. *Acta Mater.* **2003**, *51*, 5775–5799. [CrossRef]
13. Chen, W. Application of advanced aluminum alloys in A380 structures. *Aviat. Maint. Eng.* **2005**, *2*, 40–41.
14. Shu, D. *Mechanical Properties of Engineering Materials*, 1st ed.; Machinery Industry Press: Beijing, China, 2007.
15. Cheng, S.; Zhao, Y.H.; Zhu, Y.T.; Ma, E. Optimizing the strength and ductility of fine structured 2024 Al alloy by nano-precipitation. *Acta Mater.* **2007**, *55*, 5822–5832. [CrossRef]
16. Liu, Z.; Li, Y.; Liu, Y.; Xia, Q. Development of Al-Cu-Mg-Ag alloys. *Chin. J. Nonferr. Met.* **2007**, *12*, 1905–1915.
17. Wang, S.C.; Starink, M.J. Precipitates and intermetallic phases in precipitation hardening Al-Cu-Mg-(Li) based alloys. *Int. Mater. Rev.* **2013**, *50*, 193–215. [CrossRef]

metals

MDPI

Article

On the Relationship between Structural Quality Index and Fatigue Life Distributions in Aluminum Aerospace Castings [†]

Hüseyin Özdeş and Murat Tiryakioğlu *

School of Engineering, University of North Florida, 1 UNF Dr. Jacksonville, FL 32224, USA; huseyinozdes@gmail.com

* Correspondence: m.tiryakioglu@unf.edu; Tel.: +1-904-620-1390; Fax: +1-904-620-1391

† An earlier version of this paper also appears in Shape Casting: 6th International Symposium, pp. 85–92, part of the 2016 TMS Annual Meeting.

Academic Editor: Nong Gao
Received: 4 March 2016; Accepted: 1 April 2016; Published: 7 April 2016

Abstract: Tensile and fatigue testing results of D357 and B201 aluminum alloy aerospace castings reported in the literature have been reanalyzed. Yield strength–elongation bivariate data have been used as a measure of the structural quality of castings, and converted into quality index. These results as well as fatigue data have been analyzed by using Weibull statistics. A distinct relationship has been observed between expected fatigue life and quality index. Moreover, probability of survival in fatigue life was found to be directly linked to the proportions of the quality index distributions in two different regions, providing further evidence about the strong relationship between elongation, *i.e.*, structural quality and fatigue performance.

Keywords: structural quality; metal fatigue; weakest link; elongation

1. Introduction

There are a number of tests used by engineers to determine the mechanical properties of structural parts used in aerospace and automotive applications. Among those tests, the most widely used one is the tensile test which is required in many industrial standards and specifications, such as MIL-A-21180D [1]. Although the root cause in 90% of all in-service failures in metallic components is fatigue [2,3], to the authors' knowledge, there is no requirement or specification for fatigue performance in any industrial or military specification.

In castings, the probability of premature failure under stress increases with increasing number density and size of structural defects such as pores and inclusions. Hence, the degradation in mechanical properties including tensile strength [4,5], elongation (e_F) [6–9], fracture toughness as well as fatigue life (N_f) [10] is directly related to the structural quality of castings.

To quantify structural quality by using tensile data, a new quality index, Q_T, has been introduced by one of the authors and his coworkers [11–13].

$$Q_T = \frac{e_F}{e_{F(\max)}} = \frac{e_F}{\beta_0 - \beta_1 \sigma_Y} \tag{1}$$

where, $e_{F(\max)}$ is the maximum elongation, alternatively referred to as the "ductility potential" of the alloy representing the defect-free condition, σ_Y is the yield strength, β_0 and β_1 are alloy dependent constants which were determined from the maximum ductility values over a wide range of yield strength, by analyzing hundreds of data from the aerospace and premium casting literature. Tiryakioğlu and Campbell [13,14] divided the Q_T space into three distinct regions and provided recommendations for quality improvement for each region. When tensile data are in Region 1 ($0 \leqslant Q_T < 0.25$), the premature failure is primarily due to "old" oxides which was the surface of re-melted castings, foundry returns and/or ingot. In this region, tensile specimens do not neck and fatigue failure starts from defects on or close to the specimen surface. Region 2 ($0.25 \leqslant Q_T < 0.70$) represents castings that are free from major "old" oxides but there is still a considerable density of "young" oxides, entrained into the casting during melt transfers and/or filling of the mold. Tensile specimens may show some necking and there will be occasional fatigue failures initiating from internal defects with facets around them [15]. In Region 3 ($0.70 \leqslant Q_T \leqslant 1.0$), tensile specimens are expected to neck and deform significantly beyond ultimate tensile strength [16]. Moreover, fatigue fracture is predominantly due to internal defects, exhibiting facets on fracture surfaces.

Recently, the tensile elongation requirement for castings in industrial and military specifications was interpreted as a *de facto* fatigue life specification [17]. Furthermore, it was shown that there is a distinct relationship between the Q_T and N_f distributions in A206-T7 castings. A similar approach is followed in this study and data from aerospace literature are reanalyzed for a potential relationship between the quality index and fatigue life in aerospace castings.

2. Materials and Methods

2.1. Experiments by Ozelton et al.

Four datasets reported by Ozelton *et al.* [18] who investigated the durability and damage tolerance for D357-T6 and B201-T7 cast aluminum alloys were reanalyzed in this study. For both alloys, two solidification rates based on the pour temperature and the chill material were used. The experimental details for "slow" and "fast" cooled specimens are given in Table 1.

Table 1. Experimental design used by Ozelton *et al.* for D357 and B201 aerospace castings.

	D357-T6		B201-T7	
Solidification Rate	Slow	Fast	Slow	Fast
Pouring T (°C)	782	748	787	732
Chill Material	Iron	Copper	Iron	Copper

2.2. Statistical Analysis

Because mechanical properties that involve fracture can be directly linked to casting defects, the Weibull distribution [19–21] based on the "weakest link" theory [22], has been used to characterize these properties. For the Weibull distribution, the cumulative probability function is expressed as:

$$P = 1 - \exp\left[-\left(\frac{\sigma - \sigma_T}{\sigma_0}\right)^m\right] \tag{2}$$

where, P is the probability of failure at a given stress (or fatigue life) at or lower, σ_T is the threshold value below which no failure is expected, σ_0 is the scale parameter and m is the shape parameter, alternatively known as the Weibull modulus. Note that when $\sigma_T = 0$, Equation (2) reduces to the 2-parameter Weibull distribution. The mean of the Weibull distribution is found by:

$$\overline{\sigma} = \sigma_T + \sigma_0 \Gamma\left(1 + \frac{1}{m}\right) \tag{3}$$

where, Γ represents the gamma function. The probability density function, f, for the Weibull distribution is expressed as;

$$f = \frac{m}{\sigma_0}\left(\frac{\sigma - \sigma_T}{\sigma_0}\right)^{m-1}\exp\left[-\left(\frac{\sigma - \sigma_T}{\sigma_0}\right)^m\right] \tag{4}$$

Tensile data were transformed to Q_T by using the β_0 and β_1 of 36 and 0.064 MPa^{-1} for D357 [12] and 34.5 and 0.047 MPa^{-1} for B201 [23], respectively. Weibull distributions with both two and three parameters have been fitted to the tensile and fatigue life data.

3. Results

Ozelton *et al.* performed tensile and fatigue tests in accordance with the ASTM B557 and ASTM E466, respectively. The geometry of the fatigue specimen was carefully selected to mimic aircraft components with holes where fatigue cracks are usually initiated due to stress concentrations. In total, 170 fatigue life and 165 tensile test results obtained by Ozelton *et al.* have been re-evaluated in the present investigation. It is significant that there were no fatigue run-outs in the datasets.

The dot-plot for elongation data for D357-T6 and B201-T7 castings is presented in Figure 1. Note in Figure 1a that the highest data for "fast" solidification is higher than in "slow" solidification, although minimum data in both datasets are similar. Hence the scatter is higher in "fast" solidification D357 castings. In Figure 1b, the elongation data for "fast" B201-T7 castings are only slightly higher than in "slow" castings. Moreover, there is an apparent gap in both datasets, as indicated in Figure 1b.

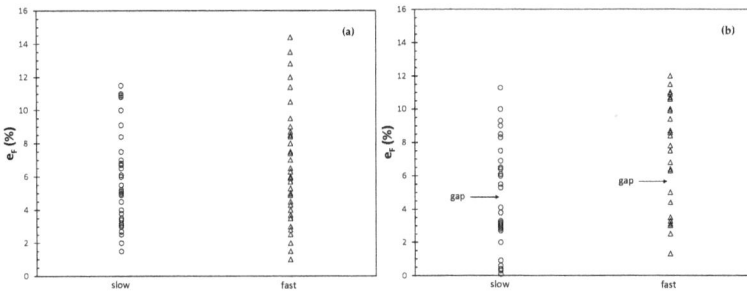

Figure 1. Dot-plot for elongation data of "slow" and "fast" specimens for (**a**) D357-T6 and (**b**) B201-T7 aluminum alloy castings.

The fatigue life data for the two aluminum alloy castings are presented in Figure 2. For D357, minimum data are almost identical for "slow" and "fast" castings, Figure 2a. However longest fatigue life is significantly higher for "fast" castings. As in elongation, Figure 1b, fatigue life data in B201 aluminum alloy castings have a significant gap, as shown in Figure 2b. The gaps in datasets are an indication that data have been collected from two distinct distributions. Hence, there is evidence that there is a mixture of at least two distributions in elongation and fatigue life data for B201-T7 aluminum alloy castings.

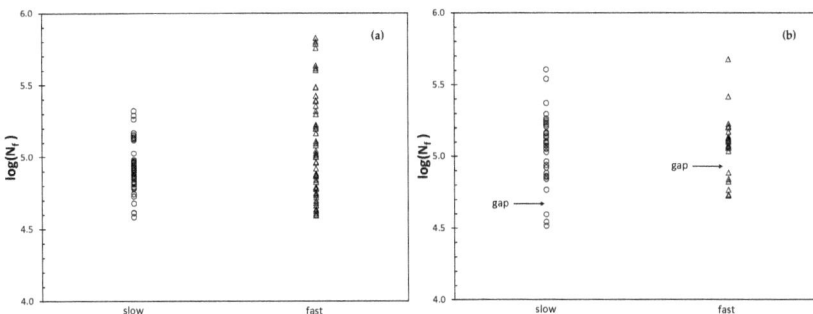

Figure 2. Dot-plot for fatigue life of "slow" and "fast" specimens for (**a**) D357-T6 and (**b**) B201-T7 aluminum alloy castings.

Weibull distributions were fitted to Q_T and N_f data by using the maximum likelihood method. The Weibull probability plots for Q_T and N_f for D357-T6 aluminum alloy castings are presented in Figure 3. Estimated Weibull parameters for each dataset are given in Table 2. Fits indicated by the two curves in Figure 3 are in close agreement with the data. The goodness-of-fit of the estimated parameters was tested by using the Anderson-Darling statistic [24]. In all cases, the hypothesis that the data come from the fitted Weibull distributions could not be rejected. Note in Figure 3a that the Q_T data for fast cooled castings fall on almost a straight line, which indicates that the threshold is close to zero (0.010), Table 2. The data for slow-cooled castings indicate a curve relationship which is indicative of a positive threshold. Moreover, both fatigue life distributions have almost the same threshold. Because lowest fatigue life in a distribution is determined by the largest defects possible in specimens [25], the size of the largest defects is almost the same in "fast" and "slow" datasets, regardless of how fast the metal solidified.

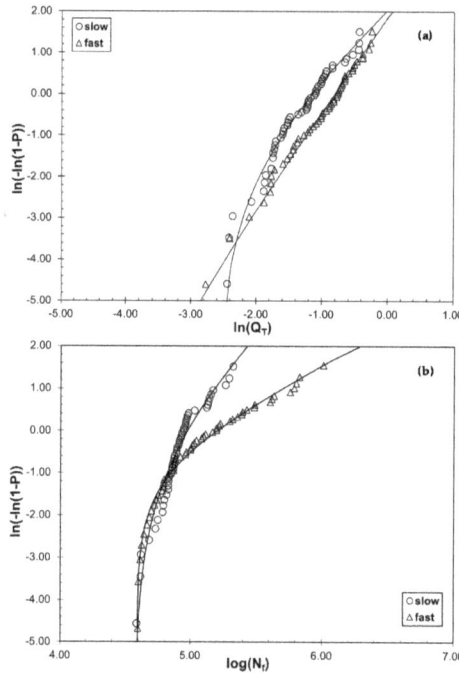

Figure 3. Weibull probability plots of (**a**) Q_T and (**b**) N_f in D357-T6 aluminum alloy castings.

Note that for B201, fatigue life and quality index data was found to have Weibull mixtures, as indicated in Table 2. In such cases, the cumulative probability is expressed as [26];

$$P = pP_L + (1-p)P_U \tag{5}$$

where, p is the fraction of the lower Weibull distribution in the mixture and subscripts L and U refer to the lower and upper Weibull distributions, respectively. The probability density function for a Weibull mixture is given as:

$$f = pf_L + (1-p)f_U \tag{6}$$

The Weibull probability plots for Q_T and N_f for B201-T7 aluminum alloy castings are provided in Figure 4. Note that for both Q_T and N_f, there are inflection points in the probability plots which are indicative of Weibull mixtures [25–27].

4. Discussion

Both Q_T and N_f data for B201 aluminum alloy castings showed Weibull mixtures is noteworthy. Analysis of fracture surfaces in A206-T7 aluminum alloy castings showed [10,17,28] that the lower distribution for elongation was attributed to the "old", coarse oxide bifilms that were generated during previous melt processing or were on the skin of the ingots. For fatigue life, the lower distribution is due to the fatigue crack initiation at surface defects. The two lower distributions are linked because the probability that a defect will be on the surface of the fatigue specimen increases with its size and number density [17]. Hence, premature fracture in fatigue has to be accompanied by low elongation, or alternatively, Q_T. As expected, increased solidification rate has a positive effect on both Q_T and N_f. It is also noteworthy that the improvement is most significant in the lower distributions. Moreover, the lower distributions remain significantly separated from the upper distributions, showing that chilling is a much less effective way to improve properties than eliminating structural defects, mainly bifilms and pores. It has been only recently understood [29,30] that the degradation of and variability in the mechanical properties of castings are related to these very defects that are incorporated into the bulk of the liquid by an entrainment process, in which the surface oxide folds over itself. In most steel castings, the oxide has a significantly lower density than the metal, and therefore floats to the surface quickly, leaving the metal relatively free of defects. In aluminum alloys, the folded oxide has practically neutral buoyancy, so that defects tend to remain in suspension. The layer of air in the folded oxide can: (i) grow into a pore as a result of the negative pressure due to contraction of the solidifying metal and/or rejection of gases, originally dissolved in liquid metal, upon solidification; or (ii) remain as an un-bonded surface, like a crack, in the solidified alloy, which usually serves as heterogeneous nucleation sites for intermetallics.

Table 2. Estimated Weibull parameters.

Alloy	Solidification Rate	Distribution Tag			Weibull parameters		
				p	σ_T	σ_0	m
D357-T6	Slow	Q_T			0.077	0.239	1.52
		N_f			37,291	57,340	1.43
	Fast	Q_T			0.010	0.433	2.27
		N_f			38,688	120,966	0.74
B201-T7	Slow	Q_T	Lower	0.250	0	0.066	1.13
			Upper		0.179	0.234	1.34
		N_f	Lower	0.114	26,351	16,963	1.62
			Upper		67,720	90,284	0.82
	Fast	Q_T	Lower	0.188	0	0.200	4.79
			Upper		0	0.623	6.26
		N_f	Lower	0.281	51,372	18,730	0.95
			Upper		113,610	58,081	0.57

Prior studies [5,31–35] have shown that there are multiple types of defects in castings, including bifilms and pores associated with bifilms. From a process viewpoint, it is not surprising to find Weibull plots for tensile data that reveal at least two populations of defects [25]:

1. the original rather fine scattering of defects remaining in suspension in the original poured liquid from the crucible or ladle (prior damage). These "old" bifilms have a typical minimum thickness of approximately 10 µm and show only coarse wrinkles.
2. the large new bifilms (new damage) that would have been produced during the melt transfer and/or pouring and filling if the filling system was not designed properly. These "young" oxides have a minimum thickness of tens of nanometers or less and show fine wrinkles on fracture surfaces of castings.

The probability density functions (Equations (4) and (6)) for Q_T and corresponding N_f distributions are presented in Figure 5. The different shapes of N_f distributions are a product of the use of the 3-parameter version of the Weibull distribution and the value of the shape parameter; when $m \leqslant 1$, the shape of the 3-parameter Weibull distribution resembles that of an exponential decay curve.

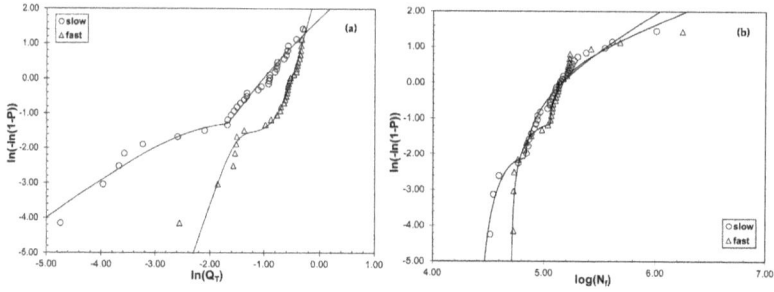

Figure 4. Weibull probability plots for (**a**) Q_T and (**b**) N_f in B201-T7 aluminum alloy castings.

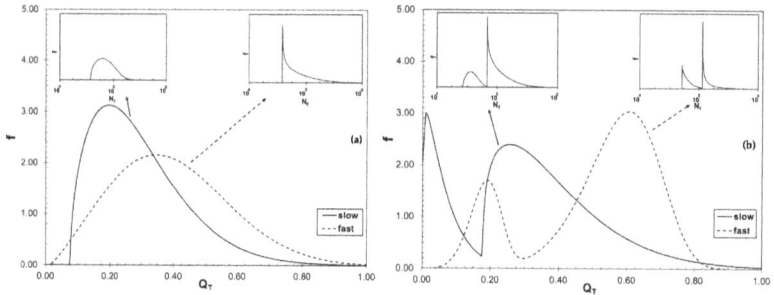

Figure 5. Probability density functions for Q_T and corresponding probability density functions for N_f for (**a**) D357 and (**b**) B201 aluminum alloy castings.

The expected (mean) values for all six distributions were calculated by using Equation (3) and the estimated Weibull parameters in Table 2. As stated above, the lower distributions in Q_T and N_f in B201 were associated with each other. The relationship between expected Q_T and N_f values is presented in Figure 6. Similar to what was reported for A206-T7 aluminum alloy castings [17], a linear relationship was observed between expected Q_T and the logarithm of expected fatigue life. The best fit line indicated with dashed lines in Figure 6 has the following equation:

$$\log(\overline{N}_f) = 4.57 + 1.46\overline{Q}_T \tag{7}$$

Note that the best fit equation estimates approximately 10^6 cycles when the specimen could reach the maximum quality (defect free condition). Therefore, fatigue life of aerospace castings can be extended by at least six times if structural defects are eliminated.

The proportions of the Q_T distributions in all three regions as well as the probability of survival after 10^5 cycles were calculated by using estimated Weibull parameters in Table 2. The results are presented in Table 3. The proportion of Region 1 ($Q_T \leqslant 0.25$) *versus* probability of $N_f > 10^5$ is plotted in Figure 7a. Clearly, probability of survival after 10^5 cycles decreases significantly with the proportion of castings in Region 1. A similarly strong relationship between the probability of survival after 10^5 cycles *versus* the proportion of castings in Region 3 ($Q_T > 0.70$) is presented in Figure 7b. As the casting quality is improved, probability of survival also increases, as can be expected. Hence there is strong evidence in Figure 7, as well as in Figure 6, that elongation (quality index) and fatigue performance are related; (i) there is a strong correlation between mean values, and (ii) the probability of survival is directly linked to the proportion of the elongation distribution in Region 1 and Region 3. Therefore, the statement that the elongation requirement in industrial specifications is a *de facto* fatigue life specification is justified. Research is underway to expand this understanding and develop predictive models for aerospace castings.

Table 3. Fraction of distributions for Q_T in each region and probability of survival at 10^5 cycles.

Distribution Tag			$P\,(Q_T \leqslant 0.25)$	$P\,(0.25 \leqslant Q_T < 0.70)$	$P\,(Q_T \geqslant 0.70)$	$P\,(N_f \geqslant 10^5)$
D357	Slow		0.458	0.529	0.014	0.321
	Fast		0.230	0.713	0.056	0.546
B201	Slow	L	0.989	0.011	0.000	0.000
		U	0.183	0.763	0.054	0.650
	Fast	L	0.946	0.054	0.000	0.084
		U	0.003	0.871	0.126	1.000

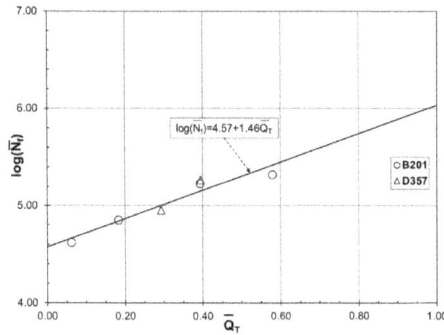

Figure 6. Relationship between the means of Q_T and N_f distributions.

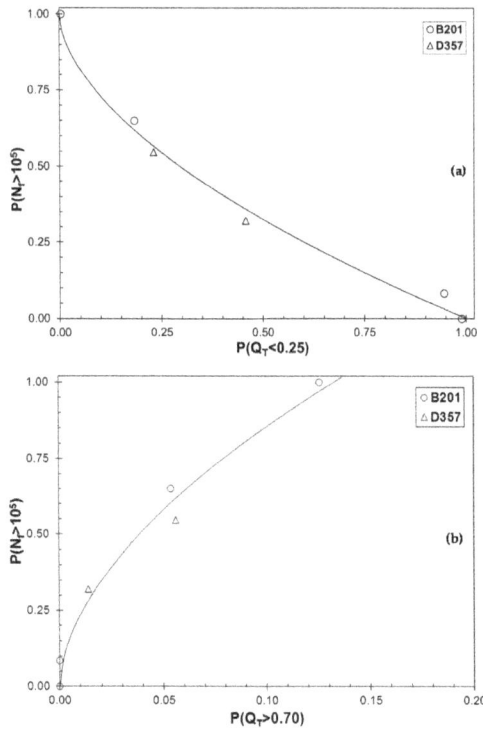

Figure 7. The change in the probability of survival after 10^5 cycles *versus* the estimated fraction of the Q_T distribution in (**a**) Region 1 and (**b**) Region 3.

5. Conclusions

- The quality index, Q_T, can be used to characterize the structural integrity of D357 and B201 aluminum alloy castings.
- Probability plots for both Q_T and N_f distributions for B201 showed strong indications of Weibull mixtures.
- There is a strong relationship between the mean Q_T and N_f values as calculated from estimated Weibull parameters.
- There is a strong negative correlation between the proportion of Q_T in Region 1 and probability of survival for 10^5 cycles. Similarly, a strong positive correlation exists between the proportion of Q_T in Region 3 and probability of survival for 10^5 cycles, providing further evidence for the strong link between elongation and fatigue performance.
- The statement that the elongation requirement in industrial specifications is a *de facto* fatigue life specification is justified.

Author Contributions: Hüseyin Özdeş and Murat Tiryakioğlu collaborated in all phases of this paper, namely the data collection, statistical analysis and interpretation of results.

Conflicts of Interest: The authors declare no conflict of interest.

Nomenclature

β_0, β_1	alloy dependent constants
f	Weibull probability density function
Γ	gamma function
m	shape parameter
σ	mean value of the Weibull distribution
N_f	fatigue life
σ_0	scale parameter
p	fraction of the lower distribution
σ_T	threshold value below which no failure is expected
P	probability of failure
σ_Y	yield strength (MPa)
P_L	probability from lower distribution
e_F	elongation
P_U	probability from upper distribution
$e_{F(max)}$	maximum elongation, ductility potential
Q_T	quality index

References

1.   MIL-A-21180 D: Aluminum-Alloy Castings. Available online: http://everyspec.com/MIL-SPECS/MIL-SPECS-MIL-A/MIL-A-21180D_6087/ (accessed on 1 April 2016).
2.   Dieter, G.E. *Mechanical Metallurgy*; McGraw-Hill: New York, NY, USA, 1976.
3.   Reed-Hill, R.E. *Physical Metallurgy Principles*; Van Nostrand: Princeton, NJ, USA, 1964.
4.   Green, N.R.; Campbell, J. Statistical distributions of fracture strengths of cast Al-7Si-Mg alloy. *Mater. Sci. Eng. A* **1993**, *173*, 261–266. [CrossRef]
5.   Nyahumwa, C.; Green, N.R.; Campbell, J. Influence of casting technique and hot isostatic pressing on the fatigue of an Al-7Si-Mg alloy. *Metall. Mater. Trans. A* **2001**, *32*, 349–358. [CrossRef]
6.   Lee, C.D.; Shin, K.S. Effect of microporosity on the tensile properties of AZ91 magnesium alloy. *Acta Mater.* **2007**, *55*, 4293–4303. [CrossRef]
7.   Lee, C.D. Effect of grain size on the tensile properties of magnesium alloy. *Mater. Sci. Eng. A* **2007**, *459*, 355–360. [CrossRef]

8. Tiryakioğlu, M. On estimating the fracture stress and elongation of Al–7%Si–0.3%Mg alloy castings with single pores. *Mater. Sci. Eng. A* **2010**, *527*, 4546–4549. [CrossRef]
9. Song, J.; Xiong, S.M.; Li, M.; Allison, J. *In situ* observation of tensile deformation of high-pressure die-cast specimens of AM50 alloy. *Mater. Sci. Eng. A* **2009**, *520*, 197–201. [CrossRef]
10. Staley, J.T., Jr.; Tiryakioğlu, M.; Campbell, J. The effect of hot isostatic pressing (hip) on the fatigue life of A206-T71 aluminum castings. *Mater. Sci. Eng. A* **2007**, *465*, 136–145. [CrossRef]
11. Tiryakioğlu, M.; Campbell, J.; Alexopoulos, N.D. Quality indices for aluminum alloy castings: A critical review. *Metall. Mater. Trans. B* **2009**, *40*, 802–811. [CrossRef]
12. Tiryakioğlu, M.; Campbell, J.; Alexopoulos, N.D. On the ductility of cast Al-7 pct Si-Mg alloys. *Metall. Mater. Trans. A* **2009**, *40*, 1000–1007. [CrossRef]
13. Tiryakioğlu, M.; Campbell, J. Quality index for aluminum alloy castings. *Int. J. Met.* **2015**, *8*, 39–42. [CrossRef]
14. Tiryakioglu, M.; Campbell, J. Quality index for aluminum alloy castings. *AFS Trans.* **2013**, *13*, 217–222. [CrossRef]
15. Tiryakioğlu, M.; Campbell, J.; Nyahumwa, C. Fracture surface facets and fatigue life potential of castings. *Metall. Mater. Trans. B* **2011**, *42*, 1098–1103. [CrossRef]
16. Alexopoulos, N.D.; Tiryakioğlu, M. On the uniform elongation of cast Al–7%Si–0.6%Mg (A357) alloys. *Mater. Sci. Eng. A* **2009**, *507*, 236–240. [CrossRef]
17. Tiryakioğlu, M. On the relationship between elongation and fatigue life in A206-T71 aluminum castings. *Mater. Sci. Eng. A* **2014**, *601*, 116–122. [CrossRef]
18. Ozelton, M.; Mocarski, S.; Porter, P. Durability and Damage Tolerance of Aluminum Castings. Available online: http://www.dtic.mil/dtic/tr/fulltext/u2/a245237.pdf (accessed on 1 April 2016).
19. Weibull, W. *A Statistical Theory of the Strength of Materials*; Generalstabens litografiska anstalts förlag: Stockholm, Sweden, 1939.
20. Weibull, W. *The Phenomenon of Rupture in Solids*; Generalstabens litografiska anstalts förlag: Stockholm, Sweden, 1939.
21. Weibull, W. A statistical distribution function of wide applicability. *J. Appl. Mech.* **1951**, *18*, 293–297.
22. Peirce, F.T. Tensile tests for cotton yarns–"the weakest link" theorems on the strength of long and of composite specimens. *J. Text. Inst. Trans.* **1926**, *17*, T355–T368.
23. Tiryakioglu, M.; Campbell, J. Ductility, structural quality, and fracture toughness of Al–Cu–Mg–Ag (A201) alloy castings. *Mater. Sci. Technol.* **2009**, *25*, 784–789. [CrossRef]
24. Anderson, T.W.; Darling, D.A. A test of goodness of fit. *J. Am. Stat. Assoc.* **1954**, *49*, 765–769. [CrossRef]
25. Tiryakioğlu, M.; Campbell, J. Weibull analysis of mechanical data for castings: A guide to the interpretation of probability plots. *Metall. Mater. Trans. A* **2010**, *41*, 3121–3129. [CrossRef]
26. Tiryakioğlu, M. Weibull analysis of mechanical data for castings ii: Weibull mixtures and their interpretation. *Metall. Mater. Trans. A* **2015**, *46*, 270–280. [CrossRef]
27. Jiang, S.; Kececioglu, D. Maximum likelihood estimates, from censored data, for mixed-weibull distributions. *IEEE Trans. Reliab.* **1992**, *41*, 248–255. [CrossRef]
28. Staley, J.T., Jr.; Tiryakioğlu, M.; Campbell, J. The effect of increased hip temperatures on bifilms and tensile properties of A206-T71 aluminum castings. *Mater. Sci. Eng. A* **2007**, *460–461*, 324–334. [CrossRef]
29. Campbell, J. *Castings*, 2nd ed.; Elsevier: Oxford, UK, 2003.
30. Campbell, J. Entrainment defects. *Mater. Sci. Technol.* **2006**, *22*, 127–145. [CrossRef]
31. Wang, Q.G.; Crepeau, P.N.; Davidson, C.J.; Griffiths, J.R. Oxide films, pores and the fatigue lives of cast aluminum alloys. *Metall. Mater. Trans. B* **2006**, *37*, 887–895. [CrossRef]
32. Nyahumwa, C.; Green, N.; Campbell, J. Effect of mold-filling turbulence on fatigue properties of cast aluminum alloys (98–58). *Trans. Am. Foundrymen's Soc.* **1998**, *106*, 215–223.
33. Zhang, B.; Poirier, D.R.; Chen, W. Microstructural effects on high-cycle fatigue-crack initiation in A356.2 casting alloy. *Metall. Mater. Trans. A* **1999**, *30*, 2659–2666. [CrossRef]
34. Eisaabadi B, G.; Davami, P.; Kim, S.K.; Tiryakioğlu, M. The effect of melt quality and filtering on the weibull distributions of tensile properties in Al–7%Si–Mg alloy castings. *Mater. Sci. Eng. A* **2013**, *579*, 64–70. [CrossRef]
35. Wang, Q.G.; Apelian, D.; Lados, D.A. Fatigue behavior of A356-T6 aluminum cast alloys. Part i. Effect of casting defects. *J. Light Met.* **2001**, *1*, 73–84. [CrossRef]

metals

MDPI

Article

Onset Frequency of Fatigue Effects in Pure Aluminum and 7075 (AlZnMg) and 2024 (AlCuMg) Alloys

Jose I. Rojas [1,*] and Daniel Crespo [2]

[1] Department of Physics-Division of Aerospace Engineering, Universitat Politècnica de Catalunya, c/ Esteve Terradas 7, 08860 Castelldefels, Spain

[2] Department of Physics, Universitat Politècnica de Catalunya, c/ Esteve Terradas 7, 08860 Castelldefels, Spain; daniel.crespo@upc.edu

* Correspondence: josep.ignasi.rojas@upc.edu; Tel.: +34-93-413-4130

Academic Editor: Nong Gao
Received: 30 December 2015; Accepted: 23 February 2016; Published: 1 March 2016

Abstract: The viscoelastic response of pure Al and 7075 (AlZnMg) and 2024 (AlCuMg) alloys, obtained with a dynamic-mechanical analyzer (DMA), is studied. The purpose is to identify relationships between the viscoelasticity and fatigue response of these materials, of great interest for structural applications, in view of their mutual dependence on intrinsic microstructural effects associated with internal friction. The objective is to investigate the influence of dynamic loading frequency and temperature on fatigue, based on their effect on the viscoelastic behavior. This research suggests that the decrease of yield and fatigue behavior reported for Al alloys as temperature increases may be associated with the increase of internal friction. Furthermore, materials subjected to dynamic loading below a given threshold frequency exhibit a static-like response, such that creep mechanisms dominate and fatigue effects are negligible. In this work, an alternative procedure to the time-consuming fatigue tests is proposed to estimate this threshold frequency, based on the frequency dependence of the initial decrease of the storage modulus with temperature, obtained from the relatively short DMA tests. This allows for a fast estimation of the threshold frequency. The frequencies obtained for pure Al and 2024 and 7075 alloys are 0.001–0.005, 0.006 and 0.075–0.350 Hz, respectively.

Keywords: aluminum alloys; AlZnMg; AlCuMg; viscoelasticity; dynamic-mechanical analysis; internal friction; loading frequency; fatigue

---

1. Introduction

Fatigue is a form of failure that may occur in structures subjected to dynamic loading, even at stress levels significantly lower than the ultimate tensile strength under static loading [1]. Failure results from a gradual process of damage accumulation and local strength reduction, which is manifested by crack initiation and propagation, after relatively long periods of dynamic loading. It is particularly dangerous in structural applications, because of its brittle, catastrophic nature and because it occurs suddenly and without warning, since very little plastic deformation is observed in the material prior to failure [1,2]. The fatigue fracture behavior of materials is dominated by the microstructure [3]. When a material is subjected to dynamic loading, energy is dissipated due to internal friction phenomena. Most of this energy manifests as heat and causes temperature increases in the material, a process termed hysteresis heating. It has been suggested that all metals, when subjected to hysteresis heating, are prone to fatigue [4].

In previous investigations, the viscoelastic response (including the internal friction behavior) of Al alloys (AA) 7075 and 2024 was measured with a dynamic-mechanical analyzer (DMA) [5,6]. In this work, experimental results on the viscoelastic response of pure Al are presented, first. Second, these results for pure Al and the aforementioned alloys are analyzed with the purpose of identifying relationships between the viscoelastic response and the fatigue behavior of these materials, in view of their mutual dependence on intrinsic microstructural effects associated with internal friction [7]. Particularly, the objective is to investigate the influence of the dynamic loading frequency and temperature on fatigue, based on the effect of these variables on the viscoelastic behavior. The results seem to support the work by Amiri and Khonsari [4], as per the correlation between the fatigue life and the initial hysteresis heating during dynamic loading. Namely, it is likely that the decrease of yield and fatigue response observed in some metals as temperature increases is associated with the increase of internal friction with temperature. Moreover, following previous investigations by other researchers, suggesting the existence of a threshold frequency marking the transition from a static-like response of the material to the advent of fatigue effects, in this work, an alternative procedure is proposed to estimate this threshold frequency based on experimental data obtained with the relatively short DMA tests. These findings are of remarkable importance, especially for the alloys, in view of their widespread use in structural applications under dynamic loading. Particularly, AA 7075 and 2024 are key representatives of the AlZnMg and AlCuMg alloy families (or 7xxx and 2xxx series, respectively), belonging to the group of age-hardenable alloys. These alloys feature excellent mechanical properties and are highly suitable for a number of industrial applications, especially in the aerospace sector and transport industry [8].

1.1. Influence of the Loading Frequency on the Fatigue Response of Metals

Fatigue may be sensitive, for instance, to the strength, the manufacturing conditions and the surface treatment of the material, but also to the loading frequency and loading environment, the displacement rates and the stress amplitude [1,9–11]. In this work, we address the effects of the dynamic loading frequency and temperature, in conjunction with the microstructure.

Much research has been devoted to ascertain whether accelerated laboratory tests (*i.e.*, with loading frequencies higher than those in service conditions) affect the fatigue response and how, but this is yet a controversial issue. This is particularly true for the study of high cycle fatigue (HCF) and very high cycle fatigue (VHCF) behavior by means of very high frequency tests. Tests in VHCF and very low crack growth rates are time consuming with conventional fatigue testing techniques, like rotating bending, with a maximum frequency of 100 Hz. Hence, accelerated laboratory tests are very interesting because a significant reduction of testing time is possible using high speed servo-hydraulic machines [12], which may work at frequencies of 600 Hz, or especially using ultrasonic equipment, which may reach frequencies of 20 kHz [13].

Zhu *et al.* [14] state that environmental effects need to be considered, and Mayer *et al.* [15] explain that this is so because the time-dependent interaction with the environment may cause an extrinsic frequency influence on fatigue properties, on top of the intrinsic strain rate effects. Furuya *et al.* [12] state that frequency generally affects high frequency fatigue tests because: (1) fatigue limits and lives decrease due to the temperature increase caused by plastic deformation [16]; (2) dislocations may not match the applied frequency because dislocation movement is slow compared to sonic velocity [17]; and (3) provided that embrittlement by hydrogen diffusion had an effect [18], fatigue lives would depend on both the number of loading cycles and time. However, Mayer [19] reported also that the HCF behavior of metallic alloys is relatively insensitive to the test frequency, provided that the ultrasonic testing procedure is appropriate (e.g., adequate cooling) and that fatigue-creep interaction and the time-dependent interaction with the environment are negligible. The reasons suggested are, on the one hand, that cyclic plastic straining is limited near the fatigue limit or the threshold of fatigue crack growth (FCG), and thus, plastic strain rates are low, even at high frequencies; and on the other hand, the fact that shear stress has little sensitivity to strain rate [20]. Mayer *et al.* [15] also commented that the influence of frequency becomes significantly smaller if the dynamic stress amplitude is lower, maybe because cyclic loading is almost perfectly elastic.

For body-centered cubic (bcc) metals and metallic alloys, the HCF behavior is reported to be more sensitive to frequency than for face-centered cubic (fcc) metals [13]. However, Furuya *et al.* [12] observed that the fatigue behavior of high strength steels is independent of frequency. The argued cause was their extremely high strength and, thus, reduced plasticity and dislocation mobility. The hysteresis energy is low in low plasticity materials, and thus, the frequency effects on fatigue associated with the temperature increase are minimized. Likewise, Yan *et al.* [21] observed very little variation of the fatigue strength of high strength steel when testing at a conventional frequency (52.5 Hz) and at an ultrasonic frequency (20 kHz).

For an Al alloy similar to AA 7075 tested in the HCF regime at room temperature (RT), samples tested at 100 Hz were reported to fail earlier than those tested at 20 kHz. However, the effect of frequency on fatigue behavior was not statistically significant [15]. On the contrary, for E319 cast Al alloy at 293, 423 and 523 K, fatigue life at 20 kHz was 5–10-times longer than that at 75 Hz [14], but this author states that fatigue crack initiation is not influenced either by temperature or frequency. Rather, the observed difference in fatigue life is attributed to environmental effects on FCG rate. The fact that the moisture of ambient air deteriorates the fatigue life of high strength Al alloys by increasing the FCG rate has also been suggested by other authors [15,22]. Namely, Menan and Henaff [23] suggest for AA 2024-T351 that fatigue and corrosion may interact, such that FCG rates are enhanced. These synergistic effects are more notorious at low frequencies, for a given number of cycles at RT. Finally, Benson and Hancock [24] observed strain rate effects on cyclic plastic deformation of AA 7075-T6, provided that cyclic stresses were close to the yield stress.

As per low frequency loading, on the one hand, Nikbin and Radon [25] proposed a method to predict the frequency region of interaction between creep and FCG using static data (obtained at 423 K for Al alloy RR58) and RT high frequency fatigue data and assuming a linear cumulative damage law. The results showed that the interaction region is 0.1–1 Hz for the Al alloy (see Figure 4 in [25]). In the intermediate (steady state) stage of cracking for static and low frequency tests, crack growth is sensitive to frequency, and the fracture mode is time dependent inter-granular in nature, suggesting that creep dominates. Conversely, for high frequency tests, crack growth is insensitive to frequency, and the fracture mode is trans-granular, suggesting that pure fatigue mechanisms dominate. The results indicated also little interaction between these processes.

On the other hand, Henaff *et al.* [26] analyzed creep crack growth (CCG) rates, FCG rates and creep-fatigue crack growth (CFCG) rates of AA 2650-T6 at 293, 403 and 448 K, for frequencies of 0.05 and 20 Hz. The objective was to enable the prediction of crack growth resistance of that alloy under very low frequency loading at elevated temperatures. It was concluded that, in the studied frequency range, frequency has only a slight effect on FCG rates at 448 K. In particular, under low frequency loading, a high increase was observed in the fracture surface fraction of the inter-granular type, similar to that corresponding to CCG. This shows that creep damage might occur during loading at low frequency, in accordance with the findings in [25]. Henaff *et al.* [26] reported also that, for a given temperature, CFCG is unaffected by frequency above a critical value of the loading frequency (see Figure 12b in [26]). Below, CFCG is inversely proportional to excitation frequency, *i.e.*, a time-dependent crack growth processes take place. This researcher suggested the existence of a creep-fatigue-environment interaction, as CFCG is affected by the environment at low frequency loading, and proposed an alternative method to predict CFCG rates at very low frequencies, using a superposition model and results obtained at higher frequencies.

1.2. Influence of Temperature on the Fatigue Response of Al Alloys

For E319 cast Al alloy tested at 293, 423 and 523 K, Zhu *et al.* [14] observed that the fatigue strength decreases with temperature and that the temperature dependence of the fatigue resistance at 108 cycles follows the temperature dependence of the yield and tensile strength for this alloy closely. Furthermore, by integration of a universal version of a modified superposition model, the effects of temperature, frequency and the environment on the S-N curve of this alloy can be predicted, and it is possible also to extrapolate ultrasonic data to conventional fatigue behavior [14]. Henaff *et al.* [26] concluded that the temperature has almost no influence on FCG rates for AA 2650-T6, after conducting tests at 293, 403 and 448 K and frequencies of 0.05 and 20 Hz. Amiri and Khonsari [4] state that the initial slope of the temperature rise due to hysteresis heating observed at the beginning of fatigue tests is a characteristic of metals. Capitalizing on this, they developed an empirical model that predicts fatigue life based on that slope, thus preserving testing time. Indeed, the correlation of the temperature evolution with fatigue has been used successfully in many ways, aside from for predicting fatigue life, as in the previous example. Namely, it has been used for providing information on FCG [27] and the endurance limit of materials [28] or for quantification of the cumulative damage in fatigue [16]. Furthermore, the heat dissipated during ultrasonic cycling can be used to calculate the cyclic plastic strain amplitude [15].

1.3. Influence of the Microstructure on the Fatigue Response of Al Alloys

There is abundant research in the literature on the effect of the microstructure on the fatigue response of materials. For example, it is proposed that the mechanisms responsible for the fatigue fracture behavior are associated with the competing and synergistic influences of intrinsic microstructural effects and interactions between dislocations and the microstructure [3]. Indeed, researchers claim that the prediction of fatigue life should be possible based on the knowledge of the microstructure prior to the beginning of service, without the need for expensive, time-consuming fatigue experiments [29]. This would enable optimization of the material properties by controlling the microstructure. Accordingly, a model based on dislocation stress was proposed to predict S-N curves using microstructure/material-sensitive parameters instead of constitutive equation parameters [29]. The model is successful for low cycle fatigue life prediction.

2. Materials and Methods

The tested specimens were machine cut from a sheet of as-received pure Al (99.5 wt. % purity according to the supplier, Alu-Stock, S.A., Vitoria-Gasteiz, Spain) in the H24 temper. The H24 temper consists of cold-working (*i.e.*, strain hardening) beyond the desired hardness, followed by a softening treatment consisting of annealing up to halfway of the peak hardness. The specimens were rectangular plates 60 mm long, 8–12 mm wide and 2 mm thick. Half of these plates were annealed at 750 K for 30 min and immediately quenched in water to RT, to remove the strain hardening. A TA Instruments Q800 DMA (TA Instruments, New Castle, DE, USA) was used to measure the viscoelastic response of the samples in N_2 atmosphere. Namely, the DMA measured the storage modulus E' (*i.e.*, the elastic-real-component of the dynamic tensile modulus, accounting for the deformation energy stored by the material), the loss modulus E'' (*i.e.*, the viscous-imaginary component of the dynamic tensile modulus, accounting for the energy dissipation due to internal friction during relaxation processes) and the loss tangent (also termed mechanical damping or tanδ) [7]. The 3-point bending clamp was used, and the DMA was set to sequentially apply dynamic loading with frequencies ranging from 1–100 Hz, at temperatures from RT to 723 K in step increments of 5 K. More details on the procedure, as well as the viscoelastic data of AA 7075-T6 and 2024-T3 used in this work, can be found in [5,6].

3. Results and Discussion

3.1. Storage Modulus

Figure 1a shows E' for pure Al in the H24 temper, from RT to 648 K, while Figure 1b shows E' for pure Al, from RT to 723 K, in both cases as obtained from DMA tests at frequencies ranging from 1–100 Hz. The behavior of E' for pure Al is similar in some aspects to that observed for AA 7075-T6 and 2024-T3 [5,6] and AA 6082 [30] (an Excel file including the values of E', E'' and loss tangent measured for pure Al in the H24 temper, pure Al, AA 7075-T6 and AA 2024-T3 is provided as Supplementary Material). For example, E' also decreases initially. The slope at low temperatures (below the beginning of the dissolution of Guinier-Preston (GP) and Guinier-Preston-Bagariastkij (GPB) zones, for the alloys) is what is most interesting in this study, as explained in Section 3.5. Furthermore, a significant decrease in E' is observed, with the beginning of this drop shifted to higher temperatures (from around 423–523 K) as the loading frequency increases. Thus, at a given temperature, the alloys show a stiffer response (*i.e.*, E' is larger) at higher frequencies, as expected. Furthermore, E' depends more significantly on frequency at high temperatures (above 423 K). The fact that the viscoelastic behavior becomes more prominent with temperature has already been observed in amorphous alloys [31], aside from AA 7075-T6 and 2024-T3 [5,6] and AA 6082 [30].

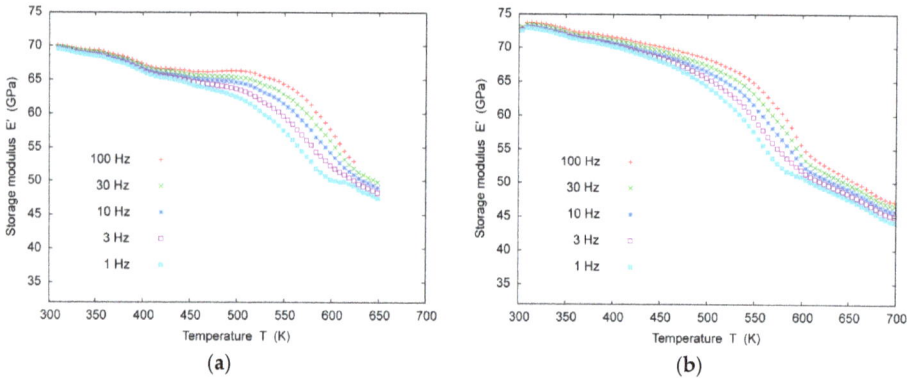

Figure 1. Storage modulus E' *vs.* temperature T from dynamic-mechanical analyzer tests at frequencies ranging from 1–100 Hz: (**a**) for pure Al in the H24 temper, from room temperature (RT) to 648 K; (**b**) for pure Al, from RT to 723 K.

3.2. Loss Modulus

Figure 2a shows E'' for pure Al in the H24 temper, from RT to 648 K, while Figure 2b shows E'' for pure Al, from RT to 723 K, in both cases as obtained from DMA tests at frequencies ranging from 1–100 Hz. In this case, the behavior of E'' for pure Al shows noticeable differences to that observed for AA 7075-T6 and 2024-T3 [5,6] and AA 6082 [30]. At low temperatures, the slopes of E'' are similar for all of the studied frequencies and not very steep (all show almost a plateau). At 393–533 K, the slopes increase sharply. This variation in the slope is shifted towards higher temperatures with increasing loading frequency. The observed behavior may be due to the viscous loss at higher frequencies competing with shorter relaxation times. Since the relaxation time decreases with temperature due to the Arrhenius-type behavior of the relaxation rate [7], this means that the temperature above which the viscous effect exceeds the relaxation is higher for higher frequency. In other words, higher frequency viscous loss curves rise at a higher temperature than lower frequency curves.

For AA 7075-T6, 2024-T3 and 6082, the sharp growth in E'' with temperature reaches very high values without showing a peak, which is usually explained by the presence of coupled relaxations [7]. On the contrary, for pure Al, E'' clearly exhibits a peak, which is achieved virtually at the same temperature for all of the frequencies (around 573 K). The peak is larger (both in width and height) as the loading frequency decreases. Previous works suggest that AlZnMg alloys, AlCuMg alloys and pure Al exhibit mechanical relaxation peaks associated with dislocations and grain boundaries [32,33]. For example, dislocation motions explain some internal friction peaks associated with semi-coherent precipitates for the alloys [34] and also the Bordoni peak, which has been extensively studied in cold-worked pure Al [7]. In this case, the observed peak corresponds to a typical internal friction peak in polycrystalline Al, related to grain boundaries [7]. In particular, the mechanism governing this relaxation is based on sliding at boundaries between adjacent grains. Upon application of stress, this process starts with the sliding of a grain over the adjacent one, caused by the shear stress acting initially across their mutual boundary. As a consequence, the shear stress is reduced gradually, and opposing stresses build up at the end of the boundary and into other adjacent grains. The process terminates when the shear stress has vanished across most of the boundary, and most of the total shearing force is sustained by the grain corners.

In addition, in Figure 2, a transition is observed around 473–513 K between the low temperature region where E'' is smaller for lower loading frequencies and the high temperature region where E'' is smaller for higher frequencies (a stiffer response in this case is expected). Finally, after the aforementioned peak, E'' seems to increase again in Figure 2b, particularly for the lower frequencies.

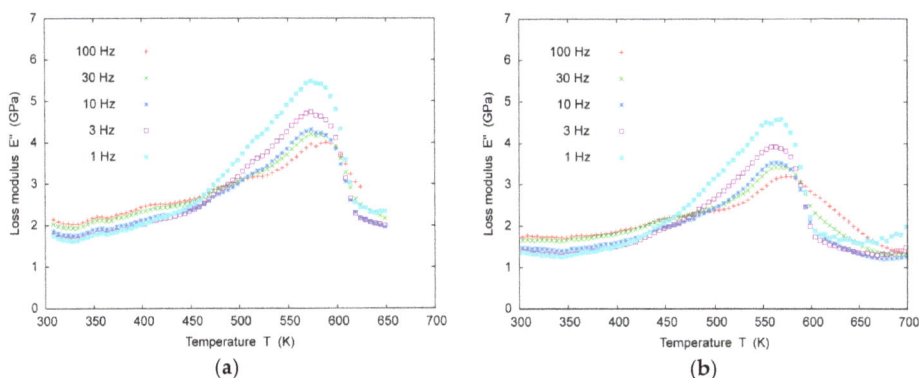

(a)   (b)

Figure 2. Loss modulus E'' *vs.* temperature T from dynamic-mechanical analyzer tests at frequencies ranging from 1–100 Hz: (**a**) for pure Al in the H24 temper, from room temperature (RT) to 648 K; (**b**) for pure Al, from RT to 723 K.

As usual, the loss tangent obtained from the DMA tests exhibits qualitatively the same behavior as E'' (the plots of loss tangent *vs.* temperature are included as Supplementary Material). The only remarkable comment is that there are no appreciable differences between the mechanical damping behavior for pure Al in the H24 temper and for pure Al, like for the E'' behavior (differences in the measured absolute values fall virtually within the instrument accuracy).

3.3. Temperature Dependence of the Storage Modulus

In absence of microstructural transformations, e.g., for pure Al and the alloys in the range RT to 373 K, E' decreases linearly with temperature. The reasons supporting this assumption are explained herein. There is abundant literature reporting a decrease with temperature of the elastic stiffness constants of metals, e.g., the static elastic modulus of pure Al and Al alloys [35–38]. This decrease can be assumed linear in a wide temperature range. Significant deviations from linearity are only observed close to 0 K and well above the temperature range of most interest for this research, *i.e.*, well above RT to 373 K, as explained in Section 3.5. Wolfenden and Wolla [39] observed also a highly quasi-linear decrease of the dynamic elastic modulus with temperature, measured at high frequency (80 kHz) from RT to 748 K, for pure Al and for AA 6061 reinforced with alumina.

Figure 3 in [6] shows a comparison of static and high frequency dynamic elastic modulus data available in the literature with data obtained with the DMA for AA 2024-T3 and for pure Al at 1 Hz. As expected, our DMA results for pure Al in the range of RT to 373 K fall between the static and high frequency dynamic values of elastic modulus in the literature, but they show a pronounced decrease of dynamic elastic modulus starting around 423–523 K. This is coincident with the observed E'' peak, as shown in Figure 2. Wolfenden and Wolla [39] assumed a linear behavior from RT to 748 K and explained this in terms of the Granato-Lücke theory for dislocation damping [40], but their data are too scattered to be explained with a single mechanism in the considered temperature range, as shown by our DMA results. Moreover, the low frequency dynamic elastic modulus of pure Al should also be sensitive to micrometric mechanisms, such as boundary migration during recrystallization and grain growth, as shown by Zhang *et al.* [41]. This mechanism is likely to be much more relevant in fatigue processes. However, the microstructural evolution and mechanical properties of AA 7075-T6 and 2024-T3 are controlled by the successive redissolution and precipitation of minority phases, and that is why it is likely that boundary migration is not the single most significant mechanism governing the behavior of their dynamic elastic moduli.

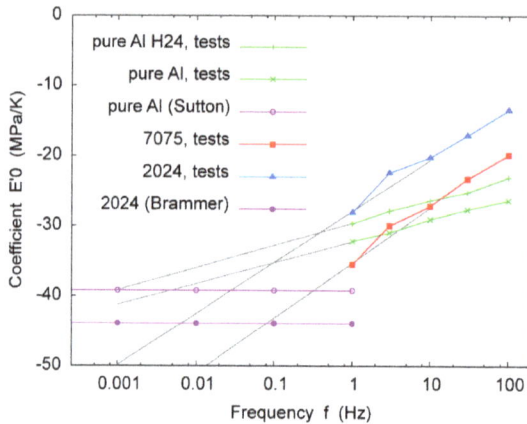

Figure 3. Temperature softening coefficient E_0' *vs.* frequency f for pure Al in the H24 temper, pure Al, Al alloy (AA) 7075-T6 and AA 2024-T3, as obtained from linear regression of test data from the dynamic-mechanical analyzer. Logarithmic tendency lines fitted to the data to extrapolate them to lower frequencies are also shown, as well as the rates of loss of static elastic modulus with temperature, as obtained by linear regression of data in the literature for pure Al [36] and AA 2024 [35].

Consequently, it can be assumed that, in absence of microstructural transformations, E' decreases linearly with temperature. Accordingly, Table 1 shows the slope of the decrease of E' with temperature for pure Al in the H24 temper and for pure Al, as obtained from linear regression of the DMA experimental results. In the following, to refer properly to this parameter on the basis of its physical meaning, it is termed "temperature softening coefficient", E_0'. Finally, it is important to note that the larger decrease of E' for low frequencies is typically explained by the Arrhenius-type behavior of the relaxation rate. That is, the mechanical relaxation time decreases with temperature [7], so that, at low frequencies, the shorter relaxation times lead to responses with larger values of E'' and smaller values of E'.

Table 1. Temperature softening coefficient E_0', for pure Al in the H24 temper and for pure Al, as obtained from linear regression of dynamic-mechanical analyzer test data.

Loading Frequency (Hz)	E_0' (Pure Al in H24 Temper) (MPa·K^{-1})	E_0' (Pure Al) (MPa·K^{-1})
100	−23.1	−26.4
30	−25.2	−27.7
10	−26.4	−29.1
3	−27.9	−31.0
1	−29.7	−32.2

3.4. Effect of Internal Friction on Fatigue Strength

Considering that viscoelasticity is linked to the fatigue and yield stress behavior [7], it is likely that the decrease of fatigue and yield stress behavior observed in metals as temperature increases is associated with the increase of internal friction. The reasons supporting this assumption are explained herein. Namely, Zhu *et al.* [14] observed that the fatigue strength of an Al alloy decreases with temperature, with a behavior that follows closely the temperature dependence of the yield and tensile strength. Furthermore, Liaw *et al.* [16] suggested that the fatigue limits and fatigue lives of steels decrease due to the temperature increase caused by plastic deformation. Finally, from the work by Amiri and Khonsari [4], it appears that the more intense the hysteresis heating of a metal during dynamic loading, the lower its fatigue resistance. Extrapolating the assumption stated above to AA 7075-T6, 2024-T3 and AA 6082, their yield and fatigue strength would decrease with temperature according to the observed increase in internal friction in the tested temperature range, as shown in [5,6] and [30], respectively. For pure Al, the yield and fatigue strength would decrease up to the E'' peak.

Furthermore, if considering the influence of loading frequency on internal friction, our results suggest that below 423 K, the yield and fatigue strength would decrease with increasing frequency, while at high temperatures, these properties would increase. These presumptions agree with some of the research reported in the literature, but still, there is controversy about the effect of frequency on the fatigue response, as shown in Section 1.1. For example, our findings agree with results from low frequency investigations for Al alloy RR58 at 423 K [25] and for AA 2650-T6 at 175 °C [26], pointing out that fatigue life is sensitive to frequency and that it increases with loading frequency. There is agreement also with the results for E319 cast Al alloy at 423 and 523 K, pointing out that fatigue life at 20 kHz was longer than at 75 Hz [14]. However, this behavior is attributed to environmental effects on the FCG rate, rather than intrinsic temperature or frequency effects.

3.5. Onset Frequency of Fatigue Effects

It is reasonable to accept that, at sufficiently low loading frequency, fatigue effects become negligible. Indeed, in a series of dynamic loading tests with loading frequency decreasing to very low values, the test time scale will eventually become much larger than the largest mechanical relaxation time for any of the possible relaxation processes for the tested material. That is, the reduction in the loading frequency is equivalent to an increase in the reaction time, and therefore, there is a threshold frequency below which the reaction time is longer than the relaxation time. Consequently, the viscous effect as the loss mechanism is negligible as compared to the relaxation. This means that eventually, E'' will decrease to a level that there will be no appreciable frictional energy loss due to relaxation effects. Recalling the statement by Amiri and Khonsari [4] that some degree of hysteresis heating (which is caused by the energy dissipated due to internal friction) is necessary for metals to experience fatigue, the conclusion is that, eventually, fatigue effects vanish. This hypothesis is in line with research by Henaff *et al.* [26] and Nikbin and Radon [25], suggesting the existence of a critical value of loading frequency below which, for a given temperature, the material exhibits a static-like response (relaxed case), crack growth is sensitive to frequency and static creep dominates instead of pure fatigue mechanisms.

Next, a procedure to estimate this threshold frequency is proposed based on the frequency dependence of the temperature softening coefficient E_0'. The procedure is illustrated in Figure 3, which shows E_0' as a function of the loading frequency for pure Al in the H24 temper and pure Al, as obtained from linear regression of DMA data, and for AA 7075-T6 and 2024-T3, as obtained in [5,6]. These values of E_0' are compared to average values of the rate of loss of static elastic modulus with temperature, as obtained by linear regression of data in the literature for pure Al [36] and AA 2024 [35]. To calculate these averages, data from Kamm and Alers [37] and from Varshni [38] were disregarded, as they correspond to temperatures below RT. Data from Wolfenden and Wolla [39] were also disregarded, since these data are too scattered in a broad temperature range, and thus, the computable slope is probably not representative.

Assuming that, at the threshold frequency, E_0' should become equal to the rate of loss of the elastic modulus with temperature under static loading conditions, this threshold frequency may be estimated by the intersection of the latter rate with a tendency line extrapolating the behavior of E_0' measured experimentally to lower frequencies. In the example shown in Figure 3, the intersection of a logarithmic tendency line with the rate of loss of the static elastic modulus for pure Al in the H24 temper and pure Al gives a threshold frequency of around 0.001 and 0.005 Hz, respectively. For AA 2024-T3, the threshold frequency obtained with the same procedure would be about 0.006 Hz. For AA 7075-T6, no data on the variation of the elastic stiffness constants with temperature were found in the literature, but using the data available for AA 2024 and pure Al, a threshold frequency of 0.075 and 0.350 Hz would be obtained, respectively. These results are similar to those in the literature: Henaff et al. [26] reported a critical frequency of 0.020 Hz for AA 2650-T6, and Nikbin and Radon [25] reported that the transition region is 0.100–1 Hz for cast Al alloy RR58 at 423 K. However, to better assess the performance of the proposed procedure, further comparison with experimental data on fatigue response at very low frequencies is necessary. Unfortunately, there is a lack of this type of test data [26], due to the extremely long testing time. It is interesting to note that, on one side, according to our results, fatigue effects would seem to appear at lower loading frequencies for pure Al, compared to the alloys. Thus, apparently, the precipitation structure in the alloys would cause not only hardening, but would also enable the alloys to be loaded at a wider range of low frequencies without experiencing fatigue. On the other side, fatigue would appear already at lower frequencies for pure Al in the H24 temper compared to pure Al. In this case, the reason may be the lower ductility (and, thus, lower resistance to FCG) associated with the H24 temper.

The proposed procedure for the determination of the threshold frequency is a major result of this work. Low frequency fatigue experiments are, by definition, very long. Furthermore, there is still controversy about the effects of the exposure to the environment on the experimental results, as it is not feasible to perform longstanding experiments in constant environment conditions. Besides, the environment during service life is likely to be different from that during the experiments, and thus, the estimations of the threshold frequency based on conventional, time-consuming fatigue experiments are likely to be inaccurate. The determination of the threshold frequency from the data obtained with the relatively short DMA tests (less than 3 h) reflects the intrinsic properties of the material only. This offers also a standardized method, which allows precise comparison between different alloys. Furthermore, it is quite insensitive to the specific instrumental range available for the tests, provided the range is large enough to allow a consistent regression fit.

4. Conclusions

It was suggested that some degree of hysteresis heating is necessary for metals to experience fatigue when subjected to dynamic loading. Hysteresis heating is caused by energy dissipated due to internal friction, and thus, an increase in the latter should have an effect on the fatigue response. In particular, the results of this research suggest that the decrease of yield and fatigue behavior reported for Al alloys as temperature increases may be associated with the increase of internal friction with temperature. Due to the Arrhenius-type behavior of relaxation processes, the relaxation time decreases with temperature. The reduction in the loading frequency is equivalent to an increase in the reaction time, and therefore, there is a threshold frequency below which the reaction time is longer than the relaxation time. Consequently, the viscous effect as the loss mechanism is negligible as compared to the relaxation. In other words, with the dynamic loading frequency decreasing to very low values, eventually there will be no appreciable frictional energy dissipation and, thus, no hysteresis heating, due to mechanical relaxation phenomena. Thus, below the threshold frequency, the material will exhibit a static-like response (relaxed case), such that creep mechanisms dominate and fatigue effects are negligible. In this work, an alternative procedure to the time-consuming conventional fatigue tests is proposed to estimate this threshold frequency, based on the frequency dependence of the slope of the initial decrease of E' with temperature, which in this work is termed the temperature softening coefficient. The interesting point of our approach comes from the fact that this coefficient is easily obtained from the relatively short DMA tests, hence allowing for a fast estimation of the threshold frequency. For pure Al, AA 2024-T3 and AA 7075-T6, the threshold frequencies obtained with this procedure are 0.001–0.005, 0.006 and 0.075–0.350 Hz, respectively. This suggests that fatigue effects start to appear at lower loading frequencies for pure Al, while the alloys may be loaded at a wider range of low frequencies without experiencing fatigue, probably due to effects related to the presence of precipitates. However, to better assess the performance of the proposed procedure, further comparison with experimental data on fatigue response at very low frequencies is necessary.

Supplementary Materials: The following supplementary material is available online at http://www.mdpi.com/2075-4701/6/3/50/s1: (1) supplementary file 1 (MS Excel file): DMA test data, which includes the measured values of storage modulus E', loss modulus E'' and loss tangent (also termed mechanical damping or tanδ) for pure Al in the H24 temper, pure Al, AA 7075-T6 and AA 2024-T3; (2) Figure S1a: mechanical damping of pure Al in the H24 temper; and Figure S1b: mechanical damping of pure Al.

Acknowledgments: Work supported by the MINECO Grant FIS2014-54734-P and the Generalitat de Catalunya/Agència de Gestió d'Ajuts Universitaris i de Recerca (AGAUR) Grant 2014 SGR 00581. As part of these grants, we received funds for covering the costs to publish in open access.

Author Contributions: Jose I. Rojas conceived of, designed and performed the experiments. Jose I. Rojas and Daniel Crespo analyzed the data and wrote the paper.

Conflicts of Interest: The authors declare no conflict of interest.

Abbreviations

The following abbreviations are used in this manuscript:

AA	Aluminum alloy(s)
bcc	Body-centered cubic
CFCG	Creep-fatigue crack growth
CCG	Creep crack growth
DMA	Dynamic-mechanical analyzer
fcc	Face-centered cubic
FCG	Fatigue crack growth
GPBZ	Guinier–Preston–Bagariastkij zones
GPZ	Guinier–Preston zones
HCF	High cycle fatigue
RT	Room temperature
VHCF	Very high cycle fatigue

References

1. Callister, W.D.; Rethwisch, D.G. *Fundamentals of Materials Science and Engineering*, 4th ed.; John Wiley & Sons, Ltd.: Singapore, Singapore, 2013; pp. 1–910.

2. Van Kranenburg, C. *Fatigue Crack Growth in Aluminium Alloys*; Technische Universiteit Delft: Delft, The Netherlands, 2010; Volume 1, pp. 1–194.

3. Srivatsan, T.S.; Kolar, D.; Magnusen, P. Influence of temperature on cyclic stress response, strain resistance, and fracture behavior of aluminum alloy 2524. *Mater. Sci. Eng. A Struct. Mater. Prop. Microstruct. Process.* **2001**, *314*, 118–130. [CrossRef]

4. Amiri, M.; Khonsari, M.M. Life prediction of metals undergoing fatigue load based on temperature evolution. *Mater. Sci. Eng. A Struct. Mater. Prop. Microstruct. Process.* **2010**, *527*, 1555–1559. [CrossRef]

5. Rojas, J.I.; Aguiar, A.; Crespo, D. Effect of temperature and frequency of dynamic loading in the viscoelastic properties of aluminium alloy 7075-T6. *Phys. Status Solidi C* **2011**, *8*, 3111–3114. [CrossRef]

6. Rojas, J.I.; Crespo, D. Modeling of the effect of temperature, frequency and phase transformations on the viscoelastic properties of AA 7075-T6 and AA 2024-T3 aluminum alloys. *Metall. Mater. Trans. A* **2012**, *43*, 4633–4646. [CrossRef]

7. Nowick, A.S.; Berry, B.S. *Anelastic Relaxation in Crystalline Solids*, 1st ed.; Academic Press: New York, NY, USA, 1972; pp. 1–677.
8. Starke, E.A.; Staley, J.T. Application of modern aluminum alloys to aircraft. *Prog. Aerosp. Sci.* **1996**, *32*, 131–172. [CrossRef]
9. Hong, Y.; Zhao, A.; Qian, G. Essential characteristics and influential factors for very-high-cycle fatigue behavior of metallic materials. *Acta Metall. Sin.* **2009**, *45*, 769–780.
10. Braun, R. Transgranular environment-induced cracking of 7050 aluminium alloy under cyclic loading conditions at low frequencies. *Int. J. Fatigue* **2008**, *30*, 1827–1837. [CrossRef]
11. Nikitin, I.; Besel, M. Effect of low-frequency on fatigue behaviour of austenitic steel AISI 304 at room temperature and 25 °C. *Int. J. Fatigue* **2008**, *30*, 2044–2049. [CrossRef]
12. Furuya, Y.; Matsuoka, S.; Abe, T.; Yamaguchi, K. Gigacycle fatigue properties for high-strength low-alloy steel at 100 Hz, 600 Hz, and 20 kHz. *Scr. Mater.* **2002**, *46*, 157–162. [CrossRef]
13. Papakyriacou, M.; Mayer, H.; Pypen, C.; Plenk, H.; Stanzl-Tschegg, S. Influence of loading frequency on high cycle fatigue properties of b.c.c. and h.c.p. metals. *Mater. Sci. Eng. A Struct. Mater. Prop. Microstruct. Process.* **2001**, *308*, 143–152. [CrossRef]
14. Zhu, X.; Jones, J.W.; Allison, J.E. Effect of frequency, environment, and temperature on fatigue behavior of E319 cast aluminum alloy: Stress-controlled fatigue life response. *Metall. Mater. Trans. A Phys. Metall. Mater. Sci.* **2008**, *39A*, 2681–2688. [CrossRef]
15. Mayer, H.; Papakyriacou, M.; Pippan, R.; Stanzl-Tschegg, S. Influence of loading frequency on the high cycle fatigue properties of AlZnMgCu1.5 aluminium alloy. *Mater. Sci. Eng. A Struct. Mater. Prop. Microstruct. Process.* **2001**, *314*, 48–54. [CrossRef]
16. Liaw, P.K.; Wang, H.; Jiang, L.; Yang, B.; Huang, J.Y.; Kuo, R.C.; Huang, J.G. Thermographic detection of fatigue damage of pressure vessel steels at 1000 Hz and 20 Hz. *Scr. Mater.* **2000**, *42*, 389–395. [CrossRef]
17. Urabe, N.; Weertman, J. Dislocation mobility in potassium and iron single-crystals. *Mater. Sci. Eng.* **1975**, *18*, 41–49. [CrossRef]
18. Murakami, Y.; Nomoto, T.; Ueda, T. Factors influencing the mechanism of superlong fatigue failure in steels. *Fatigue Fract. Eng. Mater. Struct.* **1999**, *22*, 581–590. [CrossRef]
19. Mayer, H. Fatigue crack growth and threshold measurements at very high frequencies. *Int. Mater. Rev.* **1999**, *44*, 1–34. [CrossRef]
20. Laird, C.; Charsley, P. *Ultrasonic Fatigue*, 1st ed.; The Metallurgical Society of AIME: Philadelphia, PA, USA, 1982.
21. Yan, N.; Wang, Q.Y.; Chen, Q.; Sun, J.J. Influence of loading frequency on fatigue behavior of high strength steel. *Prog. Fract. Strength Mater. Struct.* **2007**, *353–358*, 227–230. [CrossRef]
22. Verkin, B.I.; Grinberg, N.M. Effect of vacuum on the fatigue behavior of metals and alloys. *Mater. Sci. Eng.* **1979**, *41*, 149–181. [CrossRef]
23. Menan, F.; Henaff, G. Influence of frequency and waveform on corrosion fatigue crack propagation in the 2024-T351 aluminium alloy in the S-L orientation. *Mater. Sci. Eng. A Struct. Mater. Prop. Microstruct. Process.* **2009**, *519*, 70–76. [CrossRef]
24. Benson, D.K.; Hancock, J.R. Effect of strain rate on cyclic response of metals. *Metall. Trans.* **1974**, *5*, 1711–1715. [CrossRef]
25. Nikbin, K.; Radon, J. Prediction of fatigue interaction from static creep and high frequency fatigue crack growth data. In Advances in Fracture Research, Proceedings of the 9th International Conference in Fracture (ICF9), Sydney, Australia, 1–5 April 1997; Karihaloo, B.L., Mai, Y.W., Ripley, M.I., Ritchie, R.O., Eds.; Pergamon Press, Ltd.: Kidlington, UK, 1997; Volume 1–6, p. 429.
26. Henaff, G.; Odemer, G.; Benoit, G.; Koffi, E.; Journet, B. Prediction of creep-fatigue crack growth rates in inert and active environments in an aluminium alloy. *Int. J. Fatigue* **2009**, *31*, 1943–1951. [CrossRef]
27. Botny, R.; Kaleta, K.; Grzebien, W.; Adamczewski, W. A method for determining the heat energy of the fatigue process in metals under uniaxial stress: Part 2. Measurement of the temperature of a fatigue specimen by means of thermovision camera-computer system. *Int. J. Fatigue* **1986**, *8*, 35–38. [CrossRef]
28. Luong, M.P. Fatigue limit evaluation of metals using an infrared thermographic technique. *Mech. Mater.* **1998**, *28*, 155–163. [CrossRef]
29. Chung, T.E.; Faulkner, R.G. Parametric representation of fatigue in alloys and its relation to microstructure. *Mater. Sci. Technol.* **1990**, *6*, 1187–1192. [CrossRef]

30. Rojas, J.I.; Lopez-Ponte, X.; Crespo, D. Effect of temperature, frequency and phase transformations on the viscoelastic behavior of commercial 6082 (Al-Mg-Si) alloy. *J. Alloys Compd.* in preparation.

31. Jeong, H.T.; Kim, J.H.; Kim, W.T.; Kim, D.H. The mechanical relaxations of a $Mm_{55}Al_{25}Ni_{10}Cu_{10}$ amorphous alloy studied by dynamic mechanical analysis. *Mater. Sci. Eng. A Struct. Mater. Prop. Microstruct. Process.* **2004**, *385*, 182–186. [CrossRef]

32. Belhas, S.; Riviere, A.; Woirgard, J.; Vergnol, J.; Defouquet, J. High-temperature relaxation mechanisms in Cu-Al solid-solutions. *J. Phys.* **1985**, *46*, 367–370. [CrossRef]

33. Riviere, A.; Gerland, M.; Pelosin, V. Influence of dislocation networks on the relaxation peaks at intermediate temperature in pure metals and metallic alloys. *Mater. Sci. Eng. A Struct. Mater. Prop. Microstruct. Process.* **2009**, *521–522*, 94–97. [CrossRef]

34. Mondino, M.; Schoeck, G. Coherency loss and internal friction. *Phys. Status Solidi A Appl. Res.* **1971**, *6*, 665–670. [CrossRef]

35. Brammer, J.A.; Percival, C.M. Elevated-temperature elastic moduli of 2024-aluminum obtained by a laser-pulse technique. *Exp. Mech.* **1970**, *10*, 245–250. [CrossRef]

36. Sutton, P.M. The variation of the elastic constants of crystalline aluminum with temperature between 63 K and 773 K. *Phys. Rev.* **1953**, *91*, 816–821. [CrossRef]

37. Kamm, G.N.; Alers, G.A. Low temperature elastic moduli of aluminum. *J. Appl. Phys.* **1964**, *35*, 327–330. [CrossRef]

38. Varshni, Y.P. Temperature dependence of the elastic constants. *Phys. Rev. B* **1970**, *2*, 3952–3958. [CrossRef]

39. Wolfenden, A.; Wolla, J.M. Mechanical damping and dynamic modulus measurements in alumina and tungsten fiber-reinforced aluminum composites. *J. Mater. Sci.* **1989**, *24*, 3205–3212. [CrossRef]

40. Granato, A.; Lucke, K. Theory of mechanical damping due to dislocations. *J. Appl. Phys.* **1956**, *27*, 583–593. [CrossRef]

41. Zhang, Y.; Godfrey, A.; Jensen, D.J. Local boundary migration during recrystallization in pure aluminium. *Scr. Mater.* **2011**, *64*, 331–334. [CrossRef]

metals

MDPI

Article

On the Improvement of AA2024 Wear Properties through the Deposition of a Cold-Sprayed Titanium Coating

Antonello Astarita [1,*], Felice Rubino [2], Pierpaolo Carlone [2], Alessandro Ruggiero [2], Claudio Leone [3,4], Silvio Genna [4], Massimiliano Merola [2] and Antonino Squillace [1]

[1] Department of Chemical, Materials and Industrial Production Engineering, University of Naples Federico II, P. Tecchio 80, Naples 80125, Italy; squillac@unina.it

[2] Department of Industrial Engineering, University of Salerno, Via Giovanni Paolo II, Fisciano, Salerno 84084, Italy; frubino@unisa.it (F.R.); pcarlone@unisa.it (P.C.); ruggiero@unisa.it (A.R.); mmerola@unisa.it (M.M.)

[3] Department of Industrial and Information Engineering, Second University of Naples, Via Roma 29, Aversa (Ce) 81031, Italy; claudio.leone@unina.it

[4] CIRTIBS Research Centre, University of Naples Federico II, P.le Tecchio 80, Naples 80125, Italy; sgenna@unina.it

* Correspondence: antonello.astarita@unina.it; Tel.: +39-081-768-2364

Academic Editor: Nong Gao
Received: 25 March 2016; Accepted: 4 August 2016; Published: 11 August 2016

Abstract: This paper deals with the study of the enhancement of the tribological properties of AA2024 through the deposition of a titanium coating. In particular two different coatings were studied: (1) untreated titanium coating; and (2) post-deposition laser-treated titanium coating. Titanium grade 2 powders were deposited onto an aluminium alloy AA2024-T3 sheet through the cold gas dynamic spray process. The selective laser post treatment was carried out by using a 220 W diode laser to further enhance the wear properties of the coating. Tribo-tests were executed to analyse the tribological behaviour of materials in contact with an alternative moving counterpart under a controlled normal load. Four different samples were tested to assess the effectiveness of the treatments: untreated aluminium sheets, titanium grade 2 sheets, as-sprayed titanium powders and the laser-treated coating layer. The results obtained proved the effectiveness of the coating in improving the tribological behaviour of the AA2024. In particular the laser-treated coating showed the best results in terms of both the friction coefficient and mass lost. The laser treatment promotes a change of the wear mechanism, switching from a severe adhesive wear, resulting in galling, to an abrasive wear mechanism.

Keywords: AA 2024; titanium; cold spray; laser treating; coating; wear; tribotest

1. Introduction

Aluminium alloys are widely used in many industrial fields (e.g., aerospace, automotive, naval, buildings and so on) due to their low weight, ease of manufacturing and good mechanical and electrochemical properties. On the other hand, for some applications (e.g., coupling with carbon fibre-reinforced plastics, applications in high corrosive environments or applications in which a superior wear resistance is required) enhanced superficial properties, in terms of both corrosion resistance and wear resistance, are required. A solution that could overcome these issues is the production of titanium coatings on aluminium components. In this way it is possible to couple the light weight and relatively low-cost nature of aluminium with the high superficial properties of titanium.

This solution is of particular interest in aeronautics, in which light weight is a primary requirement for all the components.

Aluminium alloys mainly used in aeronautics, e.g., the alloys of 2xxx and 7xxx series, are heat-treatable alloys and are age hardened. Thus, they require a deposition technique in which the substrate remains at a low temperature during the whole process. Cold gas dynamic spray (CGDS) is an innovative deposition technology matching these requirements. CGDS is an additive manufacturing process used to create a coating layer by means of high velocity impacts of metallic particles dispersed in supersonic gas flows. In this process, coating deposition occurs at relatively low temperatures when compared to other spray technologies, allowing sprayed particles to preserve their solid state. Appropriate flow conditions are enforced, combining a high pressure-heated gas with a converging/diverging nozzle (De Laval nozzle). Metallic particles are propelled to supersonic velocity (i.e., 500–1000 m/s), and the high kinetic energy causes the impingement of the particles onto the substrate [1]. The macroscopic plastic deformation, induced by the high velocity impact, was indicated by some researchers as the main bonding mechanism [2,3]. Two different scenarios have been observed, depending on material pairing and particle velocity: (a) particles rebounding from the substrate or (b) particles bonding to the substrate [4]. The velocity at which bonding is achieved is referred to as the critical velocity and relies on the particle size and distribution as well as the substrate material. The coatings exhibit high density and conductivity, good corrosion resistance and high hardness, due to the cold worked microstructure [5]. Recent studies discussed the capabilities of post-deposition laser treatments to compensate for some process drawbacks, reducing the porosity, improving the superficial hardness and increasing the anti-corrosion and tribological properties of the coating [6–8]. Several post-deposition treatments were successfully used to modify the microstructure and to improve the properties of the cold-sprayed titanium layer. Marrocco et al. (2011) proved that post-deposition laser treatment on titanium coating eliminates the residual micro-porosity and forms a high quality corrosion barrier layer [3]. Other authors studied the laser treatments in order to improve the superficial properties of both coatings and bulk materials [9–11]. The authors' previous work showed the optimal process parameters of the laser treatment to obtain a compact titanium dioxide layer on cold-sprayed titanium coating [12]. In literature, other authors investigated the deposition of titanium coatings in order to improve the tribological properties [13,14], where different coating techniques were taken into account.

This paper presents a study of the enhancement of the wear properties of an AA2024 rolled sheet through the deposition of a cold-sprayed titanium coating.

Two different coatings were studied, (1) untreated coating; (2) post-deposition laser-treated coating, in order to assess the influence of the laser treatment on the wear and tribological properties of titanium cold spray coating. Current developments in Ti processing predict the accessibility of lower-cost Ti and have revived interests in exploring the tribo-behaviour of Ti alloys as bearing materials [15–18]. Titanium and its alloys exhibit low tribological properties but their wear resistance can be improved by surface treatments promoting growth of the surface hardness, leading to changes of the wear mechanism and a lowering of the wear rate [19]. Frictional contact of titanium alloys both against other materials and titanium alloys themselves, especially under pure sliding conditions, quickly damages the contact surface area and results in the transfer of some particles of the material to the counter face [20,21]. Titanium and its alloys are characterised by a high and unstable friction factor and strong adhesive wear degradation through an elevated propensity to seizure (localised damage due to the diffusion welding between sliding surfaces [22]) and scuffing (damage characterised by surface unevenness and asperities called hillocks [23]). A couple of papers dealing with the tribological properties of cold-sprayed titanium coatings [24,25] are also available in the literature, but, to date, the coupling between the cold-sprayed coating and laser treatment has not been studied and understood.

Tribological characteristics of titanium alloys can be improved by different kinds of treatments. Cassar et al. [26] proposed a triode plasma oxidation treatment on Ti-6Al-4V, finding a reduction in the wear rate under a range of different wear test regimes. Wang et al. [27] studied the time-dependant effects of thermal oxidation in aqueous atmospheres performed on Ti-6Al-4V alloy which affect its surface roughness and wear properties. The fretting tests performed revealed an improvement in the fretting wear resistance after oxidation and an optimum time treatment equal to 4 h. Bell and Dong (2000) [6] proved that the oxidation treatment of Ti6Al4V led to a significant improvement in wear resistance, but there is no detailed analysis focusing on the application of the laser treatments on cold-sprayed coatings. Friction and wear tests were performed on untreated aluminium sheets, titanium grade 2 sheets, as-sprayed titanium powders and the laser-treated coating layer to compare the wear properties of different materials and investigate the effectiveness of the treatments.

2. Experimental Section

Commercially pure grade 2 titanium particles with a mean size of approximately 40 µm were deposited onto 3-mm-thick AA2024-T3 plates. The chemical composition and the main properties of both grade 2 titanium and AA2024 are available in literature [28,29] and not here reported in the interest of brevity. A low pressure cold spray facility equipped with a round-shaped exit nozzle, with a final diameter of 4.8 mm, was used for spraying. Helium, used as carrier gas, avoided the oxidation of titanium particles during the deposition process. The particles were sprayed at a velocity of 680 m/s with helium as carrier gas, the gas temperature and pressure were kept at 600 °C and at 12 bars. A superficial layer of cold sprayed titanium powder 5 mm thick was obtained. The successive milling process allowed to obtain the surface finishing required for the following laser treatment and tribo-tests. After the milling process, a 2.5-mm-thick titanium coating with a good superficial finishing was obtained. Laser treatments were carried out using a diode laser (IPG DLR-200-AC, IPG Photonics, Tokyo, Japan). The laser power is transferred via an optical fibre to the laser head and focused/defocused on the plate surface by means of a focusing lens (corresponding to a spot diameter of about 2 mm). Previous work [12] gave necessary information about optimal process parameters to obtain a fully dense and compact titanium oxidised layer on cold-sprayed coating surface: scan speed and power of laser beam were 50 mm/min and 200 W.

Friction and wear tests were performed with a ball-on-flat testing apparatus on a TR-BIO 282 Reciprocatory Friction Monitor (Ducom Instruments, Bangalore, India), following a consolidate protocol [30,31]. The tribopair was made up by a titanium ball and a flat specimen of the analysed materials, aluminium alloy, titanium grade 2, titanium as-sprayed coating and laser treated layer. The Ti6Al4V ball had a diameter of 3 mm, the specimens were realised with dimensions of $25 \times 25 \times 5$ mm^3. The reciprocating movement was imposed to the sphere by a stepper motor. The frictional force was monitored by a load cell, and the evolution of the coefficient of friction, during the test, was recorded. A gravimetric analysis was accomplished to evaluate the wear, expressed as mass loss. Before each test, the sample and the ball were cleaned with ethanol and compressed air, subsequently weighted on a precision scale (accuracy of 0.01 mg). These tests were made in partial compliance with the prescriptions of Test Method ASTM G133, Procedure A. The normal force was 12.0 N, instead of 25.0 N, and the test time was 20 min, instead of 16 min and 40 s as prescribed by the norm. All other advices of Test Method G133 were followed: the stroke length and the frequency oscillation were respectively 10 mm and 5 Hz (alternative motion). Considering the smaller radius of the ball—1.5 mm instead of the 4.7 mm radius recommended—a lower load was applied, the resulting pressure (Hertzian contact stress) was around 1.4 GPa, which is a suitable stress for such tribocouples. A longer test was required to look for a stable value of the friction coefficient, this choice brought a total sliding distance of 120 m instead of 100 m. Every test was realised at room temperature, in laboratory air at controlled levels of relative humidity. A topographic surface acquisition was carried out with a 3D non-contact optical profilometer, PLµ neox (Sensofar, Terrassa, Spain), which operates either as a confocal microscope or as a white light interferometer. The worn surfaces, previously cleaned from debris, were scanned using a confocal lens with magnification of 5×. The scans provided 3D and contour images, which gave qualitative information on the wear process and quantitative information, such as the maximum wear depth.

Microscopical analysis were also performed, employing Scanning Electron Microscopy (Hitachi TM 3000, Tokyo, Japan), to observe macroscopic features (depth and width) of the laser treated zone as well as to investigate microstructural aspects. Kroll's reagent (2 mL HF, 6 mL HNO_3 and H_2O up to 100 mL) was adopted to unveil microstructure, grain boundaries and phase distribution. Finally, Vickers microhardness tests were conducted on the laser treated specimen to characterise the mechanical properties of material.

3. Results and Discussion

Micrographies of a laser treated specimen (Figure 1) show a dense titanium oxide layer produced on the surface of the titanium coating. In the centre of the treated sample the oxide penetrates the coating up to about 30% of its thickness. Below the oxide layer, the heat produced by the laser source caused the formation a heat-affected zone, characterised by a microstructure with a coarser grain than the material base. In the discussion, the following terms will be used to identify the different metallurgical zones:

(1) Oxidised zone: the region of the cold-sprayed layer that was oxidised by the laser treatment;
(2) Heat-affected zone: similarly to what happens in laser beam welding, this is the region of the cold-sprayed layer that was affected by the heat generated during the treatment;
(3) Base material: the region of the cold-sprayed layer that was not affected by the laser treatment;
(4) Substrate: the aluminium plate used as a substrate for the deposition process.

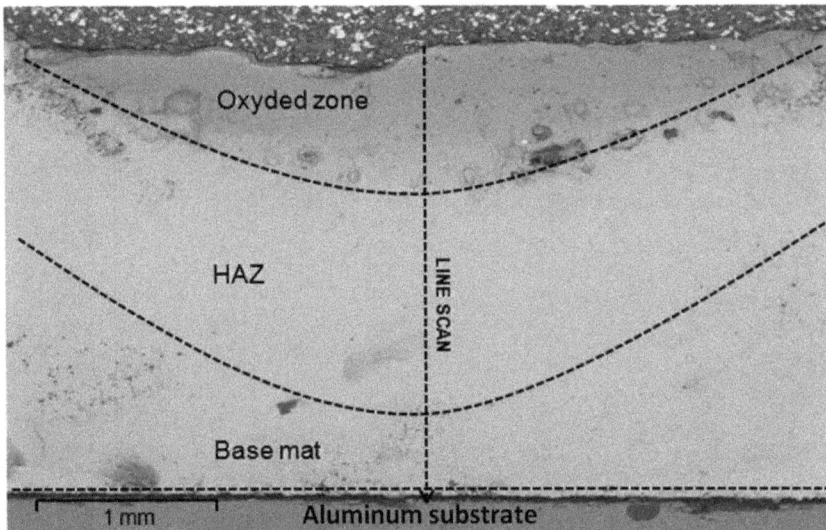

Figure 1. Cross-section of laser treated specimen (magnification 60×).

Figure 2 shows the different microstructures produced by the laser treatment within the cold-sprayed layer. Three different zones are visible in the sections of the treated specimen: the oxidised zone (Figure 2a), in which the typical microstructure of the titanium oxide is exhibited, the heat-affected zone (Figure 2b), which is characterised by lamellae of coarser dimensions than the as-sprayed titanium due to the heat input occurring during the laser process, and the base material (Figure 2c), characterised by lamellae of coarser dimensions. EDS analyses confirmed the existence of a rutile (TiO_2

titanium dioxide) layer. Line scan EDS analysis (Figure 3), performed in the center of the cross-section following the pattern labelled "line scan" represented in Figure 1, revealed the concentration of titanium and oxygen from the top of the surface to the interface between the coating and substrate.

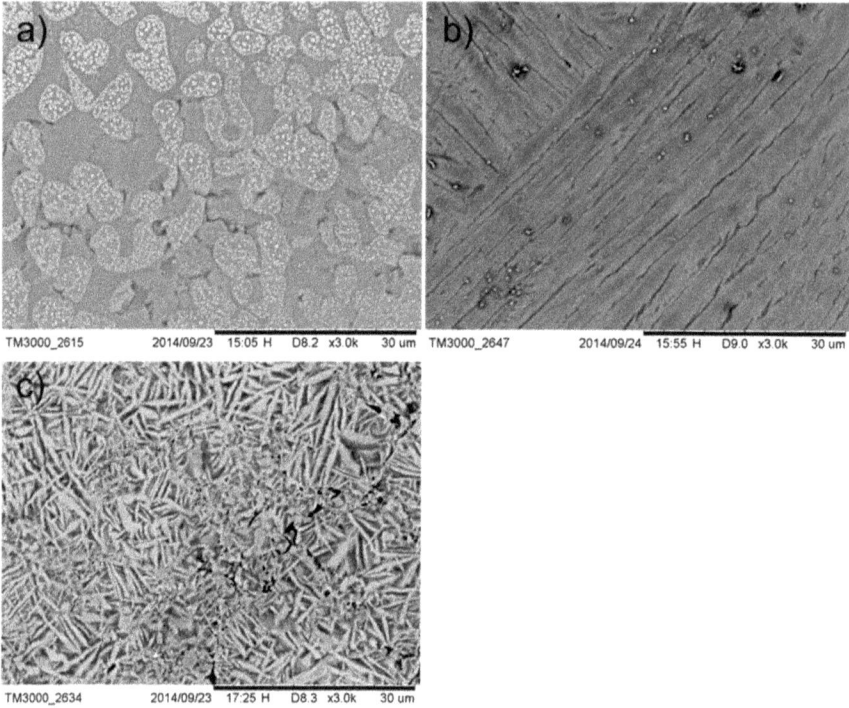

Figure 2. Different microstructures observed in the cold-sprayed coating after the laser treatment: (**a**) titanium oxidised zone; (**b**) heat-affected zone; (**c**) base material.

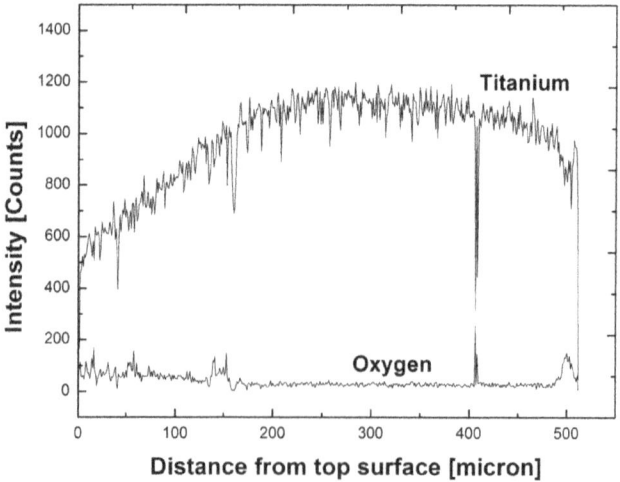

Figure 3. Titanium and oxygen intensity along the scan line indicated in Figure 1.

The relative concentration of two elements (Ti, O), in the topmost layer of the cross-section, is approximately equal to the stoichiometric one for the titanium dioxide; moving to the interface with the aluminium substrate, the oxygen percentage decreases to characteristic values of titanium grade 2. It should be noted that the XRD analysis of the treated surface, presented by the authors in a previous paper [32], confirmed the formation of a rutile layer as well as a thick oxygen diffused zone underlying the oxidised zone. This result will be useful in the further discussion of the results.

The micro-hardness values of treated materials on the coating surface were about 1000 HV, much greater than the values of cold-sprayed titanium (about 200 HV) reported in the literature [33] (Figure 4). The tests were performed following the same pattern of the EDS analysis.

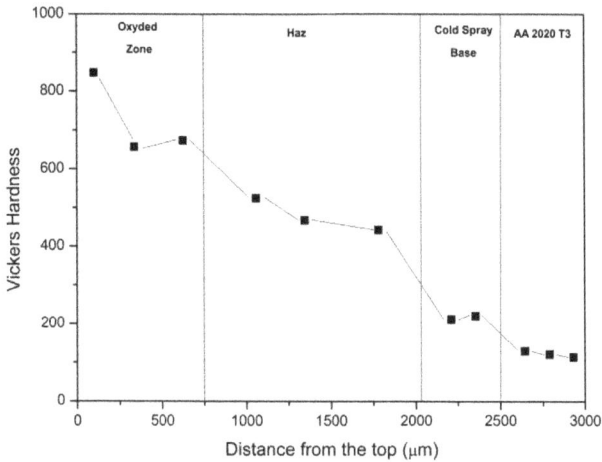

Figure 4. Microhardness measurements taken through the plate cross-section.

Microhardness measurements strongly indicated the formation of a very hard surface layer. This observation is in agreement with the results of the tribological test (Figure 2), which showed the major mass loss occurred in the titanium ball rather than in the laser-treated titanium. In addition, microhardness tests on the substrate indicated a microhardness close to 150 HV (the pre-deposition microhardness of the AA 2024 T3 alloy) which suggests that the aluminium substrate was mechanically unaffected by the deposition process. Thus, it can be conjectured that both the deposition process and the laser treatment do not affect the temper state of the aluminium substrate, which is an extremely important result of this work.

The main aim of the tribo-tests was to evaluate the wear resistance of the materials. Tribological tests have highlighted that the laser treatment can significantly improve the friction and wear properties of titanium. Figure 5 compares the development of the friction coefficient (CoF) of the laser-treated coating with the cold-sprayed coating, as well as samples of the as-received titanium and aluminium alloy.

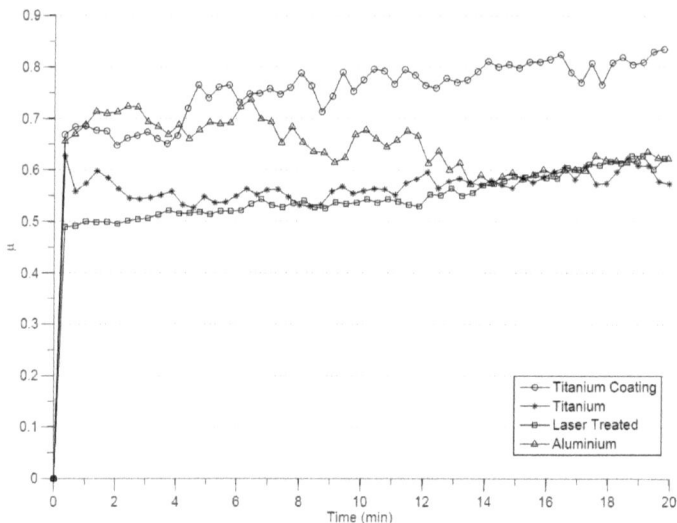

Figure 5. Evolution of friction coefficients for the processed plates with time.

The evolution of the friction coefficient is unsteady and its variation derives from two elements: the trend and the fluctuations around the mean value. The trend is due to several phenomena such as the reciprocal adjustment of surface asperities leading to an alteration of the real contact area, the inclusion of (micro) wear debris [34]—third body wear—and the variability of the contact temperature; the fluctuation around the mean is typical of the reciprocating test where the relative velocity is variable. The aluminium plate vs. titanium ball tribopair couple (the latter, hereinafter, omitted for simplicity) indicates a slightly increasing trend only after 14 min of testing, where μ is near 0.60. For the titanium plate μ was in the range of 0.50–0.60 for the duration of the test, ending with a constant trend. This result is consistent with the values proposed by Miller and Budinski [35,36], who, in their investigations on the tribological properties of titanium and its alloys, analysed several tribo-pairings and wear mechanisms. In the case of the titanium cold spray coating, after 14 min, fluctuations were still noticeable due to the surface unevenness (a characteristic of the as-sprayed coating) and the friction coefficient, μ, varied in the range of 0.75–0.85, with an increasing trend. This trend is consistent with the results of Khun et al. [25]. They observed a higher friction coefficient of Ti cold spray coating compared to the Ti bulk material, with a very irregular profile and large fluctuations. The measured value CoF, after an initial peak, reached a steady-state value of approximately 0.78. Finally, the development of the laser-treated coating presented a rising trend, reaching—with small fluctuations—the value of 0.62. The overall rising trend in the final phase of the tests is supported by the results of Wang et al. from their tribological analysis on rutile coating [27], where the formation of debris is held accountable for this phenomenon. From our analysis a similar behaviour emerged from the different tribocouples in the last six minutes of the test, excluding the titanium coating for which higher μ values were found. In Figure 6, the results from the gravimetric analysis are presented. Both the values of the specimen wear and the sphere wear are shown and summed up. Concerning the aluminium case, the wear of the ball was almost zero, whereas the flat sample lost a considerable quantity of mass. Laser-treated coating returned the highest wear of the ball with respect to the different tribopairs. On the other hand, this specimen lost very little mass. The overall wear of the tribopair for the laser-treated coating case is comparable to the one relative to the titanium plate; in the latter case, however, the main loss was detected on the specimen side. The oxidation of titanium indicated significant improvements in its tribological properties. The TiO$_2$ showed a smoother friction coefficient profile with less fluctuations and reached a lower value of μ

compared to the titanium plate, which is characterised by an unstable CoF with evident fluctuations. The same behaviour was observed for the cold spray coating. The improvements are also highlighted by the limited wear rate of the oxidised coating (see Figure 6), which is much lower than the as-sprayed titanium and titanium bulk material. Krishna et al. [37] confirmed this behaviour. They proposed that the improvement in wear resistance is due to the high hardness of the oxide layer and its strong adhesion with the underlying material. Dalili et al. [38] confirmed the role played by the enhanced surface hardness, resulting from the formation of a harder oxide layer and a thick oxygen diffusion zone, in improving the wear resistance of titanium which resulted in a lower friction coefficient and negligible weight loss. They proposed that the oxide layer prevents extensive plastic deformation of the titanium, thus changing the nature of the contact area from metallic/metallic pair to ceramic/metallic tribo-pair [39,40]. This improvement of the wear properties of the titanium layer is fundamentally linked to the presence and contribution of rutile. In this regard, in his papers [41] on the tribological properties of titanium, Sun proved that the improvement of the tribological properties of the titanium after the thermal oxidation treatment can be attributed to the formation of a rutile layer and also highlighted the correlation between the tribological properties and the characteristics of the rutile [42].

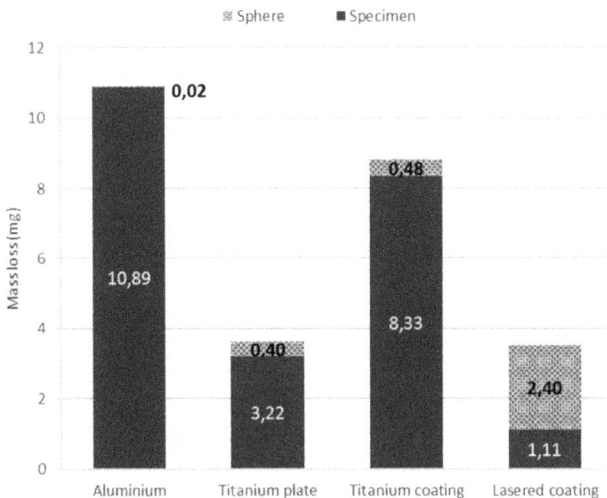

Figure 6. Wear comparison in term of mass loss for the four processing methods.

The tribotest instrument recorded the evolution of the frictional force during the sliding of the sphere along the test, and the force was recorded with an acquisition frequency equal to 100 times the oscillating frequency.

In Figure 7, the typical evolution of the frictional force during sliding is represented for each specimen type. The fitted points represent that two cycles took every 10,000 points, assumed to be exemplificative of the total evolution during the test. The X-axis plots the acquisition points, corresponding to the time progress, in a logarithmic scale. In Figure 7a the aluminium case is shown, and the cycles are quite irregular; this unpredicted behaviour, in the first analysis, can be attributed to an inhomogeneous specimen density in the proximity of the centre line of the stroke. In Figure 7b, the force evolution cycles for the titanium plate are shown. Again, the force has an irregular gait, but exhibits the typical hysteresis shape typical of this evolution. It has two peaks along the two central axes, where the ball diverts its motion and, consequently, the friction switches from the kinetic to the static phase. The loop cycles of the last two instances, namely the titanium coating (Figure 7c) and the laser-treated coating (Figure 7d), are outlined by a smooth hysteresis. These cycles exhibit a regular and flat shape during the sliding of the ball, i.e., the kinetic phase. The values of the energy ratio evaluated all along the test confirmed that the gross slip regime (wear-dominated regime) is always greater than 0.2, i.e., from a minimum of 0.3 to the maximum 0.9.

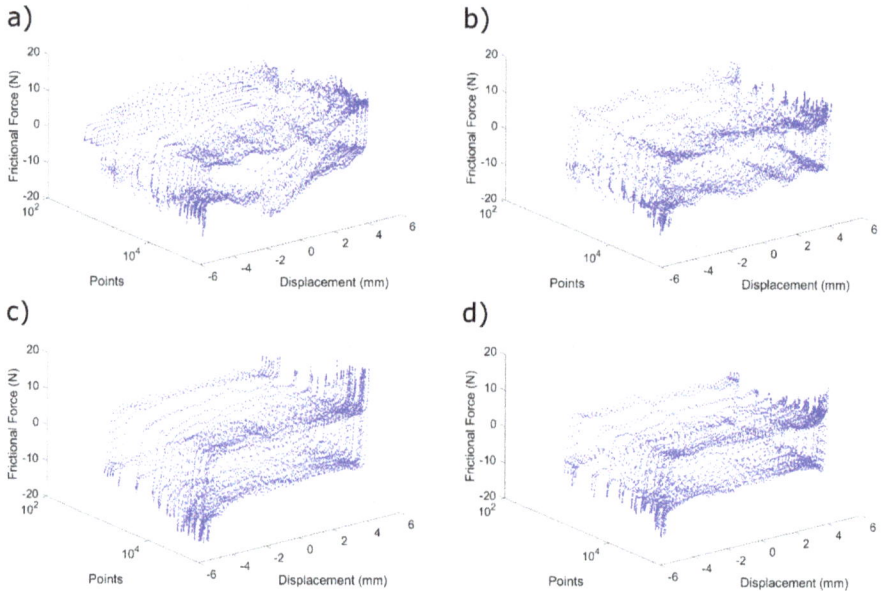

Figure 7. Loop cycles of the frictional force for: (**a**) aluminium; (**b**) titanium plate; (**c**) titanium coating; (**d**) laser-treated coating cases.

In Figure 8, the images obtained from the topography acquisitions are displayed. The wear track on the aluminium plate (Figure 8a) displays an uneven shape: it is non-symmetrical in respect to the *Y*-axis (perpendicular to the sliding direction). This observation is congruent with the friction loop cycles, which also exhibited irregular profiles. The wear track is deeper in proximity to the right central test-axis where the frictional force was the highest: the maximum wear depth along this midline profile is equal to 470 μm—the deepest recorded among the specimens. The titanium plate also exhibits an irregular shape (Figure 8b), with the deepest wear in the proximity of the two central axes of the ball—120 μm for the left and 130 μm for the right. It is also noticeable, as red points on the contour image, that there is an area of strong plastic deformation along the border of the wear track. In Figure 8c, the worn track of the titanium coating is displayed, and the profile is regular and smooth, presenting a maximum wear of 251 μm. This confirmed the poor resistance of cold spray coating against the counter-sphere, which results in a higher weight loss of the coating itself compared to both the titanium plate and oxidised layer (see Figure 6). Figure 8d shows the topography images on the laser-treated coating worn surface. As expected, the topography did not provide an unequivocal estimation of the wear depth, due to the irregular profile produced by the laser operation. As a result, the surface has an indented profile and the worn area is only distinguishable due to the presence of a large boundary incision.

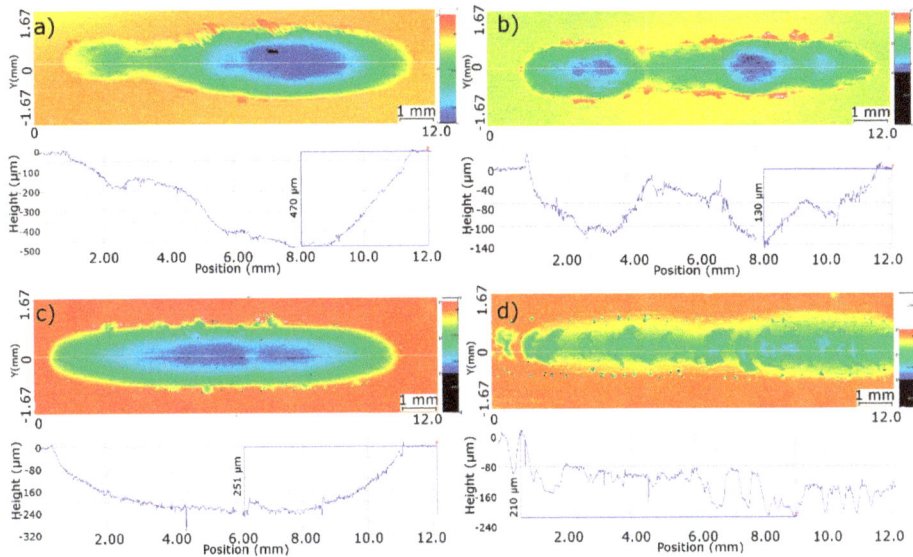

Figure 8. Wear tracks: contours and midline profiles: (**a**) aluminium plate; (**b**) titanium plate; (**c**) as-sprayed titanium coating; (**d**) laser-treated titanium coating.

4. Conclusions

The above presented and discussed results are summarised as follows. Both coatings under investigation, the treated one and the untreated one, showed better wear performances relative to the untreated AA2024 alloy. In particular the laser-treated coating showed the best wear resistance, better than the bulk titanium. These enhanced properties can be attributed to the presence of a superficial hard oxidised layer produced by the laser treatment. The presence of this oxide layer was confirmed by the EDX analysis in which an increase in the oxygen content within the treated area was observed. Three different metallurgical zones, with different microstructures, were visible in the areas of the treated tracks: the base material, the heat-affected zone and the treated zone. On the other hand, the temper state of the aluminium substrate was not affected by the process. The laser treatment influenced both the friction coefficient and the wear mechanism of the coating. The untreated coating showed an adhesive wear with a noticeable mass loss from the coating itself and a negligible wear of the ball. Abrasive wear mechanisms were observed testing the treated coating, with a negligible wear of the coating but a severe wear and mass lost on the sphere.

Acknowledgments: The authors gratefully thank Dermot Brabazon of the School of Mechanical and Manufacturing Engineering of the Dublin City University for the fruitful discussion concerning laser processing and treatments.

Author Contributions: All the authors contributed equally to this research activity. In particular: Claudio Leone and Silvio Genna carried out the laser treatment; Alessandro Ruggiero and Massimiliano Merola carried out the tribological tests; Antonello Astarita and Antonino Squillace carried out the cold spray deposition and also did some experimental tests; Pierpaolo Carlone and Felice Rubino carried out the experimental characterization. All the authors contributed at the writing of the paper.

Conflicts of Interest: The authors declare no conflict of interest.

References

1. Schmidt, T.; Gaertner, F.; Kreye, H. New Developments in Cold Spray Based on Higher Gas and Particle Temperatures. *J. Therm. Spray Technol.* **2006**, *15*, 488–494. [CrossRef]
2. Morgan, R.; Fox, P.; Pattison, J.; Sutcliffe, C.; O'Neill, W. Analysis of cold gas dynamically sprayed aluminium deposits. *Mater. Lett.* **2004**, *58*, 1317–1320. [CrossRef]
3. Marrocco, T.; McCartney, D.; Shipway, P.; Sturgeon, A. Production of titanium deposits by cold-gas dynamic spray: Numerical modeling and experimental characterization. *J. Therm. Spray Technol.* **2006**, *15*, 263–272. [CrossRef]
4. Astarita, A.; Durante, M.; Langella, A.; Montuori, M.; Squillace, A. Mechanical characterization of low pressure cold-sprayed metal coatings on aluminium. *Surf. Interface Anal.* **2013**, *45*, 1530–1535. [CrossRef]
5. Irissou, E.; Legoux, J.G.; Arsenault, B.; Moreau, C. Investigation of Al-Al$_2$O$_3$ Cold Spray Coating Formation and Properties. *J. Therm. Spray Technol.* **2007**, *16*, 661–668. [CrossRef]
6. Bell, T.; Dong, H. Enhanced wear resistance of titanium surfaces by a new thermal oxidation treatment. *Wear* **2000**, *238*, 131–137.
7. Wang, Y.; Li, C.; Guo, L.; Tian, W. Laser remelting of plasma sprayed nanostructured Al$_2$O$_3$-TiO$_2$ coatings at different laser power. *Surf. Coat. Technol.* **2010**, *204*, 3559–3566. [CrossRef]
8. Sidhu, B.S. Laser surface remelting to improve the erosion-corrosion resistance of nickel-chromium-aluminium-yttrium (NiCrAlY) plasma spray coatings. In *Laser Surface Modifications of Alloys for Corrosion and Erosion Resistance*; Kwok, C.T., Ed.; Woodhead Publishing: Cambridge, UK, 2012; pp. 355–366.
9. Chikarakara, E.; Aqida, S.; Brabazon, D.; Naher, S.; Picas, J.A.; Punset, M.; Forn, A. Surface modification of HVOF thermal sprayed WC-CoCr coatings by laser treatment. *Int. J. Mater. Form.* **2010**, *3*, 801–804. [CrossRef]
10. Chikarakara, E.; Naher, S.; Brabazon, D. High speed laser surface modification of Ti-6Al-4V. *Surf. Coat. Technol.* **2012**, *206*, 3223–3229. [CrossRef]
11. Chikarakara, E.; Naher, S.; Brabazon, D. Process mapping of laser surface modification of AISI 316L stainless steel for biomedical applications. *Appl. Phys. A Mater. Sci. Process.* **2010**, *101*, 367–371. [CrossRef]
12. Rubino, F.; Astarita, A.; Carlone, P.; Genna, S.; Leone, C.; Minutolo, F.M.C.; Squillace, A. Selective Laser Post-Treatment on Titanium Cold Spray Coatings. *Mater. Manuf. Process.* **2015**. [CrossRef]
13. Zhang, E.; Xu, L.; Yang, K. Formation by ion plating of Ti-coating on pure Mg for biomedical applications. *Scr. Mater.* **2005**, *53*, 523–527. [CrossRef]
14. Zhang, X.H.; Liu, D.X.; Tan, H.B.; Wang, X.F. Effect of TiN/Ti composite coating and shot peening on fretting fatigue behaviour of TC17 alloy at 350 °C. *Surf. Coat. Technol.* **2009**, *203*, 2315–2321. [CrossRef]
15. Qu, J.; Blau, P.J.; Watkins, T.R.; Cavin, O.B.; Kulkarni, N.S. Friction and wear of titanium alloys sliding against metal, polymer, and ceramic counterfaces. *Wear* **2005**, *258*, 1348–1356. [CrossRef]
16. Nieslony, P.; Cichosz, P.; Krolczyk, G.M.; Legutko, S.; Smyczek, D.; Kolodziej, M. Experimental studies of the cutting force and surface morphology of explosively clad Ti-steel plates. *Measurement* **2016**, *78*, 129–137. [CrossRef]
17. Ruggiero, A.; D'Amato, R.; Gómez, E.; Merola, M. Experimental comparison on tribological pairs UHMWPE/TIAL6V4 alloy, UHMWPE/AISI316L austenitic stainless and UHMWPE/Al$_2$O$_3$ ceramic, under dry and lubricated conditions. *Tribol. Int.* **2016**, *96*, 349–360. [CrossRef]
18. Ruggiero, A.; D'Amato, R.; Gómez, E. Experimental analysis of tribological behaviour of UHMWPE against AISI420C and against TiAl6V4 alloy under dry and lubricated conditions. *Tribol. Int.* **2015**, *92*, 154–161. [CrossRef]
19. Bloyce, A.; Morton, P.H.; Bell, T. *ASM Handbook*; ASM Int.: Materials Park, OH, USA, 1994; Volume 5, p. 835.
20. Bell, T.; Dong, H. Tribological enhancement of titanium alloys. In Proceedings of the First Asian Conference on Tribology, Beijing, China, 12–15 October 1998; pp. 421–427.
21. Dong, H.; Bell, T. Towards designer surfaces for titanium components. *Ind. Lubr. Tribol.* **1998**, *50*, 282–289. [CrossRef]
22. Friction, Lubrication, and Wear Technology. In *ASM Handbook*; Blau, P.J., Ed.; ASM Int.: Materials Park, OH, USA, 1994; Volume 18, p. 1879.
23. Astarita, A.; Durante, M.; Langella, A.; Squillace, A. Elevation of tribological properties of alloy Ti-6% Al-4%V upon formation of a rutile layer on the surface. *Met. Sci. Heat Treat.* **2013**, *54*, 662–666. [CrossRef]

24. Kataria, S.; Kumar, N.; Dash, S.; Tyagi, A.K. Tribological and deformation behaviour of titanium coating under different sliding contact conditions. *Wear* **2010**, *269*, 797–803. [CrossRef]

25. Khun, N.W.; Li, R.T.; Loke, K.; Khor, K.A. Effects of Al-Cr-Fe Quasicrystal Content on Tribological Properties of Cold-Sprayed Titanium Composite Coatings. *Tribol. Trans.* **2015**, *58*, 616–624. [CrossRef]

26. Cassar, G.; Wilson, J.C.A.-B.; Banfield, S.; Housden, J.; Matthews, A.; Leyland, A. Surface modification of Ti-6Al-4V alloys using triode plasma oxidation treatments. *Surf. Coat. Technol.* **2012**, *206*, 4553–4561. [CrossRef]

27. Wang, S.; Liao, Z.; Liu, Y.; Liu, W. Influence of thermal oxidation duration on the microstructure and fretting wear behaviour of Ti6Al4V alloy. *Mater. Chem. Phys.* **2015**, *159*, 139–151. [CrossRef]

28. Davis, J.R. (Ed.) *ASM Specialty Handbook: Aluminium and Aluminium Alloys*; ASM Int.: Materials Park, OH, USA, 1993.

29. Hanson, B. *The Selection and Use of Titanium: A Design Guide*; Institute of Materials: London, UK, 1995.

30. Ruggiero, A.; Valasek, P.; Merola, M. Friction and Wear Behaviours of Al/Epoxy Composites during Reciprocating Sliding Tests. *Manuf. Technol.* **2015**, *15*, 684–689.

31. Ruggiero, A.; Merola, M.; Carlone, P.; Archodoulaki, V.-M. Tribo-mechanical characterization of reinforced epoxy resin under dry and lubricated contact conditions. *Compos. B Eng.* **2015**, *79*, 595–603. [CrossRef]

32. Carlone, P.; Astarita, A.; Rubino, F.; Pasquino, N.; Aprea, P. Selective Laser Treatment on Cold-Sprayed Titanium Coatings: Numerical Modeling and Experimental Analysis. *Metall. Mater. Trans. B* **2016**. [CrossRef]

33. Wong, W.; Rezaeian, A.; Irissou, E.; Legoux, J.-G.; Yue, S. Cold Spray Characteristics of Commercially Pure Ti and Ti-6Al-4V. *Adv. Mater. Res.* **2010**, *89–91*, 639–644. [CrossRef]

34. Cassar, G.; Wilson, J.C.A.-B.; Banfield, S.; Housden, J.; Matthews, A.; Leyland, A. A study of the reciprocating-sliding wear performance of plasma surface treated titanium alloy. *Wear* **2010**, *269*, 60–70. [CrossRef]

35. Miller, P.D.; Holladay, J.W. Friction and Wear Properties of Titanium. *Wear* **1958**, *2*, 133–140. [CrossRef]

36. Budinski, K.G. Tribological properties of titanium alloys. *Wear* **1991**, *151*, 203–217. [CrossRef]

37. Krishna, D.S.R.; Brama, Y.L.; Sun, Y.R. Thick rutile layer on titanium for tribological applications. *Tribol. Int.* **2007**, *40*, 329–334. [CrossRef]

38. Dalili, N.; Edrisy, A.; Farokhzadeh, K.; Li, J.; Lo, J.; Riahi, A.R. Improving the wear resistance of Ti-6Al-4V/TiC composites through thermal oxidation (TO). *Wear* **2010**, *269*, 590–601. [CrossRef]

39. Kim, K.; Geringer, J. Analysis of energy dissipation in fretting corrosion experiments with materials used as hip prosthesis. *Wear* **2012**, *296*, 497–503. [CrossRef]

40. Fouvry, S.; Kapsa, P.; Vincent, L. Analysis of sliding behaviour for fretting loadings: Determination of transition criteria. *Wear* **1995**, *185*, 35–46. [CrossRef]

41. Sun, Y. Thermally oxidised titanium coating on aluminium alloy for enhanced corrosion resistance. *Mater. Lett.* **2004**, *58*, 2635–2639. [CrossRef]

42. Sun, Y. Tribological rutile-TiO_2 coating on aluminium alloy. *Appl. Surf. Sci.* **2004**, *233*, 328–335. [CrossRef]

metals

MDPI

Article

Effect of Friction Stir Welding Parameters on the Mechanical and Microstructure Properties of the Al-Cu Butt Joint

Sare Celik [1],* and Recep Cakir [2]

[1] Department of Mechanical Engineering, Faculty of Engineering and Architecture, Balikesir University, Balikesir 10145, Turkey
[2] Personnel Recruitment Resources, Turkish Land Forces, Ankara 06590, Turkey; cakirbey2006@hotmail.com
* Correspondence: scelik@balikesir.edu.tr; Tel.: +90-266-612-9495

Academic Editor: Nong Gao
Received: 7 April 2016; Accepted: 23 May 2016; Published: 31 May 2016

Abstract: Friction Stir Welding (FSW) is a solid-state welding process used for welding similar and dissimilar materials. FSW is especially suitable to join sheet Al alloys, and this technique allows different material couples to be welded continuously. In this study, 1050 Al alloys and commercially pure Cu were produced at three different tool rotation speeds (630, 1330, 2440 rpm) and three different tool traverse speeds (20, 30, 50 mm/min) with four different tool position (0, 1, 1.5, 2 mm) by friction stir welding. The influence of the welding parameters on the microstructure and mechanical properties of the joints was investigated. Tensile and bending tests and microhardness measurements were used to determine the mechanical properties. The microstructures of the weld zone were investigated by optical microscope and scanning electron microscope (SEM) and were analyzed in an energy dispersed spectrometer (EDS). Intermetallic phases were detected based on the X-ray diffraction (XRD) analysis results that evaluated the formation of phases in the weld zone. When the welding performance of the friction stir welded butt joints was evaluated, the maximum value obtained was 89.55% with a 1330 rpm tool rotational speed, 20 mm/min traverse speed and a 1 mm tool position configuration. The higher tensile strength is attributed to the dispersion strengthening of the fine Cu particles distributed over the Al material in the stir zone region.

Keywords: Friction Stir Welding; AA1050; Cu; mechanical properties; microstructure

1. Introduction

Friction Stir Welding (FSW), was invented and patented by The Welding Institute UK (TWI) in 1991 [1]. FSW as a solid-state process has gained a lot of importance due to its advantages such as providing good mechanical properties, especially with aluminum alloy, and quality joints [2,3]. This method has advantages compared to conventional welding methods since there is no distortion, porosity and cracks during the application [4,5]. Very good quality welds have been obtained using FSW in joining aluminum, magnesium, titanium, copper and steel materials. Recently, studies on joining dissimilar materials have been carried out [6–8]. The accurate joining of dissimilar materials is very important in terms of its use in important fields including the chemical, nuclear, aerospace, transportation, power generation, and electronics industries [9,10].

Copper and aluminum are important metals for the electrical industry due to their good electrical and thermal conductivity as well as high corrosion resistance and mechanical properties. Many studies for different welding methods have been conducted in order to joint these two materials in high-voltage, direct-current distribution lines; and the different techniques of joining copper/aluminum has become a research subject [11]. However, the welding of aluminum to copper by fusion welding is generally difficult because of the wide difference in their physical, chemical and mechanical properties and the tendency to form brittle intermetallic compounds (IMCs). Therefore, solid-state joining methods such as friction welding, roll welding and explosive welding have received much attention. These methods, however, have a few drawbacks. For example, friction welding and roll welding lack versatility, and there are safety problems involved in explosive welding [12].

Several studies have been carried out on the effects of dissimilar aluminum and copper welding parameters on the microstructure and mechanical properties in the weld zone and the detection of intermetallic phases that occurs in the weld zone [5,13–17]. In fact, several works have already addressed the dissimilar friction stir welding of these materials, in both butt and lap joint configurations. However, Al-Cu lap joining has been much more explored than friction stir butt welding, for which, so far, only a small number of studies have been conducted [8]. The studies have concluded with different results and could not achieve high strengths, yet very few studies have addressed tool positioning parameters. In particular, the effect of the tool positioning on the complex material flow pattern and the resultant properties have not yet been revealed in detail for Al-Cu materials.

In this study, AA1050 with a thickness of a 4 mm is friction stir welded to pure copper sheets at three different tool rotation speeds (630, 1330, 2440 rpm), three different tool traverse speeds (20, 30, 50 mm/min), and four different tool positions (0, 1, 1.5, 2 mm); finally, the mechanical and microstructural properties of the joint are evaluated.

2. Materials and Methods

Pure copper (99.9%) and 1050 aluminum alloy plates with a thickness of 4 mm were joined by FSW. Aluminum and copper plates are prepared in 100×150 mm dimensions. The mechanical properties of aluminum and copper that are used in this study is shown in tab:metals-06-00133-t001.

Table 1. Mechanical properties of Al and Cu.

Properties	Aluminum (Al)	Copper (Cu)
Tensile Strength (MPa)	111.20	231.38
Elongation (%)	14.98	41.03
Hardness (HV)	41	88

Two materials are positioned on the fixture and it is ensured that they do not draw apart; Cu is leaned to the advancing side, while Al is leaned to the retreating side as shown in Figure 1.

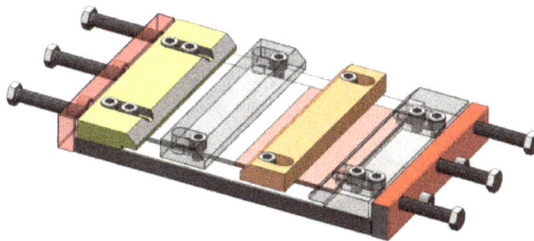

Figure 1. Schematic representation of the fixture.

The tool material selected is high-speed steel in order to keep the hardness resistance and avoid corrosion on the stir pin during the process. Heat treatment is applied to the stir pin and a 62HRc value is achieved. A cylindrical tool of M4 $\times$ 3.87 mm with a shoulder of 18 mm is used. The welding parameters are determined by preliminary studies and literature. The constant parameters are as follows:

- Direction of rotation of the tool: Clockwise
- Tilt Angle: $1.5°$
- Standby Time: 60 s.

Experiments were performed with different sets of rotational and traverse speeds in order to achieve high strength in the welded parts. In these experiments, stir pin was positioned at "0" (zero) on both aluminum and copper plates. Although the welding surface appearance seems proper, gaps in microstructure were formed, as shown in Figure 2. The gaps and welding that was not fully formed cause low mechanical values in the welded parts, and low tensile strength. It was therefore concluded that the welding of the materials was not fully performed. Afterwards, the studies were continued by changing the position of the stir pin. It was positioned to the Al side from the butt center line since it is a softer material compared to Cu. After preliminary trials, with the understanding of the significant importance of the tool positioning, the welding parameters were determined as shown in tab:metals-06-00133-t002. The nomenclature adopted in the text for labelling the different welds will identify the welding condition, *i.e.*, 630/20/1 means 630 rpm of rotational speed, 20 mm· min^{-1} of traverse speeds and 1 mm of pin positions, respectively.

Figure 2. Weld cross section with "0" tool position.

Table 2. Al-Cu Welding Parameters in Friction Stir Welding (FSW).

Tool Rotational Speed (rpm)	Tool Traverse Speed (mm/min)	Tool Positioning (to the Al Side (mm))
630–1330–2440	20	1
		1.5
		2
	30	1
		1.5
		2
	50	1
		1.5
		2

The tensile specimens were extracted from the weld joint and tested using an electromechanical controlled universal testing machine as per ASTM E8 M-04 guidelines. Three tensile tests have been performed for every welding sample and the average value has been obtained. The strain rate was 2 mm/min. Bending test specimens were prepared perpendicular to the welding direction in accordance with the ASTM E855-08 standard. Two rows of microhardness measurement were made from both the lower and the upper surface of specimens that were perpendicular to the welding section. The first measurement was taken at 0.5 mm below the surface, and the second measurement was taken at 0.5 mm above the lower surface. A sanding process with grit No. from 220 to 1200 according to CAMI grit designation sandpapers was performed on the samples that were taken from the cross section perpendicular to the welding direction in order to detect the microstructural changes at the weld zones after joining. The welded area was polished with 3 μm and 1 μm diamond paste and etched. In the etching process; 100 mL of distilled water, 4 mL of saturated sodium chloric, 2 g of potassium dichromate and an etching reagent consisting of 5 mL sulfuric acid were used for the Cu side; Keller's solution was used for Al side, and the results were examined with a Nikon Eclipse MA100 optical microscope (Nikon, Tokyo, Japan) in the laboratories of Turkish Land Forces which is located Balikesir, Turkey. In addition, point and linear energy dispersed spectrometer (EDS) analyses were carried out after the examination of the weld zones with a scanning electron microscope (SEM) in the Scientific and Technological Research Council of Turkey (TUBITAK) that is located in Gebze, Turkey. X-ray diffraction (XRD) analysis was conducted to examine the phase occurring in the weld zone.

3. Results and Discussions

Cross sections perpendicular to the welding direction, and the bottom and top surfaces of joints that are formed with dissimilar welding parameters were photographed. Images from the welded parts are given in Figures 3 and 4. By comparing the surface photographs in Figures 3a,b and 4a,b, the differences in surface finishing can be easily observed. Welding defects such as gaps, holes and joint failure were not registered when the bottom and top surface of the welded part were examined. In fact, whereas the 1330/20/1 weld presents a very smooth surface composed of regular and well-defined striations, similar to those obtained in similar copper friction stir welding by Galvão *et al.* [8], signs of significant tool submerging and the formation of massive flash are observed at the surface of the 630/50/1 weld. It is important to stress that, although both welds have been carried out under the same welding conditions, the 630/50/1 weld surface presents defects usually associated with excessive heat input during friction stir welding. This result is in good agreement with Leitão *et al.* [18], who studied the influence of base materials properties on defect formation during AA5083 and AA6082 FSW.

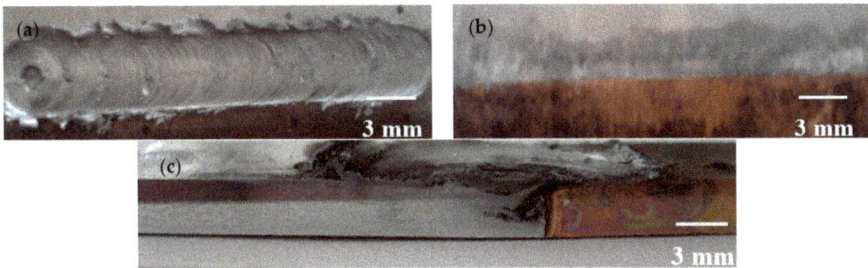

Figure 3. Macrograph of the welded part under 630/50/1 conditions: (**a**) Upper surface; (**b**) Lower surface; (**c**) Cross section.

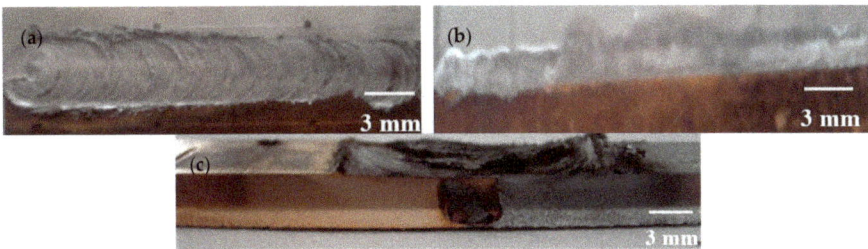

Figure 4. Welded part macro-images under 1330/20/1 condition (**a**) Upper surface; (**b**) Lower surface; (**c**) Cross section.

Comparing the cross section macrographs of both welds, displayed in Figures 3c and 4c, important differences in the structure and morphology of the bonding area can also be observed. The image of the cross section of the 630/50/1 weld shows that the Al-Cu interaction zone of this weld is restricted to the pin influence zone. Minor evidence of the material stirred by the pin can be observed in Figure 3c, that the total inefficient mixing between the aluminum and copper gave rise to a large discontinuity between both base materials, preventing the effective joining of the plates. In fact, according to Figure 3, the coupling between the two materials only occurred at the advancing side of the tool where the aluminum was pushed into the copper. The cross section macrographs of the 1330/20/1 weld are shown in Figure 4c. From the pictures, it can be concluded that the Cu/Al interaction volume for the 1330/20/1 weld is significantly larger than that observed for the 630/50/1 weld.

A full mixture could not be reached in "0" positioned Al-Cu joining; however, 1, 1.5 and 2 mm tool shifting led to a homogeneous mixture, increasing the mechanical values. Tensile specimens that were friction stir welded with tool shifting are given Figure 5, and the tensile strength test depending on the rotational speed results are given in Figures 6–8. The tensile strength of Al and Cu were found to be 111.20 MPa and 231.38 MPa, respectively. As seen in the strength chart, the 1330/20/1 specimen has the highest tensile strength at 99.58 MPa, and the lowest tensile strength is 27.59 MPa in the 630/50/1 specimen. Analyzing the graph in Figure 6, an increment in tensile strength was observed when the tool shifted from 1 mm to 1.5 mm with the same tool speed (630 rpm) and traverse speed (20 mm/min). On the other hand, a slight decrease in strength value was seen when the tool was shifted to 2 mm from the center. Additionally, it is concluded that tensile strength values were increased with the increase of tool positioning in 30 mm/min and 50 mm/min tool speeds. Higher strength values were obtained in conditions with low speeds, high traverse rates and tool positioning since they lead to sufficient welding temperature and weld width.

The highest strength values in welded parts were reached in 1330 rpm rotational speeds as shown in Figure 7. Ideal temperatures occurred in Al-Cu FSW at this rotation speed, so that a thinly dispersed and homogeneous mixture is obtained. The strength of intermetallic phase increases with the effect of heat during FSW, however, it will not be brittle, and this conforms with the literature [13,14].

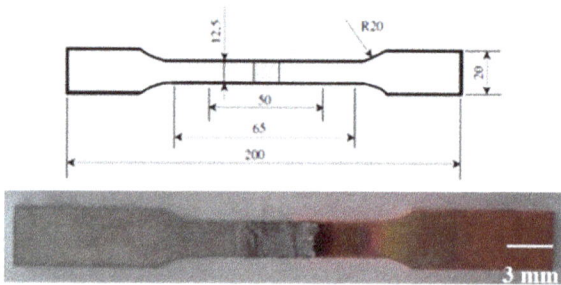

Figure 5. Dimension and macro imagine of the tensile specimen.

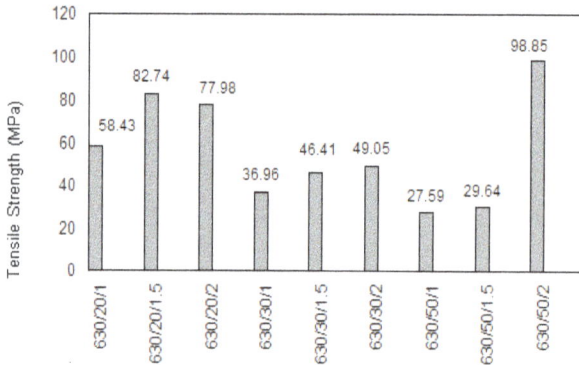

Figure 6. Tensile test results of 630 rpm.

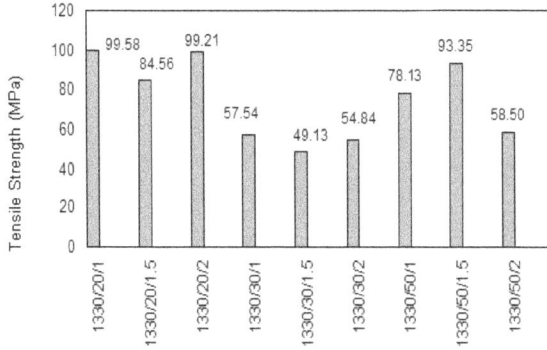

Figure 7. Tensile test results of 1330 rpm.

Figure 8 shows the trials with the highest rotation speed (2440 rpm), and it is observed that the tensile strength of the welded part is increases as the traverse speeds and tool positioning increase. High tensile strength was obtained as can be seen in Figure 8, and 92.91 MPa of tensile strength is reached with 30 mm/min traverse speed and 1 tool shifting condition. However, it is seen that the tensile strength value is decreased under the highest traverse speed (50 mm/min) and tool positioning (2 mm). The reasons for this are the lack of formation of any homogeneous mixture area in the weld zone and the fact that the adequate temperature is not supplied to the joint. Additionally, it is considered that the thickness of intermetallic phases is increased due to high heat input under low traverse speeds (20 mm/min).

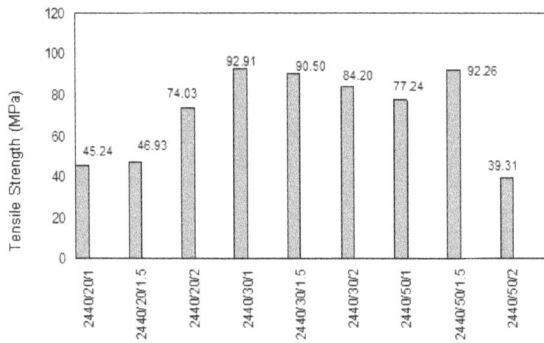

Figure 8. Tensile test results of 2440 rpm.

The higher tensile strength of the Al-Cu weld joints mainly depends on the distribution of fine particles and the low intermetallic thickness formation and grain boundary strengthening in the nugget zone. Due to the stirring of the tool, the Cu particles were fragmented from the Cu side and distributed in the stir zone. These fine Cu particles were completely transformed into hard brittle intermetallic due to the interfacial reaction with the Al matrix [5,19]. The tensile tests as a whole shows that there is adequate temperature during FSW and so the homogeneous mixture conditions leading to an Al-Cu reaction are reached. As a result of tensile tests, ruptures usually occur in weld zone and heat affected zone (HAZ) in aluminum welds. In the literature, the reason for the rupture occurrences in Al side is explained with two factors; the first is that the formation of the weld zone happened to be on the Al side, and the second factor is that the tensile strength of the base material Al is lower than the other base material Cu [11]. Ruptured surfaces of the specimens that have the highest and the lowest tensile strength are considered for the evaluation. SEM images of the ruptured surfaces are shown in Figures 9 and 10. When the SEM images are examined, it is concluded that the ruptured surface of specimens (Figure 9) that have higher mechanical properties are ductile, while the others' surface (Figure 10) are brittle. Many dimples in Al side of the rupture surface are found in the 1330/20/1 specimen, and a small amount of dimples are found on the ruptured surface of the 630/50/1 specimen.

Three point bending tests are carried out on the specimens that are cut with a water jet from the welded joints in 20 × 100 mm dimensions. Additionally, base materials have been tested; images can be seen in Figure 11a. The welded specimens are loaded until they take a U-shape or a failure is observed. As shown in Figure 10b, no failure is found on the 1330/20/1 specimen after the bending test. On the other hand, fractures and failures are found in HAZ and welded zones, especially on the specimens that have low tensile strength.

Hardness values are evaluated on the transverse cross section of welded parts. Hardness results measured from the top and bottom plates of the weldments under different parameters are illustrated in Figures 12–14. The microhardness values of the base metals were found to be 88 HV for Cu, and 41 HV for Al.

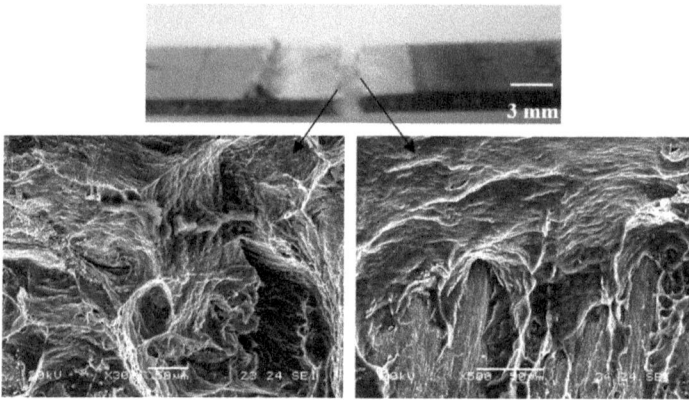

Figure 9. Surface images after tensile tests and scanning electron microscope (SEM) images of ruptured surface of welded joints in the 1330/20/1 specimen.

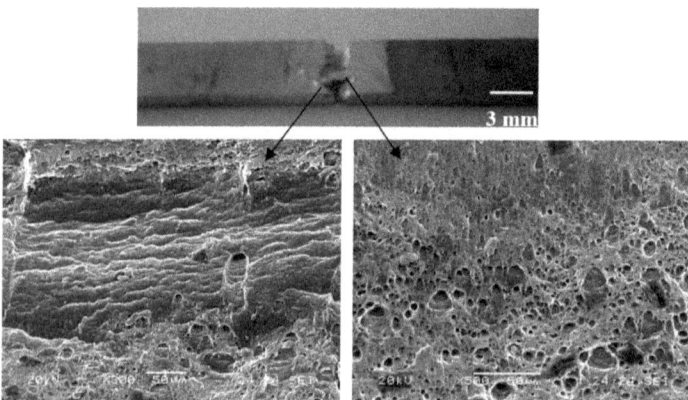

Figure 10. Surface images after tensile tests and SEM images of ruptured surface of welded joints in the 630/50/1 specimen.

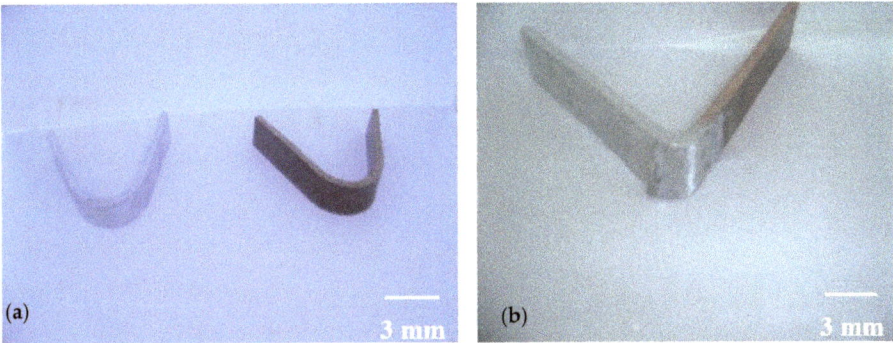

Figure 11. Bending test results of (**a**) base materials, (**b**) welded parts.

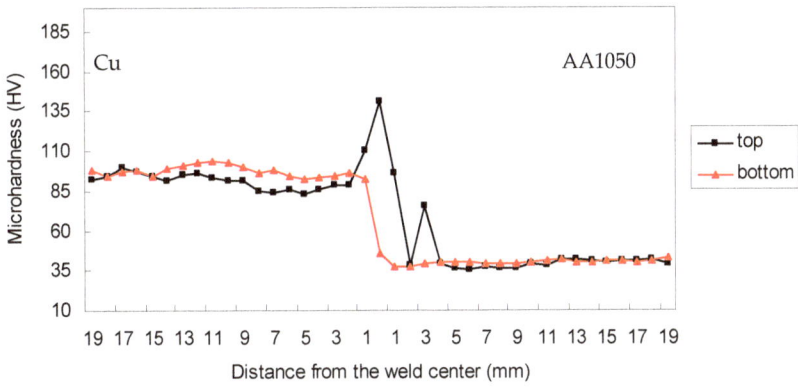

Figure 12. Hardness profile on the transverse cross section of the 630/50/1 specimen.

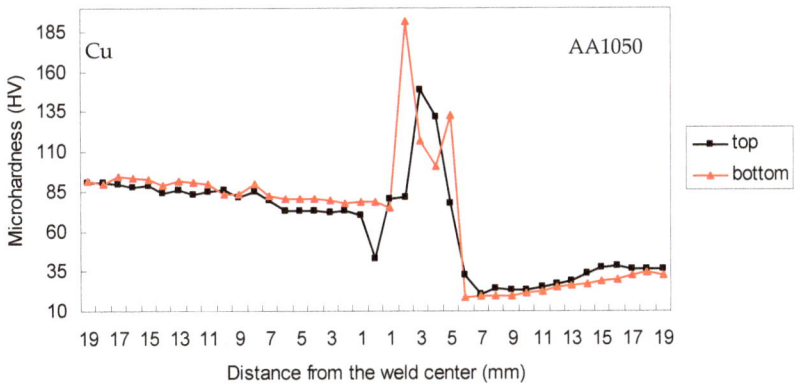

Figure 13. Hardness profile on the transverse cross section of the 1330/20/1 specimen.

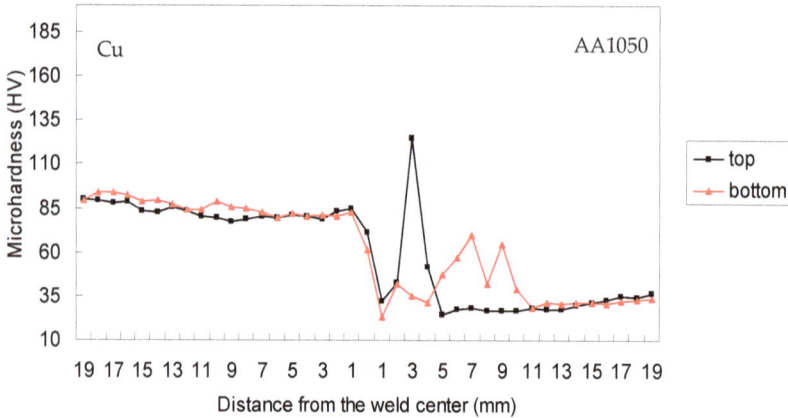

Figure 14. Hardness profile on the transverse cross section of the 2440/30/1 specimen.

In Figure 12, in analyzing the hardness changes of the specimen 630/50/1, which has low rpm and high traverse speed, it is observed that the weld zone formed is considerably narrow. Similarly, in Figures 13 and 14 the data show that specimens that have medium and high traverse speeds (1330/20/1 and 2440/30/1) have higher tensile strengths and formed larger weld zones compared to the 630/50/1 specimen. A wide weld zone shows the existence of the full mixture of materials. The sudden increase in hardness value in the weld zone, especially on the top plate, is considered to happen because of the intermetallic phases between Al-Cu under the influence of heat during welding. The hardness values in the composite structure were much higher than those of the Al side. This enhanced hardness of the Al matrix should be mainly attributed to the strengthening from the ultrafine grains. Moreover, the hardness of the layered structures was measured as high as 185 HV which was higher than that of the Cu bulk. Previous studies indicated that the hardness of the Al-Cu IMCs was very high compared to that of the Cu, and the maximum hardness value could reach 760 HV [14]. Therefore, the high hardness value of the layered structure originated mainly from the Al-Cu IMCs.

In this study, the microstructures of HAZ on the Al side, Cu side, and weld zones of all specimens are studied in details. Through these studies, it is found that the weld zone is formed on the Al side since the stir pin was positioned to the Al side in specific values (1, 1.5, 2 mm). Moreover, the composite structure between the aluminum and copper is remarkable in Al-Cu FSW joining.

The microstructure of the specimens that have the highest and lowest tensile strength are given in order to compare and evaluate the changes in strength and the structural changes in weld zones. The microstructure of base materials are illustrated in Figure 15a,b, 630/50/1 specimen's microstructure is given in Figures 16 and 17 represents the 1330/20/1 specimen's microstructure.

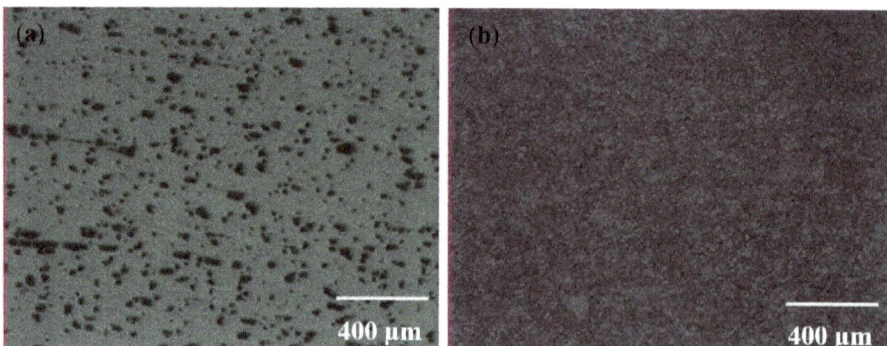

Figure 15. Microstructures of base materials: (a) Al-1050; (b) Cu.

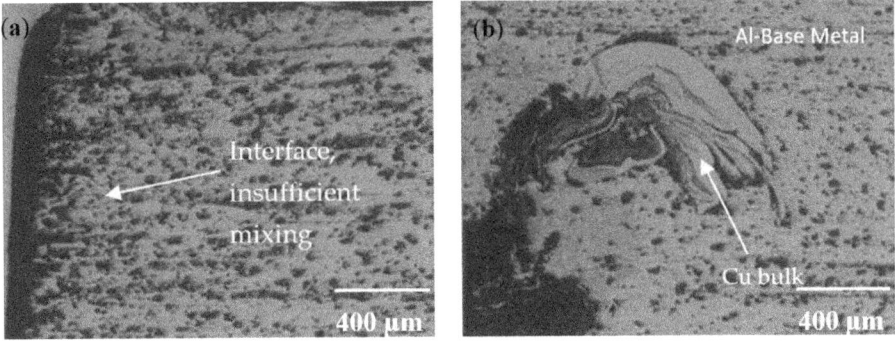

Figure 16. Welded zone of the 630/50/1 specimen: (**a**) Al side; (**b**) Nugget Zone.

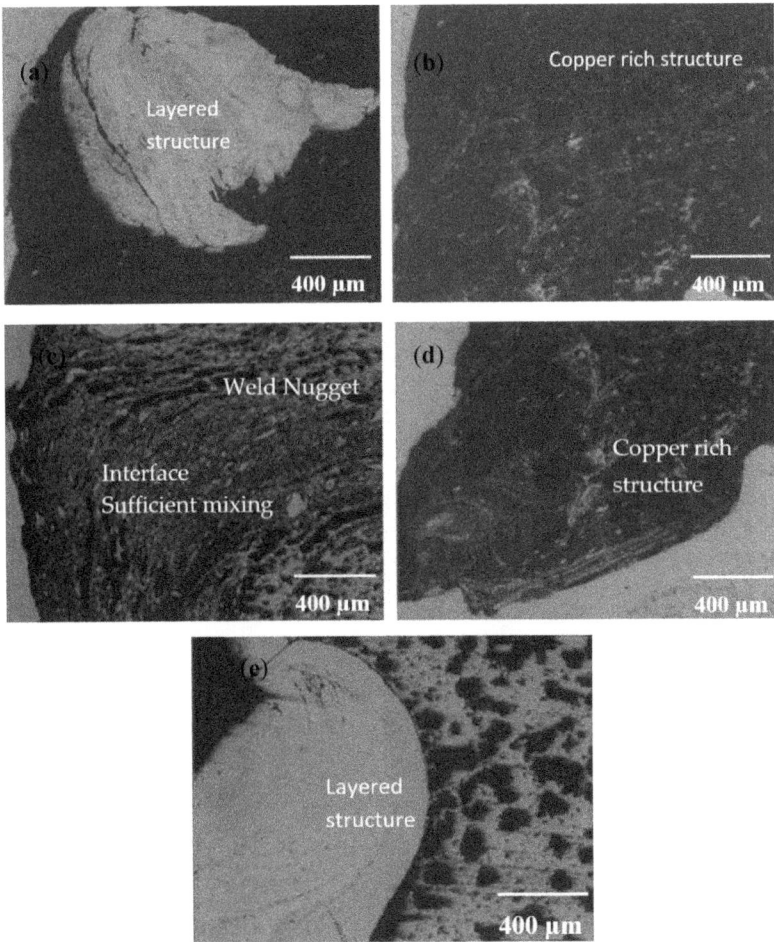

Figure 17. Welded zone of the 1330/20/1 specimen: (**a**) Al side top area; (**b**) Al side mid-area; (**c**) Weld Nugget; (**d**) Al side bottom area; (**e**) Al base material transition.

The material is flowing from the advancing side to the retreating side at the front end of the tool. This creates a vacancy in the advancing side. At the rear end, the materials are transported from the retreating side to the advancing side. When the material transported is not large enough to fill the vacancy, a tunnel defect occurred. Under the 630/50/1 parameters, low material flow is observed due to less heat input. Cavities and insufficient mixture are observed as can be seen from Figure 16, and these are the reasons that explain the low strength values.

The microstructure image of the interface between the Al and the Cu is shown in Figure 17. The optimum range of heat was enough to plasticize the Cu material near the area of the interface. Thus, the fine discontinuous Cu particles were detached and distributed in the stir zone. An obvious interface existed between the Al matrix and the Cu bulk, and a layered structure could be observed in the Cu bulk under the Al-Cu interface. Figure 17a,d shows the magnified view of the interface between the Al matrix and the Cu bulk. As shown in Figure 17, a clearer nugget zone occurred which differs from the low tensile 630/50/1 specimen. Additionally, the homogenous distribution of Cu bulks in Al increased the mechanical properties of the 1330/20/1 specimen.

When the SEM images of welded zones are evaluated, as given in Figure 18a, the mixture was not fully formed, and only a very small portion of it occurred in the Al side for the 630/50/1 specimen. On the other hand, Figure 18b confirms that the mixture occurred at the desired level in the Al side for the 1330/20/1 specimen, which has a higher tensile strength value. After a linear EDS analysis shown in Figure 19, it is observed that Al and Cu concentration is low in 630/50/1 at the zone no. 1, which is shown in Figure 18a. In contrast with this, the concentration of Al and Cu was found to be dense in the 1330/20/1 specimen at zone no. 1, which is shown in Figure 18b, and EDS analysis is illustrated in Figure 20. Comparing the EDS analysis of the 630/50/1 and the 1330/20/1 specimens, it is observed that the amount of copper was less and the blend of materials was not sufficient in the 630/50/1 specimen, which has a lower tensile strength. The lack of a full blend between Al-Cu and the low heat input are the reasons for the low tensile strength that was obtained from the joints with a 630 rpm rotational speed, compared to other tool rotational speeds (1330 and 2440 rpm). Moreover, adequate heat input and the generation of a composite structure between Al-Cu are the arguments for achieving a high tensile strength value after the welding with 1330 rpm tool rotational speed, compared to tensile values that were obtained from welding with speeds of 630 and 2440 rpm. The mechanical properties that resulted from 2440 rpm rotational speed are slightly lower compared to the 1330 rpm speed. Due to heat input incrementation and the formation of more intermetallic components at the Al-Cu interface, the brittleness is enhanced and it is considered that this caused a reduction in tensile strength. As introduced in other studies [5,12], a decrease in the tensile strength of the joint happens with the increase in the thickness of the intermetallic phases.

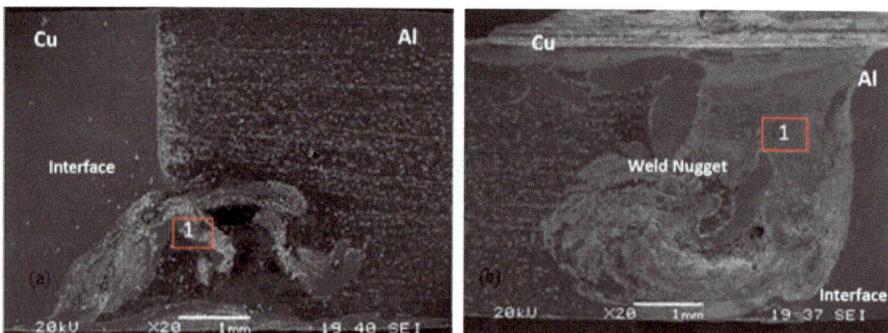

Figure 18. SEM images of (**a**) 630/50/1 specimen; (**b**) 1330/20/1 specimen.

The literature shows that intermetallic phases such as Al_2Cu, Al_4Cu_9, $CuAl$, Al_2Cu_3 and $AlCu_4$ will occur with the increase in temperature between the aluminum and copper. Al_2Cu phases occur at 150 °C, while Al_4Cu_9 phases occur at 350 °C. When the intermetallic phase reaches 10 μm in thickness, the strength of the bond indicates a sharp decrease [5,20]. XRD analysis was conducted in order to determine the intermetallic phases that may occur in the weld zone due to the high mechanical properties. The thickness of the intermetallic compound layer is a function of temperature and holding time. The atomic diffusion of Cu and Al through the intermetallic compound is the main controlling process for the intermetallic compound growth [12,21]. The analysis results in Figure 21 are analyzed and, in accordance with the literature, the $CuAl_2$ and Al_4Cu_9 intermetallic phases are determined in the mixture region.

During the friction stir welding process, the average temperatures measured from the welding zones ranged between 300 and 461 °C, depending on welding parameters. In the majority of parameters, these temperature values are sufficient for the formation of Al_2Cu and Al_4Cu_9 phases, as determined by XRD analysis. Changes in the strength values of welded specimens are explained by the temperature differences in the weld zone depending on welding parameters. The elasticity of the material at low temperatures cannot be achieved, so that a homogeneous mixture zone also cannot be formed. On the other hand, in high temperatures brittleness is formed due to the increase of intermetallic phases. In accordance with the literature, the lowest tensile strengths obtained under the 630/50/1 and 630/50/1.5 parameters which have the lowest temperature value (300 °C) at the welding zone. It is observed that adequate heat is not generated for the formation of Al_4Cu_9 phase. Additionally, a decrease in tensile strength is observed since the thickness of the intermetallic phases is enlarged under the parameter of 2440/50/2, which reaches the highest temperature (461 °C).

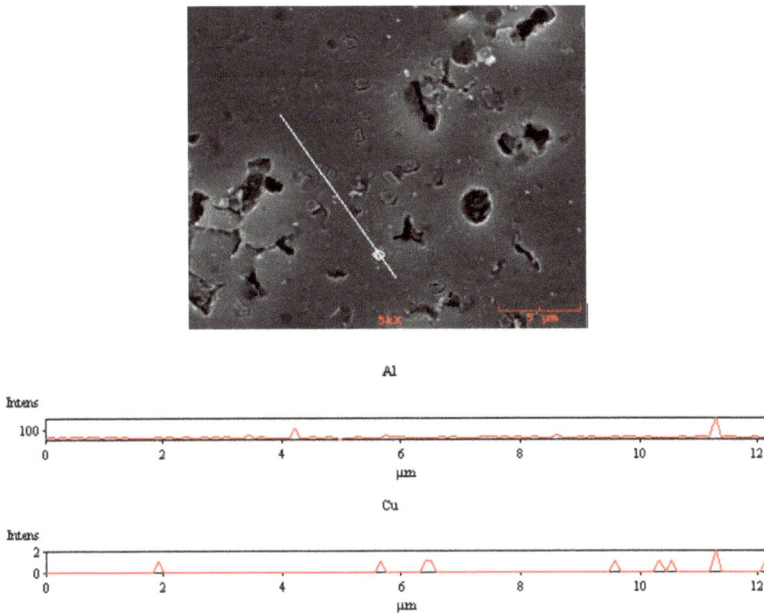

Figure 19. Energy dispersed spectrometer (EDS) linear analysis (630/50/1 specimen, zone No. 1).

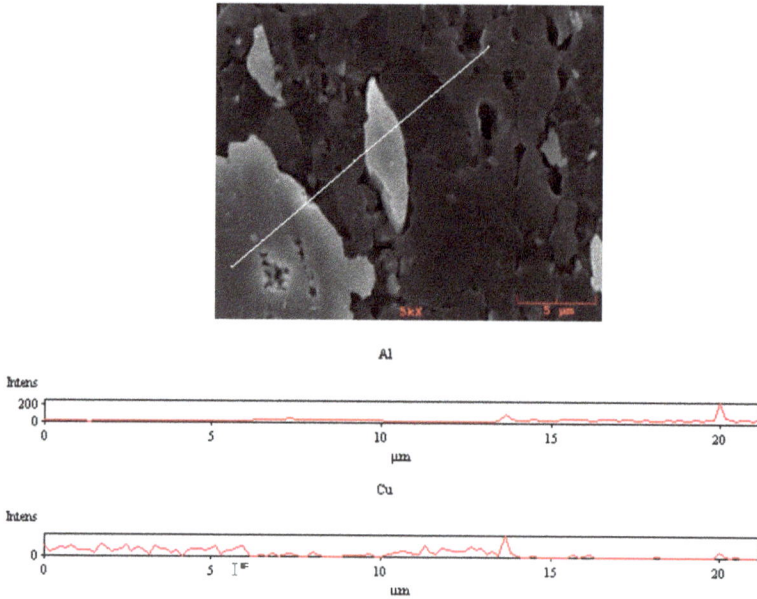

Figure 20. EDS linear analysis (1330/20/1 specimen, zone No. 1).

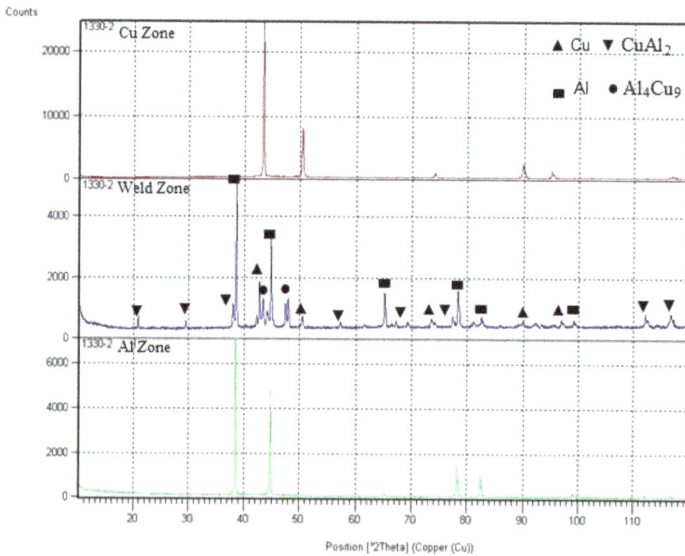

Figure 21. X-ray diffraction (XRD) graphs of base materials and weld zone.

4. Conclusions

1. In this study, the friction stir butt weldability of pure Cu and 1050 Al alloy was examined, and it was successfully accomplished under different parameters by using a cylindrical pin tool. Failures were observed in the weldings that has none tool shifting (zero positioned tool). Macro-level welding defects were not observed on the welded surfaces in the case of joints for which the stir

pin was positioned at 1, 1.5 and 2 mm to the Al side. However, micro-level gaps were observed in low tensile strength specimens.

2. Tensile and bending tests, as well as hardness measurements were made in order to determine the mechanical properties of joints. When the welding performance of joints was evaluated, the maximum value was found to be 89.5% with a 1330 rpm tool rotational speed, a 20 mm/min traverse speed and a 1 mm tool position configuration. As a result of the tensile test it was observed that ruptures usually occurred in joint zones and heat-affected zones of aluminum.

3. Due to the Al-Cu layered structure in the weld center and intermetallic phases, a hardness increase in weld zone was observed. This had the effect of mixing particles that break off from the copper in the advancing side being moved into the aluminum matrix in the retreating side. Since the weld zone was formed on the Al side, the Cu bulk in the Al matrix and intermetallic phases increased in hardness. In high tensile strength specimens, the weld zones were observed to be larger.

4. Microstructural analysis showed that the blending area happened to be on the Al side since the end of the stir pin was shifted to the Al side in proper values (1, 1.5, 2 mm). Higher strength values were obtained in a homogeneous composite structure.

5. According to linear and point EDS analysis, Al and Cu were detected on the cross sections and fracture surfaces of joints that were obtained after tensile tests. It was observed that the Cu content in the weld zones was less in specimens with a low tensile strength compared to high tensile strength specimens.

6. $CuAl_2$ and Al_4Cu_9 intermetallic phases were determined in the phase analysis that was performed using X-ray diffraction (XRD). The increase of the intermetallic phase had a lowering effect on the fragility and strength.

Acknowledgments: This work was supported by the Balikesir University under Scientific Research Projects Program grant No. BAP.2012/49.

Author Contributions: S. Celik conceived, designed the experiments; R. Cakir performed the experiments under the supervision of S. Celik; both S. Celik. and R. Cakir analyzed the data; the microstructure analyses were performed in TUBITAK of Gebze Office (The Scientific and Technological Research Council of Turkey). S. Celik wrote the paper.

Conflicts of Interest: The authors declare no conflict of interest.

Abbreviations

The following abbreviations are used in this manuscript:

FSW	Friction Stir Welding
EDS	Energy Dispersed Spectrometer
SEM	Scanning Electron Microscope
XRD	X-ray Diffractometer
IMCs	Intermetallic Compounds
HAZ	Heat Affected Zone

References

1. Thomas, W.M.; Nicholas, E.D.; Needham, J.C.; Murch, M.G.; TempleSmith, P.; Dawes, C.J. International Patent Application No. PCT/GB92/02203 and GB Patent Application No. 9125978.8, 6 December 1991.

2. Lee, W.B.; Jung, S.B. The joint properties of copper by friction stir welding. *Mater. Lett.* **2004**, *58*, 1041–1046. [CrossRef]

3. Jata, K.V.; Semiatin, S.L. Continuous Dynamic Recrystallization during Friction Stir Welding of High Strength Aluminum Alloys. *Scr. Mater.* **2000**, *43*, 743–749. [CrossRef]

4. Hwang, Y.M.; Fan, P.L.; Lin, C.H. Experimental study on Friction Stir Welding of copper metals. *J. Mater. Process. Technol.* **2010**, *210*, 1667–1672. [CrossRef]

5. Muthu, M.F.X.; Jayabalan, V. Tool travel speed effects on the microstructure of friction stir welded aluminum–copper joints. *J. Mater. Process. Technol.* **2015**, *217*, 105–113. [CrossRef]

6. Abdollah-Zadeh, A.; Saeid, T.; Sazgari, B. Microstructural and mechanical properties of friction stir welded aluminum/copper lap joints. *J. Alloy. Compd.* **2008**, *460*, 535–538. [CrossRef]

7. Mubiayi, M.P.; Akinlabi, E.T. Friction Stir Welding of Dissimilar Materials between Aluminium Alloys and Copper—An Overview. In Proceedings of the World Congress on Engineering 2013 Vol III, WCE 2013, London, UK, 3–5 July 2013.

8. Galvão, I.; Verdera, D.; Gesto, D.; Loureiro, A.; Rodrigues, D.M. Influence of aluminium alloy type on dissimilar friction stir lap welding of aluminium to copper. *J. Mater. Process. Technol.* **2013**, *213*, 1920–1928. [CrossRef]

9. Saeid, T.; Abdollah-zadeh, A.; Sazgari, B. Weldability and mechanical properties of dissimilar aluminum-copper lap joints made by friction stir welding. *J. Alloy. Compd.* **2010**, *490*, 652–655. [CrossRef]

10. Scialpi, A.; de Filippis, L.A.C.; Cavaliere, P. Influence of shoulder geometry on microstructure and mechanical properties of friction stir welded 6082 aluminium alloy. *Mater. Des.* **2007**, *28*, 1124–1129. [CrossRef]

11. Barlas, Z.; Uzun, H. Sürtünme karıştırma kaynağı yapılmış Cu/Al-1050 alın birleştirmesinin mikroyapı ve mekanik özelliklerinin incelenmesi. *Gazi Üniversity Mühendislik Mimarlik Fakküllttes Dergisi* **2010**, *25*, 857–865.

12. Xue, P.; Xiao, B.L.; Wang, D.; Ma, Z.Y. Achieving high property friction stir welded aluminium/copper lap joint at low heat input. *Sci. Technol. Weld. Join.* **2011**, *16*. [CrossRef]

13. Xue, P.; Ni, D.R.; Wang, D.; Xiao, B.L.; Ma, Z.Y. Effect of friction stir welding parameters on the microstructure and mechanical properties of the dissimilar Al-Cu joints. *Mater. Sci. Eng. A* **2011**, *528*, 4683–4689. [CrossRef]

14. Ouyang, J.; Yarrapareddy, E.; Kovacevic, R. Microstructural evolution in the friction stir welded 6061 aluminum alloy (T6-temper condition) to copper. *J. Mater. Process. Technol.* **2006**, *172*, 110–122. [CrossRef]

15. Liu, P.; Shi, Q.; Wang, W.; Wang, X.; Zhang, Z. Microstructure and XRD analysis of FSW joints for copper T2/aluminium 5A06 dissimilar materials. *Mater. Lett.* **2008**, *62*, 4106–4108. [CrossRef]

16. Xue, P.; Xiao, B.L.; Ni, D.R.; Ma, Z.Y. Enhanced mechanical properties of friction stir welded dissimilar Al-Cu joint by intermetallic compounds. *Mater. Sci. Eng. A* **2010**, *527*, 5723–5727. [CrossRef]

17. Genevois, C.; Girard, M.; Huneau, B.; Sauvage, X.; Racineux, G. Interfacial Reaction during Friction Stir Welding of Al and Cu. *Miner. Met. Mater. Soc. ASM Int.* **2011**, *42*. [CrossRef]

18. Leitão, C.; Loureiro, A.; Rodrigues, D.M. Influence of Base Material Properties and Process Parameters on Defect Formation during FSW. In Proceedings of the International Congress on Advances in Welding Science and Technology for Construction, Energy & Transportation Systems, Antalya, Turkey, 24–25 October 2011; pp. 177–184.

19. Venkateswaran, P.; Reynolds, A.P. Factors affecting the properties of frictionstir welds between aluminum and magnesium alloys. *Mater. Sci. Eng. A* **2012**, *545*, 26–37. [CrossRef]

20. Çelik, S. An Investigation of Diffusion Welding Parameters for Pure Aluminum and Copper in Inert Gas. Ph.D. Thesis, Balikesir University, Balikesir, Turkey, 1996.

21. Kim, H.G.; Kim, S.M.; Lee, J.Y.; Choi, M.R.; Choe, S.C.; Kim, K.H.; Ryu, J.S.; Kim, S.; Han, S.Z.; Kim, W.Y.; *et al.* Microstructural evaluation of interfacial intermetallic compounds in Cu wire bonding with Al and Au pads. *Acta Mater.* **2014**, *64*, 356–366. [CrossRef]

metals

MDPI

Article

Effect of Surface States on Joining Mechanisms and Mechanical Properties of Aluminum Alloy (A5052) and Polyethylene Terephthalate (PET) by Dissimilar Friction Spot Welding

Farazila Yusof [1,2,*], **Mohd Ridha bin Muhamad** [1,2], **Raza Moshwan** [1,2], **Mohd Fadzil bin Jamaludin** [1,2] and **Yukio Miyashita** [3]

1   Department of Mechanical Engineering, University of Malaya, Kuala Lumpur 50603, Malaysia;
     ridha.muhamad@gmail.com (M.R.M.); raza_moshwan@yahoo.com (R.M.);
     ibnjamaludin@um.edu.my (M.F.J.)
2   Centre of Advanced Manufacturing and Material Processing, University of Malaya,
     Kuala Lumpur 50603, Malaysia
3   Department of System Safety, Nagaoka University of Technology, Nagaoka 940-2188, Japan;
     miyayuki@mech.nagaokaut.ac.jp
*   Correspondence: farazila@um.edu.my; Tel.: +60-3-7967-7633

Academic Editor: Nong Gao
Received: 10 February 2016; Accepted: 18 March 2016; Published: 28 April 2016

Abstract: In this research, polyethylene terephthalate (PET), as a high-density thermoplastic sheet, and Aluminum A5052, as a metal with seven distinct surface roughnesses, were joined by friction spot welding (FSW). The effect of A5052's various surface states on the welding joining mechanism and mechanical properties were investigated. Friction spot welding was successfully applied for the dissimilar joining of PET thermoplastics and aluminum alloy A5052. During FSW, the PET near the joining interface softened, partially melted and adhered to the A5052 joining surface. The melted PET evaporated to form bubbles near the joining interface and cooled, forming hollows. The bubbles have two opposite effects: its presence at the joining interface prevent PET from contacting with A5052, while bubbles or hollows are crack origins that induce crack paths which degrade the joining strength. On the other hand, the bubbles' flow pushed the softened PET into irregularities on the roughened surface to form mechanical interlocking, which significantly improved the strength. The tensile-shear failure load for an as-received surface (0.31 µm R_a) specimen was about 0.4–0.8 kN while that for the treated surface (>0.31 µm R_a) specimen was about 4.8–5.2 kN.

Keywords: friction spot welding; surface roughness; dissimilar welding; polyethylene terephthalate (PET); aluminum alloy; bubble

1. Introduction

To solve environmental and energy problems, light materials are increasingly being adopted as structural constituents in different transportation industries, such as the automotive, aeronautic and train industries. The various parts of light materials are combined and joined to assemble such structures and components for cars, airplanes and trains. There are several joining technologies [1] including welding, bolt joining adhesive joining, and so on. Tungsten inert gas (TIG) welding and laser welding are commonly used to join similar and dissimilar light metal sheets.

Friction stir welding (FSW) has been recently developed and successfully applied to join light materials such as aluminum alloys and titanium alloys. The welding method is a promising ecological welding method that enables workers to diminish material waste and avoid radiation and harmful gas emissions that often occur from fusion welding [2,3]. The main process parameters in controlling the quality of joints are tool rotation speed, tool traverse speed, vertical pressure on the specimen, the tilt angle of the tool, tool geometry, and others [4,5]. During welding, a non-consumable tool attached with a specially designed pin rotates and pushes the butting edges of the two plates to be joined. The friction heat causes the material to soften, allowing the tool to penetrate into the material surface. The tool shoulder sits on the specimen's surface during penetration. Under this condition, the rotating tool traverses along the joining line. Thus, generated frictional heat causes both materials to soften under the tool where joining is achieved. This process is suitable for joining plates and sheets. However, it can also be employed for joining pipes and hollow sections [6]. Although the FSW process was initially developed for aluminum alloys [7–9], it also has a great potential for the welding of copper [10], titanium [11], steel [12], magnesium [13], metal matrix composites [14], and different material combinations [15].

Recently, studies have shown that applying FSW welding to thermoplastics is successful and various factors influencing its joinability were investigated. The effect of tool tilt angle and welding speed on the tensile strength in FSW of polyethylene [16] was studied using different tool dimensions and pre-heated pins [17]. Taguchi's approach to parameter design and analysis of variance were utilized with further experimental confirmation for FSW of polyethylene. It was shown that the optimum welding parameters were tool rotational speed of 3000 rpm, traverse of 115 mm/min and tilt angle of three degrees [18]. In a novel study by Vijendra, a new hybrid friction stir welding process, i-FSW, was designed. In i-FSW, the tool is heated during welding by an induction coil, and the temperature is precisely maintained through feedback control [19]. Another study on tool design where a new self-reacting tool with a convex pin was utilized showed the greatest effects on tensile strength in welding of acrylonitrile butadiene styrene (ABS) sheets [20]. Recent research that combines experimental and analysis models for optimization is popular to reduce experimental cost and for the ability to predict optimal conditions precisely. Factors that influence FSW were optimized using a factorial design and analyzed using Artificial Neural Networks (ANN) to compare the experimental and the model analysis, which has demonstrated that tool plunge rate, dwell time and waiting time, plunging force, and torque were discussed as the most influential factors [21,22]. A study of an ABS sheet optimized by Analysis of Variance (ANOVA) and Response Surface Methodology (RSM) demonstrated high diameter ratio and low rotational speed, which are optimal. A comparison indicated the more accurate prediction of a corresponding model for a conical pinned tool than a cylindrical probe tool. The recommended conditions identified are two degrees for tilt angle, 900 rpm rotational speed, a tool with diameter ratio of 20/6 and linear speed of 25 mm/min, which generated a weld joint with equal yield strength to the base material [23]. In another study, Simoes analysed the material flow and thermo-mechanical phenomena taking place during FSW of polymers. Polymethylmethacrylate (PMMA) was used owing to the high transparency so that polarization during tool penetration could be observed clearly. It has been reported that due to the polymers' rheological and physical properties, the thermo-mechanical conditions during FSW are very different from those registered during the welding of metals. The material flow and temperature distribution between metallic and polymeric materials were compared based on the Arbegast flow-partitioned deformation zone model for FSW in metals. The formation of discontinuities was indicated as one of the main weldability problems for polymers [24].

There are very few reports on the friction stir spot welding of polymer-polymer as well as polymer-metal combinations, which have great applicability and are in high demand, especially in the automotive industry [25,26]. In the author's previous work [27], the dissimilar joining of aluminum alloys (A5052) and polyethylene terephthalate (PET) was attempted using the frictional energy generated from friction spot welding. In the joining conditions shown for plunge depth of 0.7 mm, the lower plunge speed exhibited higher tensile strength, which was the result of longer contact time and more generated heat. The process yielded the dissimilar joining of the two materials despite the low joining strength. In the dissimilar friction stir butt joining of aluminum and Polycarbonate

(PC), the feasibility was achieved, but the concerns remain regarding lower tensile strength due to fracture induced voids [28]. The dissimilar joining of aluminum and thermoplastics by adopting the hole-clinching method has been reported in several studies. Lee investigated tool shapes such as punch diameter, punch corner radius and die depth on hole-clinching for dissimilar materials [29]. Studies on the joinability of rigid thermoplastic polymers with aluminum AA6082-T6 alloy sheets by mechanical clinching have revealed that fracture at the metal or polymer sheet was the main factor contributing to unsuccessful joinability. Joinability has been examined by studying mechanical interlocking manipulated by tool geometry [30], tool shapes [31] or temperature [32]. These studies focused on tool shapes that directly influence mechanical interlocking at the microstructural level. An analysis study by Wirth on the bonding behavior and joining mechanism of aluminum and thermoplastics recommended optimal conditions such as holding time, axial force, *etc.* [33]. High lap joint quality with shear strength of 5–8 MPa was reported in a case study of aluminum and laser transmission joints of nylon [34] and PMMA [35] where in both cases the temperature reached the melting temperature of the thermoplastics.

In the present study, the effect of surface roughness on the joining strength of an FSW-ed PET-A5052 dissimilar joint is investigated with the aim to increase the joining strength. The joining mechanism and effects of surface roughness are discussed in detail.

2. Materials and Methods

Aluminum alloy (A5052) and polyethylene terephthalate (PET) specimens were machined to dimensions of 40 × 100 mm and 3 mm thickness. The chemical composition of A5052 is shown in Table 1. The physical and mechanical properties of A5052 and PET are shown in Table 2. The aluminum alloy A5052 specimens were prepared in the as-received condition, and six different surface roughness values were prepared by wire brushing and etching (Etchant: Hydrochloric acid and Aluminium chloride), as shown in Table 3. The surface roughness of the A5052 specimens was measured by means of a profilometer (Mitutoyo, Kawasaki, Japan, Model: SJ-201) with a diamond stylus. The arithmetic mean surface roughness values, R_a, obtained by averaging three measurements for each specimen are shown in Table 3. In this study, a lap joint arrangement is investigated, with an A5052 specimen positioned on top and a PET specimen on the bottom.

Table 1. Chemical composition of A5052 (mass percentage) specified by American Society for Testing and Materials (ASTM)

Material	Al	Si	Fe	Cu	Mn	Mg	Cr	Others
A5052	Balance	<0.25	<0.4	<0.1	0.15-0.35	2.2-2.8	<0.1	<0.15

Table 2. Physical and mechanical properties of polyethylene terephthalate (PET) and A5052.

Properties	PET	A5052
Density, (g/cm^3)	1.45	2.68
Glass transition temperature, (°C)	80	-
Melting temperature, (°C)	200–225	607–649
Specific heat capacity, (J/g°C)	1.00	0.88
Thermal conductivity, (W/cm°C)	0.0024	1.38
Ultimate tensile strength, (Mpa)	55	193
Yield stress, (Mpa)	-	89.6
Modulus of elasticity, (Gpa)	2.7	70.3
Elongation at break, (%)	125	25

Table 3. Surface roughness, R_a for different surface treatments.

No.	Surface Treatments	Surface Roughness, (µm R_a)
1	As-received (AR)	0.31
2	Wire brushing 80 times (WB)	1.04
3	Etching 30 s (E0.5)	0.47
4	Etching 1 min (E1)	0.61
5	Etching 2 min (E2)	1.42
6	Etching 6 min (E6)	3.40
7	Etching 10 min (E10)	4.16

Figure 1 displays the schematic diagram of lap joint specimen positioning (left) and rotating tool dimensions (right). The dark region at the center of the assembly represents the welding area. Two alignment pads with 3 mm thickness were attached to the end of the joint specimens to align the specimens on the machine. The rotating tool consists of a shoulder with 8 mm diameter and a conical center probe with 3 mm diameter. In the FSW process, the rotating tool partially penetrated the A5052 surface to a certain depth and then moved up after a dwell time of 2 s. Table 4 lists the welding parameters, including rotating speed, plunge speed, plunge depth and dwell time. A single lap shear test of the FSW joints was performed using a tensile test machine (Shimadzu, Kyoto, Japan, Model: AGS-X, 10 kN) with a loading rate of 0.5 mm/min. Scanning electron microscope (SEM) were used to study and observe the micro-structural interlocking, joining structure and material flow at cross-sectional of the joint interface.

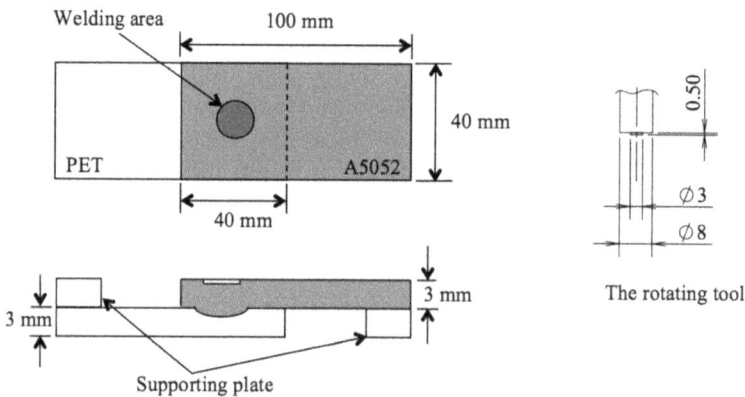

Figure 1. Schematic diagram of lap joint specimen positioning details (**left**) and the rotating tool dimension (**right**).

Table 4. List of welding parameters during the friction stir spot joining.

Parameters	Value
Spindle speed, (rpm)	3000
Plunge depth, (mm)	0.4 and 0.7
Plunge speed, (mm/min)	10, 20 and 40
Dwell time, (s)	2

3. Results and Discussion

3.1. *Joining Mechanism*

Adhesion between metal and polymer is a complicated process. Therefore, it is crucial to understand the bonding mechanism between both materials. For polymer and metal interfaces, the principal mechanisms include physical adsorption, mechanical interlocking and chemical adhesion (covalent bonds) [36]. In this study, different surface conditions were utilized in order to improve the adhesive behavior of A5052 and PET. According to Qizhou Yao and Jianmin Qu [37], metal surfaces are microscopically rough. Thus, when a liquid is applied to a rough surface, it conforms to the rough surface and fills the irregularities of the substrate surfaces, such as microgrooves, holes and dips. Similar behavior is predicted during the joining of A5052 and PET.

In specimen preparation, the A5052 surface was roughened by pre-treating. It is believed that higher surface roughness will increase the adhesive properties between two materials [38–40]. For A5052-PET joints with treated surface roughness, the PET would soften and conform to the A5052 irregularities, creating an interfacial region. This region would have the intermediate physical properties of PET and A5052. Therefore, higher joining strength is achievable for specimens with higher surface roughness.

3.2. *Effect of Surface Roughness on Welded Area*

As mentioned in the previous report [27], the welded area could be clearly observed through the transparent PET side after FSW. In this region, the molten or softened PET is strongly attached to the A5052. In the present study, the welded area was also measured for all FSW joints. The results are presented in Figure 2a,b. The welded area ranged from 450 to 900 mm^2 for the specimens welded at a plunge depth of 0.4 mm. For the specimens welded at plunge depth of 0.7 mm, the welded area ranged from 550 to 1100 mm^2. Therefore, a larger welded area was obtained for 0.7 mm plunge depth compared to 0.4 mm. It is speculated that the deeper tool penetration generated a greater amount of heat, thus producing a larger welded area. This effect is in line with Oliveira *et al.*'s report [1] that deeper tool penetration induces high heat input and greater joining strength.

In the current study, it was observed that the joined area of all specimens was larger when welded at lower plunge speed. A lower plunge speed induces substantial heat at the tool tip due to the longer spin duration, hence increasing the joined area. It was also noticed that the joined area for the A5052(AR)-PET joint converged and declined at higher plunge speed, even with the different surface roughness values of the A5052 specimens. This implies that the heat generated from rapid plunge speeds is similar to the variance of pre-treated surface roughness specimens used in the experiment.

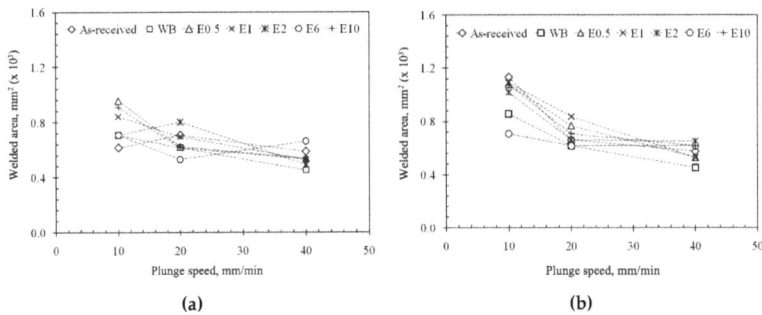

Figure 2. Effect of plunge speed on the welded area for plunge depth (**a**) plunge depth 0.4 mm and (**b**) plunge depth 0.7 mm.

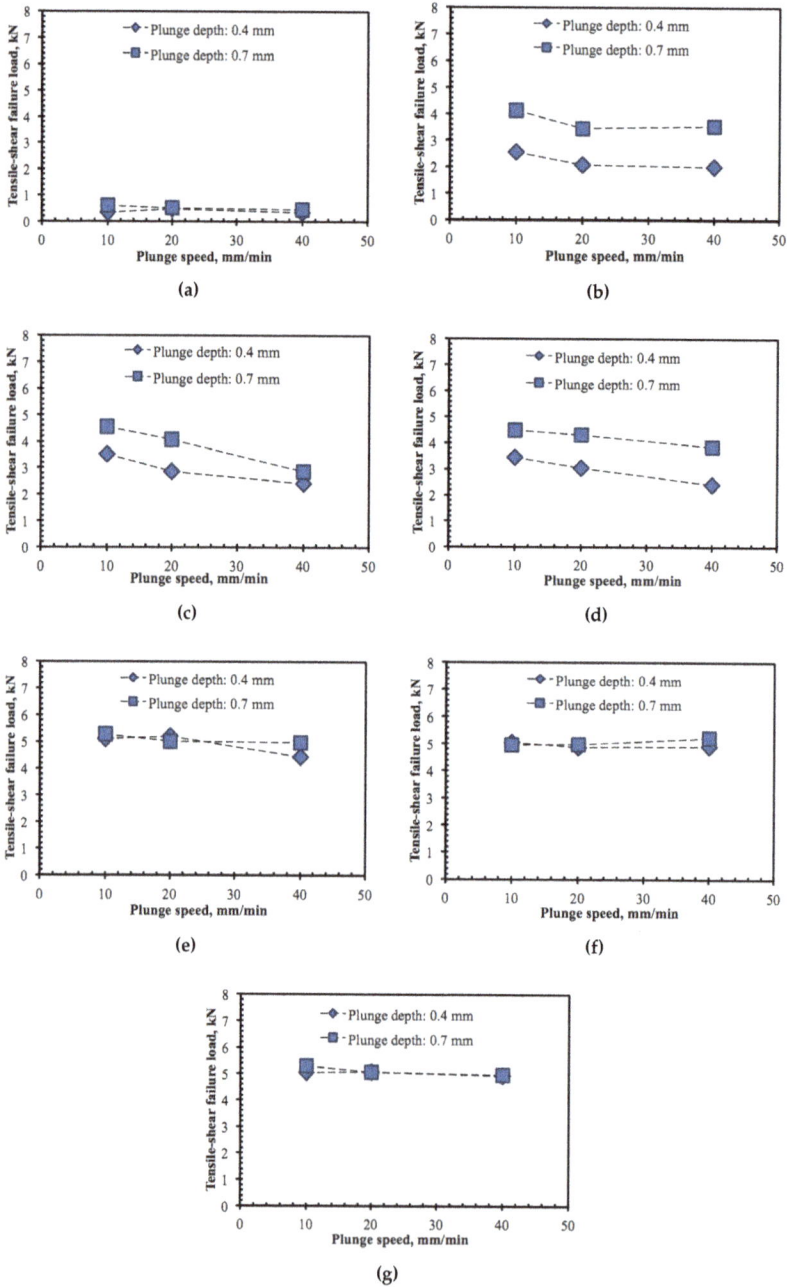

Figure 3. Effect of plunge speed on the tensile-shear failure load for different plunge depth. (**a**) As-received; (**b**) wire brushing; (**c**) etching 30 s; (**d**) etching 1 min; (**e**) etching 2 min; (**f**) etching 6 min; and (**g**) etching 10 min.

3.3. Effect of Surface Roughness on Tensile-Shear Strength

Figure 3a–g show the relationships between plunge speed and tensile shear failure load for the specimens welded at plunge depths of 0.4 and 0.7 mm, respectively. Although there is a slight decrease in tensile-shear failure load as the plunge speed increases for WB, E0.5 and E1, generally failure load was constant regardless of the plunge speed. Additionally, the plunge depth 0.7 mm yielded greater failure load than 0.4 mm, apparently from more heat induced. The results E2, E6 and E10 exhibited that higher surface roughness produced a higher tensile-shear failure load. Notably at a higher plunge speed, the difference could be significantly observed.

The fracture surface from the single lap shear test provides insight into the behavior of the failure load observed. Figure 4 displays images of fracture surfaces in the single lap shear test. Three types of failures were identified, and it was observed that surface roughness influenced the type of failure in the single lap shear test. Type 1: Failure is caused by separation at the welded interface without damage to the parent materials. It includes the A5052(AR)-PET joint at all plunge depths and plunge speeds, and the A5052(WB)-PET joint at 0.4 mm plunge depth and all plunge speeds. Type 2: Failure starts at the edge of the bubble formation region in the center of the welded area. It includes the A5052(E0.5)-PET and A5052(E1)-PET joints at 10 mm/min plunge speed and at all plunge depths. Type 3: Failure occurs from the edge of the welded area following the large plastic deformation of the PET specimen. It includes the A5052(E6)-PET and A5052(E10)-PET joints at all plunge speeds and plunge depths and the A5052(E2)-PET joint at a plunge depth of 0.7 mm and all plunge speeds. In failure Type 1 (T1), the upper (A5052) and lower (PET) specimens were separated from each other, and no severe damage occurred on either side. For failure Type 2 (T2), the upper and lower specimens were separated from each other, but the lower specimen was broken. For failure Type 3 (T3), the upper and lower specimens did not easily separate, and the lower specimen was elongated before fracturing.

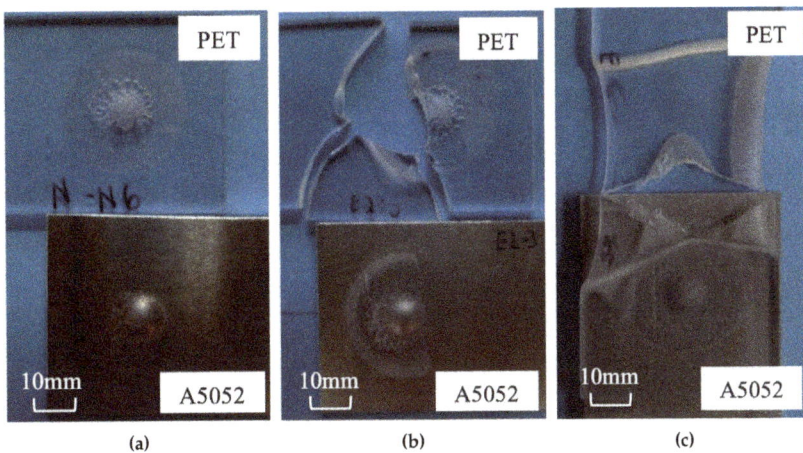

(a)   (b)   (c)

Figure 4. Three types of failure mode occurred in the A5052-PET joints. (**a**) Separation at welded interface; (**b**) fracture at the edge of bubbles or hollow structures; and (**c**) fracture at the edge of welded area after PET deformation.

It was found that the tensile shear failure load was relatively low when the fracture was initiated at the interface (Type 1), which is clearly observed on the A5052(AR)-PET joint. It is believed that the joint was weak due to bubble formation. From the fracture path identified, it was found that the path appeared at the edge of the bubbles as observed in Figure 5. This indicates that the bubbles near the joining interface may have induced the fracture and directly affected the joining strength. In addition, the bubbles formed in the A5052(AR)-PET joint were relatively larger than the specimens with treated surfaces. The observation suggests that bubble size affects the failure load. In the case of Type 2 and Type 3, the tensile shear failure load was relatively high, with a number of specimens having reached the maximum load when the fracture was initiated in the PET sheet. However, in Type 3, the PET was elongated, and necking occurred at the PET side near the edge of the joined area. This phenomenon shows that a strong joint was achieved with the materials. Strong interfacial bonding between PET and A5052 was attained due to higher surface roughness treated by etching. Further investigations of the cross-sectional observation at the joined interface could explain these behaviors.

The relationships between plunge speed and shear strength are shown in Figure 6a,b, respectively. The shear strength was calculated from the ratio of shear failure load to joined area. The shear strength for specimens that could not be separated (Type 3) is not included in Figure 6. As seen from Figure 6a,b, the specimens with surface modifications exhibited higher shear strength than the A5052 (AR)-PET joint. The highest shear strength of 9.03 MPa was achieved for the A5052(E2)-PET joint (2 min etching, 40 mm/min plunge speed, and 0.4 mm plunge depth). The joined specimens with etching-roughened surfaces had higher surface roughness than A5052(AR).

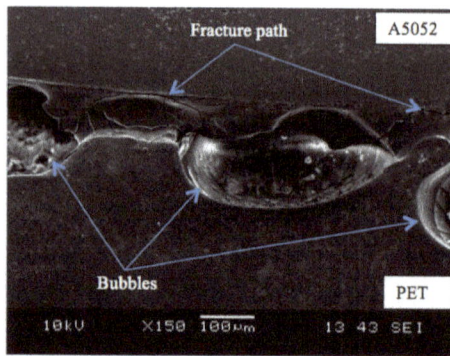

Figure 5. SEM photograph showing fracture path propagated at the bubbles.

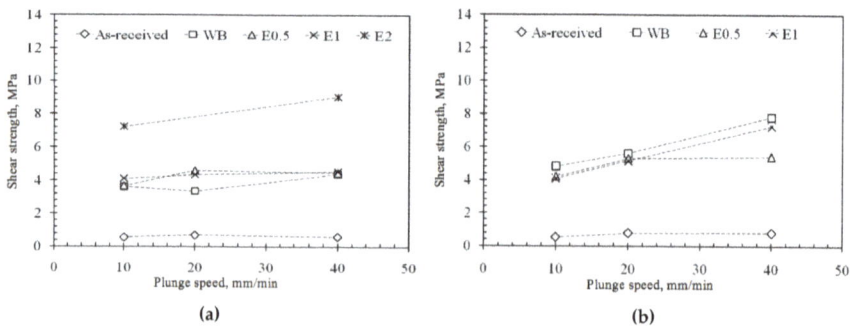

Figure 6. Effect of plunge speed on shear strength for different plunge depth. (**a**) Plunge depth: 0.4 mm and (**b**) plunge depth: 0.7 mm.

Therefore, stronger interfacial bonding between A5052 and PET occurred, thus improving the shear strength. The molten PET flows into irregularities such as pores, holes and crevices due to the pushing of bubbles' internal pressure. This phenomenon effectively conforms PET to the surface.

Further investigation of the joined interface cross section could clarify the mechanical interlocking effect. It was also observed that the shear strength increased with increasing plunge speed. Similarly, plunge speed has a pushing effect or impact on the molten PET flow, which promotes the mechanical interlocking of materials.

3.4. Cross-Sectional Observations of the Joined Interface

Cross-sectional observations of the joined regions for three typical joint specimens corresponding to Types 1, 2 and 3 are shown in Figures 7–9, respectively. Figure 7 indicates significant bubble formation with a maximum bubble size of approximately 1 mm in the center region of the welded area in the A5052(AR)-PET joint with surface roughness of 0.31 μm R_a. These bubbles formed from evaporated PET due to heat generated by the tool. The high internal pressure in the bubbles drove the molten PET into the aluminum surface irregularities, as seen in the first region in Figure 7. However, these large bubbles induced fractures and thus degraded the failure strength of the joined interface. Here, the material flow initiated by the protruded part of the aluminum towards the PET. The heat is transferred by conduction, softens the PET and dislocates it away from the plunged area. As it cools, the PET shrinks towards the plunge area simultaneously adhered to the rough aluminum surface by formed mechanical interlockings.

For the Type 2 specimens with surface roughness of 0.47 and 0.61 μm R_a, the size and number of bubbles were smaller compared to the Type 1 specimen, as seen in Figure 8. Therefore, the tensile shear failure load for Type 2 was higher than for the Type 1 specimen. For the Type 3 specimens with surface roughness of 1.42, 3.41 and 4.16 μm R_a, the size and number of bubbles were much smaller compared to the Type 1 and 2 specimens, as seen in Figure 9. It was observed that the molten PET flowed into the craters of the A5052 roughened surface. During tool penetration, the molten PET conformed to the rough surface and filled up the irregularities to form mechanical interlocking. Therefore, a rougher A5052 surface generates an anchoring effect between A5052 and PET, thus significantly improving the bonding strength of the joint.

As a result, the fracture was induced outside the bonding interface, which was at the PET in this case. According to the preceding results, the FSW joining of A5052 and PET contributed by the bonding of molten PET to A5052 when it conformed to the rough surface of A5052 with the aid of the inner pressure of the bubbles. This behavior created mechanical interlocking, which resulted in high tensile shear failure load. The effect of surface roughness, which can induce mechanical interlocking, was significant for the present FSW joining of A5052 and PET. The tensile shear failure load for the as-received surface roughness of 0.31 μm R_a was 0.5 kN while that for the treated specimen with a surface roughness of over 1.4 μm R_a was 5 kN. Notably, specimens with higher surface roughness have more bubbles than specimens with lower surface roughness. This is due to the presence of large amounts of microgrooves, concavities and holes in the rough surface that get filled by molten PET during the process. This significantly promotes structural interlocking and enhanced strength.

The experimental data demonstrated that bubbles pushing PET to enter the irregularities comprise a critical factor for improving joint strength in the process. However, the cooling of PET leaves bubbles in the hollow structure, which adversely reduces the strength. Preventing these bubbles from flowing away from the surface becomes difficult when the cooling process initiates. Another factor that promotes joint strength was high plunge speed. Therefore, FSW conditions with a lower temperature than PET vaporization which produces bubbles and higher plunge speed is expected to achieve superior failure strength without the adverse effects of the hollow structure.

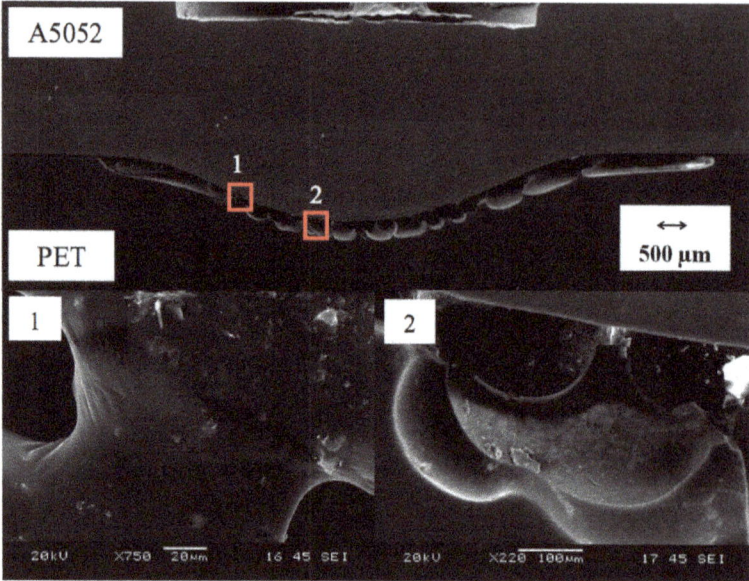

Figure 7. Cross-sectional observation of A5052(AR)-PET specimen joint at plunge speed of 20 mm/min and plunge depth of 0.7 mm.

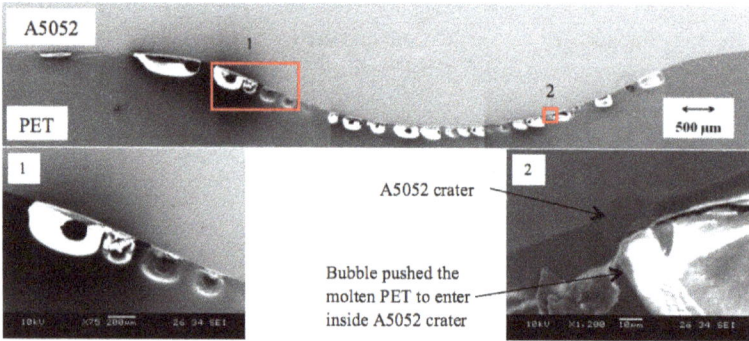

Figure 8. Cross-sectional observation of A5052(E0.5)-PET specimen joint at plunge speed of 10 mm/min and plunge depth of 0.7 mm.

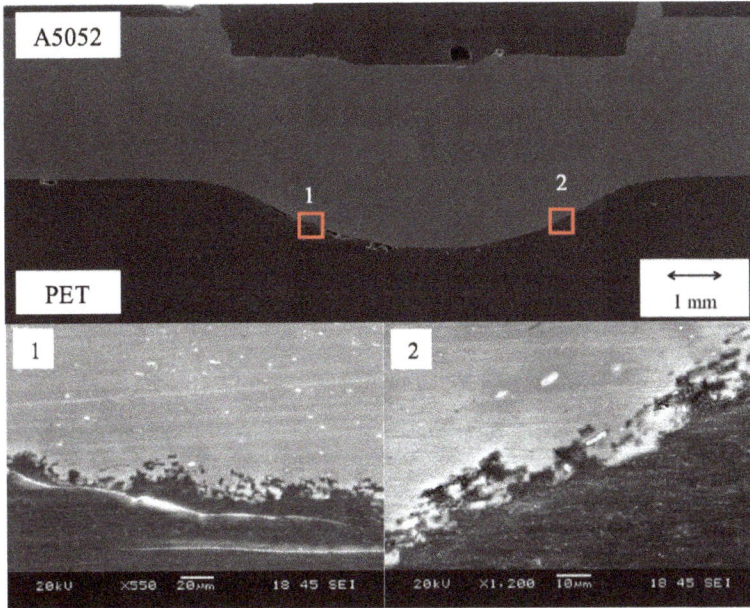

Figure 9. Cross-sectional observation of A5052(E6)-PET specimen joint at plunge speed of 20 mm/min and plunge depth of 0.7 mm.

4. Conclusions

FSW joining of A5052 and PET with various surface roughness values was carried out to investigate the joining mechanism and welding strength improvement. The PET joining interface melted due to heat induced by the FSW process and it adhered to the A5052 joining surface during the cooling process. Due to the low PET vaporizing temperature, part of the molten PET vaporized to form bubble-like holes beneath the PET joining interface. The internal pressure of the bubbles contributed to pushing the molten PET against the A5052 joining interface. Additionally, higher plunge speed also promoted a similar effect.

The cross-sectional observations showed that bubbles formed on the PET side in all joined specimens. The size and distribution of the bubbles significantly affected the shear strength of the welded joint. Smaller bubbles formed in the pre-treated A5052-PET joints compared to the A5052(AR) joints. The molten PET conformed to the roughened surface irregularities to form mechanical interlocking.

In the present FSW joining between A5052 and PET, a significant improvement of tensile-shear failure load was achieved when the surface roughness was larger than 0.31 µm R_a(as-received). The tensile-shear failure load was 0.4–0.8 kN for lower surface roughness specimens (0.31 µm R_a) and 4.8–5.2 kN for roughened surface specimens (>0.31 µm R_a).

Acknowledgments: This research was funded by University Malaya Research Grant (UMRG) through grant No. RP035A-15AET. The authors thank Nobushiro Seo from Nippon Light Metal Co. Ltd. for preparing the specimens.

Author Contributions: Farazila Yusof performed the research and wrote the article. Mohd Ridha bin Muhamad performed data and micrograph analysis, and manuscript revision. Raza Moshwan and Mohd Fadzil bin Jamaludin assisted in conducting the experiments and data analysis and Yukio Miyashita provided guidance and advise.

Conflicts of Interest: The authors declare no conflict of interest.

References

1. Oliveira, P.H.F.; Amancio-Filho, S.T.; Dos Santos, J.F.; Hage, E. Preliminary study on the feasibility of friction spot welding in PMMA. *Mater. Lett.* **2010**, *64*, 2098–2101.
2. Rodrigues, D.M.; Loureiro, A.; Leitao, C.; Leal, R.M.; Chaparro, B.M.; Vilaça, P. Influence of friction stir welding parameters on the microstructural and mechanical properties of AA 6016-T4 thin welds. *Mater. Des.* **2009**, *30*, 1913–1921.
3. Feng, A.H.; Xiao, B.; Ma, Z. Grain boundary misorientation and texture development in friction stir welded SiCp/Al–Cu–Mg composite. *Mater. Sci. Eng. A* **2008**, *497*, 515–518.
4. Cam, G. Friction stir welded structural materials: Beyond Al-alloys. *Int. Mater. Rev.* **2011**, *56*, 1–48.
5. Leal, R.M.; Leitão, C.; Loureiro, A.; Rodrigues, D.M.; Vilaça, P. Material flow in heterogeneous friction stir welding of thin aluminum sheets: Effect of shoulder geometry. *Mater. Sci. Eng. A* **2008**, *498*, 384–391.
6. Ghosh, M.; Kumar, K.; Kailas, S.V.; Ray, A.K. Optimization of friction stir welding parameters for dissimilar aluminum alloys. *Mater. Des.* **2010**, *31*, 3033–3037.
7. Colligan, K. Material Flow Behavior during Friction Stir Welding of Aluminum. *Weld. J.* **1999**, *78*, 229–237.
8. Murr, L.; Liu, G.; McClure, J. Dynamic recrystallization in friction-stir welding of aluminum alloy 1100. *J. Mater. Sci. Lett.* **1997**, *6*, 1801–1803.
9. Dawes, C.J.; Thomas, W.M. Friction stir process welds aluminum alloys. *Weld. J.* **1996**, *75*, 41–50.
10. Barlas, Z.; Uzun, H. Microstructure and mechanical properties of friction stir butt welded dissimilar Cu/CuZn30 sheets. *J. Achiev. Mater. Manuf. Eng.* **2008**, *30*, 182–186.
11. Ramirez, A.J.; Juhas, M.C. Microstructural Evolution in Ti-6Al-4V Friction Stir Welds. *Mater. Sci. Forum* **2003**, *426–432*, 2999–3004.
12. Sato, Y.S.; Nelson, T.W.; Sterling, C.J.; Steel, R.J.; Pettersson, C.O. Microstructure and mechanical properties of friction stir welded SAF 2507 super duplex stainless steel. *Mater. Sci. Eng. A* **2005**, *397*, 376–384.
13. Commin, L.; Dumont, M.; Masse, J.E.; Barrallier, L. Friction stir welding of AZ31 magnesium alloy rolled sheets: Influence of processing parameters. *Acta Mater.* **2009**, *57*, 326–334.
14. Nami, H.; Adgi, H.; Sharifitabar, M.; Shamabadi, H. Microstructure and mechanical properties of friction stir welded Al/Mg2Si metal matrix cast composite. *Mater. Des.* **2011**, *32*, 976–983.
15. Yan, Y.; Zhang, D.T.; Qiu, C.; Zhang, W. Dissimilar friction stir welding between 5052 aluminum alloy and AZ31 magnesium alloy. *Trans. Nonferrous Metals Soc. China* **2010**, *20*, s619–s623.
16. Squeo, E.A.; Bruno, G.; Guglielmotti, A.; Quadrini, F. Friction stir welding of polyethylene sheets. In *The Annals of "DUNĂREA DE JOS" University of Galati Fascicle V, Technologies in Machine Building*; Galati University Press: Galati, Romania, 2009; pp. 241–146.
17. Bilici, M.K.; Yükler, A.I.; Kurtulmuş, M. The optimization of welding parameters for friction stir spot welding of high density polyethylene sheets. *Mater. Des.* **2011**, *32*, 4074–4079.
18. Arici, A.; Selale, S. Effects of tool tilt angle on tensile strength and fracture locations of friction stir welding of polyethylene. *Sci. Technol. Weld. Join.* **2007**, *12*, 536–539.
19. Vijendra, B.; Sharma, A. Induction heated tool assisted friction-stir welding (i-FSW): A novel hybrid process for joining of thermoplastics. *J. Manuf. Process.* **2015**, *20*, 234–244.
20. Pirizadeh, M.; Azdast, T.; Ahmadi, S. Friction stir welding of thermoplastics using a newly designed tool. *Mater. Des.* **2014**, *54*, 342–347.
21. Lambiase, F. Mechanical behaviour of polymer-metal hybrid joints produced by clinching using different tools. *Mater. Des.* **2015**, *87*, 606–618.
22. Paoletti, A.; Lambiase, F.; Di Ilio, A. Optimization of Friction Stir Welding of Thermoplastics. *Procedia CIRP* **2015**, *33*, 563–568.
23. Sadeghian, N.; Besharati Givi, M.K. Experimental optimization of the mechanical properties of friction stir welded Acrylonitrile Butadiene Styrene sheets. *Mater. Des.* **2015**, *67*, 145–153.
24. Simões, F.; Rodrigues, D. Material flow and thermo-mechanical conditions during Friction Stir Welding of polymers: Literature review, experimental results and empirical analysis. *Mater. Des.* **2014**, *59*, 344–351.
25. Arici, A.; Mert, S. Friction Stir Spot Welding of Polypropylene. *J. Reinf. Plast. Compos.* **2008**, *27*, 2001–2004.
26. Amancio-Filho, S.; Bueno, C.; dos Santos, J.; Huber, N.; Hage, E. On the feasibility of friction spot joining in magnesium/fiber-reinforced polymer composite hybrid structures. *Mater. Sci. Eng. A* **2011**, *528*, 3841–3848.

27. Yusof, F.; Miyashita, Y.; Seo, N.; Mutoh, Y.; Moshwan, R. Utilising friction spot joining for dissimilar joint between aluminum alloy (A5052) and polyethylene terephthalate. *Sci. Technol. Weld. Join.* **2012**, *17*, 544–549.

28. Moshwan, R.; Rahmat, S.M.; Yusof, F.; Hassan, M.A.; Hamdi, M.; Fadzil, M. Dissimilar friction stir welding between polycarbonate and AA 7075 aluminum alloy. *Int. J. Mater. Res.* **2015**, *106*, 258–266.

29. Lee, S.H.; Lee, C.J.; Kim, B.H.; Ahn, M.S.; Kim, B.M.; Ko, D.C. Effect of Tool Shape on Hole Clinching for CFRP with Steel and Aluminum Alloy Sheet. *Key Eng. Mater.* **2014**, *622–623*, 476–483.

30. Lambiase, F. Joinability of different thermoplastic polymers with aluminum AA6082 sheets by mechanical clinching. *Int. J. Adv. Manuf. Technol.* **2015**, *80*, 1995–2006.

31. Lambiase, F.; Paoletti, A.; Di Ilio, A. Mechanical behaviour of friction stir spot welds of polycarbonate sheets. *Int. J. Adv. Manuf. Technol.* **2015**, *80*, 301–314

32. Lambiase, F.; Di Ilio, A. Mechanical clinching of metal-polymer joints. *J. Mater. Process. Technol.* **2015**, *215*, 12–19.

33. Wirth, F.X.; Zaeh, M.F.; Krutzlinger, M.; Silvanus, J. Analysis of the bonding behavior and joining mechanism during friction press joining of aluminum alloys with thermoplastics. *Procedia CIRP* **2014**, *18*, 215–220.

34. Liu, F.C.; Liao, J.; Nakata, K. Joining of metal to plastic using friction lap welding. *Mater. Des.* **2014**, *54*, 236–244.

35. Hussein, F.I.; Akman, E.; Genc Oztoprak, B.; Gunes, M.; Gundogdu, O.; Kacar, E.; Hajim, K.I.; Demir, A. Evaluation of PMMA joining to stainless steel 304 using pulsed Nd:YAG laser. *Opt. Laser Technol.* **2013**, *49*, 143–152.

36. Ho, P. Chemistry and adhesion of metal-polymer interfaces. *Appl. Surf. Sci.* **1990**, *41–42*, 559–566.

37. Yao, Q.; Qu, J. Interfacial versus cohesive failure on polymer-metal interfaces in electronic packaging—effects of interface roughness. *J. Electron. Packag.* **2002**, *124*, 127.

38. Hay, K.M.; Dragila, M.I. Physics of fluid spreading on rough surfaces. *Int. J. Numer. Anal. Model.* **2008**, *5*, 85–92.

39. Nakae, H.; Inui, R.; Hirata, Y.; Saito, H. Effects of surface roughness on wettability. *Acta Mater.* **1998**, *46*, 2313–2318.

40. Chen, Y.; Nakata, K. Effect of the Surface State of Steel on the Microstructure and Mechanical Properties of Dissimilar Metal Lap Joints of Aluminum and Steel By Friction Stir Welding-Redorbit. *Sci. Technol. Weld. Join.* **2010**, *15*, 293–298.

metals

MDPI

Article

Study on the Surface Integrity of a Thin-Walled Aluminum Alloy Structure after a Bilateral Slid Rolling Process

Laixiao Lu [1], Jie Sun [1,*], Xiong Han [2] and Qingchun Xiong [2]

[1] Key Laboratory of High Efficiency and Clean Mechanical Manufacture of Ministry of Education, School of Mechanical Engineering, Shandong University, Jinan 250061, China; lulaixiao@163.com
[2] CAC Chengdu Aircraft Industrial (Group) Co., Ltd., Chengdu 610000, China; xiong-han@163.com (X.H.); xiongqingchun@163.com (Q.X.)
* Correspondence: sunjie@sdu.edu.cn; Tel.: +86-531-88394593 (ext. 8)

Academic Editor: Nong Gao
Received: 3 March 2016; Accepted: 21 April 2016; Published: 26 April 2016

Abstract: For studying the influence of a bilateral slid rolling process (BSRP) on the surface integrity of a thin-walled aluminum alloy structure, and revealing the generation mechanism of residual stresses, a self-designed BSRP appliance was used to conduct rolling experiments. With the aid of a surface optical profiler, an X-ray stress analyzer, and a scanning electron microscope (SEM), the differences in surface integrity before and after BSRP were explored. The internal changing mechanism of physical as well as mechanical properties was probed. The results show that surface roughness (Ra) is reduced by 23.7%, microhardness is increased by 21.6%, and the depth of the hardening layer is about 100 μm. Serious plastic deformation was observed within the subsurface of the rolled region. The residual stress distributions along the depth of the rolling surface and milling surface were tested respectively. Residual stresses with deep and high amplitudes were generated via the BSRP. Based on the analysis of the microstructure, the generation mechanism of the residual stresses was probed. The residual stress of the rolling area consisted of two sections: microscopic stresses caused by local plastic deformation and macroscopic stresses caused by overall non-uniform deformation.

Keywords: bilateral slid rolling process (BSRP); surface integrity; aluminum alloys; thin-walled structure

---

1. Introduction

Aluminum alloys are widely used in the aviation industry due to its significant advantages, such as high specific strength, specific stiffness, anti-fatigue, *etc*. Aluminum alloys, titanium alloys, and composite materials are the three most important kinds of structural materials in modern aircraft. However, machining distortion is inevitable in machining processes for the aerospace monolithic components of aluminum alloys. The causes are numerous and diversified, including work blank residual stresses, machining stresses, stiffness variations in machining processes, and so on. To guarantee the high precision of parts, distortion correction processes are unavoidable for distorted aerospace monolithic components. A rolling method, which can introduce high-amplitude residual compressive stresses, can be used to correct the distorted components. The stresses will be redistributed after the rolling operations. Accordingly, the component configuration can be changed [1,2].

In rolling processes, a smooth wheel, roller, or ball with a high hardness is adopted to act on the part of the surfaces with a certain pressure for surface hardening, achieving residual compressive stresses. In 1929, this technology was firstly used in axles, crankshafts, and other parts for surface strengthening in Germany. Nowadays, after decades of development, it is widely used in aerospace, automotive motorcycle, precision machinery, bio-medicine, *etc.* [3,4]. Typical rolling technologies include burnishing, deep rolling (DR), and low plasticity burnishing (LPB). The main purpose of burnishing is to achieve a low surface roughness (Ra). However, it does not have much influence on residual stress [5]. The operation parameters and executing components of DR and burnishing are different. Since the rolling depth in DR is larger, a high compressive residual stress and a deep work hardening layer are obtained. Therefore, it is generally implemented at the outer surface of the key parts for strengthening, such as the shaft shoulder, crankshaft fillets, *etc.* [6,7]. In LPB processes, the most distinctive feature is that a carbide alloy or ceramic ball is motivated by the high-pressure liquid. As a result, a better Ra and deeper residual compressive stress can be obtained. Further, the defect of hardened material was also solved [8].

Recently, with the development of assistive technology, a variety of new burnishing methods have been proposed, such as the ultrasonic surface rolling process (USRP) [9], laser-assisted burnishing (LAB) [10], and cryogenic burnishing (CB) [11]. In the LAB processes, the workpiece surface layer is temporarily and locally softened by a controllable laser beam and then immediately processed by a conventional burnishing tool. LAB processes can produce a much better surface finish, a higher surface hardness, and a similar residual compressive stress compared with other conventional technologies used in hard materials, while the CB process is suited for a relatively soft metal like aluminum alloys. Research for rolling techniques mainly includes experimental, theoretical, and finite element simulation.

For the needs of practical application, substantial experimental research has been carried out to study the influence of parameters on surface integrity, including rolling depth, feed speed, rolling speed, the diameter of the wheel or ball, and the number of rolling passes [7]. Furthermore, multifarious prediction models have also been established based on experimental results. For example, Aysun *et al.* proposed a roughness prediction model and an optimization strategy of the ball burnishing process based on a desirability function approach and response surface methodology [12]. Yu *et al.* investigated the influences of parameters on surface integrity and established a prediction model for Ra and residual stress [13].

A surface strengthening mechanism was revealed through an analysis of microstructures. Avilés *et al.* and Juijerm *et al.* found that the fatigue lifetimes of rolled specimens at room and high temperatures were increased [8,14]. Zhu *et al.* found that the fatigue crack sources shifted from the surface to the second surface, which increased fatigue resistance ability [15]. Wang *et al.* found that the breakdown and corrosion potential of 40Cr were positively moved after the USRP, which indicated an improvement in corrosion resistance [9].

With the development of finite element technology, finite element modeling and a simulation method have been used in rolling analysis. Skalski *et al.* firstly proposed a two-dimensional finite element model of the rolling processes to analyze the changes of penetration depth under different pressures and roller diameters [5]. The problem is that a two-dimensional model cannot properly simulate the situation of the roller pressed into the workpiece. A three-dimensional model was firstly used to analyze the roll forming process of an annular groove by Kim *et al.* to study the influence of feed rate on residual stress [16]. However, the wheel or roller is often assumed as a rigid body in the finite element model in order to simplify the calculation, which affected the accuracy of the analysis. Rodríguez *et al.* considered the ball as a linear, elastic, and isotropic material, whereas the workpiece was an isotropic, plastic-hardening material with bilinear behavior, and the residual stress distribution accorded well with the experimental results [17].

Most of the research has focused on the outer cylindrical surface, the end surface, and the horizontal surface. Normally, an appropriative tool was installed on a lathe or machining center. However, there are structural restrictions when those rolling technologies are directly applied to monolithic components. The overall structures of these components are generally thin-walled parts with low structural rigidity. Unnecessary distortion will be introduced when the pressure is applied to one side. Therefore, bilateral slid rolling technology should be adopted in the correction processes of aerospace monolithic components.

In the USRP, double rollers with symmetrical pressure act on both sides of the thin-wall structure to ensure that no additional torque is introduced. Wang *et al.* have tested the surface qualities (including Ra, hardness, and the residual stresses of the surface) after the bilateral slid rolling process (BSRP) [1], but the variation mechanism of the thin-walled structure was not discussed. Since the particularity application of the BSRP, there has been little direct research. However, some surface treatment methods for aerospace monolithic components can still provide reference and guidance. Nam *et al.* presented a response surface methodology to optimize the surface properties of microhardness and residual stress in a shot peening process for aerospace structural components [18]. Nie *et al.* investigated the influence on high cycle fatigue resistance of laser shock peening (LSP) for the compressor blade made of TC11 titanium alloy. The relationship between fatigue characteristics and effects on residual stress and microstructural changes was established to reveal the strengthening mechanism of LSP [19]. Hennig *et al.* used the shot peening method with steel media in aerofoil, fillet, and annulus to introduce compressive residual stresses to increase the high cycle fatigue. The special designed caliper nozzles have been successfully utilized for different engine types in Rolls-Royce automobiles for five years [20].

Therefore, a self-designed BSRP appliance that is suitable for aerospace monolithic components was used to conduct the rolling experiments with a milled thin-walled aluminum alloy structure. The impact of rolling processes on surface integrity is discussed, addressing surface topography, microhardness, microstructure, and residual stress. The internal changing mechanism of physical as well as mechanical properties is revealed. Additionally, the generation mechanisms of residual stresses were probed by analyzing the microscopic plastic deformation and macroscopic non-uniform deformation.

2. Materials and Methods

2.1. Experimental Materials

The experiments were performed on a 60-mm-thick 7050-T7451 aluminum alloy sheet, which was manufactured by Kaiser Aluminum & Chemical Corp, Oakland, CA, USA. The original residual stress was measured with the crack-compliance method. Figure 1 shows the locations of the samples and the stress distribution of the transverse profile along the thickness direction in the blank. Since the initial surface conditions have a critical influence on surface integrity, one of the end faces (*XOY* plane) of the sheet was milled first to imitate the condition of a processed surface. Then, two equally thin-walled structures, E1 and E2, were cut via wire electrical discharge machining to reduce the machining stresses. The E1 sample was used to conduct the tests of the surface profile and microstructure, and E2 sample was used to study the residual stress distribution. The sample size is $40 \times 20 \times 3$ mm, and the top half of the sample was rolled, which is marked with blue in Figure 1a.

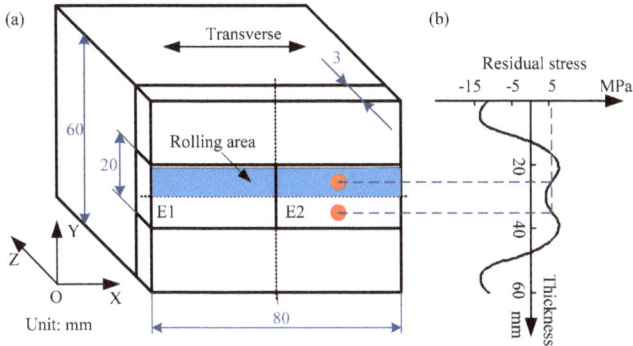

Figure 1. Experimental materials: (**a**) schematic of sampling location; (**b**) initial residual stress distribution of experimental blank.

2.2. The BSRP Appliance and Experimental Design

Double rollers must be implemented on both sides of the rib with equal pressure due to the fact that the stiffness of aerospace monolithic components is low and easy to be distorted. To meet this requirement, the BSRP appliance was designed and manufactured, as shown in Figure 2a. The rollers with smooth surfaces were made of rolling bearings and mounted on two sliding blocks through bolts, respectively. One block was fixed to basal body, while another can be driven in a chute by a rotating preload nut. In the rolling process, the target component was first placed between the rollers. Then, the proposed preload force was applied by adjusting the preload nut. At the end, the rolling processes could be conducted by pulling the workpiece from one side to another. Figure 2b shows the typical rolling processes of beam structures.

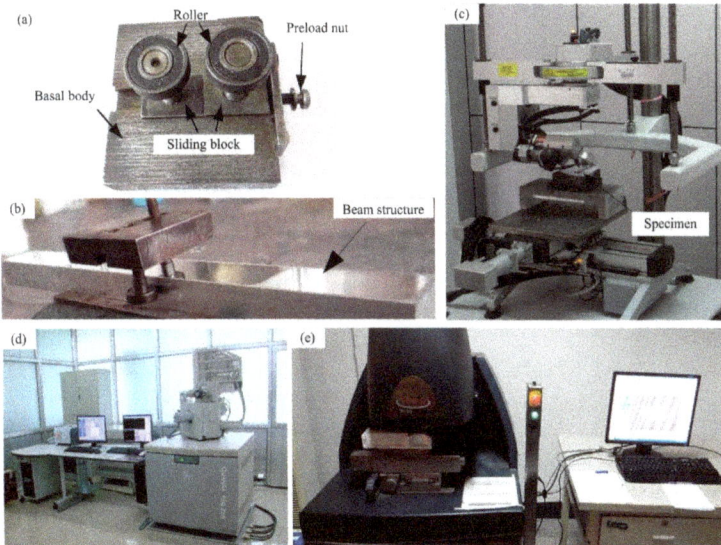

Figure 2. BSRP appliance and experimental processes: (**a**) BSRP appliance; (**b**) typical rolling processes of beam structures; (**c**) residual stresses test; (**d**) SEM system; (**e**) surface topography test.

In the rolling processes, the rolling depth is directly determined by the rotation angle of the preload nut. The materials of rollers bore steel GCr15 with an elastic modulus of 207 GPa, while the material of the target component was aluminum alloy 7050-T7451 with an elastic modulus of 71.7 GPa. In order to effectively control the rolling depth, a corresponding relation between the rotation angle of the preload nut and the rolling depth was established based on a large number of measurements. The relationship between the rotation angle and rolling depth is shown in tab:metals-06-00099-t001.

Table 1. Quantitative relationship between the rotation angle and rolling depth.

Rotation Angles (°)	Rolling Depth (mm)
45	0.03
90	0.08
135	0.11
180	0.13
225	0.15
270	0.17

In this experiment, the preload nut was rotated at 180°, so the rolling depth was 0.13 mm. Each component was bi-directionally rolled to eliminate the effect of unevenness caused by single rolling. The diameter and thickness of the roller were 28 mm and 8 mm, respectively. Since the rolling speed in correction operations was set in a very low level. There is no obvious change in surface integrity when the rolling speed is under such a low level. As a result, the rolling speed was set to 0.15 m/min.

2.3. Measurement Equipment and Methods

The milled side of the thin-walled structure, including the milling surface and the rolling surface, was selected to conduct the following tests. Surface topographies were observed by the NT9300 surface profiler (Veeco, Plainview, NY, USA). The MH-06 type microhardness tester (Everone Enterprisks Ltd., Shanghai, China) was used to measure the distribution of the microhardness. The experimental load was 10 g, and the hold time was 5 s. The microstructures of the cross sections were observed by the DMLM microscope system (Leica, Wetzlar, Germany) and the QF-250 scanning electron microscope (SEM) system (Bionand, Málaga, Spain). The X-Stress 3000 stress analyzer (Stresstech Oy, Vaajakoski, Finland) was used to measure the residual stress with a Cr target and diffraction angle of 139.3°. In order to test the residual stress distribution along the depth, the electrolytic polishing method was adopted to remove surface material step by step. The detailed tests processes are shown in Figure 2.

3. Results

3.1. Surface Profile

The surface profile of the junction region between rolled and milled regions is shown in Figure 3a. To compare the surface topography visually and take the impact of milling marks into consideration, two paths, as indicated by the black line in Figure 3a, were selected for the rolling surface and the milling surface, respectively. The profile data are plotted in Figure 3b. On the milling surface, the milling marks could be clearly differentiated, while the heights of the peaks were reduced in the rolling region. The material at the peaks was squeezed into valleys by the pressure of the rollers. As a result, shallow marks were filled. However, deeper marks could not be eliminated completely due to the restrictions of metal flow. The range of milling and rolling surfaces was reduced from 4.908 μm to 3.142 μm, by 40.0%. The roughness of the milling surface was 1.01 μm and reduced to 0.77 μm after the BSRP, by 23.7%.

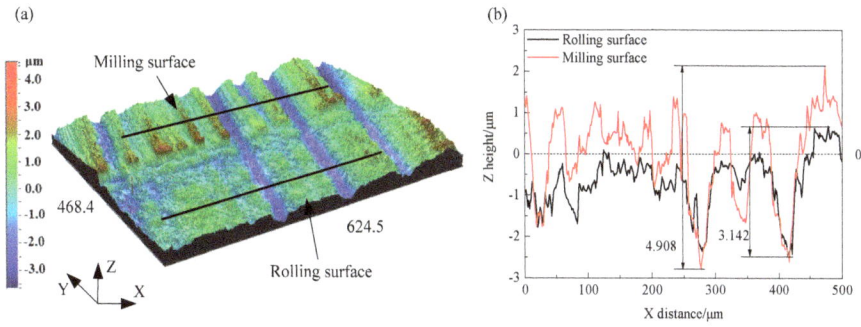

Figure 3. Surface topography: (**a**) surface topography comparison of rolling and milling surfaces; (**b**) morphological comparison of selected paths.

3.2. Microhardness

Microhardness distributions along the depths of the milling and rolling regions were tested three times, and the average value was calculated for further study. Figure 4 shows the microhardness distribution and the depth of the hardened layer. The microhardness of the milling surface was 104.6 $HV_{0.01}$. It improves with increases in depth and reaches a matrix value of 118 $HV_{0.01}$, indicated in Figure 4 by a dotted line, at a depth of 80 μm, while the surface microhardness of the rolling region is 127.2 $HV_{0.01}$ and decreases with increases in depth. It reaches the matrix hardness at approximately 100 μm deep below the surface.

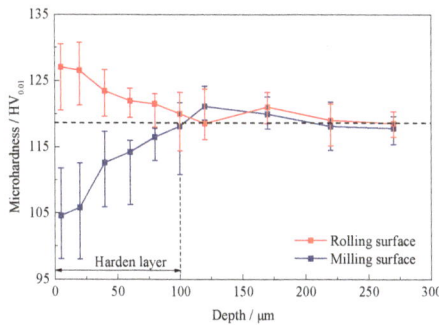

Figure 4. Distribution of microhardness along the depth.

For the milling surface, due to the influence of thermal and other factors during the machining processes, surface materials were softened. Therefore, the hardness of the milling surface is lower than that of the matrix material. However, in the rolling operation, severe plastic deformation and high amplitude stress were introduced to the rolling surface. As a result, the surface material was strengthened by rolling processes. The surface microhardness increased by 21.6%, compared to the milling surface.

3.3. Microstructure

Figure 5 shows the positional relation of the microstructure for observing samples. The normal direction of the M2 section is perpendicular to the rolling direction, and the M1 sample is compared for the milling region. On the other hand, the normal directions of the M3 and M4 sections are parallel to the direction of the rolling region.

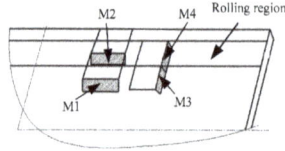

Figure 5. Schematic of samples' positional relation.

Figure 6 shows the variation of the microstructure after the rolling processes. In Figure 6b, the milling surface is not flat and has a lot of defects. However, surface defects are significantly reduced after rolling processes, as shown in Figure 6b. The rolling surface is smoother than the milling surface, which is indicated by the fact that it has a low friction coefficient and higher anti-corrosion ability. In rolling processes, the micro-peaks of the milling surface contacted with the harder rollers first. Then, the materials of micro-peaks were compacted, and flowed to valley areas due to severe plastic deformation caused by the differences in material hardness and rolling pressure. However, some deeper valleys could not be eliminated completely.

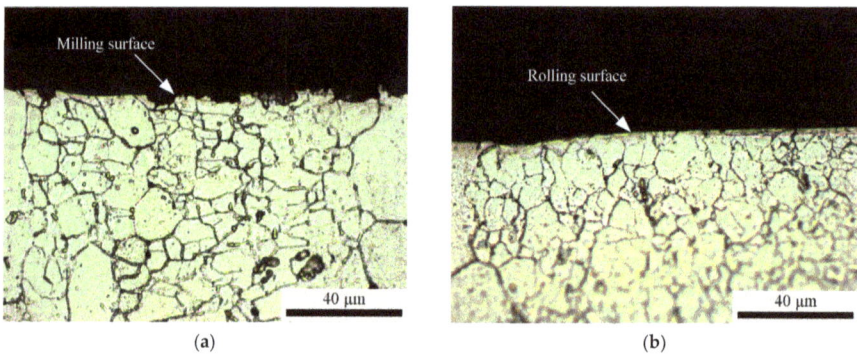

(a)

(b)

Figure 6. Microstructure comparison of specimens ($\times$500): (**a**) M1 and (**b**) M2.

3.4. Residual Stresses

Residual stresses tests were conducted with the E2 sample. The direction of X axis is defined in parallel to the rolling direction. Figure 7 shows the distributions of residual stresses along the depth.

From Figure 7, residual stresses distributions of the milling surface show a "spoon" shaped in both directions. The amplitude of residual compressive stresses exists on the milled surface, and it increases first and then decreases as depth increases. The residual stress at the surface is -78.5 MPa in the X direction and reaches a maximum value of -141.6 MPa at a depth of 21 μm, while the value in the Y direction is -19.5 MPa at the surface and reaches a maximum value of -89.1 MPa at a depth of 34 μm. The stresses of the X direction recover to the levels of the matrix at a depth of 75 μm, while it is 110 μm in the Y direction.

The distribution characteristic and amplitude of residual stresses in the rolling region are significantly different from that of the milling region. For the rolling region, maximum compressive stress values were obtained at the surface with -253.8 MPa and -217.1 MPa in the X and Y directions, respectively. The value of the residual compressive stress decreases as depth increases. According to the slope of the curve, the residual stress layer can be divided into three areas: the significant impact region, the smooth transition region, and the region in which the effect disappeared. The significant impact region was located from the surface to 150 μm deep. In this region, residual stress was significantly affected by rolling processes and increased as depth increased, while the residual stress was relatively stable at approximately -100 MPa from 150 μm deep to 500 μm deep below the surface. This layer was defined as the smooth transition region. Next is the disappeared effect region. In this region, the residual stress decreased rapidly and reached the matrix stress status at depths of 600–700 μm.

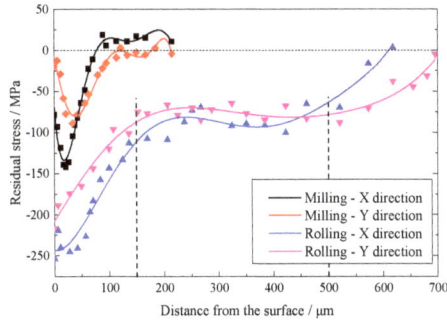

Figure 7. Distribution of residual stresses along the depth.

4. Discussion

4.1. Variation Mechanisms of the Rolling Surface

The surface strengthening mechanism was further discussed based on the variation in the microstructure. Figure 8 shows the typical microstructures of the rolling and milling region. On the milling surface, the defects can be clearly observed, whereas the rolling surface is smooth, and the boundary line of the rolling surface is more regular than that of the milling surface. Due to the plastic deformation of the surface materials, the milling trace, which is one of the major factors that affect surface quality, was reduced by the rolling processes. This corresponds to the vibration of the surface profile. It is suggested that the rolling processes could be used in improving the surface quality for aerospace monolithic components.

In conventional rolling processes, with a high rolling speed and pressure, surface grains are pronounced deformed, and metal flow traces can be clearly observed on the shallow surface. The surface region for rolling-treated specimens can then be divided into two zones from the surface to the core: the refined-grain zone and the coarse-grain zone [21]; however, in this study, the rolling surface had only a very tiny metal flow traces after the BSRP, as shown in Figure 8b,d. The region with metal flow traces was defined as the refined-microstructure zone. The depth of the refined-microstructure zone is approximately determined on the basis of visual analysis of the microstructure [22]. The average depth of the refined-microstructure zone induced by the BSRP is about 5 μm, as shown in Figure 8b. However, the refined-microstructure zone could not be clearly distinguished in Figure 8d. Comparing samples M2 and M4, the cross section of sample M2 is perpendicular to the rolling direction. It is suggested that the flow of material caused by rolling processes is mainly along the rolling direction, while plastic strain is hardly noted after the milling process in Figure 8a,c.

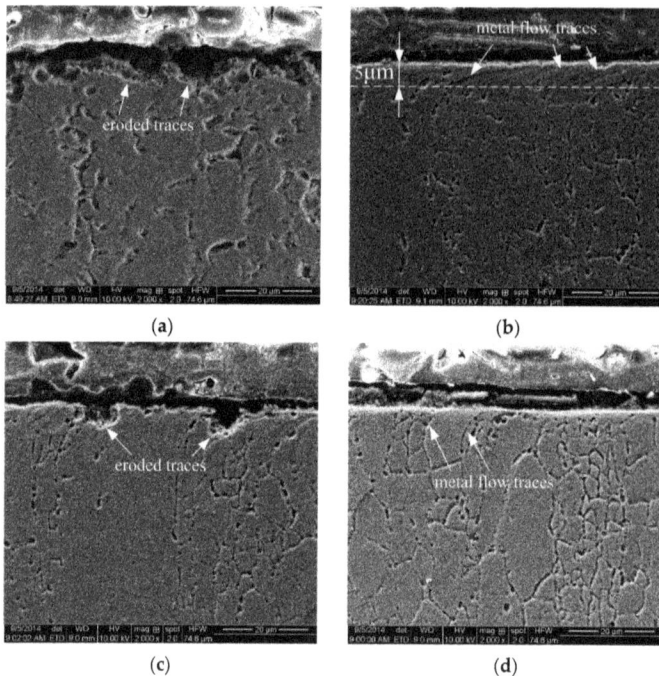

Figure 8. SEM photographs of specimens (×2000): (**a**) M1; (**b**) M2; (**c**) M3; and (**d**) M4.

In the refined-microstructure zone, slipping is hard for the grains with "hard" orientations, leading to the formation of dislocations in the interior of the grains [23]. The dislocation proliferation happened on the surface after the rolling led to the increase of dislocation density. Thus, work hardening resulted from the increased plastic deformation and dislocation. Figure 4 shows that, as a consequence, the profile of the microhardness distribution beneath the surface is altered.

These refined surface layers with higher hardness and compressive residual stress due to the mechanism of severe plastic deformation and dislocation would provide benefits to enhance the mechanical and physical properties of the components, such as improving their corrosion/wear resistance and increasing their fatigue lives [11]. During the preparation operations of the samples, surfaces were corroded to different extents, although the corrosion time was equal. It can be observed in Figure 8a,c that the remarkable eroded traces could be found. However, the rolling surface shows no discernible evidence of corrosion. The superficial area was decreased after the rolling processes, so the corrosion quantity was also reduced. It is indicated that the surface corrosion resistance ability was improved by the rolling processes.

4.2. Generation Mechanism of Residual Stresses

Residual stresses in the rolling region were closely related with the rolling effect. The distribution of full-width-at-half-maximum (FWHM) along the depth was also obtained by the X-ray diffraction method, as shown in Figure 9. The FWHM describes the width of the diffraction peaks. It is related to the heat treatment, surface hardening, and other microscopic residual stresses caused by the grain size and the value of deformation and dislocation density [14,24]. In Figure 9b, the FWHM of the rolling surface is higher than the milling surface. The depth of the affected layer is up to 50 μm, which is indicated by the fact that severe plastic deformation and dislocation were generated by rolling processes in the surface and subsurface grains. With the emergence of dislocations, the microscopic residual stresses were introduced to the surface material.

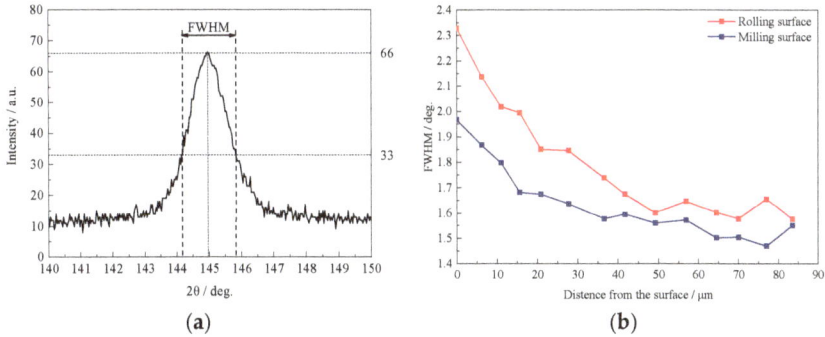

Figure 9. Distribution of full-width-at-half-maximum (FWHM) along the depth: (**a**) intensity profile; (**b**) FWHM distribution.

This phenomenon can be explained by analyzing the microstructures. After rolling processes, the materials in the surface region were flattened and elongated under the pressure of rollers, as shown in Figure 10. The original shape and positional relationship of the grains were changed by this uneven deformation. As a result, interacting forces between distorted grains were generated, as shown in Figure 10b. This force was presented as residual compressive stresses, which exist in the entire rolling region. These phenomena were consisted of severe plastic deformations and metal flow traces in the surface and subsurface, as shown in Figure 7b,d.

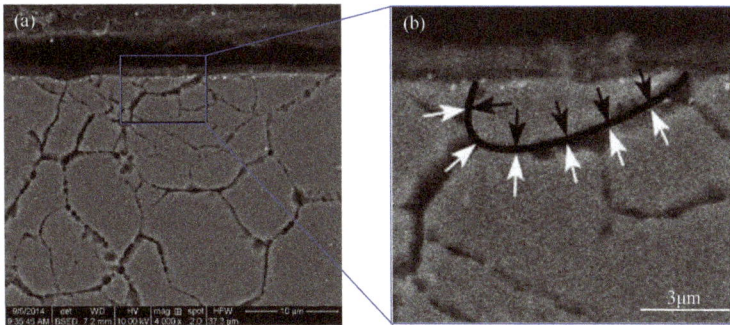

Figure 10. Distribution of intergranular forces: (**a**) Specimen M2 (×4000) and (**b**) schematic of intergranular forces.

Since only half of the structure was rolled, there were two kinds of macroscopic deformation in the thin-walled structure after the BSRP. For the rolled region, the surface materials were plastically deformed, and the center materials were elastically deformed. For the whole structure, it was bent under the effect of residual stresses and micro-plastic deformation.

Under the pressure of rollers, the materials at the peaks flowed into nearby valleys. In the rolling processes, the materials of the surface layer were plastically deformed, while materials in the deep layer were only passively elastically deformed and would recover after rolling, as shown in Figure 11. As a result, macroscopic residual compressive stresses were developed in the surface materials, while residual tensile stresses were obtained in the elastically deformed region.

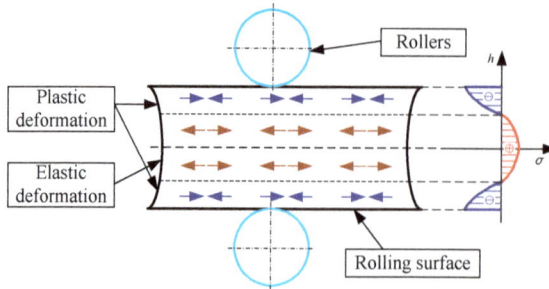

Figure 11. Schematic of deformation and residual stresses caused by rolling processes.

In the rolling process, only the upper half part was rolled. According to the results of the residual stresses, deep and high amplitude compressive stresses were generated in the rolling region via this kind of local rolling process. If the effect of the structure depth was ignored, the residual stresses were equivalent to an average value, as shown in Figure 12a. The thin-walled structure was easily distorted due to its low stiffness, since the introduced residual stresses and micro-plastic deformation in the whole structure were unbalanced. The structure was bent under the effect of residual stress and micro-plastic deformation. As a result, macroscopic stress was generated, as shown in Figure 12b. Essentially, the residual stresses introduced by the rolling processes was the result of the re-equilibrium of stresses caused by micro-plastic deformation and macro-bending deformation in the thin-walled structure, as shown in Figure 12c.

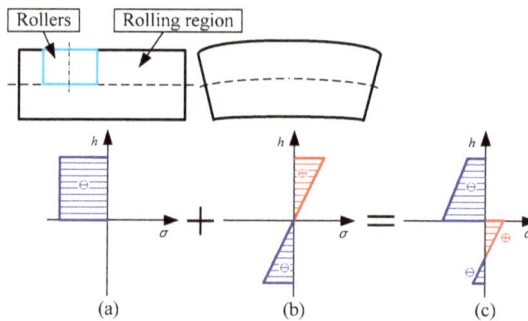

Figure 12. Residual stresses caused by local rolling processes: (**a**) rolling stress; (**b**) bending stress; (**c**) combined stress.

5. Conclusions

Self-designed bilateral slid rolling equipment was employed in aluminum alloy thin-walled structure rolling processes. The surface integrity as well as the strengthening mechanism of the rolled surface was analyzed after the BSRP.

(1) The surface integrity of the aluminum alloy 7050-T7451 was significantly improved by the BSRP. After the rolling processes, surface defects prominently reduced. Specifically, Ra reduced by 23.7% from 1.01 to 0.77 μm, surface microhardness increased by 21.6%, and the hardened layer depth was up to 100 μm.

(2) Flowing traces of surface materials were observed in the rolling region within the sample cross section. It was indicated that the surface grains were plastically deformed under the pressure of the rollers, which is also considered to be the main reason for surface hardening and the generation of residual stresses.

(3) Deep and high amplitude residual compressive stresses were obtained in the rolling processes. According to the influence degree, the residual stress layer can be divided into three regions: the significant impact region, the smooth transition region, and the region in which the effect disappeared. Based on the distribution of the FWHM and the surface microstructures, generation mechanisms of residual stresses are revealed.

Acknowledgments: The authors want to thank the National Natural Science Foundation of China (No. 51275277) for the support.

Author Contributions: L. Lu contributed to the experimental testing and wrote the manuscript under the guidance of J. Sun. X. Han contributed to the design and manufacture of the BSRP appliance. Q. Xiong conducted the residual stresses tests and revised the manuscript.

Conflicts of Interest: The authors declare no conflicts of interest.

References

1. Wang, Z.Q. Study on Theory and Approach for Correcting Aerospace Monolithic Component Due to Machining Distortion Using Rolling Method. Ph.D. Thesis, Shandong University, Jinan, China, 2009.
2. Wang, Z.Q.; Li, J.F.; Sun, J.; Li, G.Y. FEM analysis of deformation correction by side-wall rolling of aircraft monolithic components. *China Mech. Eng.* **2009**, *10*, 612–616.
3. Uddin, M.S.; Hall, C.; Murphy, P. Surface treatments for controlling corrosion rate of biodegradable Mg and Mg-based alloy implants. *Sci. Technol. Adv. Mat.* **2015**. [CrossRef]
4. Zhao, J.; Xia, W.; Li, F.L.; Zhou, Z.Y. Research status and developing tendency of burnishing mechanism. *Tool Eng.* **2010**, *44*, 3–8.
5. Skalski, K.; Morawski, A.; Przybylski, W. Analysis of contact elastic-plastic strains during the process of burnishing. *Int. J. Mech. Sci.* **1995**, *37*, 461–472. [CrossRef]
6. Altenberger, I. Deep Rolling-the past, the present and the future. In Proceedings of the 9th International Conference for Shot Peening, Paris, France, 6–9 September 2005.
7. Beghini, M.; Bertini, L.; Monelli, B.D.; Santus, C.; Bandini, M. Experimental parameter sensitivity analysis of residual stresses induced by deep rolling on 7075-T6 aluminium alloy. *Surf. Coat. Technol.* **2014**, *254*, 175–186. [CrossRef]
8. Avilés, R.; Albizuri, J.; Rodríguez, A.; de Lacalle, L.N.L. Influence of low-plasticity ball burnishing on the high-cycle fatigue strength of medium carbon AISI 1045 steel. *Int. J. Fatigue* **2013**, *55*, 230–244. [CrossRef]
9. Wang, T.; Wang, D.; Liu, G.; Gong, B.; Song, N. Investigations on the nanocrystallization of 40Cr using ultrasonic surface rolling processing. *Appl. Surf. Sci.* **2008**, *255*, 1824–1829. [CrossRef]
10. Tian, Y.; Shin, Y.C. Laser-assisted burnishing of metals. *Int. J. Mach. Tool Manu.* **2007**, *47*, 14–22. [CrossRef]
11. Huang, B.; Kaynak, Y.; Sun, Y.; Jawahir, I.S. Surface layer modification by cryogenic burnishing of Al 7050-T7451 alloy and validation with FEM-based burnishing model. *Procedia CIRP* **2015**, *31*, 1–6. [CrossRef]
12. Sagbas, A. Analysis and optimization of surface roughness in the ball burnishing process using response surface methodology and desirabilty function. *Adv. Eng. Softw.* **2011**, *42*, 992–998. [CrossRef]
13. Yu, X.; Sun, J.; Li, S.T.; Li, J.F. Influences of burnishing process on surface quality integrity of EA4T axles and establishment of prediction model. *China Surf. Eng.* **2014**, *27*, 87–95. [CrossRef]
14. Juijerm, P.; Altenberger, I. Fatigue behavior of deep rolled Al-Mg-Si-Cu alloy at elevated temperature. *Scr. Mater.* **2006**, *55*, 943–946. [CrossRef]
15. Zhu, Y.L.; Li, L.; Wang, K.; Huang, Y.L. An integrated ultrasonic deep rolling and burnishing technology for anti-fatigue manufacturing. *J. Mech. Eng.* **2009**, *45*, 183–186. [CrossRef]
16. Kim, W.; Kawai, K.; Koyama, H.; Miyazaki, D. Fatigue strength and residual stress of groove-rolled products. *J. Mater. Process. Technol.* **2007**, *194*, 46–51. [CrossRef]
17. Rodríguez, A.; de Lacalle, L.N.L.; Celaya, A.; Lamikiz, A.; Albizuri, J. Surface improvement of shafts by the deep ball-burnishing technique. *Surf. Coat. Technol.* **2012**, *206*, 2817–2824. [CrossRef]
18. Nam, Y.; Jeong, Y.; Shin, B.; Byun, J. Enhancing surface layer properties of an aircraft aluminum alloy by shot peening using response surface methodology. *Mater. Des.* **2015**, *83*, 566–576. [CrossRef]

19. Nie, X.; He, W.; Zang, S.; Wang, X.; Zhao, J. Effect study and application to improve high cycle fatigue resistance of TC11 titanium alloy by laser shock peening with multiple impacts. *Surf. Coat. Technol.* **2014**, *253*, 68–75. [CrossRef]

20. Hennig, W.; Feldmann, G.; Haubold, T. Shot peening method for aerofoil treatment of blisk assemblies. *Proced. CIRP* **2014**, *13*, 355–358. [CrossRef]

21. Cheng, M.; Zhang, D.; Chen, H.; Qin, W. Development of ultrasonic thread root rolling technology for prolonging the fatigue performance of high strength thread. *J. Mater. Process. Technol.* **2014**, *214*, 2395–2401. [CrossRef]

22. Wu, B.; Wang, P.; Pyoun, Y.; Zhang, J.; Murakami, R. Study on the fatigue properties of plasma nitriding S45C with a pre-ultrasonic nanocrystal surface modification process. *Surf. Coat. Technol.* **2013**, *216*, 191–198. [CrossRef]

23. Zhao, J.; Xia, W.; Li, N.; Li, F.L. A gradient nano/micro-structured surface layer on copper induced by severe plasticity roller burnishing. *Trans. Nonferr. Met. Soc. China* **2014**, *24*, 441–448. [CrossRef]

24. Li, Y.; Li, J.Y.; Liu, M.; Ren, Y.Y.; Chen, F.; Yao, G.C.; Mei, Q.S. Evolution of microstructure and property of NiTi alloy induced by cold rolling. *J. Alloy. Compd.* **2015**, *653*, 156–161. [CrossRef]

metals

MDPI

Article

The Establishment of Surface Roughness as Failure Criterion of Al–Li Alloy Stretch-Forming Process

Jing-Wen Feng [1,2,3], Li-Hua Zhan [1,2,3,*] and Ying-Ge Yang [1,2,3]

[1]   School of Mechanical and Electrical Engineering, Central South University, Changsha 410083, Hunan, China; fengjingwen1@csu.edu.cn (J.-W.F.); 133711022@csu.edu.cn (Y.-G.Y.)
[2]   State Key Laboratory of High Performance Complex Manufacturing, Central South University, Changsha 410083, Hunan, China
[3]   2011 Collaborative Innovation Center, Central South University, Changsha 410083, Hunan, China
*   Correspondence: yjs-cast@csu.edu.cn; Tel.: +86-731-8883-0254

Academic Editor: Nong Gao
Received: 17 December 2015; Accepted: 28 December 2015; Published: 7 January 2016

Abstract: Taking Al–Li–S4–T8 Al–Li alloy as the study object, based on the stretching and deforming characteristics of sheet metals, this paper proposes a new approach of critical orange peel state characterizations on the basis of the precise measurement of stretch-forming surface roughness and establishes the critical criterion for the occurrence of orange peel surface defects in the stretch-forming process of Al–Li alloy sheet metals. Stretching experiments of different strain paths are conducted on the specimens with different notches so as to establish the Al–Li–S4–T8 Al–Li alloy, forming limit diagram and forming limit curve equation, with the surface roughness of characteristic critical orange peel structure as the stretch-forming failure criterion.

Keywords: Al–Li–S4 Al–Li alloy; stretch-forming; orange peel; forming limit; surface roughness

1. Introduction

Aircraft skin serves as the shape part of an aircraft and constructs the aerodynamic configuration of the aircraft, featuring big size and direct contact with air. Therefore, it requires an accurate shape, smooth streamline, and no surface defects, *etc.* [1]. As a relatively common aircraft skin-forming approach in the field of aeronautics and astronautics manufacturing, the technology of stretch-forming is widely used in the manufacturing of large-scale aircraft skin as part of aircraft aerodynamic configuration [2–4]. In the stretch-forming process, the clamps of the stretch-forming machine clamp both ends of the sheet metal and move along a certain track, or the die goes up to make the sheet metal contact the stretch-forming die, creating uneven plane stretching strain to make the metal sheet conform to the stretch-forming die so as to obtain the required part shape [5–7]. Compared with other approaches to aircraft skin forming, the stretch-forming technology might cause surface defects such as orange peel. Orange peel not only affects the appearance of the aircraft skin, but also damages its surface integrity. Especially in the case of mirror skin, orange peel easily appears due to mirror skin's internal structure and polished surface, seriously affecting its service life, which is usually the main reason for the scrapping of parts [8].

The defect of orange peel is a kind of rough orange peel-like morphology found on the surface of shaped products. In general, coarse and unevenly structured grains on the alloy surface are considered the reason of the orange peel defect appearing on the alloy surface during the stretch-forming process, while the grain size of the alloy has a certain relation to the extent of the deformation. At a certain temperature, when there are relatively small deformations, recrystallization usually does not appear in the alloy and the grains maintain their original state; however, when deformation reaches a certain degree, recrystallized grains will become very coarse. In the manufacturing process of aluminum alloy skin sheet metal, there are usually multiple hot rolling and cold rolling processes as well as several heat treatments including annealing, solution and aging. Because of the imperfection in the control over cold deformation and the choice of heat treatment technology in the manufacturing process of aluminum alloy sheet metal, recrystallized grain structure tends to be coarse and show different sizes, resulting in the piling up of dislocation on large grains and the rapid increases of stress in the follow-up stretch-forming process. Thus, the areas of large grains reach and exceed the elastic limit in advance; the non-synchronous deformation between large grains and small ones gives rise to minor cracks on the surface of the material, manifesting as the orange peel structure at the macro-level [9].

The forming limit of sheet metal is a criterion used to describe whether the sheet metal fails to form or not. To identify the concept of forming limit, one should first determine the failure criterion of sheet metal's forming process. Al–Li alloy as a new type of aluminum alloy has shown a wide application prospect in the fields of aeronautics and astronautics due to its good qualities of low density, high specific strength, and high specific stiffness, and it has become a hot subject in the research field of aluminum alloy materials, being regarded as one of the important candidate materials of modern aeronautic and astronautic structures [10,11]. However, due to the high cost, poor cold plasticity at ambient temperature, evident anisotropy and easy cracking in cold working compared with other regular aluminum alloys, Al–Li alloy at present can only be processed into relatively simple parts and faces great difficulty in the processing and manufacturing of more complex structural parts. Therefore, some of its own attributes also limit Al–Li alloy's application in structural components [12–15]. Relevant researchers have conducted corresponding research on the formability of Al–Li alloy, which, however, mainly focus on the study of the Al–Li alloy's structure property and the aspect of hot forming. Literature [16] explored the 2397-T87 Al–Li alloy with a thickness of 130 mm for the microstructure, stretching property and fracture toughness of layers with different thicknesses and at different orientations. By the uniaxial tension test under different hot deformation conditions, the forming limit test with cracking as a failure criterion, and the stretch-forming tests of 5A90 Al–Li alloy sheet metal, literature [17] built the Al–Li alloy forming limit model and determined the technological parameters for Al–Li alloy to acquire good deformation performance.

Nevertheless, for aviation aircraft skin materials, cracking is not often taken as the criterion for judging whether the skin fails to form or not, especially for Al–Li alloy, the reason of which usually lies in that the forming failure is caused by the appearance of orange peel structure in the stretch-forming process. Currently, there are relatively few studies on the forming limit caused by the defect of orange peel in the stretch-forming process, and a forming failure criterion targeting the orange peel has not yet taken shape. As the occurrence of the orange peel phenomenon is a slowly changing and accumulated process, and all of the previous research on such a phenomenon was conducted by eye measurement, the results have certain randomness.

In this paper, experimental research on the problem of orange peel in the stretch-forming process of Al–Li alloy is conducted, and a novel approach of critical orange peel characterization is proposed on the basis of the precise measurement of stretch-forming surface roughness. Furthermore, the judgment criterion of the orange peel defect is analyzed and established, the stretching tests on specimens of different strain paths are conducted combining the technology of optical deformation measurement, and, ultimately, the stretch-forming limit diagram and its forming limit curve equation are obtained with the surface roughness of orange peel structure with critical characteristics such as the stretch-forming failure criterion.

2. Experiments

2.1. Instrument and Methods

In order to conduct synchronization tests on stress-strain and orange peel surface defect in the stretch-forming process of Al–Li alloy, a stretch-forming test and testing system are established to obtain the critical strain state of orange peel of the product, in which the optical deformation measurement instrument and the universal testing machine operate in collaboration, as shown in Figure 1.

The optical deformation measurement instrument used in the experiment adopts the deformation measurement system based on computer vision technology developed by the German company GOM for three-dimensional deformation analysis. By controlling the synchronized operation of the optical deformation measurement instrument and the universal testing machine, the complete monitoring of the total stretch-forming deformation process of the specimens can be achieved and the computer image processing system can be further utilized to obtain the true strain change rules at different positions on the specimen surface throughout the stretch-forming process.

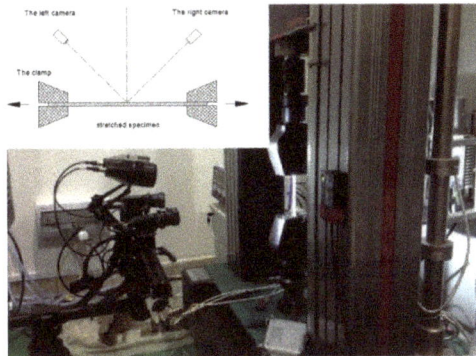

Figure 1. The stretching and testing system.

Compared with the stretch-forming mechanism, this experimental mechanism is to some extent simplified, in particular with the top die removed. However, in the actual stretch-forming process of the sheet metal, with the stretch-forming die going up, the sheet metal gradually bends to attach to the die as shown in Figure 2. Due to the existence of friction, the material flow of the attaching segment AA' tends to be limited to a relatively small deformation, whereas the free segments without contact with the die AB and A'B' tend to have relatively greater deformation without the restriction of friction. Therefore, the orange peel structure usually appears first on the free segments between the chucks and the die corners, and then slowly spreads toward the forming surface that attaches to the die. Thus, it can be seen that the deformation of the free segments on the sheet metal is the principal factor affecting the appearance of orange peel structure. As the deformation of the free segments on the sheet metal is similar to its stretch-forming, the conventional stretching test of sheet metal can be used in the research on the influence of the orange peel phenomenon on the stretch-forming of the materials.

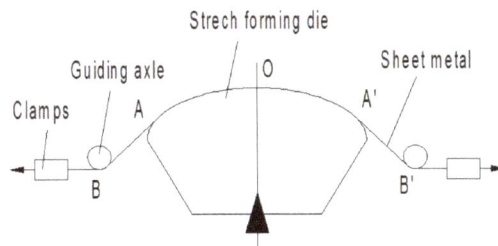

Figure 2. The schematic diagram of stretch-forming die.

2.2. Experimental Design

2.2.1. Experiment Design of Critical State Criterion of Orange Peel Defect

The experimental material was the Al–Li–S4–T8 Al–Li alloy which was used for aircraft skin with a thickness of 2 mm. See Figure 3 for the structure and size of the stretched specimens. The experimental specimens were polished to make the surfaces bright and without obvious scratches.

Meanwhile, in order to establish the same initial surface state for all specimens, the roughness of polished specimens was measured on the optical surface profilometer to ensure similar polishing effects for all the specimens.

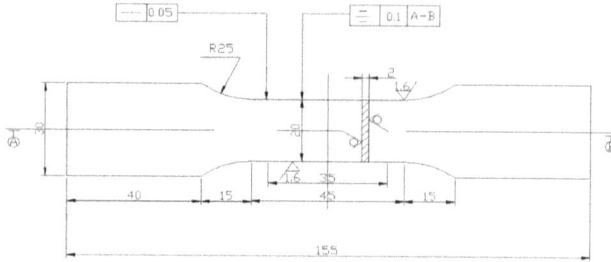

Figure 3. The shape and size of stretched specimen.

Stretching tests of different strains were conducted on polished specimens to investigate the evolution situation of the orange peel phenomenon on material surfaces under the condition of different deformations. See stretching strains in Table 1. Additionally, tests were also conducted at four different strain speeds on T8 alloy to investigate the impacts of different deforming speeds on the orange peel phenomenon on specimen surfaces in the experiment. By contrasting the surface morphology and true strain of the specimen at different speeds, when tiny orange peel phenomena appear on the surface at different strain speeds, the true strain of the Al–Li–S4–T8 Al–Li alloy metal sheet can be measured with the results shown in Table 2. It is observed from the table that when the strain speed is 0.0005/s, the strain at the sampling point is the greatest when tiny orange peel structure on the specimen surface appears. Thus, in actual stretch-forming treatment of aircraft skin, the said speed can be employed to alleviate the appearance of the orange peel phenomenon.

Table 1. Stretching strains of samples.

Experiment Batch No.	1	2	3	4	5
Stretching Strain	0%	3%	6%	9%	12%

Table 2. Critical strain and roughness of samples.

Specimen No.	Strain Speed/s^{-1}	Critical Strain/%	Roughness R_a/nm
1	0.0001	1.78	841
2	0.0005	1.87	827
3	0.001	1.14	973
4	0.0015	1.19	1002

After the stretching tests, both sides of the specimens were cleaned again, and the side used for surface morphology observation was placed under the optical profilometer for surface analysis, obtaining the morphology, nephogram and roughness of the specimen surfaces. Combined with the criterion of the orange peel defect obtained by the experiment, the areas in which critical orange peel structure occurred were found on the specimens, and the strain state of the areas can be also found at the same positions on the other side of the specimens, which actually is the critical strain state of the appearing orange peel defect.

2.2.2. Experiment Design of Stretch Forming Limit Diagram

During the stretch-forming process of aircraft skin, the strain state of sheet metals mainly fell between uniaxial stretching and plane strain with approximately linear strain paths [18]. Thus, stretch-forming specimens that could implement different strain paths were designed with the typical specimen structure and size given in Figure 4, among which the straight-edge stretched specimen was close to the uniaxial stretching state of strain, while the $R = 10$ notched specimen was close to the plane state of strain [19].

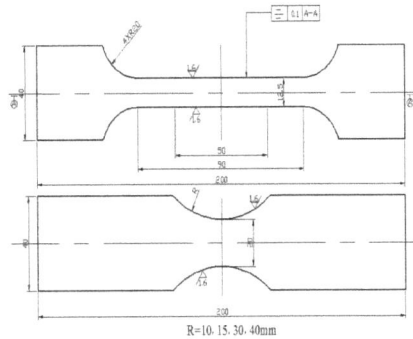

R=10, 15, 30, 40mm

Figure 4. Size of samples with different strain paths.

In order to conduct observation and analysis of the surface morphology and strain state of the specimens, both sides of the specimens were polished until without an obvious scratch, and a profilometer was used to measure the surface roughness to achieve similar polishing effects for each specimen. The specimens with ethanol were cleaned after polishing. As shown in Figure 5, the black and white speckle patterns were sprayed on half of one side of the specimens, so as to analyze the true strain on specimen surfaces during the stretching process; another half was used to observe the evolvement of the orange peel phenomenon on specimen surfaces during the stretching process. Meanwhile, another side, only polished, was used to measure surface morphology and roughness with the profilometer.

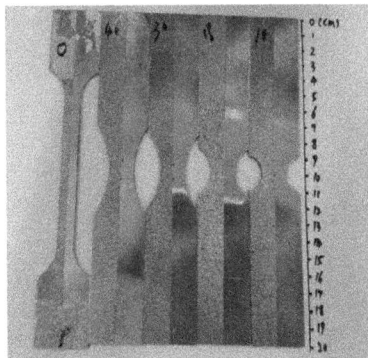

Figure 5. Sample appearances along different strain paths after surface treatment.

3. Results and Analysis

3.1. The Equilibrium Diagram of Tensile Specimen before Deformation

The grain size grade of the specimen can be measured in accordance with the Metal Methods for Estimating the Average Grain Size, as shown in Table 3 below. From the measured grain size grades, as shown in Figures 6 and 7 it is observed that the average grain size of the specimen is 60–80 μm, suggesting that the specimen has already fallen into the category of open grain structure. Meanwhile, judging from the equilibrium diagram, the grain structure of the specimen is extremely uneven with a large difference in size between the large grains and small grains. It is also discovered from the measurement that the size of the large grains is over 100 μm, but that of the small ones is merely around 10 μm. Thus, the equilibrium diagram can explain the appearance of orange peel in the process.

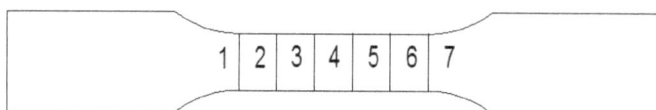

Figure 6. The exact location on the sample.

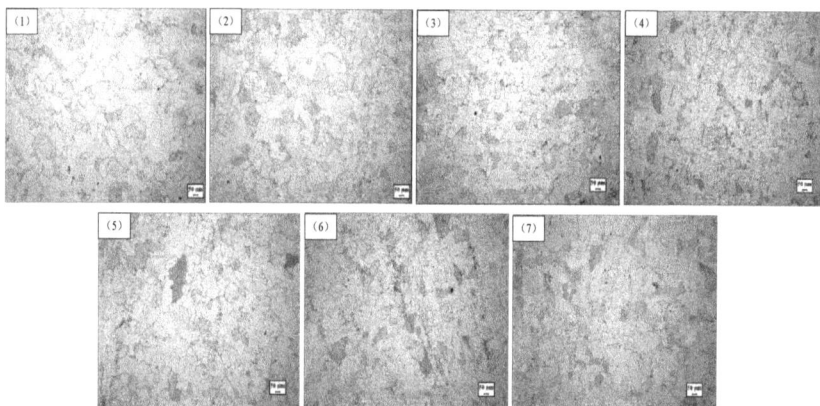

Figure 7. The equilibrium diagram of tensile specimen before deformation. (**1**) is correspondence with 1 in Figure 6; (**2**) is correspondence with 2 in Figure 6; (**3**) is correspondence with 3 in Figure 6; (**4**) is correspondence with 4 in Figure 6; (**5**) is correspondence with 5 in Figure 6; (**6**) is correspondence with 6 in Figure 6; (**7**) is correspondence with 7 in Figure 6.

Table 3. Grain size analysis of tensile specimen.

Heat Treatment Condition	Type of Analysis	Sampling Point							Average
		1	2	3	4	5	6	7	
T8	Grain Size Grade	5.02	5.13	4.51	4.38	4.98	4.62	4.56	4.7
	Average Diameter/μm	62	59	76	77	70	75	75	71

3.2. The Establishment of the Criterion for Critical State of Orange Peel Defect

The specimens were cleaned after stretch processing with ethanol, and surface observation and analysis on the optical profilometer were conducted. The photographed morphology can be seen in Figure 8, which shows that when the strain is 3%, the specimen surface starts to grow rough and form a kind of evenly frosted surface morphology, but without evident minor cracks or significant orange peel phenomenon; when the strain reaches 6%, light black strip areas appear on the surface, which are shallow grooves, and an embryonic form of orange peel structure can be roughly observed. When the strain reaches 9%, there are clear cracks appearing on the specimen surface, which is rather severe despite the fact that they are relatively decentralized and independent; at the macro-level, a very significant orange peel phenomenon is manifested. At this stage, the orange peel structure on the material surface not only affects the appearance of the stretch-forming parts, but also exerts great influence on the performance and service life of the specimen, whereas the minor cracks on the surface can easily give rise to stress concentration and gradually evolve into greater cracks after stress. When the strain amounts to 12%, the surface morphology of the specimen further deteriorates with more and deeper cracks, having already formed gully-like shapes.

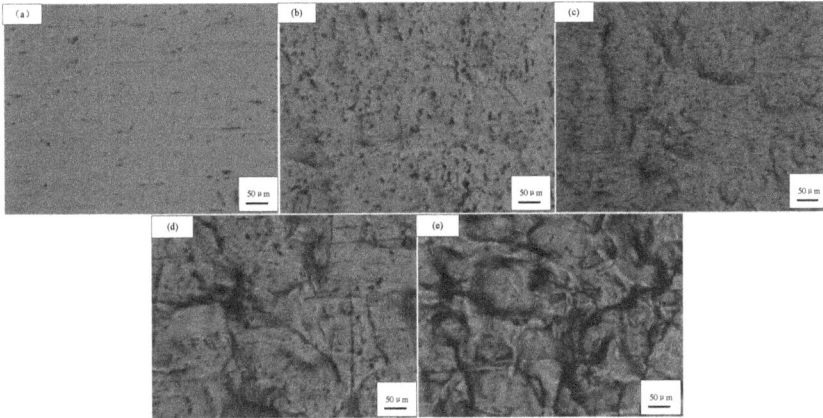

Figure 8. Surface morphology of stretched specimens with different strain variables: (a) 0%; (b) 3%; (c) 6%; (d) 9%; (e) 12%.

It can be seen from the analysis results that when the strain reaches around 6%, the specimen surface starts to form the orange peel defect in a real sense, but as the gap between the strain variables is relatively large, it is impossible to determine that the specimen is in the critical orange peel state when the strain is 6%. Thus, on the basis of the previous research, strain variables of 5% and 7% were added to a new test to observe their surface morphology after stretching, with the results given in Figure 9. When the strain was 5%, the specimen surface had not yet formed clear grooves, but through comparing the stretched specimens of 6% and 7%, it could be found that, though with roughly the same surface morphology, the scanning image of the stretched specimen with 7% strain clearly showed darker grooves with relatively deep cracks starting to develop.

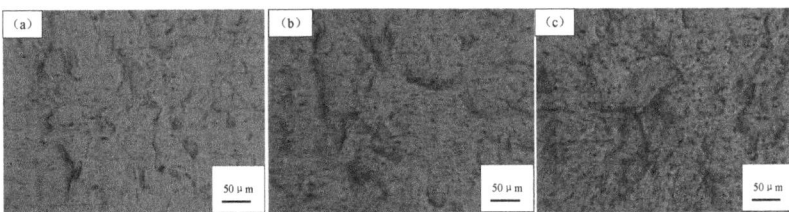

Figure 9. Surface morphology of stretched specimens with different strain variables: (a) 5%; (b) 6%; (c) 7%.

On this basis, the roughness of specimen surfaces with different strain variables is measured, the results of which are as shown in Figure 10, where R_a is the arithmetic mean of the absolute value of the distance between the dot on the profile and the baseline; R_q is the root mean square error of the profile, *i.e.*, the root mean square value of the profile's offset distance within the sampling range; R_z is the micro-irregularity on the material surface, *i.e.*, the sum of the average of the five greatest profile peaks and the average of the five smallest profile valleys within the sampling length. Combined with the surface morphology measured by the profilometer, it can been seen that, at the stage of 0%~5% strain, the main change was that the specimen surface began to wrinkle, with dispersed convexes and concaves appearing as well as a rapid change of surface roughness; at the stage of 5%~6% strain, there was little change in the surface roughness, but dispersed convexes and concaves started to gather to form lumps while grooves appeared and started to develop into cracks, which is also the major stage of qualitative change appearing on the material surface; at the stage of 6%~12% strain, cracks rapidly developed, gradually grew deeper, and joined with each other to form the gully-like morphology, resulting in the rapid increase in R_z of the specimens.

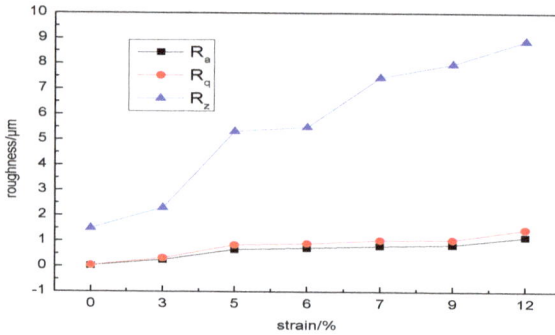

Figure 10. Changing trend of stretched sample surface roughness.

Based on the above experimental analyses, it can be assumed that when a morphology similar to that of the stretched specimen with 6% strain appears on the specimen surface, the orange peel phenomenon comes to a critical state. Thus, a roughness of $R_a = 700 \pm 50$ nm and $R_z = 5.5 \pm 0.5$ μm can be taken as a reference standard for judging whether the orange peel structure on a material surface reaches a critical state.

3.3. The Establishment of Stretch-Forming Limit Diagram

On the basis of obtaining the corresponding strain state and surface roughness of the orange peel structure on the standard specimens, the research on the stretch-forming limit under different strain paths is further conducted, through which strain data of different strain paths is measured with the surface roughness of the characteristic critical orange peel structure as the forming failure criterion, as shown in Table 4.

Table 4. Strain state of critical orange peel structure of specimens under different strain paths.

Sample	Acquisition Point 1		Acquisition Point 2		Acquisition Point 3		Acquisition Point 4		Acquisition Point 5	
	$\varepsilon_1/\%$	$\varepsilon_2/\%$	$\varepsilon_1/\%$	$\varepsilon_2/\%$	$\varepsilon_1/\%$	$\varepsilon_2/\%$	$\varepsilon_1/\%$	$\varepsilon_2/\%$	$\varepsilon_1/\%$	$\varepsilon_2/\%$
Straight edge	6.29	−1.94	6.35	−1.98	6.31	−2.08	6.25	−1.84	6.33	−1.88
$R = 10$	3.17	−0.21	3.17	−0.18	2.65	−0.15	2.76	−0.13	3.17	−0.15
$R = 15$	2.95	−0.19	3.04	−0.13	3.27	−0.12	3.31	−0.23	3.09	−0.08
$R = 30$	3.31	−0.55	3.76	−0.58	3.64	−0.77	3.72	−0.57	3.30	−0.55
$R = 40$	4.49	−1.20	4.82	−1.59	4.66	−1.27	3.92	−1.04	4.20	−1.23

Quadratic-multinomial fitting is conducted on measured data so as to establish the forming limit diagram of the orange peel phenomenon on the aluminum alloy surface with the results shown in Figure 11 and the quadratic-multinomial fitting results given in Table 5. It can be seen from the figure that the orange peel structure on the aluminum alloy surface is correlated with the strain state of the materials; meanwhile, the materials at the plane state of strain more easily develop orange peel structure than those at the uniaxial stretching strain state.

Table 5. Binomial fitting results.

Expression	A	B	C
$y = A + Bx + Cx^2$	2.99	−0.389	0.652

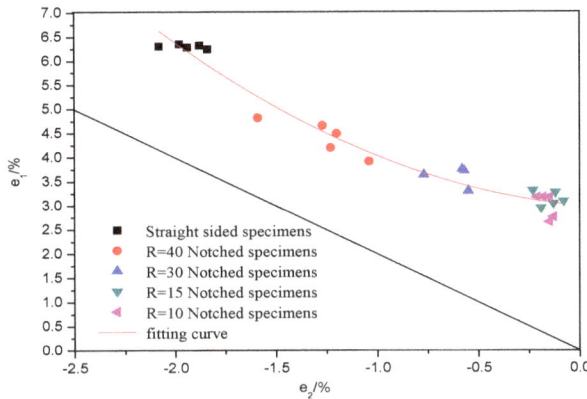

Figure 11. Stretch-forming limit diagram of Al–Li alloy at the condition of Al–Li–S4–T8.

4. Conclusions

(1) A stretch-forming experiment and testing system with the optical deformation measurement instrument and the universal testing machine operating in collaboration are constructed, the surface morphology change rule of stretched specimens with different strain variables is analyzed, and the corresponding relation between critical orange peel defect and the surface roughness of specimens is obtained. It is discovered that when critical orange peel defect appears on Al–Li alloy sheet metal at the condition of Al–Li–S4–T8, the surface roughness is $R_a = 700 \pm 50$ nm and $R_z = 5.5 \pm 0.5$ μm.

(2) By processing different notched specimens, the stretch-forming limit tests with different strain paths are conducted to obtain the forming limit diagram and forming limit curve equation $\varepsilon_1 = 2.99 - 0.389\varepsilon_2 + 0.652\varepsilon_2^2$ for Al–Li–S4–T8 Al–Li alloy, with the surface roughness of characteristic critical orange peel structure as the forming failure criterion.

Acknowledgments: This research was financially supported by the National Basic Research Program of China (2014CB046602) and the National Natural Science Foundation of China (51235010).

Author Contributions: Jing-Wen Feng and Li-Hua Zhan conceived and designed the experiments; Jing-Wen Feng and Ying-Ge Yang performed the experiments; Jing-Wen Feng analyzed the data; Jing-Wen Feng wrote the paper.

Conflicts of Interest: The authors declare no conflict of interest.

References

1. Wang, K.; Wan, M.; Hua, C.; Shao, X.F. Determination and application of coarse-grain critical pre-strain curve to aluminum alloy stretch forming. *J. Beijing Univ. Aeronaut. Astronaut.* **2013**, *39*, 508–511.
2. Han, Z.R.; Dai, L.J.; Zhang, L.Y. Current status of large aircraft skin and panel manufacturing technologies. *Aeronaut. Manuf. Technol.* **2009**, *25*, 64–66.
3. Araghi, B.T.; Manco, G.L.; Bambach, M.; Hirt, G. Investigation into a new hybrid forming process: Incremental sheet forming combined with stretch forming. *CIRP Ann. Manuf. Technol.* **2009**, *58*, 225–228. [CrossRef]
4. Kurukuri, S.; Miroux, A.; Wisselink, H.; Boogaard, T.V.D. Simulation of stretch forming with intermediate heat treatments of aircraft skins: A physically based modeling approach. *Int. J. Mater. Form.* **2011**, *4*, 129–140. [CrossRef]
5. General Editorial Board of "The Manual of Aeronautical Manufacturing Engineering". In *The Aviation Manufacturing Engineering Handbook—Aircraft Sheet Metal Process*; Aviation Industry Press: Beijing, China, 1992.
6. Hu, S.G.; Chen, H.Z. *Manufacturing Technology of Aircraft Sheet Metal Parts*; Beijing University of Aeronautics and Astronautics Press: Beijing, China, 2004.
7. Chang, R.F. *Sheet Metal Parts Manufacturing Technology*; National Defence Industry Press: Beijing, China, 1992.
8. Wan, M.; Zhou, X.B.; Li, X.X.; Wu, H. Process parameters in stretch forming of mirror skins. *Acta Aeronaut. Astronaut. Sin.* **1999**, *20*, 326–330.
9. You, Z.H.; Huang, Y.S.; Wu, R.H. A study of orange-like roughness on front side on Airplane. *Aviat. Maint. Eng.* **2001**, *6*, 46–47.
10. Zhang, R.X.; Zeng, Y.S. Development, technological characteristics and application status abroad of aluminum-lithium alloys (In Chinese). *Aeronaut. Manuf. Technol.* **2007**. [CrossRef]
11. Yin, D.F.; Zheng, Z.Q. History and current status of aluminum-lithium alloys research and development. *Mater. Rev.* **2003**, *17*, 18–20.
12. Huo, H.Q.; Hao, W.X.; Geng, G.H.; Da, D.A. Development of the new aero-craft material—Aluminum–lithium alloy. *Vac. Low Temp.* **2005**, *11*, 63–69.
13. Lyttle, M.T.; Wert, J.A. The plastic anisotropy of an Al–Li–Cu–Zr alloy extrusion in unidirectional deformation. *Metall. Mater. Trans. A* **1996**, *27*, 3503–3512. [CrossRef]
14. Li, H.; Tang, Y.; Zeng, Z.; Zheng, Z.; Zheng, F. Effect of ageing time on strength and microstructures of an Al–Cu–Li–Zn–Mg–Mn–Zr alloy. *Mater. Sci. Eng. A* **2008**, *498*, 314–320. [CrossRef]
15. Huang, J.C.; Ardell, A.J. Addition rules and the contribution of δ' precipitates to strengthening of aged Al–Li–Cu alloys. *Acta Metall.* **1988**, *36*, 2995–3006. [CrossRef]
16. Fan, C.P.; Zheng, Z.Q.; Jia, M.; Zhong, J.F.; Cheng, B.; Li, H.P.; Wu, Q.P. Microstructure, tensile property and fracture toughness of 2397 Al–Li alloy. *Rare Met. Eng.* **2015**, *44*, 91–96.
17. Ma, G.S. *Hot Forming Technology of Complex Aluminum Lithium Alloy Parts*; Chemical Industry Press: Beijing, China, 2011.
18. Jin, H.X. *Basic Experimental Research and Simulation on Aluminum Alloy Mirror Skin Tensile Forming*; Beijing University of Aeronautics and Astronautics: Beijing, China, 2009.
19. Wan, M.; Han, J.Q.; Jin, H.X.; Wu, H. Determination of strain criterion of portevin–Le chatelier effect for aluminum alloy sheets. *Trans. Nonferrous Met. Soc. China* **2006**, *16*, 1499–1503.

metals

MDPI

Article

Fabrication of Corrosion Resistance Micro-Nanostructured Superhydrophobic Anodized Aluminum in a One-Step Electrodeposition Process

Ying Huang, Dilip K. Sarkar * and X.-Grant Chen

Centre Universitaire de Recherche sur l'Aluminium (CURAL), Université du Québec à Chicoutimi,
555 Boulevard de l'Université, Chicoutimi, QC G7H 2B1, Canada; ying.huang@uqac.ca (Y.H.);
xgrant_chen@uqac.ca (X.-G.C.)
* Correspondence: dsarkar@uqac.ca; Tel.:+1-418-5455-011 (ext. 2543)

Academic Editor: Nong Gao
Received: 23 December 2015; Accepted: 22 February 2016; Published: 29 February 2016

Abstract: The formation of low surface energy hybrid organic-inorganic micro-nanostructured zinc stearate electrodeposit transformed the anodic aluminum oxide (AAO) surface to superhydrophobic, having a water contact angle of $160°$. The corrosion current densities of the anodized and aluminum alloy surfaces are found to be 200 and 400 nA/cm^2, respectively. In comparison, superhydrophobic anodic aluminum oxide (SHAAO) shows a much lower value of 88 nA/cm^2. Similarly, the charge transfer resistance, R_{ct}, measured by electrochemical impedance spectroscopy shows that the SHAAO substrate was found to be 200-times larger than the as-received aluminum alloy substrate. These results proved that the superhydrophobic surfaces created on the anodized surface significantly improved the corrosion resistance property of the aluminum alloy.

Keywords: superhydrophobic aluminum; corrosion; anodized aluminum oxide (AAO); organic-inorganic; micro-nanostructure; zinc stearate (ZnSA); potentiodynamic polarization; electrochemical impedance spectroscopy (EIS)

1. Introduction

Aluminum (Al) and its alloys are naturally-abundant engineering materials with extensive applications in daily life. In recent years, nanoporous anodic aluminum oxide (AAO) prepared by electrochemical anodization has found a multitude of applications, such as catalysis [1], drug delivery [2], biosensing [3], template synthesis [4], molecular and ion separation [5], corrosion resistance [6], and so forth. The formation of AAO on the aluminum alloy surface would act as the corrosion barrier. The formation of AAO was limited to a certain extent due to its hydrophilic behavior. Therefore, it is necessary to transform the AAO to be superhydrophobic in order to improve the corrosion resistance properties.

Superhydrophobicity, exhibiting an excellent water-repellent property, is characterized by a contact angle above $150°$. Creating a rough surface, as well as reducing the surface energy is attributed to the modification of superhydrophobicity [7]. In the last few decades, a large effort has been devoted to the realization of superhydrophobic surfaces, due to their applications in biology [8], anti-corrosion [9–12], anti-icing [13], self-cleaning [14], *etc.*

Recent publications show that the anodized surface can be made superhydrophobic by passivation with organic molecules [15–18]. In the study of Liu *et al.*, the superhydrophobic anodized surfaces were fabricated by polypropylene (PP) coating after anodizing [17]. Li *et al.*, used a very complex process to engineer superhydrophobic anodized aluminum alloy surfaces [16]. After anodizing with sulfuric acid, the anodized sample was firstly immersed in the mixing solution containing $M(NO_3)_2$ salt (M = Mg, Co, Ni and Zn) and NH_4NO_3, followed by immersion in (Heptadecafluoro-1,1,2,2-tetradecyl)trimethoxysilane (n-$CF_3(CF_2)_7CH_2CH_2Si(OC_2H_5)_3$). Vengatesh and Kulandainathan fabricated superhydrophobic anodic aluminum oxide (SHAAO) surfaces by passivation with organic molecules and show these surfaces having corrosion resistance properties indicated by potentiodynamic polarization [15]. Despite this, the fabrication, as well as corrosion resistance properties of the electrodeposited superhydrophobic surfaces on AAO have not been shown in the literature.

In this work, we have described the method to prepare the superhydrophobic surfaces on AAO by the electrodeposition process and describe their corrosion resistance properties, both using potentiodynamic polarization and electrochemical impedance spectroscopy (EIS).

2. Experimental Procedure

After pretreatment with 0.01 M NaOH at 55 °C for 3 min, the AA6061 aluminum alloy substrates were anodized using 3 vol. % H_3PO_4 aqueous solution at 10 °C at 0.01 A/cm^2 (Ametek Sorensen DCS 100-12E, Chicoutimi, QC, Canada) for 2 h. The superhydrophobic anodic aluminum oxide (SHAAO) surface was prepared by the electrodeposition process in an ethanolic solution containing 0.01 M stearic acid (SA) and 0.01 M zinc nitrate ($Zn(NO_3)_2$) by applying 20 V (Ametek Sorensen DCS 100-12E) for 10 min. Microstructural examination was conducted using a scanning electron microscope (SEM, JEOL JSM-6480 LV, Chicoutimi, QC, Canada). Surface roughness was measured using an optical profilometer. The X-ray diffraction (XRD) analyses of the samples were carried out using a Bruker D8 Discover system (Chicoutimi, QC, Canada). The chemical composition of the samples was analyzed by means of Fourier transform infrared spectroscopy (FTIR, Perkins Elmer Spectrum One, Chicoutimi, QC, Canada) and an energy dispersive X-ray spectrometer (EDX, JEOL JSM-6480 LV, Chicoutimi, QC, Canada). Wetting characteristics of sample surfaces were evaluated by measuring static contact angles (CA) using a first ten Angstrom contact angle goniometer at five positions on each substrate using a 10-µL deionized water drop. In the case of rolling-off surfaces, the contact angle was measured by holding the water drop between the needle and the surface, as presented in the inset image of Figure 1c. Electrochemical experiments were performed using a PGZ100 potentiostat and a 300-cm^3-EG & G PAR flat cell (London Scientific, London, ON, Canada), equipped with a standard three-electrode system with an Ag/AgCl reference electrode, a platinum (Pt) mesh as the counter electrode (CE) and the sample as the working electrode (WE). Before the test, the open circuit potential (OCP) was monitored for more than 20 h for stabilization by immersing the sample surface in 3 wt. % NaCl aqueous solution. The electrochemical impedance spectroscopy (EIS) was tested in the frequency range between 10 MHz and 100 kHz with a sine-wave amplitude of 10 mV. For the potentiodynamic polarization experiments, the potential was scanned from −250 mV to +1000 mV with respect to the OCP voltage at a scan rate of 2 mV/s.

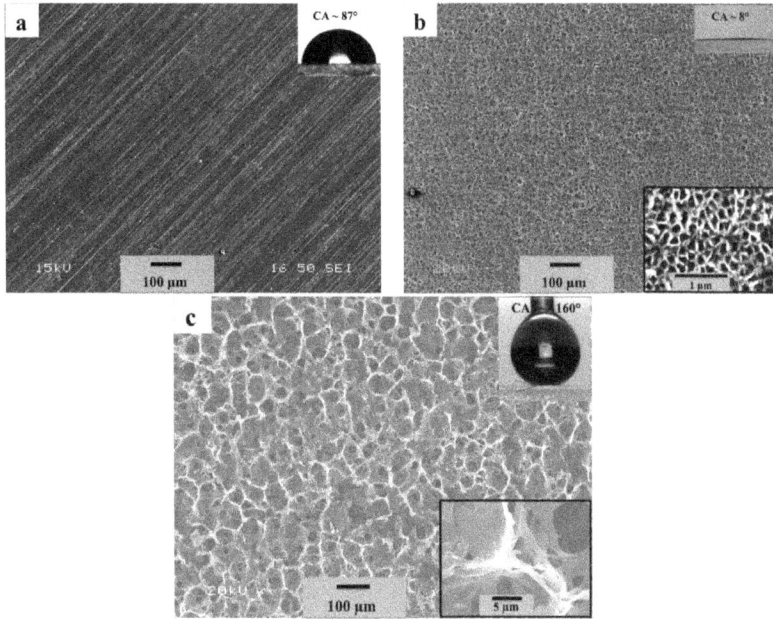

Figure 1. Secondary electron SEM micrographs showing the top surface of (**a**) the as-received aluminum alloy; (**b**) the anodized and (**c**) the electrodeposited anodized substrates. The inset top-right images present the water drop on the surfaces, and the bottom-right ones show magnified microstructures.

3. Results and Discussion

SEM images in Figure 1 reveal the evolution of the different morphologies of: (a) the as-received the aluminum alloy surface; (b) the anodized (AAO) surface; and (c) the electrodeposited hybrid organic-inorganic anodized surface using the ethanolic electrolytic solution containing stearic acid (SA) and zinc nitrate ($Zn(NO_3)_2$) at 20 V DC. The as-received aluminum alloy surface was characterized by parallel lines resulting from the rolling process, corresponding to a surface roughness of 0.45 ± 0.03 μm and a contact angle of $87° \pm 3°$ (Figure 1a).

After anodizing in phosphoric acid for a duration of 2 h, uniformly-distributed nanopores with an average diameter of approximately 100 nm and an inter-pore distance of ~137 nm are observed on the surface of the anodized substrate (Figure 1b), with a roughness of 0.68 ± 0.02 μm, while the contact angle was measured to be $8° \pm 1°$, indicating a superhydrophilic property. It is however evident from Figure 1c that the electrodeposition process resulted in the appearance of a porous network microstructure on the anodized substrate. This substrate was built by nanofibre clusters connected with each other, as presented in the inset of Figure 1c. It can be also observed that these micro-nanoporous structures are distributed uniformly on the anodized surface, resulting in a micro-nanorough surface having a roughness of 6.85 ± 1.02 μm. Interestingly, this surface shows superhydrophobic properties with a contact angle of $160° \pm 1°$ having a contact angle hysteresis of $2° \pm 1°$.

Low angle X-ray diffraction (XRD), energy dispersive X-ray spectroscopy (EDX) and Fourier transform infrared spectroscopy (FTIR) have been carried out to determine the composition of the electrodeposited micro-nanostructure thin films on the anodized substrate, as shown in Figure 2. Figure 2a(a3) shows four distinct peaks at $4.2°$, $6.26°$, $8.3°$ and $10.4°$, which correspond to zinc stearate ($(CH_3(CH_2)_{16}COO)_2Zn$) (abbreviated as ZnSA,). The possible mechanism of the formation of ZnSA has been presented as follows:

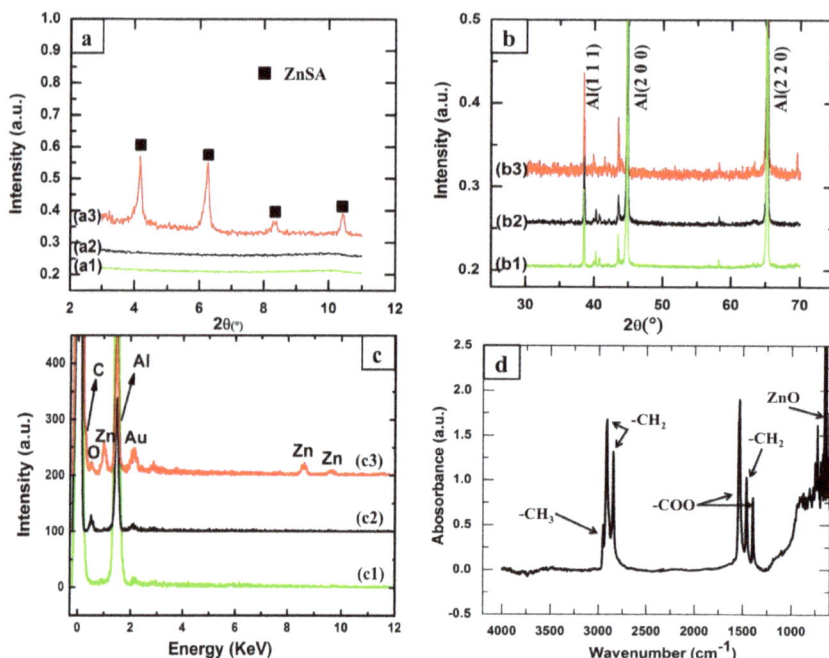

Figure 2. (a) Low angle XRD patterns of (**a1**) as-received aluminum alloy, (**a2**) AAO, (**a3**) electrodeposited anodized substrate; (**b**) high angle XRD patterns of (**b1**) as-received aluminum alloy, (**b2**) AAO and (**b3**) electrodeposited anodized substrate; (**c**) EDX spectra of (**c1**) the as-received aluminum alloy, (**c2**) anodized and (**c3**) electrodeposited anodized substrate (ZnSA = zinc stearate); (**d**) FTIR spectrum of the electrodeposited anodized substrate.

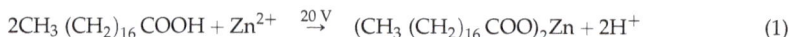

$$2CH_3 (CH_2)_{16} COOH + Zn^{2+} \xrightarrow{20 \, V} (CH_3 (CH_2)_{16} COO)_2 Zn + 2H^+ \qquad (1)$$

This reaction mechanism is very similar to that mentioned by Liu *et al.* for the electrodeposition of cerium myristate by electrodeposition on magnesium substrates to obtain superhydrophobicity [19]. When the DC voltage was applied on the electrodes, the Zn^{2+} ions close to the cathodic electrode reacted with SA, forming ZnSA and H^+ ions, as presented in Equation (1). Meanwhile, some of the H^+ ions obtained an electron and formed H_2 gas. It is noted that the as-received aluminum alloy and anodized substrates of Figure 2a(a1,a2) do not show any characteristic peaks. Furthermore, in Figure 2b, it shows the peaks at 38.47°, 44.72° and 65.1°, respectively, which are in good agreement with the characteristic peaks of Al(111), (200) and (220). This arises from the substrate of the aluminum alloy.

EDX analysis (Figure 2c) revealed that the chemical composition of the anodized surface consisted of O and Al, resulting from the formation of aluminum oxide (Al_2O_3), whereas only the Al peak was seen in the spectrum of the as-received aluminum alloy surface. However, C, Zn, O and Al are observed in the spectrum of electrodeposited anodized surface, confirming the formation of ZnSA complementary with the XRD pattern of the electrodeposited anodized substrate. It is worth mentioning that the Au peaks appearing in Figure 2c are due to the thin layer of gold on the electrodeposited thin films for improving the resolution by eliminating the charging effect of non-conducting samples during EDX analyses.

In the FTIR spectrum of the electrodeposited hybrid organic-inorganic anodized substrate (Figure 2d), the appearance of the –CH group (–CH$_3$ at 2962 cm^{-1}, as well as –CH$_2$ at 1459 cm^{-1}, 2850 cm^{-1} and 2919 cm^{-1}), as well as –COO at 1395 and 1550 cm^{-1} indicated the formation of ZnSA on the surface [20], which is in good agreement with the XRD and EDX results. These results support the formation of low surface energy methylated (–CH$_3$ and –CH$_2$) components on the electrodeposited anodized surface and make it superhydrophobic. In addition, the ZnO peaks at 560 cm^{-1} may come from the bonding of –COOZn, which further verifies the reaction production of ZnSA formed on the modified surface in a one-step electrodeposition process.

Figure 3 illustrates the variation of open circuit potential (OCP) on the superhydrophobic AAO (SHAAO) surface. The OCP value shifted from −177 to −708 mV with an average of −281 ± 87 mV into a 5-h immersion time in the salt solution. It is quite unstable during this period. From 5 to 16 h of immersion time, the OCP value varied from −533 to −686 mV with an average of −610 ± 26.8 mV. With the prolongation of immersion time to 20 h, the OCP value varied from −708 to −730 mV, with an average of −717 ± 2.8 mV. It can be seen that the fluctuation of OCP voltage reduces with time and stabilized nearly after 20 h. In contrast, the surface of the aluminum alloy substrate gets stabilized within 30 min of immersion time in the salt solution. The OCP fluctuations of the SHAAO substrate may be due to the poor wetting, as well as protective properties of the SHAAO surface with the salt solution. Therefore, the EIS and polarization experiments were performed after 20 h of immersion of the superhydrophobic surface in the salt solution while monitoring the OCP continuously.

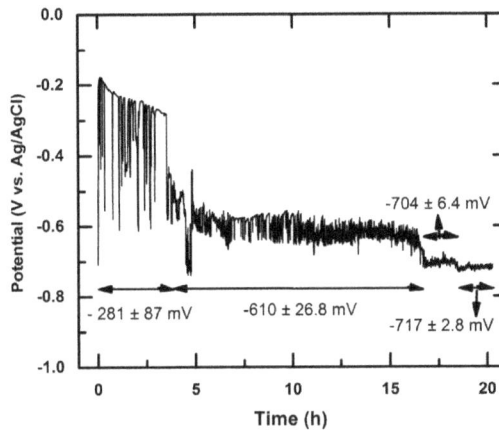

Figure 3. The variation of open circuit potential (OCP) with time on the superhydrophobic AAO (SHAAO) surface.

To evaluate the corrosion resistance performance of the fabricated anodized and SHAAO substrates, potentiodynamic polarization and electrochemical impedance spectroscopy (EIS) tests were carried out. Figure 4a shows the polarization curves of the as-received aluminum alloy, anodized and SHAAO substrates, respectively. The as-received aluminum alloy substrate exhibited a corrosion current density (I_{corr}) of 400 nA/cm^2 and a polarization resistance (R_p) of 50 kΩ·cm^2. The R_p value was calculated by the Stern-Geary equation, as shown in Equation (2).

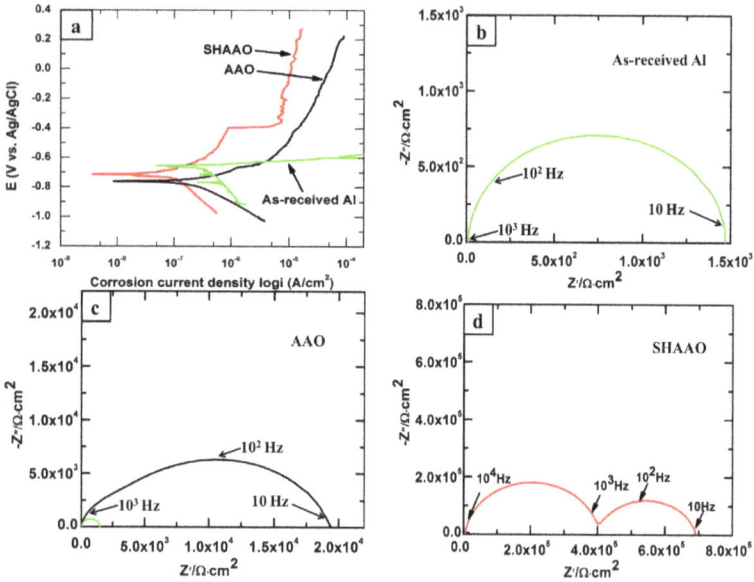

Figure 4. (**a**) Polarization curves of the as-received aluminum alloy, AAO and SHAAO substrates in 3 wt. % NaCl corrosive solution; (**b**–**d**) the Nyquist plots of (**b**) as-received aluminum alloy; (**c**) AAO (as-received aluminum alloy (a small semicircle in green close to the origin) also shown for comparison) and (**d**) SHAAO substrate (as-received aluminum alloy and AAO also shown inside for comparison, very small semicircle, nearly visible, close to the origin).

$$Rp = \frac{\beta_a \beta_c}{2.3 I_{corr}(\beta_a + \beta_c)} \tag{2}$$

When the aluminum alloy substrate was anodized, the I_{corr} was reduced to 200 nA/cm^2, and R_p increased to 87 kΩ· cm^2, indicating an improved corrosion resistance compared to the as-received substrate. This was due to a barrier layer formed on the anodized surface. On the other hand, the I_{corr} of the hybrid organic-inorganic SHAAO substrate decreased remarkably to 88 nA/cm^2, while the R_p increased up to 441 kΩ· cm^2. This shows an even better anti-corrosion performance of the SHAAO substrate relative to the anodized substrate, likely attributed to the superhydrophobic ZnSA coating formed on the surface. It is noticed that the current density of the SHAAO substrate was increased sharply at around −0.4 V *vs.* Ag/AgCl, which might have been related to the dissolution of the superhydrophobic film with the prolongation of corrosion time (more than 20 h). It is also found that the current density of the SHAAO substrate at −0.4 V to 0.2 V *vs.* Ag/AgCl was parallel to, but smaller than, that of the anodized substrate. This might be due to the superhydrophobic material ZnSA filling the anodized pore structure. As a result, the SHAAO substrate after dissolution of the surface material still presents a good corrosion resistance property.

The I_{corr} of the electrodeposited cerium stearate superhydrophobic Mg surfaces was reported to be 142 nA/cm^2 with a 30-min immersion to perform OCP [19]. Experiments performed by Vengatesh and Kulandainathan with a 30-min immersion before polarization show that the corrosion current varies between 2 and 1050 nA/cm^2, depending on the passivated molecules [15]. Moreover, in the study of He *et al.*, a 1-h immersion time is used for the stabilization [21]. In our experiment, the samples were exposed to NaCl solution for 20 h to stabilize under OCP. This indicates that our superhydrophobic film displays a better stability and durability in the corrosion test as compared to the reported values in the literature.

Metals **2016**, *6*, 47

Nyquist plots of the as-received aluminum alloy, anodized and hybrid organic-inorganic SHAAO substrate are presented in Figure 4b–d. It is well known that R_{ct} is the charge transfer resistance, describing the difficulty of the corrosion occurring on the substrate. R_{ct} of anodized substrate is found to be 14 k$\Omega \cdot$ cm^2 in Figure 4c, which is about 10-times higher than that of the as-received aluminum alloy substrate (1.5 k$\Omega \cdot$ cm^2) in Figure 4b. However, the corrosion protection of the SHAAO substrate is the combined effect of R_{ct} of 284 k$\Omega \cdot$ cm^2 (semicircle at low frequency) along with the resistance of the superhydrophobic films of 405 k$\Omega \cdot$ cm^2 (semicircle at high frequency) in Figure 4d. This indicates a significant enhancement of corrosion resistance, which is complementary to the result from the polarization curves, where the polarization resistance of SHAAO is higher than the anodized substrate, as well as for the Al substrate.

In the study of Liu *et al.* [19], the R_{ct} of the superhydrophobic magnesium substrate covered with cerium myristate was found to be 13 kΩ, which is comparable to our anodized substrate, but much smaller than our zinc stearate-covered SHAAO substrate [19].

4. Conclusions

A superhydrophobic anodic aluminum oxide (SHAAO) surface was prepared by the electrodeposition process using the ethanolic solution of stearic acid (SA) mixed with zinc nitrate (Zn(NO$_3$)$_2$) at a constant voltage of 20 V. The hybrid organic-inorganic SHAAO surface having a micro-nanoporous structure of low surface energy zinc stearate (ZnSA) exhibits a water contact angle (CA) of 160$^\circ \pm 1^\circ$. The SHAAO substrate has a polarization resistance (R_p) and charge transfer resistance (R_{ct}) of 441 and 284 k$\Omega \cdot$ cm^2, respectively, much higher than that of the as-received aluminum(R_p of 50 k$\Omega \cdot$ cm^2 and R_{ct} of 1.46 k$\Omega \cdot$ cm^2) and anodized aluminum substrate (R_p of 87 k$\Omega \cdot$ cm^2 and R_{ct} of 14 k$\Omega \cdot$ cm^2) in the corrosion. This indicates that the SHAAO substrate displays a much better corrosion resistance property as compared to the as-received aluminum alloy substrate, as well as the anodic aluminum oxide (AAO) substrate.

Acknowledgments: We acknowledge the financial support provided by the Natural Science and Engineering Research Council of Canada (NSERC) and the Aluminium Research Centre—REGAL.

Author Contributions: Ying Huang performed experiments, analysed data and wrote the draft of manuscript; D.K. Sarkar and X.-Grant Chen supervised the works. All authors took part in the discussion, the preparation and revision of the manuscript.

Conflicts of Interest: The authors declare no conflict of interest.

Abbreviations

AAO	Anodized aluminum oxide
SHAAO	Superhydrophobic anodized aluminum oxide
SA	Stearic acid
ZnSA	Zinc stearate
EIS	Electrochemical impedance spectroscopy
OCP	Open circuit potential
CA	Contact angle
SEM	Scanning electron microscope
XRD	X-ray diffraction
FTIR	Fourier transform infrared spectroscopy
EDX	Energy dispersive X-ray spectrometer

References

1. Dotzauer, D.M.; Dai, J.; Sun, L.; Bruening, M.L. Catalytic membranes prepared using layer-by-layer adsorption of polyelectrolyte/metal nanoparticle films in porous supports. *Nano Lett.* **2006**, *6*, 2268–2272. [CrossRef] [PubMed]

2. Simovic, S.; Losic, D.; Vasilev, K. Controlled drug release from porous materials by plasma polymer deposition. *Chem. Commun.* **2010**, *46*, 1317–1319. [CrossRef] [PubMed]

3. Wang, M.; Meng, G.; Huang, Q.; Xu, Q.; Chu, Z.; Zhu, C. FITC-modified PPy nanotubes embedded in nanoporous AAO membrane can detect trace PCB20 via fluorescence ratiometric measurement. *Chem. Commun.* **2011**, *47*, 3808–3810. [CrossRef] [PubMed]

4. Liu, Y.; Goebl, J.; Yin, Y. Templated synthesis of nanostructured materials. *Chem. Soc. Rev.* **2013**, *42*, 2610–2653. [CrossRef] [PubMed]

5. Jani, A.M.M.; Anglin, E.J.; McInnes, S.J.P.; Losic, D.; Shapter, J.G.; Voelcker, N.H. Nanoporous anodic aluminium oxide membranes with layered surface chemistry. *Chem. Commun.* **2009**, *21*, 3062–3064. [CrossRef] [PubMed]

6. Bouchama, L.; Azzouz, N.; Boukmouche, N.; Chopart, J.P.; Daltin, A.L.; Bouznit, Y. Enhancing aluminum corrosion resistance by two-step anodizing process. *Surf. Coat. Technol.* **2013**, *235*, 676–684. [CrossRef]

7. Huang, Y.; Sarkar, D.K.; Chen, X.-G. A one-step process to engineer superhydrophobic copper surfaces. *Mater. Lett.* **2010**, *64*, 2722–2724. [CrossRef]

8. Ciasca, G.; Papi, M.; Chiarpotto, M.; de Ninno, A.; Giovine, E.; Campi, G.; Gerardino, A.; de Spirito, M.; Businaro, L. Controlling the cassie-to-wenzel transition: An easy route towards the realization of tridimensional arrays of biological objects. *Nano-Micro Lett.* **2014**, *6*, 280–286. [CrossRef]

9. Zhang, F.; Chen, S.; Dong, L.; Lei, Y.; Liu, T.; Yin, Y. Preparation of superhydrophobic films on titanium as effective corrosion barriers. *Appl. Surf. Sci.* **2011**, *257*, 2587–2591. [CrossRef]

10. Yin, B.; Fang, L.; Tang, A.-Q.; Huang, Q.-L.; Hu, J.; Mao, J.-H.; Bai, G.; Bai, H. Novel strategy in increasing stability and corrosion resistance for super-hydrophobic coating on aluminum alloy surfaces. *Appl. Surf. Sci.* **2011**, *258*, 580–585. [CrossRef]

11. Huang, Y.; Sarkar, D.K.; Chen, X.-G. Superhydrophobic aluminum alloy surfaces prepared by chemical etching process and their corrosion resistance properties. *Appl. Surf. Sci.* **2015**, *356*, 1012–1024. [CrossRef]

12. Huang, Y.; Sarkar, D.K.; Gallant, D.; Chen, X.-G. Corrosion resistance properties of superhydrophobic copper surfaces fabricated by one-step electrochemical modification process. *Appl. Surf. Sci.* **2013**, *282*, 689–694. [CrossRef]

13. Zuo, Z.; Liao, R.; Guo, C.; Yuan, Y.; Zhao, X.; Zhuang, A.; Zhang, Y. Fabrication and anti-icing property of coral-like superhydrophobic aluminum surface. *Appl. Surf. Sci.* **2015**, *331*, 132–139. [CrossRef]

14. Watson, G.S.; Green, D.W.; Schwarzkopf, L.; Li, X.; Cribb, B.W.; Myhra, S.; Watson, J.A. A gecko skin micro/nano structure—A low adhesion, superhydrophobic, anti-wetting, self-cleaning, biocompatible, antibacterial surface. *Acta Biomater.* **2015**, *21*, 109–122. [CrossRef] [PubMed]

15. Vengatesh, P.; Kulandainathan, M.A. Hierarchically ordered self-lubricating superhydrophobic anodized aluminum surfaces with enhanced corrosion resistance. *ACS Appl. Mater. Interfaces* **2015**, *7*, 1516–1526. [CrossRef] [PubMed]

16. Li, Y.; Li, S.; Zhang, Y.; Yu, M.; Liu, J. Fabrication of superhydrophobic layered double hydroxides films with different metal cations on anodized aluminum 2198 alloy. *Mater. Lett.* **2015**, *142*, 137–140. [CrossRef]

17. Liu, W.; Luo, Y.; Sun, L.; Wu, R.; Jiang, H.; Liu, Y. Fabrication of the superhydrophobic surface on aluminum alloy by anodizing and polymeric coating. *Appl. Surf. Sci.* **2013**, *264*, 872–878. [CrossRef]

18. Fujii, T.; Aoki, Y.; Habazaki, H. Superhydrophobic hierarchical surfaces fabricated by anodizing of oblique angle deposited Al–Nb alloy columnar films. *Appl. Surf. Sci.* **2011**, *257*, 8282–8288. [CrossRef]

19. Liu, Q.; Kang, Z. One-step electrodeposition process to fabricate superhydrophobic surface with improved anticorrosion property on magnesium alloy. *Mater. Lett.* **2014**, *137*, 210–213. [CrossRef]

20. Huang, Y.; Sarkar, D.K.; Chen, X.-G. Superhydrophobic nanostructured zno thin films on aluminum alloy substrates by electrophoretic deposition process. *Appl. Surf. Sci.* **2015**, *327*, 327–334. [CrossRef]

21. He, T.; Wang, Y.; Zhang, Y.; Lv, Q.; Xu, T.; Liu, T. Super-hydrophobic surface treatment as corrosion protection for aluminum in seawater. *Corros. Sci.* **2009**, *51*, 1757–1761. [CrossRef]

metals

MDPI

Article

Differential Scanning Calorimetry and Thermodynamic Predictions—A Comparative Study of Al-Zn-Mg-Cu Alloys

Gernot K.-H. Kolb [1], Stefanie Scheiber [1], Helmut Antrekowitsch [1], Peter J. Uggowitzer [2], Daniel Pöschmann [3] and Stefan Pogatscher [1,*]

[1]   Institute of Nonferrous Metallurgy, Montanuniversitaet Leoben, Franz-Josef-Str. 18, Leoben 8700, Austria; gernot.kolb@unileoben.ac.at (G.K.-H.K.); stefanie.scheiber@unileoben.ac.at (S.S.); helmut.antrekowitsch@unileoben.ac.at (H.A.)
[2]   Laboratory of Metal Physics and Technology, Department of Materials, ETH Zurich, Vladimir-Prelog-Weg 4, Zürich 8093, Switzerland; peter.uggowitzer@mat.ethz.ch
[3]   AMAG rolling GmbH, P.O. Box 32, Ranshofen 5282, Austria; daniel.poeschmann@amag.at
*   Correspondence: stefan.pogatscher@unileoben.ac.at; Tel.: +43-3842-402-5228

Academic Editor: Nong Gao
Received: 31 May 2016; Accepted: 29 July 2016; Published: 3 August 2016

Abstract: Al-Zn-Mg-Cu alloys are widely used in aircraft applications because of their superior mechanical properties and strength/weight ratios. Commercial Al-Zn-Mg-Cu alloys have been intensively studied over the last few decades. However, well-considered thermodynamic calculations, via the CALPHAD approach, on a variation of alloying elements can guide the fine-tuning of known alloy systems and the development of optimized heat treatments. In this study, a comparison was made of the solidus temperatures of different Al-Zn-Mg-Cu alloys determined from thermodynamic predictions and differential scanning calorimetry (DSC) measurements. A variation of the main alloying elements Zn, Mg, and Cu generated 38 experimentally produced alloys. An experimental determination of the solidus temperature via DSC was carried out according to a user-defined method, because the broad melting interval present in Al-Zn-Mg-Cu alloys does not allow the use of the classical onset method for pure substances. The software algorithms implemented in FactSage®, Pandat™, and MatCalc with corresponding commercially available databases were deployed for thermodynamic predictions. Based on these investigations, the predictive power of the commercially available CALPHAD databases and software packages was critically reviewed.

Keywords: aluminium alloys; Al-Zn-Mg-Cu alloys; CALPHAD; differential scanning calorimetry

1. Introduction

Increasing standards and demands for high strength aluminium alloys for aircraft and automotive applications require the continuous improvement of heat treatment procedures and alloy chemistry to optimise critical properties such as strength, toughness and corrosion resistance. Al-Zn-Mg-Cu alloys (7xxx) are age-hardenable and favourable because of their high strength-to-weight ratio [1,2]. Their simplified precipitation sequence is generally known as [3,4]:

$$\text{SSSS} - \text{metastable GP-zones (GP I, GP II)} - \text{metastable } \eta' - \text{stable } \eta$$

where SSSS represents the supersaturated solid solution after solution treatment and quenching. Cluster and GP zones are formed during natural ageing and in early stages of artificial ageing. The metastable phase η' is commonly responsible for the main hardening process, whereas the equilibrium phase η is characterized by coarse particles and is typical of overaged conditions [5,6].

The determination of critical parameters for heat treatment procedures is often done via differential scanning calorimetry (DSC), which is a powerful technique for studying the thermodynamics and kinetics of phase changes and measures the heat flow rates in dependence on temperature and/or time [3,7]. In industrial Al-Zn-Mg-Cu alloys DSC has been especially useful to study the precipitation sequence and the possible temperature range for solution heat treatments [3,6].

In addition to experimental determination of the evolution of phases most constitutional quantities can also be predicted computationally via the CALPHAD (CALculation of PHAse Diagrams) approach [8]. CALPHAD uses different semi-empirical models to calculate the Gibbs free energy. These models are mostly generated from experimental findings. Excess Gibbs free energy contributions of non-ideal solutions are also included via semi-empirical models (e.g., Redlich-Kister polynomials) [9–13]. However, in many cases multi-component systems are not fully assessed and are only extrapolated from binary, or in special cases higher order, boundary systems. With this approach, the thermochemical properties of alloys can be described sufficiently [14]. However, the absolute accuracy for detailed alloy systems is largely unknown.

This paper illustrates how state-of-the-art thermodynamic predictions using different software packages and the corresponding databases accord with DSC results from experimentally produced Al-Zn-Mg-Cu alloys. It also illustrates a possible method for estimating the solidus temperature of 7xxx alloys, which show a wide solidification interval. Finally, it discusses the usability of thermodynamic predictions in finding optimized compositions and temperature regimes for successful solution heat treatment procedures.

2. Materials and Methods

For this study model alloys were prepared with Al 0.995 (mass fraction) and binary Al-X master alloys (X = Cu, Mn, Fe, Cr, and Ti) and pure Si, Mg, and Zn, respectively, as starting materials using an inductive melting furnace (ITG Induktionsanlagen GmbH, Hirschhorn/Neckar, Germany). To check the chemical composition, optical emission spectrometry (SPECTROMAXx from SPECTRO, Kleve, Germany) was applied during the alloying procedure and to the final products. All 7xxx example alloys are roughly variations of AA 7075 alloys; their chemical compositions are listed in Table 1. The alloy ingots were homogenized in a Nabertherm N60/85 SHA circulating air furnace at 455 °C for 4 h, and 10 h at 465 °C. The additional higher temperature was chosen in case of insufficient effectiveness at the lower temperature. Finally, the alloys were hot compressed to convert the cast structure into a wrought microstructure.

DSC measurements were performed on a Netzsch DSC 204 F1 Phoenix (Netzsch Gerätebau GmbH, Selb/Bayern, Germany) at a heating rate of 10 K/min for specimens of 4 mm × 2 mm × 0.5 mm. Samples were put into an Al_2O_3 pan in the DSC apparatus at room temperature and cooled to −40 °C at highest possible rate and equilibrated for 10 min while employing a nitrogen gas flow of 20 mL·min^{-1}. Thus, levelling of the DSC apparatus occurred at the low starting temperature and not at the interesting region above room temperature. Measurements were performed between −40 °C and 700 °C; baseline correction was performed during experiments which comprised a single DSC run using two empty Al_2O_3 pans (one as reference, the other for measuring test alloys).

Thermodynamic equilibrium calculations were performed using FactSage® 7 software [15] together with the FACT FTlite light alloy database (2015). Calculations with the MatCalc program were carried out with MC_AL_V2.029 database (2015). Pandat calculations were performed using the PanAl2013 [16] database.

Table 1. Chemical composition of the alloys measured with emission spectrometry.

Alloy #	Composition (Mass Fraction $\times 10^2$)								
	Si	Fe	Cu	Mn	Mg	Cr	Zn	Ti	Al
1	0.40	0.40	1.69	0.26	1.37	0.24	4.15	0.13	Bal.
2	0.39	0.37	1.70	0.25	1.35	0.23	6.20	0.14	Bal.
3	0.35	0.34	1.47	0.25	2.07	0.23	5.81	0.13	Bal.
4	0.18	0.10	1.26	0.10	1.87	0.16	4.73	0.05	Bal.
5	0.18	0.10	1.29	0.10	2.01	0.17	5.30	0.04	Bal.
6	0.18	0.09	1.19	0.10	2.17	0.18	5.67	0.05	Bal.
7	0.18	0.10	1.29	0.10	2.22	0.16	5.96	0.04	Bal.
8	0.15	0.10	1.17	0.09	2.12	0.20	5.24	0.04	Bal.
9	0.15	0.11	1.21	0.10	1.81	0.20	5.28	0.04	Bal.
10	0.16	0.11	1.31	0.10	1.88	0.20	5.64	0.04	Bal.
11	0.16	0.11	1.28	0.10	1.83	0.20	5.86	0.04	Bal.
12	0.16	0.11	1.27	0.10	1.80	0.21	5.98	0.04	Bal.
13	0.14	0.02	1.19	0.10	2.36	0.16	4.97	0.04	Bal.
14	0.13	0.02	1.16	0.10	2.25	0.16	5.11	0.04	Bal.
15	0.13	0.02	1.13	0.10	2.12	0.16	5.32	0.04	Bal.
16	0.13	0.02	1.16	0.10	2.14	0.16	5.73	0.04	Bal.
17	0.13	0.02	1.22	0.10	2.16	0.16	5.91	0.04	Bal.
18	0.13	0.12	1.33	0.11	2.22	0.21	5.10	0.05	Bal.
19	0.13	0.12	1.32	0.11	1.83	0.21	5.28	0.05	Bal.
20	0.14	0.12	1.33	0.10	1.84	0.21	5.65	0.05	Bal.
21	0.13	0.11	1.34	0.10	1.85	0.21	6.01	0.05	Bal.
22	0.13	0.11	1.30	0.10	1.79	0.21	6.29	0.06	Bal.
23	0.13	0.12	1.34	0.10	1.80	0.21	6.51	0.05	Bal.
24	0.12	0.12	1.21	0.11	3.04	0.18	5.47	0.06	Bal.
25	0.13	0.12	1.30	0.11	2.89	0.17	5.75	0.05	Bal.
26	0.12	0.12	1.37	0.11	2.64	0.16	6.30	0.05	Bal.
27	0.12	0.12	1.25	0.11	2.41	0.18	6.33	0.05	Bal.
28	0.14	0.13	1.37	0.11	2.47	0.16	6.79	0.05	Bal.
29	0.44	0.27	1.55	0.29	2.40	0.22	5.00	0.12	Bal.
30	0.07	0.14	1.31	0.11	2.07	0.20	5.64	0.11	Bal.
31	0.08	0.14	1.19	0.10	2.93	0.17	5.97	0.11	Bal.
32	0.06	0.03	1.33	0.10	2.38	0.17	6.02	0.08	Bal.
33	0.06	0.03	1.39	0.10	2.32	0.17	6.74	0.08	Bal.
34	0.06	0.03	1.30	0.10	2.20	0.17	6.65	0.09	Bal.
35	0.07	0.13	1.24	0.11	2.09	0.20	5.57	0.11	Bal.
36	0.07	0.15	1.27	0.11	2.05	0.20	6.04	0.10	Bal.
37	0.06	0.13	1.19	0.11	2.65	0.20	5.57	0.10	Bal.
38	0.07	0.13	1.21	0.11	2.50	0.20	5.85	0.10	Bal.

3. Results and Discussion

Figure 1 represents the thermodynamically calculated freezing range of alloy #38 (arbitrarily chosen). Obviously the coincidence of the calculations is rather low. For the solidus temperature the results differ quite strongly, ranging from 510 °C (Pandat™) to 549 °C (MatCalc). At higher temperatures, a nearly identical result is seen for all three thermodynamic programs. The temperatures where the fcc phase vanishes range from 636 °C (FactSage®) to 639 °C (MatCalc). The difference between solidus temperature and full melting of the fcc phase may stem from the small fraction of liquid formed over a large temperature interval near the solidus temperature and a large fraction of liquid formed at higher temperatures over a small temperature interval.

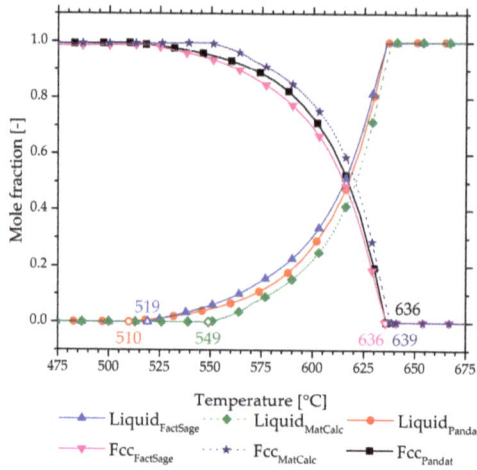

Figure 1. Thermodynamic calculation of solidus and liquidus temperatures of experimental alloy #38 using FactSage®, MatCalc, and Pandat™.

The data are plotted in Figure 2 to compare the CALPHAD results obtained for the solidus temperature. Figure 2a represents a plot of FactSage® vs. MatCalc. The R^2-value of 0.730 determined indicates an acceptable correlation between the data of these two tools. The deviation between the programs mentioned is probably caused by different databases. In Figure 2b, the calculated solidus temperatures correlate with 0.923 for Pandat™ vs. FactSage®. This correlation between FactSage® and Pandat™ is excellent by a similar database. In Figure 2c, the results of thermodynamic calculation via MatCalc vs. those via Pandat™ are shown. The correlation R^2 is about 0.719 and is slightly lower than that of FactSage® and MatCalc. This value can be expected from the excellent correlation of FactSage® and Pandat™ and the low correlation of FactSage® and MatCalc. From the correlation comparison of all three tools, it may be concluded that FactSage® and Pandat™ show quite similar trends and exhibit only small differences in the absolute solidus temperature. MatCalc calculates a slightly different trend and a higher absolute deviation in the solidus temperature compared with the two other tools.

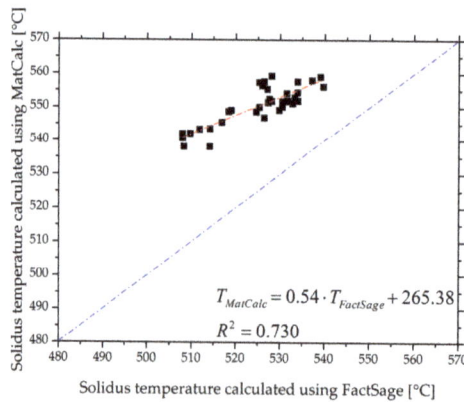

$$T_{MatCalc} = 0.54 \cdot T_{FactSage} + 265.38$$

$$R^2 = 0.730$$

(a)

Figure 2. *Cont.*

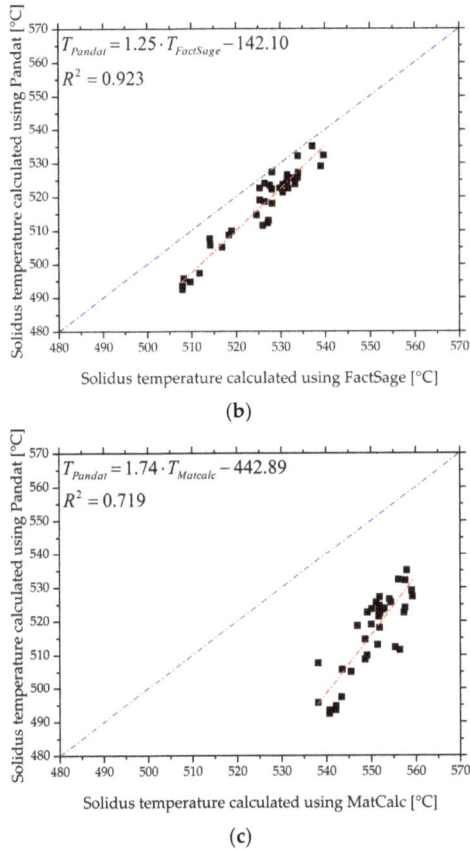

$$T_{Pandat} = 1.25 \cdot T_{FactSage} - 142.10$$
$$R^2 = 0.923$$

Solidus temperature calculated using FactSage [°C]

(b)

$$T_{Pandat} = 1.74 \cdot T_{Matcalc} - 442.89$$
$$R^2 = 0.719$$

Solidus temperature calculated using MatCalc [°C]

(c)

Figure 2. Comparison of solidus temperatures predicted by CALPHAD programs: (**a**) FactSage® vs. MatCalc; (**b**) FactSage® vs. Pandat™; and (**c**) MatCalc vs. Pandat™.

In addition to simulations, experimental investigations were carried out via DSC. Melting of pure substances at a single temperature generates broad peaks over temperature due to the thermal lag of the DSC device. Consequently, the classical onset method is used to determine the melting point. A symbolic DSC curve is shown in Figure 3. The onset (correct: extrapolated onset) is defined as the beginning of the thermal effect. Within a peak, i.e., during a transition or reaction, the baseline is defined as the curve between the region of a peak, which would have been recorded if all c_p changes had occurred but no heat of transition had been released [7]. This means that at the deviation of the base line (inflection point) tangents can be applied along the (imaginary or extrapolated) base line and the occurring peak. The intersection point of the two straight lines can be used as the onset temperature. The area under the curve is typically proportional to the enthalpy of this event [17,18]. This construction is physically useful only on the DSC melting curve of a pure substance, where it can be deployed for graphical determination of the sample temperature during melting. For alloys, the method for determining the onset is not trivial. Unfortunately, there is a broad melting interval in 7xxx alloys and the start of equilibrium melting is only associated with a low fraction of liquid formed (compare with Figure 1). Because the associated peak area in DSC scales with this fraction of liquid, the onset is smeared out. In this case the peak resulting from the real temperature interval of melting where heat is consumed and the thermal lag overlaps. Here, to account for this, we use the first deviation from the baseline (small detail in Figure 3) as a user-defined measure for the solidus temperature. The base line is defined here as the part of the DSC curve that shows no transformations (after the dissolution of precipitates but directly before melting) and is taken as a straight line. This method, however, depends on the viewed scale; we therefore define this for all measurements here as $\Delta T = C_1 = 200\,°C$ and $\Delta DSC = C_2 = 0.07\,W/g$ (see insert to Figure 3).

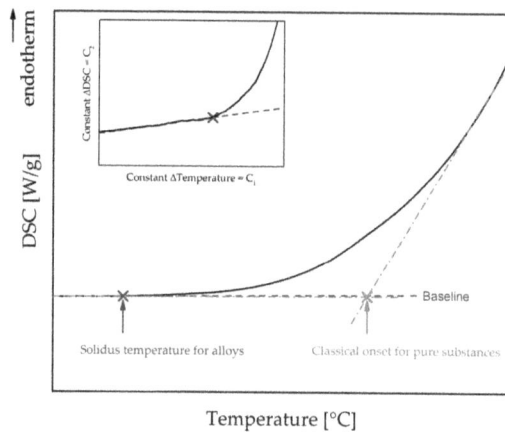

Figure 3. Comparison of a user-defined determination of solidus temperatures for alloys with large melting intervals with the classical onset technique for evaluation of the melting point of pure substances based on a DSC curve.

As an example Figure 4 shows the measured curve of the experimental alloy #38 in the interval from 470 °C to 670 °C. The DSC curve exhibits an endothermic peak at 477 °C prior to the main melting peak. The example alloy is in as-cast condition. It may be assumed that the occurrence of the first peak is caused by segregations and by the melting of the T-phase ($Al_2Mg_3Zn_3$) [6,19,20]. After a first homogenization, the alloy was measured again in the DSC. The first observed melting peak was shifted to higher temperatures (492 °C). The area of the peak was also significantly less than that in the as-cast condition. However, no complete homogenization occurred. The small peak is an indicator of the S-phase (Al_2CuMg) [6,19]. After the second homogenization at higher temperature a closer look at the DSC trace also revealed no endothermic peak due to low melting non-equilibrium segregations or undissolved phases. We assume that the alloy reached a near-equilibrium state. Now the large peak may be considered the main melting peak of the experimental alloy #38. It corresponds to the solidus temperature and is determined to be 529 °C via the user-defined method described in Figure 3.

Table 2 summarizes the solidus temperatures of the corresponding alloys and the values calculated. In Table 2 it can be seen that the predicted values agree with the results of the experimental measurements within the scatter of the various software packages (49 °C). However, a result which only deviates 31 °C between experiment and calculation can be judged as acceptable in terms of predictive character. The accuracy of our DSC results can be judged via discussion of the thermal lag of the setup and the standard deviation of the measurements. The thermal lag of the measurements at 10 K/min was quite acceptable, but can result in a systematic error of approximately 4 °C (estimated from an extrapolation of measurements with different heating rates of 5, 10, and 20 K/min to 0 K/min). The standard deviation of our solidus estimation method for repeated DSC measurements was around ±1 °C.

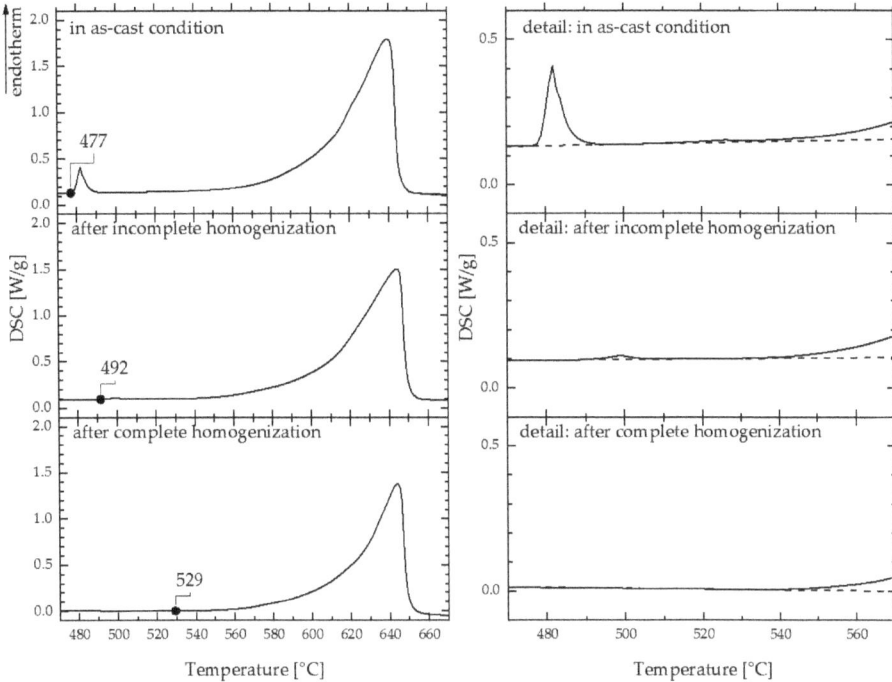

Figure 4. Change in DSC signal of experimental alloy #38 alloy during heating of three samples in conditions as-cast, incomplete homogenization and complete homogenization at constant heating rate of 10 K/min.

Figure 5 represents a parity plot of the DSC measurements and thermodynamic predictions of the solidus temperatures. In Figure 5a it can be seen that FactSage® matches acceptably with an R^2 of 0.625. At temperatures around 530 °C the coincidence is good. However, the slope of the trend line is somewhat off. Pandat™ provides a result close to that of FactSage®, with an R^2 of 0.606. The optimum overlap is in the temperature range between 530 °C and 540 °C and the trend is also slightly off. The similarity between Pandat™ and FactSage® is not surprising because these packages have been shown to deliver well-correlated predictions (see Figure 2). MatCalc supplies a slightly lower R^2, 0.540, than FactSage® and Pandat™. However, the slope of the trend line fits better. The reason for the discrepancy between the predictions and the calculated values may be due to the limitations of thermodynamic data regarding multicomponent systems, which always includes extrapolation from binary, ternary, or in special cases quaternary optimized systems. Moreover, thermodynamic data themselves are generated from experimental observations and include measurement errors. As a last point, our own measurements via DSC may also include measurement and systematic errors like the thermal lag described above. A further systematic error is possible due to the difficulties of measuring the exact melting temperature described in Figure 3. Nevertheless, thermodynamic calculations can be seen as a useful tool for predicting the solidus temperatures of a certain alloy composition to optimise 7xxx alloys. However, one should be aware of the possible inaccuracy of CALPHAD predictions. For example, a maximum absolute temperature deviation of more than 30 °C between prediction and measurement is too high for the design of new heat treatments, where adjustments are usually within a much narrower temperature range to avoid partial melting. Despite this limitation, fine-tuning of the alloy composition and heat treatment temperatures can be guided via thermodynamic predictions because trends are predicted correctly for all software-packages.

Table 2. Measured and calculated solidus temperatures.

Alloy #	T_s DSC"homogenized" (°C)	T_s FactSage® (°C)	T_s MatCalc (°C)	T_s Pandat™ (°C)
1	542	539	559	529
2	532	527	551	513
3	530	526	556	512
4	536	540	556	532
5	531	533	553	524
6	529	531	552	521
7	527	525	549	515
8	533	534	554	526
9	533	537	558	535
10	533	532	552	523
11	532	531	552	524
12	531	532	552	525
13	536	531	550	524
14	538	533	551	526
15	541	534	552	527
16	538	530	549	523
17	530	526	547	519
18	532	528	552	518
19	534	534	558	532
20	533	532	554	527
21	531	528	552	522
22	530	528	552	524
23	533	525	550	519
24	526	512	543	497
25	523	510	542	495
26	521	508	541	493
27	518	517	546	505
28	519	508	541	493
29	536	527	555	512
30	540	526	558	520
31	520	508	542	494
32	528	514	544	506
33	525	508	538	496
34	523	514	538	508
35	536	528	559	527
36	527	525	557	523
37	532	518	549	509
38	529	519	549	510

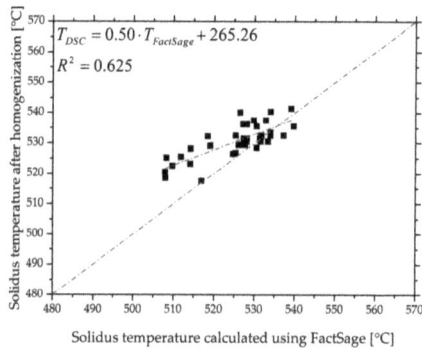

$T_{DSC} = 0.50 \cdot T_{FactSage} + 265.26$

$R^2 = 0.625$

Solidus temperature after homogenization [°C]

Solidus temperature calculated using FactSage [°C]

(a)

Figure 5. *Cont.*

$$T_{DSC} = 0.74 \cdot T_{MatCalc} + 123.70$$
$$R^2 = 0.540$$

Solidus temperature calculated using MatCalc [°C]

(b)

$$T_{DSC} = 0.38 \cdot T_{Pandat} + 333.83$$
$$R^2 = 0.606$$

Solidus temperature calculated using Pandat [°C]

(c)

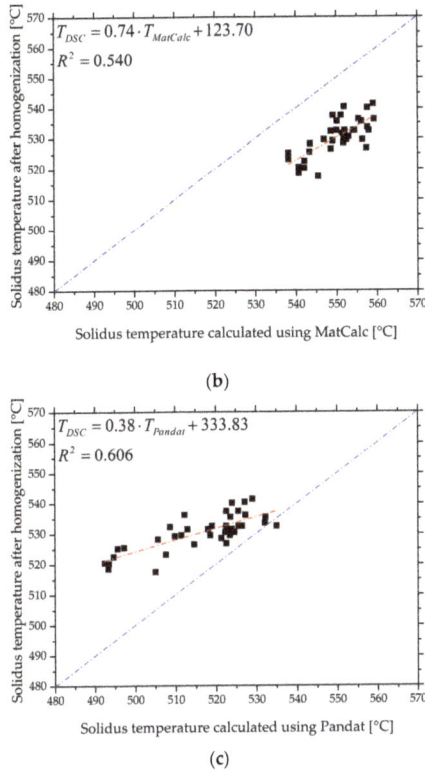

Figure 5. Representation of DSC measurement results compared with thermodynamically predicted forecasts: (**a**) FactSage® vs. DSC; (**b**) MatCalc vs. DSC; and (**c**) Pandat™ vs. DSC.

4. Conclusions

In this study we compared DSC measurements and thermodynamic predictions for Al-Zn-Mg-Cu alloys from different CALPHAD tools: the software packages FactSage®, Pandat™, and MatCalc and the corresponding databases. We showed that the simulations deliver useful information about solidus temperatures, which is commonly only available through extensive experimental work. Based on an evaluation of the quality of the predictions, we illustrated the extent to which thermodynamic predictions can help to identify optimized alloy compositions, excluding prohibited areas with low melting phases in the temperature field of solution treatment procedures. The main findings of the study are summarized as follows:

- The CALPHAD tools FactSage®, Pandat™ and MatCalc predict correlated solidus temperature values, although within a maximum observed absolute temperature deviation of 49 °C for various Al-Zn-Mg-Cu alloys.
- To compare simulated solidus temperatures to data from DSC measurements, a user-defined method for estimating the solidus temperature for alloys with a broad melting interval was introduced.
- Experimentally determined solidus temperatures agree with the predictions and deviate no more than the predictions of different CALPHAD tools themselves.

Thermodynamic tools based on the CALPHAD approach are very efficient for optimizing alloys and heat treatments, but our results show that it is critical to be aware of the boundaries of prediction accuracy.

Acknowledgments: The authors thank the Austrian Research Promotion Agency (FFG) (Grant No. 850427) and AMAG rolling GmbH for financial support of this work.

Author Contributions: G.K.-H.K. and S.P. conceived the study. G.K. and S.S. produced the alloys and performed the DSC tests. G.K.-H.K. and S.S. conducted the DSC measurements. S.P., H.A., P.J.U. and D.P. supervised the work. All authors contributed extensively to the data analysis and discussion. G.K.-H.K., S.S. and S.P. wrote the paper.

Conflicts of Interest: The authors declare no conflict of interest.

References

1. Dursun, T.; Soutis, C. Recent developments in advanced aircraft aluminium alloys. *Mater. Des.* **2014**, *56*, 862–871. [CrossRef]
2. Heinz, A.; Haszler, A.; Keidel, C.; Moldenhauer, S.; Benedictus, R.; Miller, W.S. Recent development in aluminium alloys for aerospace applications. *Mater. Sci. Eng. A* **2000**, *280*, 102–107. [CrossRef]
3. Lang, P.; Wojcik, T.; Povoden-Karadeniz, E.; Falahati, A.; Kozeschnik, E. Thermo-kinetic prediction of metastable and stable phase precipitation in Al-Zn-Mg series aluminium alloys during non-isothermal DSC analysis. *J. Alloy. Compd.* **2014**, *609*, 129–136. [CrossRef]
4. Deschamps, A.; Livet, F.; Bréchet, Y. Influence of predeformation on ageing in an Al-Zn-Mg alloy—I. Microstructure evolution and mechanical properties. *Acta Mater.* **1998**, *47*, 281–292. [CrossRef]
5. Berg, L.; Gjønnes, J.; Hansen, V.; Li, X.; Knutson-Wedel, M.; Waterloo, G.; Schryvers, D.; Wallenberg, L. GP-zones in Al-Zn-Mg alloys and their role in artificial aging. *Acta Mater.* **2001**, *49*, 3443–3451. [CrossRef]
6. Lim, S.T.; Eun, I.S.; Nam, S.W. Control of Equilibrium Phases (M,T,S) in the Modified Aluminum Alloy 7175 for Thick Forging Applications. *Mater. Trans.* **2003**, *44*, 181–187. [CrossRef]
7. Höhne, G.W.H.; Hemminger, W.F.; Flammersheim, H.-J. *Differential Scanning Calorimetry*; Springer Berlin Heidelberg: Berlin/Heidelberg, Germany, 2003.
8. Schmitz, S.M. Phasenseparation und Einfuss von Mikrolegierungselementen in Systemen mit metallischer Glasbildung. Ph.D. Thesis, Technische Universität Dresden, Dresden, Germany, April 2012.
9. Hillert, M. Partial Gibbs energies from Redlich-Kister polynomials. *Thermochim. Acta* **1988**, *129*, 71–75. [CrossRef]
10. Kattner, U.R. The thermodynamic modeling of multicomponent phase equilibria. *JOM* **1997**, *49*, 14–19. [CrossRef]
11. Das, K.N.; Habibullah, M.; Ghosh, M.; AkberHossain, N. Regression alternative to the redlich-kister equation in the determination of the excess partial molar volumes of the constituents in a binary mixture. *Phys. Chem. Liq.* **2004**, *42*, 89–94. [CrossRef]
12. Dos Santos, I.A.; Klimm, D.; Baldochi, S.L.; Ranieri, I.M. Thermodynamic modeling of the LiF-YF$_3$ phase diagram. *J. Cryst. Growth* **2012**, *360*, 172–175. [CrossRef]
13. Dos Santos, I.A.; Klimm, D.; Baldochi, S.L.; Ranieri, I.M. Experimental evaluation and thermodynamic assessment of the LiF-LuF$_3$ phase diagram. *Thermochim. Acta* **2013**, *552*, 137–141. [CrossRef]
14. Luo, A.A. Material design and development: From classical thermodynamics to CALPHAD and ICME approaches. *Calphad* **2015**, *50*, 6–22. [CrossRef]
15. Bale, C.W.; Chartrand, P.; Degterov, S.A.; Eriksson, G.; Hack, K.; Mahfoud, R.B.; Melançon, J.; Pelton, A.D.; Petersen, S. FactSage thermochemical software and databases. *Calphad* **2002**, *26*, 189–228. [CrossRef]
16. CompuTherm LLC. Software package for calculating phase diagrams and thermodynamic properties of multi-component alloys. Available online: http://www.computherm.com (accessed on 21 May 2016).
17. Riesen, R. Wahl der Basislinien. Available online: https://at.mt.com/dam/mt_ext_files/Editorial/Simple/0/basislinien_ta_usercom25ds0106.pdf (accessed on 29 July 2016).
18. Hydrate web. DSC. Available online: http://www.hydrateweb.org/dsc (accessed on 9 May 2016).
19. Rometsch, P.A.; Zhang, Y.; Knight, S. Heat treatment of 7xxx series aluminium alloys—Some recent developments. *Trans. Nonferr. Met. Soc. China* **2014**, *24*, 2003–2017. [CrossRef]
20. Saunders, N.; Miodownik, A.P. *CALPHAD (Calculation of Phase Diagrams). A Comprehensive Guide*; Pergamon: Oxford, UK, 1998.

metals

MDPI

Article

Evaluating the Applicability of GTN Damage Model in Forward Tube Spinning of Aluminum Alloy

Xianxian Wang, Mei Zhan *, Jing Guo and Bin Zhao

State Key Laboratory of Solidification Processing, School of Materials Science and Engineering, Northwestern Polytechnical University, Xi'an 710072, China; wangxianx07@163.com (X.W.); guojing_pipi@126.com (J.G.); bzhaonpu@163.com (B.Z.)
* Correspondence: zhanmei@nwpu.edu.cn; Tel.: +86-029-8846-0212-805

Academic Editor: Nong Gao
Received: 20 March 2016; Accepted: 1 June 2016; Published: 6 June 2016

Abstract: Tube spinning is an effective plastic-forming technology for forming light-weight, high-precision and high-reliability components in high-tech fields, such as aviation and aerospace. However, cracks commonly occur in tube spinning due to the complexity of stress state, which severely restricts the improvement of the forming quality and forming limit of components. In this study, a finite element (FE) model coupled with Gurson-Tvergaard-Needleman (GTN) damage model for forward tube spinning of 3A21-O aluminum alloy is established and its applicability is evaluated by experiment. Meanwhile, the GTN damage model is employed to study the damage evolution for forward tube spinning of 3A21-O aluminum alloy. The results show that the FE model is appropriate for predicting the macroscopic crack appearing in uplift area for forward tube spinning, while the damage evolution in deformation area could not be predicted well due to the negative stress triaxiality and the neglect of shear deformation. Accumulation of damage in forward tube spinning occurs mainly in the uplift area. Void volume fraction (VVF) in the outer surface of the tube is higher than that in the inner surface. In addition, it is prone to cracking in the outer surface of tube in the material uplift area.

Keywords: forward tube spinning; 3A21-O aluminum alloy; Gurson-Tvergaard-Needleman (GTN) damage model; finite element (FE) model; crack

1. Introduction

Tube spinning, also known as flow forming, is one of spinning processes widely used to produce cylindrical components with thin-walled section and high precision [1,2]. In this process, the metal is displaced axially along a mandrel, while a continuous and localized plastic deformation is applied by the feeding movement of one or more rollers and rotational motion of the mandrel to reduce the thickness of components [3,4]. According to the relationship between the direction of material flow and roller traversing, the process can be classified as forward and backward tube spinning, as shown in Figure 1 [5]. 3A21 aluminum alloy is one of the most commonly used alloys for aviation, aerospace and automotive industries because of its versatile properties, economical benefit and no need for-heat-treatment advantages [6]. However, due to the highly non-linear feature and complicated stress state during tube spinning, it is prone to cracking in 3A21 aluminum spun parts, which severely restricts the improvement of the forming quality and forming limit of components. Therefore, it is necessary to study the damage evolution of 3A21 aluminum alloy to guide the actual production of spun components.

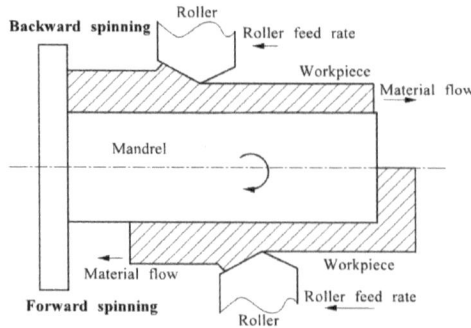

Figure 1. Schematic of forward and backward tube spinning.

Researches on the damage evolution in spinning process coupled with ductile fracture criteria have been reported in recent years. Ma *et al.* [7] investigated the damage evolution in tube spinnability of TA2 titanium tube with nine types of un-coupled ductile fracture criteria in detail. Their result indicated that except for the Freudenthal, Rice and Tracey (R-T) and Ayada models, all the other models can correctly predict the damage distribution on TA2 titanium tube in spinnability test. Cockcroft-Latham (C-L) criterion provided the highest prediction accuracy on the spinnability of TA2 titanium tube, which was only 9% less than the measured experimental value. Zhan *et al.* [8] predicted the failure occurring in shear spin-forming, splitting spin-forming of LF2M aluminum alloy by embedding the Lemaitre and Cockcroft-Latham (C&L) criteria into the finite element (FE) model. The results showed that the Lemaitre criterion was better than the C&L criterion at accurately predicting the position at which damage will occur. A thermal damage model for tube spinning process of Ti-6Al-2Zr-1Mo-1V combining the Oyane ductile fracture criterion with the relationship among damage threshold, temperature and strain rate was also established by Zhan *et al.* [9]. Their results indicated that the inner surface of the spinning region was the zone most prone to damage due to positive stress triaxiality and large strain rate.

Recently, studies on expansion and accumulation of cavities in ductile material have been carried out. It is often believed that, the ductile failure process of metal material consists of three stages in mesoscopic scale: micro-voids nucleation, growth and coalescence, respectively [10,11]. The typical model to describe these three stages is the Gurson-Tvergaard-Needleman (GTN) damage model, which is proposed by Gurson [10] and further modified by Tvergaard and Needleman [11]. Compared with other ductile fracture criteria, GTN damage model is a coupled ductile fracture criterion, incorporating the void evolution into the constitutive equations. Generally, GTN damage model and modified GTN damage model are applied to predict void initiation, propagation, and final rupture [12,13] combined with FE simulation in some process. Chen and Dong [12] predicted the damage in deep drawing test of AA6111 aluminum alloy well with a modified GTN yield criterion based on a quadratic anisotropic yield criterion and an isotropic hardening rule. Butcher *et al.* [13] predicted the burst pressure, formability and failure location in tube hydroforming of dual phase (DP600) steel using GTN constitutive model and interpreted the influence of void shape and shear on coalescence. Sun *et al.* [14] analyzed the ductile damage and failure behavior of steel sheet with edge defects under multi-pass cold rolling based on the shear GTN damage model proposed by Nahshon and Hutchinson [15]. Li *et al.* [16] indicated that the GTN damage model can predict the damage in tube bending process. However, it cannot predict the damage evolution due to the negative stress triaxiality in split spinning. It can be found that most of these studies about GTN damage model concentrate on the simple stress state. Researches on GTN damage model applied in other complicated deformation, such as tube spinning process, are limited. In tube spinning process, many researches on damage evolution have been studied based on other ductile fracture criteria. However, there are few researches on damage evolution based on GTN damage model in this process. Thus, the applicability and limitation of GTN damage model in tube spinning should be evaluated in detailed.

To investigate the applicability of GTN damage model to predict fracture in tube spinning process, an FE model coupled with GTN damage model for forward tube spinning of 3A21-O aluminum alloy is established based on ABAQUS/Explicit platform. Then the applicability of GTN damage model in spinning process is evaluated by experiment. Distributions of the stress triaxiality, the maximum principal stress, and the void volume fraction (VVF) are analyzed to reveal damage evolution in forward tube spinning finally.

2. Material and Methods

The material used in this study is 3A21-O aluminum alloy. The chemical composition of the alloy is listed in Table 1. Parameters of the material are shown in Table 2.

Table 1. Chemical composition of 3A21-O aluminum alloy.

Position	Si	Fe	Cu	Mn	Mg	Zn	Ti	Al
Mass fraction (%)	0.6	0.7	0.2	1.3	0.05	0.10	0.15	Bal.

Table 2. Material parameters of 3A21-O aluminum alloy.

Parameters	Values
Elastic modulus (GPa)	69.98
Poisson's ratio	0.33
Yield strength (MPa)	52
Strength coefficient (MPa)	188.76
Hardening exponent	0.194

Based on the ABAQUS/Explicit platform, uniaxial tensile simulations under the same condition as experiments are conducted to determine main parameters in GTN damage model. Tensile tests are carried out on a CMT5205 electronic universal testing machine (MTS Systems Corporation, Eden Prairie, MN, USA) at a maximum load of 200 KN. Dimensions of uniaxial tensile test specimen are shown in Figure 2. Experiments of forward tube spinning are carried out on a CZ900/CNC spin forming machine (Sichuan Space Industry Company, China). Taking the large size of the blank in experiment into consideration, theory of similarity [17] is adopted to improve the efficiency of calculation in simulation. The ratio of similitude is 4 in this study. The main process parameters in experiment and simulation for forward tube spinning are listed in Table 3. In order to analyze the damage evolution in forward tube spinning, the tube is divided into four areas, namely un-deformed area, uplift area [9], deformation area, and deformed area, respectively (Figure 3).

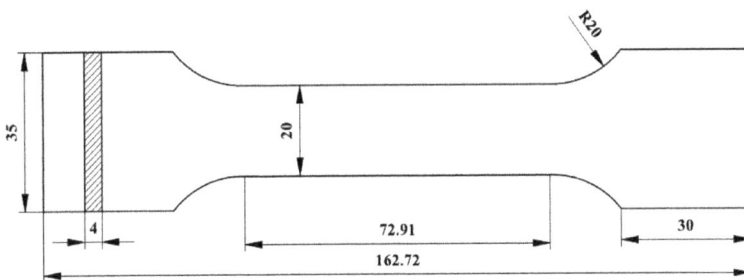

Figure 2. Dimensions of uniaxial tensile specimen (Unit: mm).

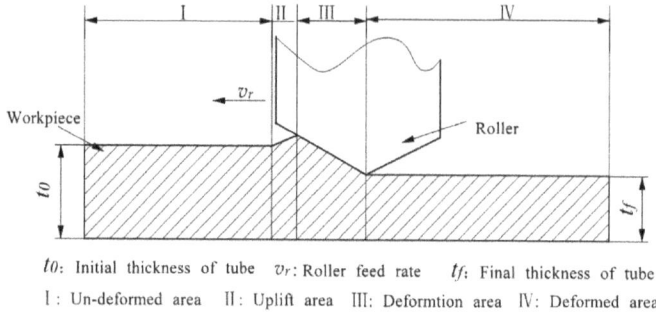

t_0: Initial thickness of tube v_r: Roller feed rate t_f: Final thickness of tube
I : Un-deformed area   II : Uplift area   III: Deformtion area   IV: Deformed area

Figure 3. Schematic of area division of the tube in forward spinning.

Table 3. Process parameters in simulation and experiment.

Parameters	Experiment	Simulation
Inner diameter of the tube d (mm)	320.6	80.15
Thickness of the tube t_0 (mm)	12	3
Initial height of the tube h (mm)	150	37.5
Roller nose radius r (mm)	5	1.25
Roller feed rate v_r (mm/r)	1.25	0.3125
Roller attack angle α (°)	30	30
Mandrel rotational speed ω (r/min)	100	100
Reduction ratio of wall thickness Ψ (%)	50	50

3. Gurson-Tvergaard-Needleman (GTN) Damage Model and Finite Element (FE) Model in Tube Spinning

3.1. GTN Damage Model

The Gurson model revised by Tvergaard and Needleman, namely the GTN damage model, can be expressed as Equation (1) [10,11]:

$$\phi(\sigma, f) = \left(\frac{\sigma_{eq}}{\sigma_y}\right) + 2q_1 f^* \cosh\left(-\frac{3q_2\sigma_m}{2\sigma_y}\right) - \left(1 + q_3 f^{*2}\right) = 0 \tag{1}$$

where, ϕ is the yield function; σ_{eq} is the von Mises equivalent stress; σ_y the mean uniaxial equivalent stress of the matrix material; σ_m is macroscopic hydrostatic pressure; f is the VVF; q_1, q_2 and q_3 are the constants for material. When $q_1 = q_2 = q_3 = 1$, Equation (1) can be degraded into the Gurson model. f^* is the modified void volume fraction that takes into account the final decrease in load when void coalescence occurs. The relationship between f^* and f is given as follows:

$$f^* = \begin{cases} f & (f \leq f_c) \\ f_c + \frac{1/q_1 - f_c}{f_F - f_c} (f - f_c) & (f_c < f < f_F) \end{cases} \tag{2}$$

where, f_c is the critical VVF when the void coalescence takes place and f_F is the VVF at the final failure of material.

The evolution of voids is characterized by the gradually growth of existing voids volume fraction (f_g) and nucleation of new voids volume fraction (f_n) in material, as shown in Equation (3):

$$df = df_g + df_n \tag{3}$$

The void growth rate is determined by the plastic incompressibility of matrix surrounding the voids with respect to the rule of mass balance in representative volume elements. It can be expressed as Equation (4):

$$df_g = (1-f)d\varepsilon^p : 1 \tag{4}$$

where, $d\varepsilon^p$ is the increment of hydrostatic plastic strain and I is a second order unit tensor.

The growth of strain controlled nucleation of new voids [18] can be expressed as Equation (5):

$$df_n = Ad\varepsilon_y^{pl} \tag{5}$$

with $A = \dfrac{f_n}{S_n \sqrt{2\pi}} \exp\left[-\dfrac{1}{2}\left(\dfrac{\varepsilon_y^{pl} - \varepsilon_n}{S_n}\right)^2\right]$, where, f_n is the total void volume fraction that can be nucleated; ε_n is the mean equivalent plastic strain for void nucleation; S_n is the corresponding standard deviation and $d\varepsilon_y^{pl}$ is the increment of equivalent plastic strain.

3.2. Determination of Parameters in GTN Damage Model

In the GTN model, q_1, q_2, q_3, ε_n and S_n can be determined by empirical values, and f_0, f_n, f_c and f_F are generally determined by scanning electron microscope (SEM) microstructures. In this study, to reflect the interaction in two void groups in GTN model, values of q_1, q_2 and q_3 are quantified as 1.5, 1 and 2.25, respectively [19]. The value of S_n is determined as 0.1 and ε_n is assigned as 0.3 [10,12,20]. Considering that there are many researches on the GTN damage model of aluminum alloy, the microstructure of 3A21 aluminum alloy can be hardly observed due to its excellent corrosion resistance, the initial void volume fracture f_0 is determined as 0.001 according to the values adopted in other aluminum alloy [12,21], and f_n, f_c and f_F are determined by anti-inference method [21] through the combination of experiment with FE simulation of uniaxial tensile test.

The true stress strain curves in experiment and in simulation with different parameters in the GTN model (Table 4) are shown in Figure 4. It can be seen that, under the same dimensions of specimen and loading condition, different parameters in the GTN damage model have significant effects on the deformation behavior of material. In experiment, ultimate tensile strength reaches to 143.13 MPa at the strain of 0.27. When the strain is larger than 0.27, there are obvious differences in stress strain curves between simulation and experiment. Considering that the specimen bears the maximum load and void coalescence begins, and the hardening properties decrease suddenly due to the coalescence of void speed which depends on the factor f_F [22], stress strain curves between simulation and experiment can be compared by critical strain (strain corresponding to tensile strength) and fracture strain. When the strain is less than 0.27, stress strain curves in simulation are similar to each other. The differences of stress strain curves between simulation and experiment in this region can be symbolized as standard deviation (*SD*), which can be expressed as:

$$SD = \sqrt{\dfrac{\sum\limits_{i=1}^{n}(x_i - x_{irel})^2}{n}} \tag{6}$$

where, x_i is the stress in simulation at a certain strain (0.02, 0.04, 0.06,...,0.26); x_{irel} is the stress in experiment at the corresponding strain; n is 13 in this study.

Standard deviation, critical strain and fracture strain under different parameters in the GTN model are shown in Table 4. It can be seen that there are minimum standard deviation under the parameters in group 5. Meanwhile, in simulation critical strain and fracture strain under these parameters are also closer to those in experiment. Thus, in this study, f_n, f_c and f_F are determined as 0.012, 0.02 and 0.04, respectively.

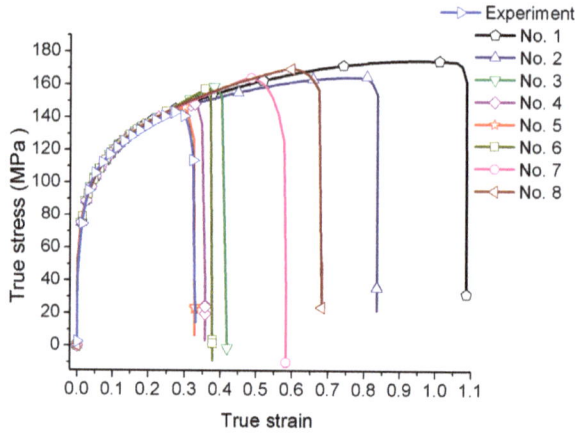

Figure 4. True stress strain curves of different parameters coupled with Gurson-Tvergaard-Needleman (GTN) damage model.

Table 4. Standard deviation, critical strain and fracture strain under different parameters in Gurson-Tvergaard-Needleman (GTN) damage model.

Group Values	1	1	3	4	5	6	7	8	Experiment
f_n	0.01	0.02	0.02	0.015	0.012	0.01	0.04	0.002	-
f_c	0.05	0.05	0.02	0.02	0.02	0.01	0.05	0.005	-
f_F	0.1	0.1	0.05	0.04	0.04	0.02	0.15	0.05	-
Standard deviation	1.47	1.30	1.70	1.26	1.25	1.75	1.36	1.72	-
Critical strain	1.01	0.81	0.38	0.32	0.30	0.35	0.48	0.59	0.27
Fracture strain	1.09	0.84	0.42	0.36	0.33	0.38	0.58	0.68	0.33

3.3. Establishment of the FE Model for Forward Tube Spinning Coupled with GTN Damage Model

Based on the ABAQUS/ Explicit platform, an FE model coupled with the GTN damage model for forward tube spinning process of 3A21-O aluminum alloy is established, as shown in Figure 5. In the model, rollers and mandrel are assumed to be analytical rigid bodies. The tube is considered as a deformable body, which is meshed by an 8-node linear brick, reduced integration and hourglass control element (C3D8R). Five layer elements are divided in the thickness direction of the tube because five layer elements are enough to get the variation information of wall thickness during spinning [17]. Three simulations with mesh sizes of 1, 1.5 and 2 mm in the axial-hoop plane are compared, as shown in Table 5. The mass scaling is 2000 to ensure the efficiency and accuracy in simulation. As seen in Table 5, there are small increases of maximum Mises stress and maximum equivalent strain with mesh size decreasing. However, central processing unit (CPU) time is increased obviously with mesh size decreasing. Considering the efficiency and accuracy of simulation, mesh size of 1.5 mm in the axial-hoop plane are adopted in this study. Coupling constraint is adopted to limit the movement of bottom surface of tube in simulation. Arbitrary Lagrangian-Eulerian (ALE) adaptive grid technique is used to avoid the distortion of the mesh and birth-and-death element is adopted to delete the element whose VVF exceeds f_F. The GTN damage model is implanted into forward tube spinning through the porous metal plasticity module in ABAQUS platform.

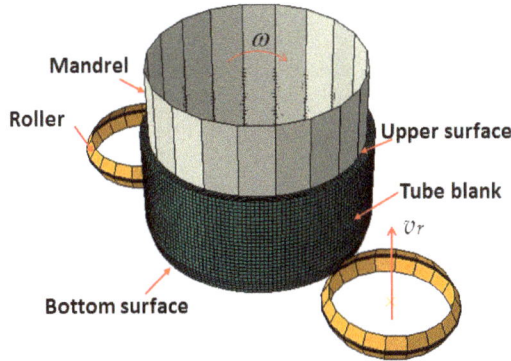

Figure 5. The finite element model for forward tube spinning.

Table 5. Simulation results under different mesh sizes.

Simulation results	Mesh Size		
	2 mm	1.5 mm	1 mm
Maximum Mises stress (MPa)	238.4	239.1	239.1
Maximum equivalent strain	2.826	3.276	3.41
Central processing unit time (h)	68	128	244

4. Evaluating the Applicability of the FE Model Coupled with GTN Damage

To verify the reliability of the FE model coupled with GTN damage model for forward tube spinning of 3A21-O aluminum alloy, the energy variation of the FE model are evaluated and corresponding experiments are carried out. Figure 6 shows the variations of kinetic energy and the ratio of kinetic energy to internal energy in forward tube spinning. It is found that, the variation of kinetic energy is stable except for the initial stage of forward tube spinning, and the ratio of kinetic energy to internal energy is less than 5% during most stage of the forward tube spinning, which indicate that the FE model is theoretically reliable [23].

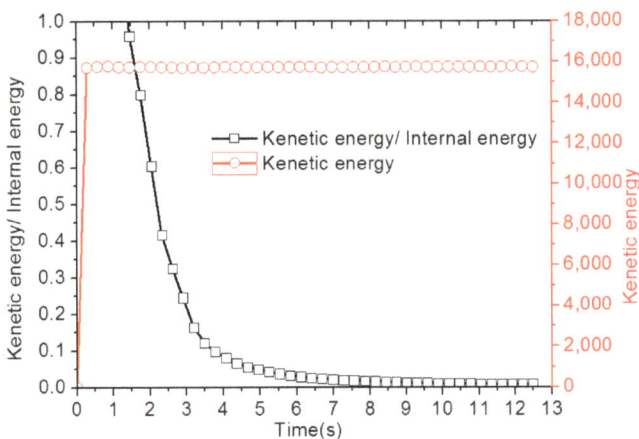

Figure 6. Energy evaluated of finite element model for forward tube spinning of 3A21-O aluminum alloy.

To evaluate the applicability of the GTN damage model in forward tube spinning in this study, experiments are conducted under the parameters as shown in Table 3. Figures 7 and 8 show the crack defects of the tube in simulation and in experiment at wall thickness reduction ratio of 50%. It can be seen that, in the outer surface of tube, some elements are deleted in the material uplift area and bottom area (Figure 7a) due to the birth-and-death element adopted when the VVF exceeds the f_F. Cracks are not observed in the inner surface of tube in simulation (Figure 7b). This indicates that in simulation, macroscopic cracks mainly occur in the material uplift area. Cracks appearing in the bottom region of tube in simulation are not emphasized in this study, because they are generated mainly due to the complicated and large force when the coupling constraint is adopted to limit the movement of the bottom surface in simulation. In experiment process, penetrating cracks (Area 1) are found in deformed area (Figure 8a,b). Except for the penetrating cracks, there is no crack in the inner surface of tube. Localized magnification image (Figure 8c) of Area 2 in Figure 8a indicates there are also many tiny cracks occurring in the material uplift area. Meanwhile, both in simulation and experiment, cracks are more serious at the peak of the material uplift area.

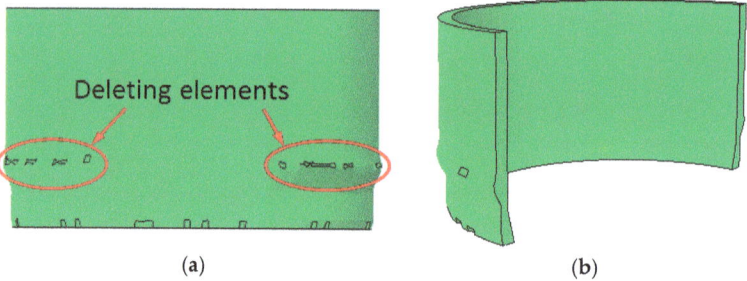

(a)   (b)

Figure 7. Crack defects in simulation in the (**a**) outer surface and (**b**) inner surface of tube at wall thickness reduction ratio of 50%.

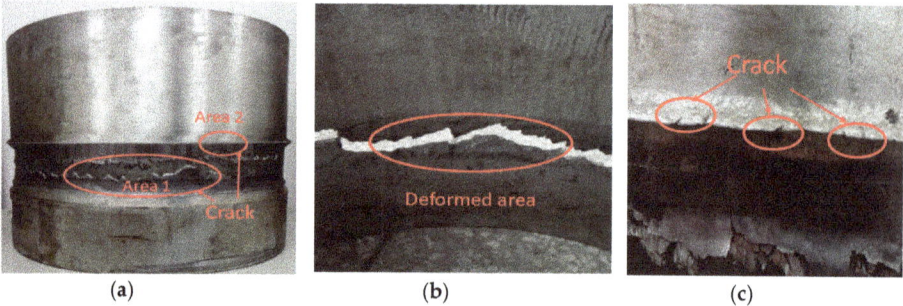

(a)   (b)   (c)

Figure 8. Crack defects of tube in experiment in the (**a**) outer surface; (**b**) inner surface and (**c**) magnified image of Area 2 in the outer surface of tube at wall thickness reduction ratio of 50%.

To analyze the discrepancy of the crack position between simulation and experiment results, Figure 9 shows the VVF distributions of two points, without crack (point A) and with crack (point B), respectively, in outer surface of spun tube at wall thickness reduction ratio of 50%. It can be seen that variations of VVF of the two points are similar to each other. Before spinning, there is no significant change of VVF. Then VVF increases sharply when the point is in the material uplift area. The maximum value of VVF of the point with crack can reach to 0.08. During spinning, VVF of the points decrease rapidly and values of it are close to zero at the end of spinning process. After spinning, values of VVF are stable and keep to zero. The similar variation of damage evolution is also found in the researches of Ma *et al.* [7] by applied Ayada damage model for tube spinning. These indicate that, in GTN damage model for forward tube spinning, the void nucleation, growth and coalescence occurs in material uplift area, then the void is annihilated during spinning process under the compression of rollers.

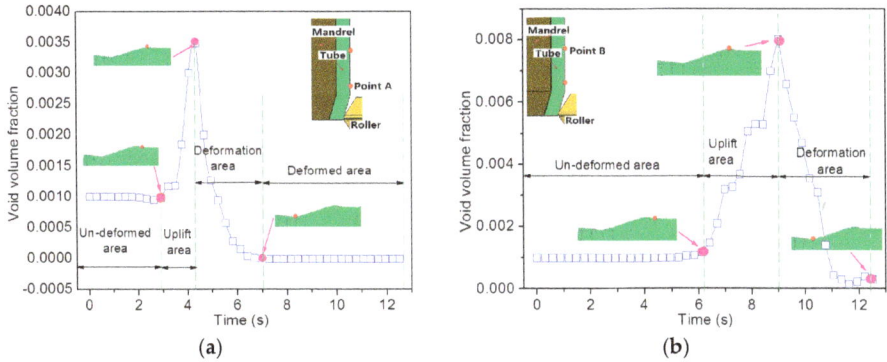

Figure 9. Distribution of void volume fraction (VVF) of the point (**a**) without crack and (**b**) with crack in forward tube spinning.

In GTN damage model, the characterization of damage evolution can be done by a combination of the stress triaxiality and Lode parameter [24]. Figure 10a,b show the variations of stress triaxiality of the two points during forward spinning process. It can be seen that, in the un-deformed area and uplift area, stress triaxiality is positive, which is good for damage accumulation. Then the stress triaxiality changes to negative in deformation area and finally changes to positive in deformed area. The variations of stress triaxiality of the two points in outer surface of spun tube are similar to researches of Ma *et al.* [7], where C-L ductile fracture criterion is employed to investigate the spinnability of tube. In general, GTN damage model or modified GTN damage model have a good performance in prediction fracture location under high stress triaxiality or shear loading [25,26]. However, the GTN damage model does not work well for uniaxial compression and plane strain compression since the void volume fraction does not increase due to the fact that void growth was suppressed in negative stress triaxiality [15]. Therefore, the GTN damage model could not precisely predict the variation of VVF in deformation area in forward tube spinning, and the variation of VVF in the subsequent deformed area is also influenced.

Meanwhile, in tube spinning, the deformed area is usually simplified as a plane strain state, namely compression deformation in radial direction and tensile deformation in axial direction. Figure 10c,d shows the variation of Lode parameter in spinning process. It can be seen that Lode parameter fluctuates widely during the whole process especially in deformation area and deformed area. In the un-deformed area, negative Lode parameter indicates there is mainly tensile deformation in this area. In uplift area, Lode parameter changes to positive gradually. Fluctuations of Lode parameter in these areas are influenced by the rapid variation of principal stresses under compatible deformation. In deformation area, the values of Lode parameter fluctuate from −0.6 to 0.4, and some values of Lode parameter are close to 0. The fluctuation of Lode parameter in deformation area is mainly resulted from the progressive deformation in spinning. When the point contacts to the roller, it undergoes shear-compression deformation; when the point is rotating to the area between the two rollers and still in the deformation area, it undergoes the addition shear-tension deformation. In addition, the degree of deformation gradually increases when the point contacts to roller again. Values of Lode parameter in deformation area indicate that the material in this area undergoes larger shear deformation than in other area of tube. In tube spinning, larger shear deformation is mainly generated in circumferential direction by the transmission of torsional moment when the blank rotates with the mandrel. In deformed area, the reason for the fluctuation of Lode parameter is similar to it in un-deformed area and material uplift area. Therefore, the shear strain in circumferential direction is also important and shouldn't be neglected especially in deformation area. Under shear loads, failure is mainly driven by the shear localization of plastic strain of the inter-voids ligaments due to void rotation and distortion [27,28]. However, the shear strain is not taken account in the classical GTN damage model, which could results in some offset of the VVF variation in deformation area as well as the subsequent deformed area.

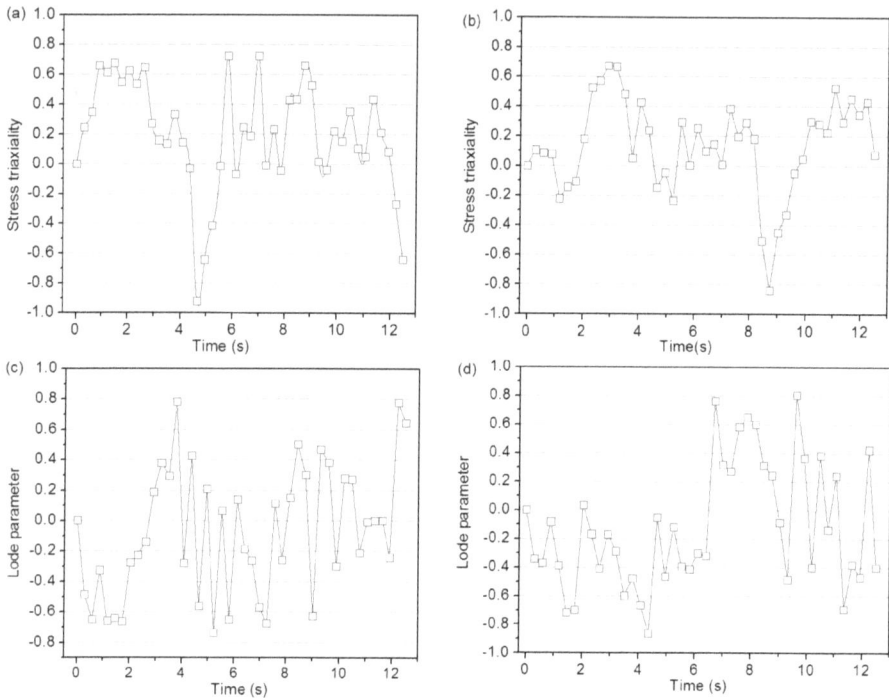

Figure 10. Distributions of stress triaxiality of the point (**a**) without crack; (**b**) with crack and lode parameter of the point (**c**) without crack and (**d**) with crack in forward tube spinning.

As explained above, the main reason for the failed prediction of the crack in deformed area is that the negative stress triaxiality and shear deformation in deformation area are not taken into consideration in the classical GTN damage model. Another reason is that, in simulation, when deformation occurs, the values of VVF are decreased and values of VVF after deformation are even smaller than the initial VVF. These mean that the macroscopic crack could be annihilated due to the dynamic change of VVF. While in experiment, once macroscopic crack appearing in tube, it is hard to eliminate but to expansion under complicated stress state. Initial crack seems to derive in the uplift area and then expansion to the deformed area leading to large cracks after severely deformation. Therefore, GTN damage model is available to predict the position of macroscopic crack appearing in the material uplift area in forward tube spinning.

5. Damage Evolution in Forward Tube Spinning

It is well known that the triaxiality stress, maximum principal stress and void volume fraction have a significance effect on damage evolution. Thus, based on the FE model coupled with GTN damage model for forward tube spinning, distributions of stress triaxiality, maximum principal stress and void volume fraction are studied, as shown in Figure 11.

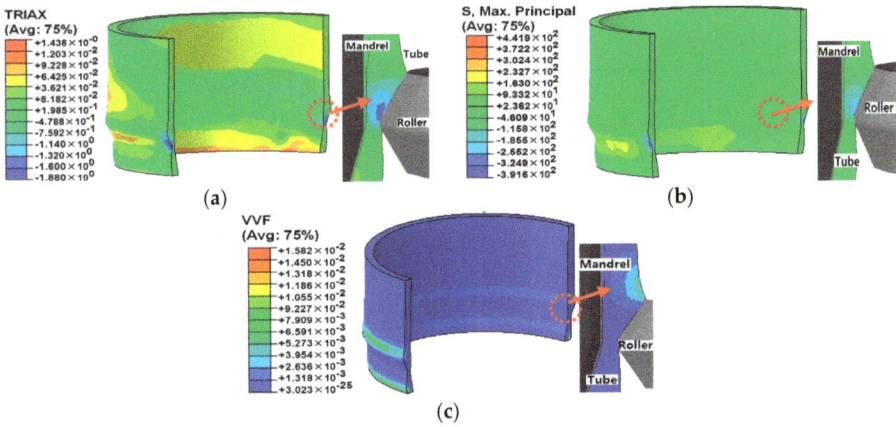

Figure 11. Distributions of (**a**) stress triaxiality; (**b**) maximum principal stress and (**c**) VVF in spun tube.

It can be seen that, in Figure 11a, the value of stress triaxiality near the upper surface of tube is close to zero due to the small compatible deformation produced in spinning; in the un-deformed area which is close to deformation area, the values of stress triaxiality change to positive. In material uplift area and deformed area, the value of stress triaxiality should be positive combined with the distributions of VVF of the points with crack and without crack (Figure 10a,b). In the deformation area contacting with roller, stress triaxiality is negative under the pressure of roller, which indicates that damage growth is restrained in this area. Neighboring areas in circumferential direction undergoes positive stress triaxiality due to the compressive deformation under the roller generating the additional tensile deformation in neighboring region. The positive stress triaxiality in the bottom surface of the tube is due to the coupling constraint applied in this surface. Characteristics of maximum principal stress in Figure 11b indicated that, in the roller contacting area (deformation area), the maximum principle stress can reach to approximately −250 MPa. The value of maximum principal stress in neighboring of roller contacting area in circumferential direction is positive due to the additional tensile deformation. Except for these areas, the maximum principle stress is close to zero because main deformation does not occur in other area of tube. Figure 11c shows the VVF distribution in the tube. It is found that: VVF in uplift area and bottom area of tube is close to 0.004 (VVF at fracture); while in other areas of the tube, the value of VVF is close to initial value of 0.001. Smaller values of VVF appear in un-deformed area because small deformation away from roller cannot lead to the increasing of VVF. In the un-deformed area near deformation area and uplift area, positive stress triaxiality results in the increasing of VVF. In the deformation area, material undergoes larger compression deformation and stress triaxiality are negative and VVF is decreased in this area. These result in the maximum VVF appearing in the uplift zone not in the deformation area. This means that GTN model could only capture the fracture occurred in material uplift area. Damage evolution in the deformation area and subsequent deformed area cannot be well predicted due to the decrease of VVF under negative deviation ratio. The distribution of lager VVF in uplift area in tube forward spinning is similar to the VVF distribution in spilt spinning in the research of Li *et al* [16]. Meanwhile, VVF in outer surface is higher than that in inner surface of spun tube. This indicates that cracks are easy to occur in the outer surface of the spun tube.

To observe the characteristics of damage evolution in the material uplift area clearly, the variation of stress triaxiality, maximum principal stress and VVF are obtained along four paths in radius direction

(Figure 12), as shown in Figure 13. It can be found that values of stress triaxiality decrease from inner surface to middle thickness section first, then increase and reach to maximum value of outer surface. This is due to the large amount of deformation in outer surface under the effect of roller and small amount of deformation under the friction between blank and mandrel in forward tube spinning. When the point is away from roller contacting area to uplift area, values of stress triaxiality gradually change from negative to positive. These indicate that in the uplift area, damage is easier to be accumulated, and cracks occur from the outer surface to inner surface on account of positive stress triaxiality. As we can see, the laws of maximum principal also show the tendency of decreasing slightly first and then increasing from the inner surface to outer surface (Figure 13b). The variations of void volume fraction along radial paths are obtained, as shown in Figure 13c. It can be found that, values of VVF in inner surface along different paths are close to each other. VVF is gradually increased when the path is away from the deformation area in outer surface. In addition, values of VVF in outer surface are greater than them in inner surface, which is resulted from accumulation of VVF under the positive stress triaxiality. This indicates that in the outer surface of material uplift area, the material is easy to crack when it is far away from rollers.

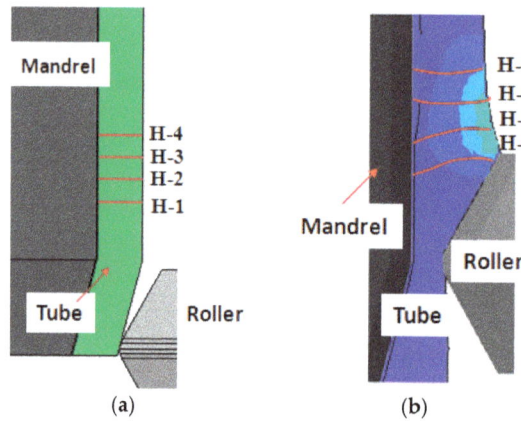

Figure 12. (a) Paths in tube blank and (b) paths in spun tube.

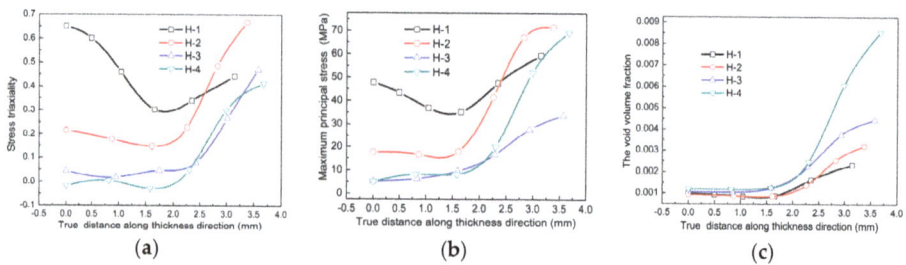

Figure 13. Distributions of (a) stress triaxiality; (b) maximum principal stress and (c) VVF in tube along the different radial paths.

According to the characteristics of stress triaxiality, maximum principal stress and void volume fraction, it can be concluded that accumulation of damage in tube spinning occurs mainly at the material uplift area. In addition, in the outer surface of material uplift area, the material is easy to crack when it is far away from rollers.

6. Conclusions

The applicability of GTN damage model in forward tube spinning of 3A21-O aluminum alloy is evaluated by experiment and the damage evolution is studied through analyzing the law of distribution of the stress triaxiality, the maximum principal stress, and the void volume fraction. The main results are as follows:

(1) Main parameters in GTN model, f_n, f_c and f_F of 3A21-O aluminum alloy, are determined 0.02, 0.012 and 0.04, respectively, by anti-inference method. A three-dimensional finite element numerical simulation model coupled the GTN damage model for forward tube spinning of 3A21 aluminum alloy is established. In addition, the GTN damage model is appropriate for predicting macroscopic crack in the material uplift area. The damage evolution in deformation area could not be predicted well with GTN damage model due to the negative stress triaxiality and the neglect of shear deformation.

(2) Accumulation of damage in forward tube spinning occurs mainly at the material uplift area through FE modeling. The VVF in the outer surface of the tube is higher than that in the inner surface in forward tube spinning. In addition, it is prone to cracking in the outer surface of tube in the material uplift area.

Acknowledgments: The authors would like to acknowledge the support from the National Science Fund for Excellent Young Scholars of China (Project 51222509), the National Natural Science Foundation of China (Project 51175429), the United Fund of Aerospace Advanced Manufacturing Technology (Project U1537203) and the Research Fund of the State Key Laboratory of Solidification Processing (Projects 97-QZ-2014 and 90-QP-2013).

Author Contributions: Xianxian Wang performed the experiments, simulations and wrote this paper under the guidance of Mei Zhan; Jing Guo and Bin Zhao assisted in performing experiments and analyzing simulation results.

Conflicts of Interest: The authors declare no conflict of interest.

Nomenclature

ϕ	Yield function
σ_{eq}	Von Mises equivalent stress
σ_y	Mean uniaxial equivalent stress of the matrix material
Σ_m	Macroscopic hydrostatic pressure
f	Void volume fraction
f^*	Modified void volume fraction
f_c	Critical void volume fraction
f_F	Void volume fraction at failure
F_n	Void volume fraction due to nucleation
F_g	Void volume fraction due to growth
S_n	Standard deviation of nucleation
q_1, q_2, q_3	Coefficients of the GTN damage model

References

1. Mohebbi, M.S.; Akbarzadeh, A. Experimental study and FEM analysis of redundant strains in tube spinning of tubes. *J. Mater. Process. Technol.* **2010**, *210*, 389–395. [CrossRef]
2. Xia, Q.X.; Xiao, G.F.; Long, H.; Cheng, X.Q.; Sheng, X.F. A review of process advancement of novel metal spinning. *Int. J. Mach. Tool. Manuf.* **2014**, *85*, 100–121. [CrossRef]
3. Music, O.; Allwood, J.M.; Kawai, K. A review of the mechanics of metal spinning. *J. Mater. Process. Technol.* **2010**, *210*, 3–23. [CrossRef]
4. Wong, C.C.; Dean, T.A.; Lin, J. A review of spinning, shear forming and flow forming processes. *Int. J. Mach. Tool Manuf.* **2003**, *431*, 1419–1435. [CrossRef]
5. Gur, M.; Tirosh, J. Plastic flow instability under compressive loading during shear spinning process. *Trans. ASME J. Eng. Ind.* **1982**, *104*, 17–22. [CrossRef]
6. Davis, J.R. *Metals Handbook*, 2nd ed.; ASM International: Russell, KS, USA, 2001.

7. Ma, H.; Xu, W.C.; Jin, B.C.; Shan, D.B.; Nutt, S.R. Damage evaluation in tube spinnability test with ductile fracture criteria. *Int. J. Mech. Sci.* **2015**, *100*, 99–111. [CrossRef]
8. Zhan, M.; Zhang, T.; Yang, H.; Li, L.J. Establishment of a thermal damage model for Ti-6Al-2Zr-1Mo-1V titanium alloy and its application in the tube rolling-spinning process. *Int. J. Adv. Manuf. Technol.* **2016**. [CrossRef]
9. Zhan, M.; Gu, C.J.; Jiang, Z.Q.; Hu, L.J.; Yang, H. Application of ductile fracture criteria in spin-forming and tube-bending processes. *Comput. Mater. Sci.* **2009**, *47*, 353–365. [CrossRef]
10. Gurson, A.L. Continuum theory of ductile rupture by void nucleation and growth. Part I. Yield criteria and flow rules for porous ductile media. *J. Eng. Mater. Technol.* **1977**, *99*, 2–15. [CrossRef]
11. Tvergaard, V.; Needleman, A. Analysis of the cup-cone fracture in a round tensile bar. *Acta Metall.* **1984**, *32*, 157–169. [CrossRef]
12. Chen, Z.Y.; Dong, X.H. The GTN damage model based on Hill' 48 anisotropic yield criterion and its application in sheet metal forming. *Comput. Mater. Sci.* **2009**, *44*, 1013–1021. [CrossRef]
13. Butcher, C.; Chen, Z.T.; Bardelcik, A.; Worswick, M. Damage-based finite-element modeling of tube hydroforming. *Int. J. Fract.* **2009**, *155*, 55–65. [CrossRef]
14. Sun, Q.; Zan, D.Q.; Chen, J.J.; Pan, H.L. Analysis of edge crack behavior of steel sheet in multi-pass cold rolling based on a shear modified GTN damage model. *Theor. Appl. Fract. Mech.* **2015**, *80*, 259–266. [CrossRef]
15. Nahshon, K.; Hutchinson, J.W. Modification of the Gurson model for shear failure. *Eur. J. Mech. A Solid* **2008**, *27*, 1–17. [CrossRef]
16. Li, H.; Fu, M.W.; Lu, J.; Yang, H. Ductile fracture: Experiments and computations. *Int. J. Plast.* **2011**, *27*, 147–180. [CrossRef]
17. Zhang, J.H.; Zhan, M.; Yang, H.; Jiang, Z.Q.; Han, D. 3D-FE modeling for power spinning of large ellipsoidal heads with variable thicknesses. *Compt. Mater. Sci.* **2012**, *53*, 303–313. [CrossRef]
18. Chu, C.C.; Needleman, A. Void nucleation effects in biaxially stretched sheets. *J. Eng. Mater. Technol.* **1980**, *102*, 249–256. [CrossRef]
19. Vadillo, G.; Fernández-Sáez, J. An analysis of Gurson model with parameters dependent on triaxiality based on unitary cells. *Eur. J. Mech. A Solid* **2009**, *28*, 417–427. [CrossRef]
20. Aravas, N. On the numerical integration of a class of pressure-dependent plasticity models. *Int. J. Numer. Methods Eng.* **1987**, *24*, 1395–1416. [CrossRef]
21. He, M.; Li, F.G.; Wang, Z.G. Forming limit stress diagram prediction of Aluminum alloy 5052 based on GTN model parameters determined by *in situ* tensile test. *Chin. J. Aeronaut.* **2011**, *24*, 378–386. [CrossRef]
22. Yan, Y.; Sun, Q.; Chen, J.; Pan, H. The initiation and propagation of edge cracks of silicon steel during tandem cold rolling process based on the Gurson-Tvergaard-Needleman damage model. *J. Mater. Process. Technol.* **2013**, *213*, 598–605. [CrossRef]
23. *ABQSUS Analysis User's Manual*; Version 6.8; ABAQUS. Inc.: Paris, France, 2008.
24. Lou, Y.S.; Huh, H. Prediction of ductile fracture for advanced high strength steel with a new criterion: Experiments and simulation. *J. Mater. Process. Technol.* **2013**, *213*, 1284–1302. [CrossRef]
25. Malcher, L.; Andrade Pires, F.M.; de Sá, J.M.A.C. An extended GTN model for ductile fracture under high and low stress triaxiality. *Int. J. Plast.* **2014**, *54*, 193–228. [CrossRef]
26. Xue, L. Constitutive modeling of void shearing effect in ductile fracture of porous materials. *Eng. Fract. Mech.* **2008**, *75*, 3343–3366. [CrossRef]
27. Chaboche, J.L.; Boudifa, M.; Saanouni, K. A CDM approach of ductile damage with plastic compressibility. *Int. J. Fract.* **2006**, *137*, 51–75. [CrossRef]
28. Kim, J.; Gao, X.S.; Srivatsan, T. Modeling of void growth in ductile solids: Effects of stress triaxiality and initial porosity. *Eng. Fract. Mech.* **2004**, *71*, 379–400. [CrossRef]

MDPI AG

St. Alban-Anlage 66

4052 Basel, Switzerland

Tel. +41 61 683 77 34

Fax +41 61 302 89 18

http://www.mdpi.com

Metals Editorial Office

E-mail: metals@mdpi.com

http://www.mdpi.com/journal/metals

www.ingramcontent.com/pod-product-compliance
Lightning Source LLC
Chambersburg PA
CBHW051709210326
41597CB00032B/5423